Self-Protection Jammer Systems

For a complete listing of titles in the
Artech House Electronic Warfare Library,
turn to the back of this book.

Self-Protection Jammer Systems

Ahmet Güngör Pakfiliz

ARTECH
HOUSE

BOSTON | LONDON
artechhouse.com

Library of Congress Cataloging-in-Publication Data
A catalog record for this book is available from the U.S. Library of Congress

British Library Cataloguing in Publication Data
A catalog record for this book is available from the British Library.

ISBN 13: 978-1-68569-011-3

Cover design by Joi Garron

© 2025 Artech House
685 Canton St.
Norwood, MA

10 9 8 7 6 5 4 3 2 1

*To my wife Serap, my daughter Ilgım, and my son Cemre,
for their support and patience during my endless studies*

Contents

CHAPTER 3
Jamming of Radar-Guided Systems

Preface

There are many valuable and informative books on electronic warfare (EW), but dedicated studies on specific EW systems are rare. This book focuses on the structure of self-protection jammer (SPJ) systems, their technical properties, operational requirements, and their utilizations. They are generally an integrated part of the self-protection EW (SPEW) systems and do not operate separately from the other subsystems. When the technical and operational properties of the SPJ systems are considered, the passive parts of the SPEW systems, such as electronic support (ES) and radar warning receivers (RWR), should also be regarded. In this context, the host platforms of the SPJ systems and the potential radio frequency (RF) threats against these platforms are the other essential topics for the SPJ systems.

This book provides general SPEW system information and a detailed technical and operational explanation of radar-guided weapon systems. It equips the readers with the theoretical and mathematical fundamentals of jamming radar systems. It emphasizes the practical jamming by discussing the RWR system architecture and Rx theory. The book then presents the theory of the SPJ systems in detail, followed by a discussion on their effective use in terms of operational and technical aspects. The practical application of the SPJ systems in airborne platforms and the decoys is also covered. The book includes a glossary of definitions of essential terms to aid understanding.

The book's primary target audience is engineers, military staff, and defense industry professionals in EW, radar, and RF-guided missile design, development, maintenance, operational support, procurement, and implementation. It would also benefit undergraduate and graduate students who aim to work in the electronic defense industry. Furthermore, the book may be an excellent supplementary source for graduate thesis studies. It aims to provide readers with the technical details and operational aspects of the SPJ systems, including the operation environments, concepts, and requirements. It also presents future projections with possible tendencies in the systems and concepts. Application-dependent problems with solutions support the theoretical parts.

The book covers theoretical information on SPJ systems and provides fundamental information for designing SPJ systems on different platforms and preparing and conducting procurement projects for them. It may be considered an essential reference book for the rapid application source. The theories are presented step by step, supported with real-life systems, and explained using applicable generic structures. The solved problems aim to clarify the physical nature of the situations. Moreover, some issues are supported with MATLAB® simulations. The book empowers the readers to apply these principles in their professional endeavors confidently by equipping them with this practical knowledge.

I hope that this book will be helpful to engineers, military personnel, and professionals in the defense industry. This book will be only one fundamental element of a booster tool for those committed to their goals and continually improving their skills through consistent study.

CHAPTER 1

Self-Protection Electronic Warfare Concept

Technological developments have significantly influenced weapon systems, increasing complexity in planning and executing operations. This complexity, although challenging, is crucial for defeating the enemy and adds a sense of accomplishment for the experienced and well-informed military staff. The battlefield has drastically changed in the last century, primarily due to the development and invention of weapon systems with electronics and computer technology. These systems not only shape the operational fields and tactics of the battles but also influence the course of the wars. Their impact on the new operational concepts is undeniable. However, these game-changer weapon systems, such as aircraft, warships, radar, and air defense systems, are not easy to sacrifice as they are the ultimate targets for adversaries.

Electronic warfare (EW) has become a cornerstone of modern battlespace, with military operations increasingly executed in a complex electromagnetic environment. The ability to control the electromagnetic spectrum has become a decisive factor in the success of military operations. The force that can disrupt the enemy's use of the electromagnetic spectrum, exploit it for its purposes, and maintain control over it gains a significant advantage. This section delves into the emergence of EW, particularly with integrating radar and communication systems in military operations. Understanding the urgency and importance of this issue in operations is necessary.

The primary objective is to safeguard highly valued and vital platforms, such as fighters, bombers, cargo aircraft, helicopters, medium and large UAVs, and naval ships, from potential threats. These threats are not to be taken lightly, and an armed force's primary responsibility is to mitigate their risks effectively. To achieve this, self-protection EW (SPEW) systems are deployed. This section provides comprehensive information about the structure and subsystems of SPEW systems. It also outlines their usage concepts and applications for different platforms, underscoring their critical role in protecting these assets.

1.1 Introduction to Self-Protection EW Systems

The search for new, astonishing methods has been an essential part of the art of war since the beginning of humanity. New weapons have been developed throughout history, and the military industry has always been the driving power for the development of technology. With electronics entering the military field, drastic changes occurred in the rules of wars and battlefields. By using electronic technology, the effectiveness of new weapons systems entering the battlefield and conventional weapons has increased. An electronic system may be a separate weapon system, such as early warning radars, military radios, and laser weapon systems. At other times, weapon systems use electronic technology, like radar-guided missiles, artillery radios, and laser designators, for missile guidance as a force multiplier. Later, electronic technology began to be supported by the software. Thus, the effectiveness of offensive weapon systems and the protection capabilities of defense systems increase.

Even though electronics usage in military systems provided general progress, it made itself felt primarily on *radio frequency* (RF). As RF importance increased, military requirements were forced to maximize RF systems exploitation for allies and reduce their use for adversaries. These technological requirements of the operation field have led to the emergence of EW. EW is a military action that exploits electromagnetic energy to provide situational awareness and create offensive and defensive effects [1].

To achieve EW goals, not only the studies in the operational field but also the studies in the technical fields gain importance. It is important to consider three EW functions at this point: *electronic warfare support* (ES), *electronic attack* (EA), and *electronic protection* (EP). The ES function includes searching for, intercepting, identifying, and locating sources of intentional and unintentional radiated *electromagnetic* (EM) energy for immediate threat recognition, targeting, planning, and conducting future operations. The EA function is related to using electromagnetic energy for offensive purposes. Furthermore, the EP function of EW involves actions to ensure effective, friendly use of the electromagnetic spectrum despite the enemy's use of electromagnetic energy. EP may be active or passive measures [2]. Equipment development, personnel training, and mission planning use these EW functions.

EW systems may be separate or part of a weapon system. A *direction finding* (DF) system for determining the location of a hostile *transmitter* (Tx) and a *stand-off-jammer* (SOJ) system are separate systems, and they conduct individual actions for supporting operations. However, a *radar warning receiver* (RWR), a *self-protection jammer* (SPJ), and a *countermeasures dispenser system* (CMDS) for chaff and flare dispensing are EW subsystems of weapon systems. Furthermore, communication or radar systems using *frequency-agile* techniques such as *frequency hopping, frequency chirp*, or *direct sequence spread spectrum* (DSSS) are EW protection techniques of weapon systems. The subsystems and special EW techniques are assumed to be force multipliers rather than separate systems. It is possible to separate them into EW functions by putting them into the intersection of multiple functions. If we consider RWR, it is a part of the ES function because of its *receiver* (Rx) structure. From another point of view, it is used to protect the mother platform and can be classified as the EP function. Similarly, the SPJ is used to jam radar-guided missiles; in this respect, it can be assumed to be a function of EA. However, SPJ systems can be considered an EP function, as they confuse the missile's radar system to protect itself.

When we look at the most potent assault weapon systems, we encounter air and naval surface platforms. Even though their roles differ, fighters, bomber aircraft, attack helicopters, and warships are the central assault units for any armed forces. Also, they have another common point regarding threats: they are targets of radar-guided missiles. There are also fatal units in land-based systems such as tanks, artillery, and *anti-aircraft artillery* (AAA), but they are not direct targets of radar-guided missiles [3]. So assault units are vulnerable to radar-guided *enemy-integrated air defense systems* (IADS) threats. These threats may be land or vessel-based *surface-to-air missile* (SAM) systems or *air-to-air missile* (AAM) systems used in *air interception* (AI) operations. Even though there are some tactical avoidances from radar-guided SAM, AAM, and AAA systems, the most effective salvation method is to use SPEW systems.

SPEW systems are subsystems of the mother platform, carrying all the EW functions on themselves. For this purpose, SPEW systems contain many EW elements. Some of these elements are used against RF-guided threats; others are counter to infrared (IR) and laser-guided threats. These elements can be changed according to the platform; however, generally used elements are listed in clusters as follows:

- *Warning receivers:* RWR, *laser warning Rx* (LWR), and *missile warning Rx* (MWR);
- *Active countermeasures:* SPJ, *directional IR countermeasures* (DIRCM), and *laser countermeasure system* (LCMS);
- *Expendable dispensers:* CMDS, disposable jammer, and smoke;
- *Central and side elements:* SPEW computer, controllers, and displays.

There are three variations of SPEW system architectures, so separating them into neatly defined categories is difficult. These general classifications are defined in [4] as follows:

- *Stand alone:* Each discrete EW system operates independently or nearly independently of every other EW system.
- *Federated:* Each EW system essentially maintains its functional boundaries. The individual EW systems commonly share data via an EW data bus, with the RWR as the bus controller. The individual EW systems communicate via the avionics data bus to receive inputs.
- *Integrated:* All EW components and other avionics systems share common processing resources and databases. Data fusion algorithms are commonly used to enhance information quality. Integrated systems can also schedule different aircraft system apertures and sensors to perform EW tasks.

Although three variants of SPEW system architecture are regularly mentioned, architectures in which different architectural features are used together for flexibility can be considered. A functional block diagram for a generic SPEW system with a federated architecture tendency is shown in Figure 1.1. Solid lines represent the main bus, and the dashed lines indicate the EW data bus. The given block diagram is assumed to be independent of the platform. In this generic architecture, all the subunits of the SPEW system operate in connection with a *central management unit* (CMU). The overall EW system is connected to the platform's central bus system through CMU. Also, the SPEW system's dedicated controllers and display have direct connections to the main bus. Using the MIL-STD-1553B bus is usual for Western military platforms. MIL-STD-1553B networking technology is used in numerous military platforms, including aircraft, ships, tanks, missiles, and satellite applications.

In Figure 1.1, cooperative subunits are clustered together as sections for clarity. The controllers and displays are directly coordinated with the CMU and related to all the sections. The operational flow for each section's processing chain is described next.

- *RF section:* The RWR detects the active and semiactive radar-guided threats. The information obtained at the output of the RWR system directly activates the appropriate countermeasures under the control of the CMU. As seen in Figure 1.1, the countermeasure may be chaff dispensing, jamming, or both. Each applied countermeasure informs the CMU about the activation. Moreover, CMU may decide to use a disposable jammer based on information taken from the RWR.
- *IR section:* This time, the sensor is a *missile warning system* (MWS). This time, the sensor is MWS. The detected threat can be IR or passive radar and cannot be distinguished from each other. However, they are treated the same if no additive indication exists. For this purpose, CMDS dispenses chaff and flare, DIRCM, or all, which are activated with the output of the MWS. Also, each countermeasure system sends activation information to the CMU.
- *Laser section:* The laser warning system (LWS) detects the laser spot of a laser designator on the platform and activates the smoke, LCMS, or both, depending on the situation. Again, the activated countermeasure system sends conditional information to the CMU.

1.2 Role of SPEW Systems

Air vehicles such as fighters, helicopters, UAVs, and warships play an offensive role when conducting assault and intelligence operations in the enemy's territory. These

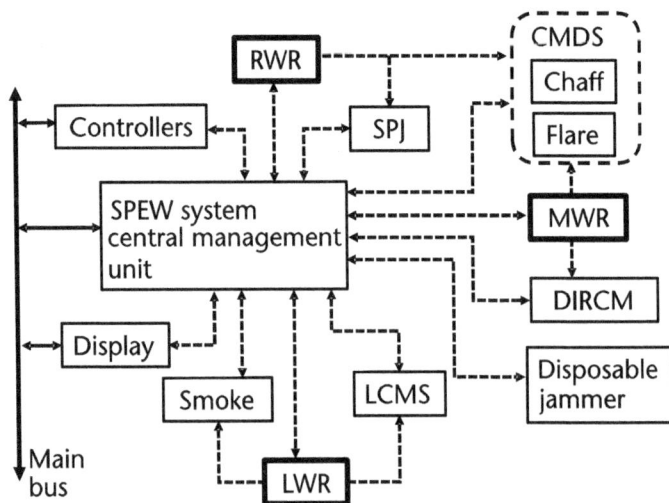

Figure 1.1 Block diagram of a general SPEW system.

platforms have increased their attack ability with technological developments. However, enemy IADS also pursues technological developments and has improved its detection and destruction abilities for adversary assault units. Even though the assault and intelligence-gathering platforms are intruders, they become targets for the IADS and vulnerable to hard-kill threats. These threats can be SAM, AAA, AAM, or ASM, some of which are RF-guided, some of which are electro-optical-guided, and some of which use both. Even though different avoidance tactics against different threats have been developed for various platforms, the most effective way to avoid them is by using the proper SPEW system. Various SPEW systems are used for different platforms, and the common property of these systems is protecting mother platforms from guided threats.

Developing an effective SPEW system depends on technology, and the success of its use depends on intelligence. In other words, the success of a self-protection system requires adapting countermeasures to the threat. The SPEW system's standard method for countering a threat starts with threat detection and identification and then executing effective countermeasures. Considering the events and actions in the SPEW countermeasure process and the main factors involved are beneficial for understanding the role of SPEW systems. For this purpose, a functional block diagram for a general SPEW system is shown in Figure 1.2. Also, the depicted SPEW block diagram is free from the mother platform, so it is valid for airborne platforms and naval vessels.

The SPEW process starts with activating the guided threats and directing the guidance to the mother platform, as shown in Figure 1.2. The threat systems may be RF, IR, or laser-guided, and the guidance methods for the threats are as follows:

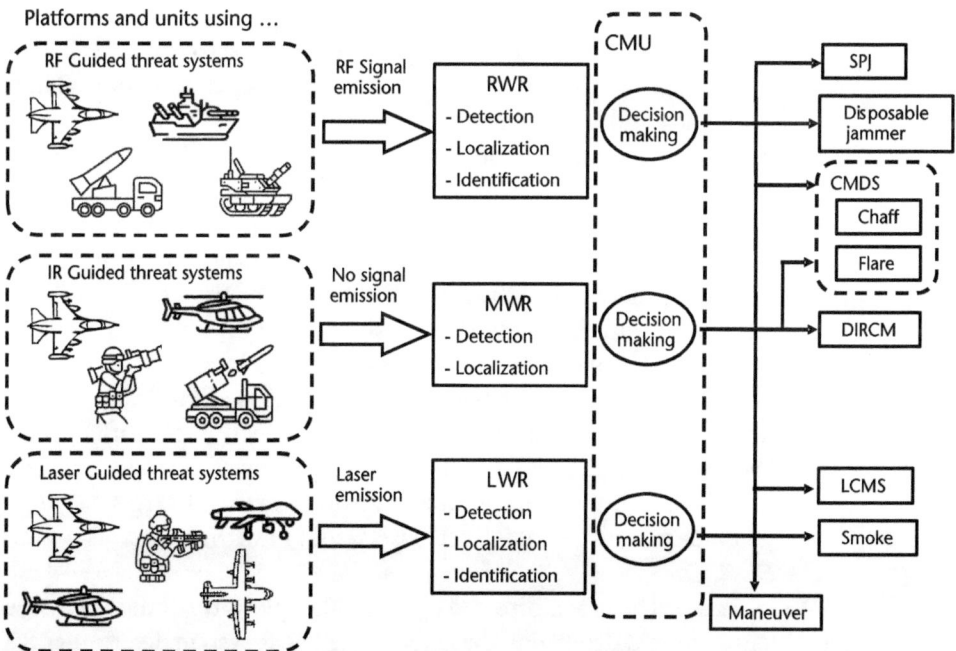

Figure 1.2 Functional block diagram of a general SPEW system.

- *RF guidance:* RF-guided threats have many applications, such as AAM, AAA, and SAM. They may be used to guide missiles and guns. They also have different guidance methods: active, semiactive, command, and passive. Their launch platform may be air, ground, or sea surface. Their targets are air vehicles and naval vessels.
- *IR guidance:* IR-guided threats are very effective and have a variety of uses for many different targets. The main reason that IR-guided missiles are so effective is that they use passive guidance, which results in difficult detection by the target platforms. They may launch from land platforms, naval vessels, air platforms, and even man-portable launchers. Their targets may also be air vehicles, naval vessels, and moving land vehicles.
- *Laser guidance:* The most common method is the *laser range finder* (LRF) for distance measurement. *Laser target designator* (LTD) and *laser beam rider* (LBR) are used for target designation. Generally, LTD and LBR are used for laser-guided weapon systems, which are mentioned separately [5].
  - LTD generates a beam of radiation designating an object to which a specific warfare agent is directed. Most often, these are missiles and bombs guided by scattered radiation of a laser spot designating the target. LTD can be placed on air, land, and sea platforms or as equipment for a single soldier. A land platform can designate a target on an air platform and vice versa. All combinations are possible, including target designation by platforms of the same type. The designators use lasers generating radiation near the IR band, a wavelength of 1,064 and about 1,500 nm.
  - LBR uses laser radiation to illuminate the back of the rocket. At the back of the missile, a laser radiation detector analyzes the received radiation and takes corrections to change the rocket's flight path. The laser beam illuminating the rocket is coded in a cross-section area to bring the rocket to the center of the laser-guided weapon systems use the semiactive guidance method. Therefore, they need a laser designator to point to the target, and the guided missile detects the return of the laser emission. The working principles and requirements for using laser-guided weapon systems may change depending on whether the designation and launching processes are carried out from the same or different platforms. Occasionally, platforms, like fighters and bombers, use LBR, and launchers of laser-guided missiles and bombs are on them. If the LTD and launcher platforms are different, this time, there are limitations to this type of threat due to the nature of the laser itself. First, the laser designator should be on a stable and low-maneuverable platform. Suitable platforms for laser designation may be helicopters or UAVs for airborne platforms, troops or land vehicles for ground, and vessels for sea surface. Second, their effectiveness differs according to their line-of-sight range and laser output power. Targets for laser-guided missiles are low-maneuverable platforms or fixed targets. These may be ground-moving targets, fixed assets, and naval vessels.

When a guided missile is directed to the mother platform of the SPEW system, the related sensor should determine the threat. The RWR system determines RF-guided threat systems. It is expected to detect the threat system's existence and identify the threat emitters using RF measurements and their technical specifications. Also, determining the threat's direction and distance are the other requirements from RWR. RWR systems fulfill these tasks via complicated signal process techniques. The RWR structure is discussed in detail in a separate section.

Determining the existence of IR-guided missile threats is more complex than deciding on RF-guided missiles since no IR emission exists in the guidance process. The launch of these missiles must be detected to apply an effective defense technique against IR-guided missiles. Unlike radar-guided missiles, IR missiles have no emission that can indicate the launch of IR-guided missiles and can be detected by detection systems such as the RWR. The only way to detect such missiles is to place an MWR on the platform. MWR systems may be active or passive. Active systems that can provide 360° horizontal detection are radar systems designed to find missiles.

Nevertheless, using an active system may cause problems, such as high weight and volume, difficulties integrating with the platform, and possibly additive maintenance. Also, an additional broadcasting system makes the platform an easily detectable target. Currently, the general trend in the world is toward passive systems. These systems are developed to detect the energy emitted by objects in IR, ultraviolet (UV), or both bands. The generally used MWRs in SPEW systems are passive warning systems for detecting, localizing, and warning of missiles approaching the mother platforms. They are utilized against SAM, ASM, or AAM threats and can detect plume and body emissions.

Laser guidance is different from radar and IR guidance in many aspects. The laser-guided threats make emissions, but not in the RF band. A laser beam radiation is formed on the target platform, typically in the visual and infrared bands. Detecting this radiation may be possible with an LWR. The LWR system is a passive defense warning system aiming to detect, track, and warn of different kinds of laser threats in the typical wavelength range between 0.5 and 1.65 μm. LWR system structure usually includes optical, sensing, and processing subsystems. The optical subsystem focuses on the incoming laser beam and transmits it toward the detector subsystem, which transduces the optical signal into a digital signal. Finally, the processor subsystem produces the warning output, which contains the characteristics of the laser radiation.

The sensors in the Rx part of the SPEW system detect different guided threats. This information is used to trigger an appropriate countermeasure against the threats. Different guidance methods and various usable countermeasures are taken place in the process. Selecting the proper countermeasure for the threat considering the situational requirements requires decision-making. This process occurs at the CMU in line with preprogramming in conventional systems. However, machine learning seems to take over decision-making shortly [6].

When an RF-guided threat is directed to the mother platform, RWR detects, localizes, and identifies the threat, and a decision process starts for the proper countermeasures. The options may be self-protection jamming, chaff dispensing, maneuvering in the appropriate direction, or a combination. The main goal of an SPJ system is to protect the mother platform. Even though the weapon radars are the primary target for sophisticated IADSs, early warning, and surveillance radars are also potential targets for SPJs. The IADS fire control systems generally have similar engagement processes: search, detect, track, and fire. A procedure is applied during the search phase, and potential targets are specified. In the detection phase, a decision procedure is conducted to determine the targets, and tracking is initiated. In the tracking phase, target tracking is maintained using a gating process called locking. In the fire phase, the lock on the target continues, but this time, the missile's intersection with the target is aimed.

Since the protection process lasts from detection to the end of avoidance of the platform from the missile, jamming techniques may vary. These techniques include noise and deceptive jamming. Noise jamming is used for concealment, which aims to prevent being a potential target for the fire control systems during the search process. Deceptive jamming creates false targets to prevent track initiation or locking. After locking or during track maintenance, different deceptive jamming methods, such as *range gate pull-off* (RGPO) or *velocity gate pull-off* (VGPO), are engaged to break the lock.

Chaff is defined as strips of frequency-cut metal foil, wire, or metalized glass fiber used to reflect electromagnetic energy expelled from shells or rockets by aircraft or ships as a radar countermeasure. Generally, from the tactical point of view, chaff is used for screening or self-protection. Screening tactics are used to establish a volume saturation and a chaff corridor to mask the platform against early warning, surveillance, and acquisition radars in the IADS. Thus, the aim is to break the chain in the command, control, communications, computers, intelligence, surveillance, and reconnaissance (C4ISR) structure of the IADS. Self-protection chaff tactics are developed to counter the *target tracking radars* (TTRs). When used with jamming and maneuvers, chaff can cause TTRs to break the lock or cause the missile to deviate from the platform.

MWR detects incoming missile threats and selects between using the DIRCM system, flare dispensing, maneuvering, or combining these after CMU's decision process. DIRCM systems use a powerful IR energy source placed on the platform. The source near the region where the energy emitted from the platform is most intense is modulated appropriately, aiming to deviate the missile from the target. The mixer output is passed through an optical filter to prevent it from being seen by the naked eye. Since there is no resource usage limitation as in flare, these systems operate continuously in the threat zone. Like how chaff is used against radars, *flare* is used against IR missiles. The flare deceives the missile by emitting more potent IR energy away from the platform on which it is launched. For the flare to be effective, the total IR energy emitted by the missile in the wavelength range that it operates must be greater than that of the protected platform in the same band. It must be fired at the right time.

If a laser beam radiation is dropped on the platform, LWR detects it with location, and after identification of the threat system, a decision process is conducted. After the decision is made, an alternative method is selected, which may be smoke, LCMS, maneuver, or a mixture. Depending on the type of laser-guided threat and the protected platform, various defense countermeasures can be implemented, and placing a smokescreen is the simplest one to avoid hitting. Smoke, dust, and debris can distort the use of laser-guided munitions. However, it is necessary to use the correct type of smoke if it is intended to give any protection against a laser threat. The effects of smoke depend upon the size of the smoke particles and the wavelength of the laser. As the size of the particles approaches the wavelength of the laser, more energy will be scattered than attenuated.

An LCMS protects high-value fixed and mobile targets from laser-guided weapon attacks. These targets may be command posts, land vehicles, naval vessels, and low-maneuverable air platforms such as helicopters and UAVs. When a hostile laser designator is aimed at the protected target, LCMS generates a shady spot by illuminating a high-power laser spot near the target, coded to mimic the hostile laser signal, thus deceiving the attacker to lock on the new spot instead of the targeted object.

1.3   Using SPEW Systems in Different Platforms

The scope of this book is RF systems since SPJ systems are related to RF threats and active countermeasures against them. Introducing the SPEW system is essential to understanding the function of the SPJ in this system and the relationship of SPJ with the other subsystems. However, after briefly discussing the SPEW system's role for all platforms, it is time to restrict the inspection for the platforms that target radar-guided weapon systems.

This part considers SPEW system requirements for airborne and naval surface platforms. Also, recently, medium and large UAVs have entered the operational fields with AI, fighter, bombing, reconnaissance, and surveillance missions. During these missions, they are under the threat of RF-guided weapons. Thus, SPEW system requirements for UAVs are also mentioned here.

1.3.1   SPEW Systems in Airborne Platforms

Airborne platforms have been an indispensable part of military power since the middle of the twentieth century and are the most critical factor determining the outcome of wars. Generally, armed forces use military airborne platforms for offensive and defensive purposes. In the offensive role, these aircraft destroy adversaries' vital installations, aircraft on the ground or wing, and ordnance depots and supplies. The defensive position provides close air support to the land-based army and intercepts enemy air strikes. Airborne platforms play a significant role in detecting and neutralizing submarines and warships in naval warfare to keep the seacoast free from enemy attack. Military aircraft also provide logistical supplies to forwarding bases, air supplies for cargo and troops, and participate in search and rescue operations. Military aviation includes transport and warcraft and consists of fixed-wing and rotary-wing aircraft. Only human-crewed aircraft are considered in this part, and unmanned air platforms are discussed separately. Military airborne platforms are categorized as follows:

- *Fighter aircraft (fighters):* Fighters are fixed-wing military aircraft, and their structure is intended to be extremely high in firepower and maneuverability. They are primarily designed for air-to-air combat but are compatible with air-to-surface missions such as ground or naval attacks and bombardments. The role of fighters is to establish air superiority in the war zone, which can be in friendly or enemy territory. If the war zone is in the allied region, which means defensive action, the role of the fighter may be an interceptor or reconnaissance near the borders. In the offensive actions, fighters may be fighter-bombers, surface attacks, surveillance, and reconnaissance. In addition, some EW missions, such as escort jamming and *suppression of enemy air defenses* (SEAD) operations, are considered offensive.

- *Bomber aircraft (bombers):* Bombers are also fixed-wing military aircraft, but their structures are more extensive, heavier, and less maneuverable than fighters are. Bombers are designed to attack ground and naval targets by dropping bombs or missiles, launching torpedoes, or deploying air-launched cruise missiles.
- *Military cargo aircraft:* This type is also called military transport aircraft. Cargo aircraft support military operations by airlifting troops and equipment over moderate or long distances. Their structure is fixed-wing, big-bodied, with large amounts of interior space for carrying military assets, and their maneuvering capability is low.
- *Military helicopters:* Helicopters are rotary-winged aircraft of different sizes and structures used for various missions. Their maneuvering capabilities and operational altitudes are limited compared with fixed-wing aircraft. The typical tasks of military helicopters are transporting troops and assets, surveillance, *combat search and rescue* (CSAR), medical evacuation, attacking ground targets, maritime patrol, and anti-submarine warfare.

Since the structures and operational usage of the above airborne platform classes are different, their SPEW requirements have some differences and some similarities. The SPEW structure with the same subcomponents is used because almost similar threats apply to each class. However, the technical specifications of subsystems change according to the following properties:

- *Radar cross-section (RCS):* RCS is a measure of power scattered in a given direction when a target is illuminated by an incident wave [7]. RCS is a function of target-dependent and RF transmission-dependent properties. The target-dependent properties are geometry and material composition. Target geometry includes size, amount of vertical structure, the method of joining the wings to the fuselage, transitions between airframe materials, and surface orientation geometry. Also, using materials that absorb electromagnetic waves reduces the RCS. Furthermore, the RCS of a target depends not only on its physical shape and composite materials but also on its subcomponents, such as pylons, bombs, antennas, and other sensors. Given these explanations, the RCS of airborne platforms can be analyzed.

    RCS reduction of airborne platforms has been studied extensively, and considerable progress has been made. This effect can be seen in the progressive RCS reduction of the third-generation, fourth-generation, and fifth-generation fighters [8]. Because of the large structure of cargo and bomber aircraft, the decrease in their RCS can be somewhat reduced. Helicopters, which are relatively close in size to fighters, have a higher RCS than fighters. The relatively high RCS of helicopters is due to the dynamic electromagnetic scattering effect caused by the principal and tail rotors. As a result, compared with fighters, cargo, bomber aircraft, and helicopters, they require a more sensitive Rx in RWR systems and higher SPJ power output.

- *Suitable space and payload capacity:* The number of avionics, radar, EW, sensors, communication, and missile systems used in aircraft has been increasing rapidly with the development of technology. Despite the increase in the number of systems, the space on the planes, the payload that they can carry, and the power that they can provide remain the same. As a result, these systems must be smaller and lighter and require less power.

  Today's airborne platforms demand high performance in small structures for every subsystem. *Size, weight, and power* (SWaP) are essential indicators for aircraft subsystems' usage efficiency. SWaP is also an important parameter for SPEW systems. However, reducing SWaP depends on the technology and design experiences. Thus, SWaP optimization for a specific platform brings challenges and impossibilities. Required SPEW system specifications can be reached with an airborne platform possessing sufficient place and power.

  However, insufficient space and power result in inadequate SPEW system requirements even with a reduced SWaP. The size of the airborne platforms directly affects meeting the needs of SPEW systems. Therefore, platforms that can allocate small space and power for SPEW systems, such as fighters and helicopters, limit the system's specifications. However, large platforms such as bombers and cargo planes are more suitable for meeting SPEW system needs.

- *Maneuvering capabilities:* Numerous factors affect the maneuvering capability of the air platforms, such as size, mass, or engine type, becoming fixed or rotary wing, and payloads. Fighters' maneuvering capability is significantly higher than cargo aircraft, bombers, and helicopters. It can increase success on the modern battlefield when adequately employed with maneuvers and countermeasure subsystems of SPEW systems such as SPJ and CMDS. Therefore, different tactics have been developed for different airborne platforms to increase survivability. Studies on maneuvering tactics using SPEW systems have focused mainly on fighters. Appropriate maneuvering of the aircraft while dispensing chaff enhances the chaff bloom rate and increases its reflection effect on radars. In addition, some tactics were developed for fighters' self-protection when chaff was employed with maneuvers and jamming.

  However, some maneuvering restrictions occur for RWR operations. RWRs provide accurate threat positioning information when the aircraft maneuvers to specific roll angle limits. Fighters' aggressive maneuvering can exceed the turn rate and these limits.

- *Operational altitudes:* The operational altitudes of aircraft are different. During the cruise, various attacking and bombing air platforms use multiple altitudes to avoid surveillance radars. Aircraft that fly at high altitudes, like cargo and bomber aircraft during the cruise, require a high-sensitivity RWR Rx and a high-power SPJ output, especially for long-range SAM systems. Helicopters fly lower profile than other aircraft and are more vulnerable to IR and radar-guided missiles. The vulnerability of helicopters against threats is not only because they fly in low profiles but also because of their low maneuverability.

For this reason, their RWR and MWS sensitivity should be very high, and their RF and IR countermeasure abilities are also powerful. When considering the fighters, their flying altitude varies over a wide range. Therefore, they must detect the threats from far distances, which means that their RWR Rx sensitivity is high, and their RWR Rx must detect the threats from short ranges without saturation, which results in a high dynamic range. In addition, their SPJ should be capable of using both linear and saturation modes.

Even though the technical specs of the subsystems vary, Figure 1.3 presents a standard SPEW system structure for airborne platforms; putting all the subsystems in a SPEW system is optional. The selection of SPEW systems' subsystems, construction, and specifications depends on the platform space, payload capacity, operational requirements, logistic issues, technical experience, and cost.

As seen in Figure 1.3, some subsystems do not fit with each platform because the related threats do not match the structure and properties of the platform. For example, laser-guided threats are unsuitable for fighters, so laser-related warning and countermeasure systems are not used on fighters. RF-guided missiles are the primary threats to airborne platforms, so RWR, SPJ, and CMDS subsystems are indispensable for airborne SPEW systems. In addition, some subsystems such as CMU, power supply, controls, and displays participate in the SPEW systems. Using a dedicated power supply for the requirements of the SPEW system is standard. A common approach is using a separate CMU for the SPEW systems to reduce the processing losses in the mission computer. CMU fuses the required data from the SPEW subsystems and conducts the decision-making and resource allocation processes. In addition, the CMU exchanges data with the mission computer via the main bus. Because the SPEW system is vital, a separate display and control group should exist. The separation here may only sometimes be meant as a different subsystem since the SPEW system's displays and controls may be in a *multifunction control and display unit* (MCDU).

It would be practical and valuable to demonstrate the layout of SPEW systems on different aircraft. For this purpose, examples of SPEW systems on fighters and helicopters are presented. The layouts for different platforms may vary, but the

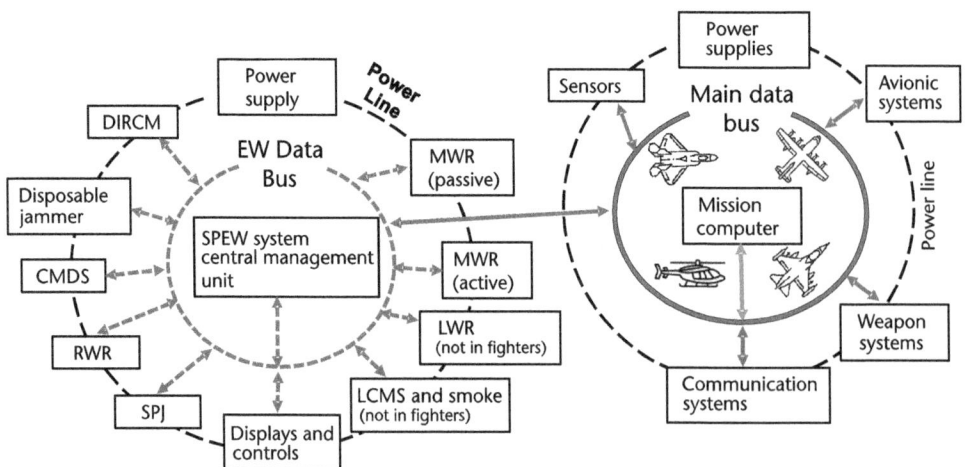

Figure 1.3   SPEW system structure for airborne platforms.

subunits' general structure and probable placements are considered in the examples. Figure 1.4 shows the design of a generic onboard SPEW system on a fighter aircraft. The subunits are shown from the top. Except for the antennas, the subunits are located inside the aircraft.

As seen in Figure 1.4, the fighter's onboard SPEW system consists of the mission computer, SPEW CMU, RF subsystems (RWR, RFJ and chaff dispenser), *electro-optical* (EO) subsystems (MWS, LWR, and flare dispenser), and disposable jammer subsystems.

After mentioning the SPEW system layout of an aircraft, it would be appropriate to give the SPEW system layout of a different air platform. Figure 1.5 demonstrates a generic layout of an onboard SPEW system on a helicopter in this context. Except for the antennas, the SPEW system's subunits are inside the chopper.

Figure 1.5 shows the helicopter's onboard SPEW system. The system comprises the mission computer, SPEW CMU, RF subsystems (RWR, RFJ and chaff dispenser), EO subsystems (MWS, LWR, DIRCM, and flare dispenser), and disposable jammer subsystems.

1.3.2   SPEW Systems in Naval Surface Platforms

Naval vessels are another potential target for RF-guided threats. This book's main scope is self-protection RF jammer systems, and this section is dedicated to the relationship between the SPEW system and SPJ. Thus, the SPEW systems used in naval vessels are considered here. The guidance methods of antiship missiles are inertial, RF, IR, and laser. Further, RF guidance methods contain active, semiactive, and passive radar guidance, jammer, and radar homing. Even though the ES (or ESM) of EW in naval ships is similar to airborne platforms, using and structure of their active parts, such as jammer and CMDS, are different. The main reason for this

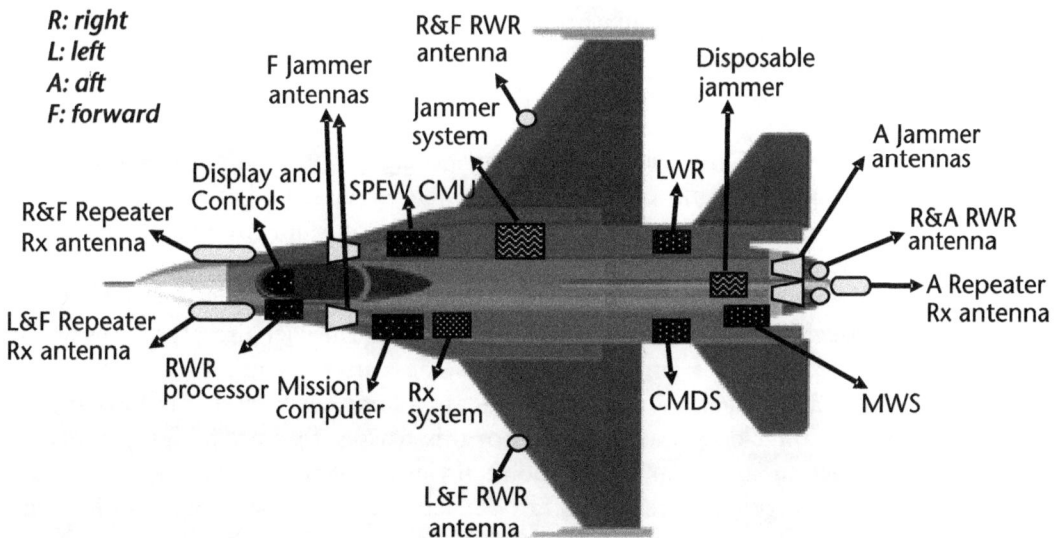

Figure 1.4 A generic SPEW system layout for a fighter aircraft.

R: right
L: left
A: aft
F: forward

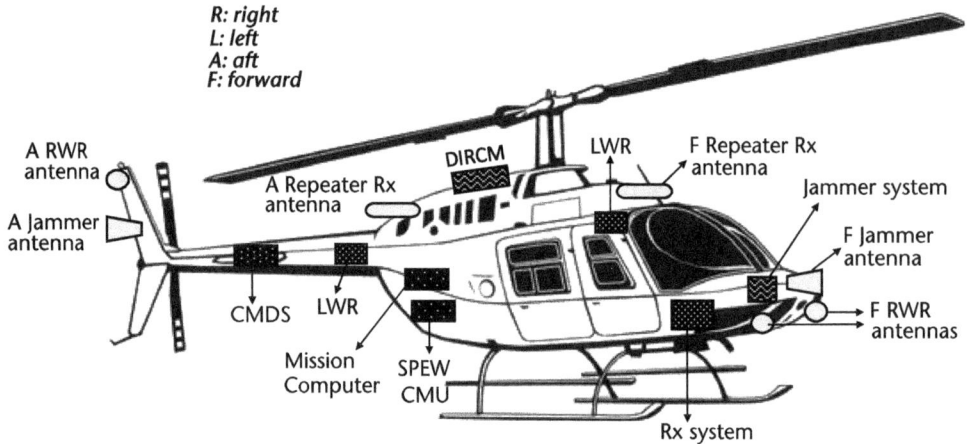

Figure 1.5 A generic SPEW system layout for a helicopter.

emerges from the limited maneuvering capabilities of vessels. The other reason is that most antiship missiles have sea-skimming properties.

Sea skimming is a technique many antiship missiles use to avoid radar and infrared detection during their approach to the target. The missiles use a trajectory about 10m above sea level. A vessel under attack can only detect a sea-skimming missile when it emerges over the horizon at approximately 15 to 25 nautical miles (28 to 46 km). According to the cruise speed, a vessel has only 25 to 60 seconds of warning time. As a result, naval ships' large bodies, low maneuvering capability, and short warning time determine the SPEW systems' structure.

Before discussing the structure of the SPEW systems and specifications of the subsystems for naval ships, it is essential to mention the types of surface warships that are often exposed to guided missiles and use SPEW systems. For this purpose, some properties of aircraft carriers, surface combatants (destroyers, frigates, and cruisers), and amphibious ships are given [9].

- *Aircraft carriers:* These warships are the largest vessels in any Navy worldwide, averaging approximately 1,100 ft (350m) long. They provide combat air support to the fleet. Aircraft carriers carry fighter aircraft, and runways allow planes to take off and land. Carriers are classified as having either conventional propulsion or nonconventional propulsion. Aircraft carriers carry numerous fighters, bombers, helicopters, and other aircraft types. However, carrying various aircraft includes ships designed to support *short-takeoff/vertical-landing* (STOVL) jet operations. While heavier aircraft such as bombers are launched from aircraft carriers, these aircraft cannot successfully land on a carrier again. Aircraft carriers are among the largest warships and the most valuable sea-based assets. For this reason, they are a high priority for protection, and from another point of view, they are the first target to attack. Another critical issue for aircraft carriers is that they are not considered surface combatants, as their attack capability comes from their air wings rather than onboard weapons.

- *Destroyers:* Destroyers are among the most used ships in navies worldwide. They are fast, maneuverable, and long-endurance warships that provide land-attack, air, water surface, and submarine defense capabilities. Their high firepower and endurance make them ideal for wars and escort operations. Today, they are the heaviest surface combatants after the cruiser.

- *Frigates:* Frigates are usually considered ships weighing over 3,000 tons. Even though frigates are smaller offensive vessels than destroyers, the line between a frigate and a destroyer is not specific. Frigates tend to focus more on antisubmarine missions. However, both classes are frequently capable of multimission. Generally, a destroyer is heavier, carries more firepower, and is slightly faster than a frigate. The role of frigates is to protect other ships of their strike group. The central part of this responsibility is to protect them from hostile submarines.

- *Cruisers:* Modern cruisers are generally the largest ships in a fleet, after aircraft carriers and amphibious assault ships, and can usually perform several roles. The cruiser is the strongest of the surface combatant class. It was built to have significant firepower and to be able to take out everything that its strike group could face. Due to some of the new destroyers having greater firepower than some cruisers, the line between cruisers and destroyers is a blur. The cruiser's role varied according to the ship's structure and the navy's operational approach. However, their missions often include air defense and land attacks.

- *Amphibious ships:* This class includes amphibious assault ships, transport dock ships, and landing ships. Amphibious ships range from 522 to 844 ft (160m to 260m) long and use landing craft and helicopters to move Marine Corps equipment and vehicles ashore. Amphibious assault ships are the biggest in this class. The amphibious assault ships provide the means for putting marines onshore using helicopters and landing craft. The role of the amphibious assault ship is fundamentally different from that of a standard aircraft carrier: its aviation facilities have the primary function of hosting helicopters to support forces ashore rather than to support strike aircraft. However, some can serve in sea control, operating as a safe base for many STOVL fighters. Amphibious transport dock ships carry Marines and landing craft for land assaults. Dock landing ships carry landing craft. They also have some logistics abilities.

Even though the self-protection concept may vary from one ship to a fleet in naval concepts, here, the self-protection of one vessel is considered. From the operational roles, structural differences, and similarities of the maritime surface platforms, some important conclusions can be drawn about them.

- Naval vessels do not have sufficient maneuvering capability to avoid guided missiles, even used in coordination with self-protection jamming and CMDS, also known as soft-kill methods.
- The time between threat detection and response is limited and has a mean of 40 seconds.
- A more extensive body means a larger RCS, and self-protection jamming may not sufficiently increase *jammer-to-signal* (J/S) to protect the vessel from a

radar-guided threat. So, especially for aircraft carriers and amphibious assault ships, self-protection jamming and CMDS usage against radar-guided threats will be ineffective.

- A strict requirement using missile defense capabilities or hard-kill techniques, such as interception missiles and *close-in-weapon systems* (CIWS), by combining data from multiple battle forces to air search sensors to obtain a real-time, composite track picture.

Considering the points mentioned above, the difficulty of defending against antiship missiles is deductible. Generally, naval vessels use a *ship self-defense system* (SSDS) rather than a SPEW system [10]. SSDS is not a solitary self-protection EW system but a self-protection system with some EW subunits—this self-protection system construction results from the warships' structure, the operational environment and threat types.

After detecting missiles, naval vessels use a three-layer defense system. In the outer layer, ships commonly use interception missiles. In the final layer, they use CIWS. In the middle layer, vessels usually use soft-kill techniques alone or combined with hard-kill techniques. Soft-kill techniques are either active *electronic countermeasures* (ECM) or passive decoys. A passive decoy system or CMDS launches chaff and flare (IR decoys) against RF and IR guidance systems.

Naval vessels' self-protection systems include ECM systems, radar and IR-based sensors, CMDS, interception missiles, CIWS, *identification friend or foe* (IFF) systems, displays, and control systems. Multiple individual processors allow each subsystem to have its computer, intending to operate at minimal CPU loading. The simplified distributed processing environment is a particularly effective way to avoid resource contention problems among multiple programs sharing the same CPU, add functionality, and allow more predictable software processing performance.

The distributed structure of a typical SSDS physical architecture is given in Figure 1.6. Each subsystem has its display and control systems and connects to a *local area network* (LAN) via an interface unit called the *LAN access unit* (LAU). The warship supplies the power requirement of each subsystem. However, they may have their DC-DC converters for specific power requirements. The subsystems take and send information to the *combat direction system* (CDS) via LAN. CDS is a centralized, automated *command-and-control* (C2) system collecting and correlating combat information.

The EO/IR sensor subsystem shown in Figure 1.6 does not only represent a 360° panoramic day and night air and surface surveillance and situational awareness system. It may also mean *IR search and track* (IRST) and *forward-looking IR* (FLIR) systems. All three EO/IR sensors, or only two or one, may exist in a vessel.

The SSDS system of a naval vessel is composed of many different subsystems and has a distributed system architecture. The distributed system architecture consists of several components spread across other computers but operates as a single network. In SSDS architecture, different countermeasure subsystems use different CMUs, which provides flexibility and decentralization for a quick response against threats. Some subsystems try to sense the dangers, while others aim to protect vessels using hard and soft-kill countermeasures. Since it comprises different subsystems, forming a centric decision for SSDS brings some troubles. The main problem is

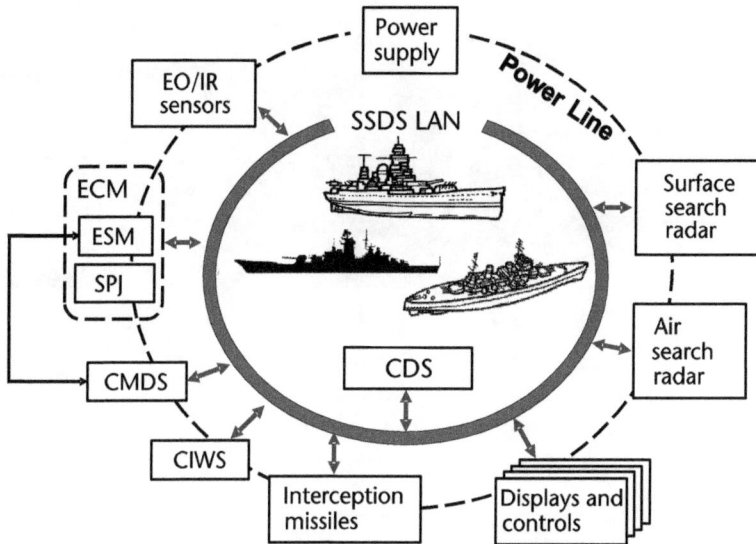

Figure 1.6 The distributed structure of a typical SSDS physical architecture.

deciding on hard-kill and soft-kill usage against threats. Preprogramming planning enables the determination of the response of SSDS subsystems to hostile systems.

The preprogramming includes the definition of target priority and optimization processes, depending on resource allocation. Resource allocation is often called the *weapon-target assignment* (WTA) or hard-kill and soft-kill deployment decision processes. The problem is optimally assigning weapons to the enemy targets so that the total expected survival value of the targets after all the engagements is minimized. Efficient solutions to this problem are of great importance to operations. So, in an engagement with the enemy, the problem must be solved in real time. The enormous combinatorial complexity of the situation implies that, even with the supercomputers available today, optimal solutions cannot be obtained in real time [11]. Thus, the response sequence generated by the system during operation may result in weak coordination between the subsystems, and sometimes the corruption of orchestration may be encountered.

It would be helpful to demonstrate the SSDS's layout on a warship. The layouts for different warships may vary, but the given example considers subunits' probable placements. Figure 1.7 shows the layout of a generic onboard SSDS on a warship. The CDS subsystem is inside the ship, but the other subsystems are outside, or only some parts are inside. However, some subsystems are inside a separate shelter.

1.3.3 SPEW Systems in Unmanned Air Vehicles

Recently, *unmanned air vehicles* (UAVs) entered the operational field in various roles and took over conventional, manned airborne platforms. Future projections show that the percentage of UAV missions will continue to increase. This increment is because using UAVs in operations eliminates the risk to a pilot's life, especially for high-risk missions. Moreover, UAVs are cheaper to procure and operate than human-crewed aircraft. Before discussing the usage of SPEW systems in UAVs,

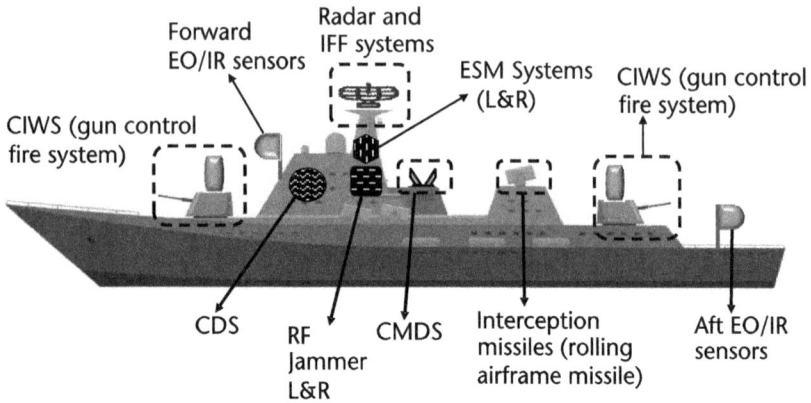

Figure 1.7 A generic SSDS layout for a warship.

brief information about UAVs' historical development and operational requirements is given.

UAVs were first tested during World War I; however, their first use in combat
was in World War II. The British constructed several UAVs during the interwar
period to be used as flying bombs and practice for anti-aircraft artillery targets. In
1942, the US Navy conducted the first experiments in adding weapons to a UAV
by attaching a television camera, transmitter, and torpedo to the aircraft. Even
though the tests succeeded, the program was canceled. At the same time, the US
Army studied the reconnaissance drone; however, it was used after World War II.
World War II's most famous use of uncrewed flight was Germany's V-1 and V-2
rocket deployment.

After World War II, development studies continued for surveillance and reconnaissance UAVs and target drones. Basic armed UAVs for naval purposes were used.
In different UAV models, piston engines, booster rockets, and jet engines were tested.
Drones with propellers were developed during this period. Reconnaissance UAVs
made their first combat appearance during the Vietnam War. During this war, an
ultrafast reconnaissance UAV capable of reaching Mach-3 was tested. Furthermore,
some drones provided live video feedback to the ship, which they engaged in directing naval gunfire. The first armed UAV was tried in this war.

After the Vietnam War, UAVs were developed in various ways, such as flying
performance, payloads, data-link packages, and remote-control systems. Israel
made a notable contribution to UAV studies during the 1970s. During this period,
they developed and used UAVs, drones' surveillance and intelligence-gathering
capabilities [12].

After the 1980s, many countries made significant investments in developing
UAV systems. Some other countries that did not have sufficient technology but
wanted to keep pace with the UAV technology made investments in procurement.
When the Cold War probability ceased, every country changed its defense policy in
the 1990s and added UAV missions to its operational strategies. After the 1990s,
UAVs provided direct support to ground forces in combat.

From the 1990s to the present, significant developments have occurred in the
light of experiences and operational requirements. The developments in UAV systems

can be examined in four subgroups. As an additional clarification, the overall system is the UAV system, and the air vehicle is the UAV. Although UAVs and drones have been mentioned separately for air vehicles up to this point, using only UAVs for both is a common approach.

1. *Air vehicle:* The central unit for a UAV system is the air vehicle or UAV. Air vehicles have developed in speed, operational altitude, payload carrying capacity, operating range, endurance, and survivability. Furthermore, avionics capabilities have improved.
2. *Payload:* There is an aim for each UAV mission; thus, payload requirements will change corresponding to the mission type. These payloads may be used for image intelligence, *communication intelligence* (COMINT), *electronic intelligence* (ELINT), *direction finding* (DF), communication relay, target designation, and jammer. Furthermore, image intelligence may be visual band, IR, or *synthetic aperture radar* (SAR) images. The jammer payload is not for self-protection jamming purposes, but it has a stand-in jamming capability. Missiles and bombs have been used recently in larger UAVs.
3. *C2 unit:* The C2 unit performs many tasks, including mission planning, payload control, and communication. A C2 unit is called a *ground control station* (GCS) or *mission control station* (MCS). Depending on the characteristics of the UAV system, GCS may be a laptop computer, a large control device, or a fixed facility. The main developments have occurred in air vehicle and MCS interoperability and commonality. In addition, enhancements have happened in the operational range between UAV and MCS and processing and timeliness properties.
4. *Data link:* Data links include any communication system between the UAV, the MCS and the required authorities. The data links of the UAV can operate either in the *line of sight* (LOS) or *beyond LOS* (BLOS). The developments have occurred in the link range, carrier signal frequency, and information signal bandwidth. With frequency agility techniques, an augmentation in transmission signal bandwidth is expected. In addition, some improvements have come about in ECM protection, interoperability, and timeliness. Another enhancement occurred in the transformation of analog data links into digital ones.

1.3.3.1 Operational Requirements of UAVs

Before discussing the operational requirements of UAV systems, consider whether their military roles will be appropriate. The general classification for military missions of UAV systems may be separated into five main task areas for present concept and future projection:

- Reconnaissance and surveillance support;
- Offensive (such as small flying aircraft as potential explosive devices and bigger UAVs in fighter and bomber roles);
- Platform simulation (such as miniature air-launched decoys-MALDs);
- Air interception;

- Special and specific missions (such as cargo, medical support, target designation and EW purposes).

Some tasks falling within the scope of their duties are currently being carried out. However, some missions are unavailable or only available in some regions worldwide. UAVs are used for reconnaissance, surveillance, platform simulation, cargo and medical support and *stand-in-jammer* (SIJ) missions. However, target designation and bomber duties are relatively new concepts, and they have started to be used in recent years. However, using UAVs in fighter and air interception roles is a conceptual approach. The military visions for the future of almost all countries predict that UAVs will appear more in the operational field, and more and more missions will be shifted to drones. In addition, the most striking future expectation is the ability of UAVs to serve in place of fighters with human control or automatically in adversaries' territory.

Before continuing with new concepts, which are the bases for using SPEW in UAVs, it would be helpful to draw a general framework about the current usage concepts of UAVs. The currently conducted UAV missions are based on forces in Table 1.1 [13].

It can be concluded that, by taking most of the existing usage concepts given in Table 1.1, UAVs do not require a SPEW system to protect themselves from RF guidance threats, or it would be an excessive investment for the UAV system. For a UAV to need a SPEW system, it must be too precious to sacrifice it, perform the mission in enemy territory, and have sufficient space for integrating the SPEW system. At this point, giving the classification for UAVs would be appropriate;

Table 1.1 The Military Missions of UAVs

Army	• Reconnaissance • Surveillance of enemy activity • Monitoring of *nuclear, biological, or chemical* (NBC) contamination • Signals intelligence (mainly COMINT) • Target designation and monitoring • Location and destruction of land mines
Navy	• Shadowing enemy fleets • Decoying missiles by the emission of artificial signatures • ELINT • Relaying radio signals • Protection of ports from offshore attack • Placement and monitoring of sonar buoys and possibly other forms of antisubmarine warfare
Air Force	• Long-range, high-altitude surveillance • Radar system jamming (SIJ) • Anti-radiation UAV for SEAD • ELINT • COMINT (mainly for *command, control, and communications* (C3) structure) • Airfield base security • Airfield damage assessment • Elimination of unexploded bombs

however, they have many different types [13, 14]. Although UAV systems include many components other than the aircraft, they are generally classified according to the capability or size of the plane. The most common classification for military UAVs is the NATO UAV classification in [14], which will be used here. Sharp lines do not separate the classification boundaries since UAVs can conduct missions in different classes. Furthermore, technological development provides smaller devices with higher capabilities. For this reason, the definitions given for UAV classes are not very strict and can be changed. The classification is shown in Table 1.2.

The propulsion system types used in UAVs are not mentioned in the classification since the functional properties of the UAVs are more important than the structure. Class specifications are more important than how they are obtained so that any proper engine can be used in each class. However, the maneuvering capabilities and mission types of UAVs vary with the propulsion method. For this purpose, the propulsion methods for UAVs are mentioned here. There are four primary types of engines used to propel UAVs. They are four-cycle and two-cycle reciprocating internal combustion engines, rotary engines, and gas turbines. Another one is the electric motor, which plays an increasing role in Class-I types of UAVs [15]. UAVs use gas turbine engines, such as turbojet, turbofan, and turboprop engines. The maneuvering capability of UAVs with turbojet and turbofan engines can be as much as the fighters' maneuvering capability or further. Furthermore, unstable designs of some UAVs might be too dangerous for humans.

As stated before, a SPEW system in a UAV must have some functional and physical properties. First, the UAV should be valuable and cannot be sacrificed. Second, it operates in the enemy territory or lethal border of the enemy IADS. The only exception is that the UAV is in the air interception role. Thus, UAVs must be in reconnaissance, surveillance, *signals intelligence* (SIGINT), fighter, bomber, or target designator roles. As another issue, the UAV must have enough space, payload, and power supply capacity. In light of these explanations, SPEW systems can be integrated into Class-II and III-type UAVs. Lastly, the mission radius or the range between MCS and UAV must be unlimited; otherwise, the UAV is not proper for missions in enemy territory. The situation may get out of control due to an operation in the enemy territory. Furthermore, the link may be lost due to over-range.

Table 1.2 The Classification of UAVs

Class	Category	Operation Altitude	Mission Radius
Class-I (<150 kg)	Small (>20 kg)	Up to 5,000 ft *above ground level* (AGL)	50 km (LOS)
	Mini (2–20 kg)	Up to 3,000 ft AGL	25 km (LOS)
	Micro (<2 kg)	Up to 200 ft AGL	5 km (LOS)
Class-II (150–600 kg)	Tactical	Up to 10,000 ft AGL	200 km (LOS)
Class-III (>600 kg)	Strike Combat	Up to 65,000 ft AGL	Unlimited (BLOS)
	High altitude long endurance (HALE)	Up to 65,000 ft AGL	Unlimited (BLOS)
	Medium altitude long endurance (MALE)	Up to 45,000 ft AGL	Unlimited (BLOS)

Considering the functional and physical properties, the UAV category should be mounted on the SPEW system, MALE, HALE, or strike combat. With the internally mounted or external pod structure, the SPEW systems used in UAVs are being conceptualized, and some trials have been realized. Furthermore, there is no official notification of using the SPEW system in UAVs in operation. As a future projection, the properties and components of UAVs' SPEW systems will be like the SPEW systems of their counterparts in piloted aircraft. The SPEW system requirements of the UAVs are defined by their operational roles in conventional airborne platforms, such as fighter, cargo, and bomber aircraft or helicopters. However, the functional and structural differences between piloted air vehicles and UAVs create variations in UAV SPEW systems' operational requirements. At this point, operational and structural similarities between them should also be considered. These variations and similarities are due to the following items:

- A piloted aircraft has at least one person, and the crew must be protected at all costs. Even if the value of a UAV is very high, it cannot be compared with human life.
- In some missions, such as reconnaissance, surveillance, SIGINT, and target designation, UAVs move with low maneuvers and are vulnerable to RF and IR-guided threats. In such tasks, the behavior of UAVs at high altitudes is like cargo-type SIGINT aircraft, while at low altitudes, their behavior is like helicopters.
- UAVs with turbojet and turbofan engines are used in fighter roles like assault, bomber, and air interceptor. In this case, the SPEW system requirements of UAVs converge with the needs of fighters.
- Generally, in the operation of remotely piloted and autonomous UAVs, the focus point is the mission. However, in conventional airborne platforms, pilots pay attention to situational awareness and continuously follow threats in the operational environments. For this reason, SPEW systems of the UAVs conducting a mission in a threat environment should be appropriately developed to adapt to the functional requirements for all components of the UAV system.
- The SPEW system must be compatible with the UAV system's air vehicle, GCS and C2 components. Furthermore, it should provide the *electromagnetic interference* (EMI) and *electromagnetic compatibility* (EMC) requirements with the avionics systems and payloads. However, it will be sufficient to consider only air vehicle compatibility with the SPEW system for a piloted aircraft. Naturally, the aircraft's SPEW system should provide EMI/EMC requirements with the avionics, sensor, and weapon systems.

Considering the explanations, a general structure of a conceptual SPEW system for a UAV system is shown in Figure 1.8. The system is conceptual since no SPEW system has been used in the operational area. The concept of operating UAVs has not required using a SPEW system thus far. However, the operational concepts prepared recently for the future projections point to the UAV missions that require SPEW systems. The conceptual structure given in the figure is that the SPEW system is a self-separate system with its data bus. The SPEW system exchanges data with the UAV *flight and mission computer* (FMC) system via the UAV's main data bus.

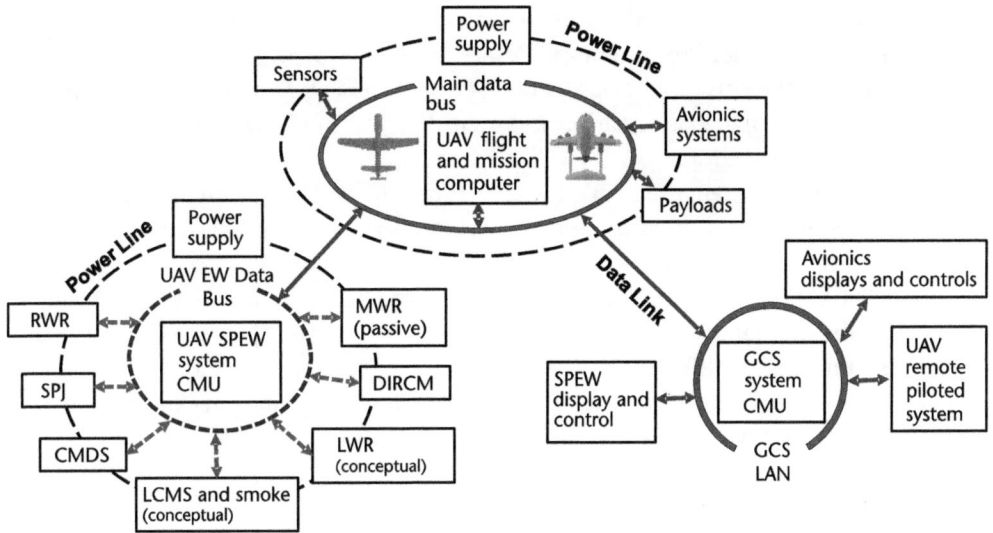

Figure 1.8   A conceptual SPEW system for a UAV system.

The SPEW subsystems of a UAV system can vary according to the role and payload capacity of the aircraft and the threats in the probable operational areas. A UAV's limited space, weight, and power capacity also define the types of SPEW subsystems and their power specifications. Thus, a detailed SWaP analysis is very important for explaining the SPEW specifications of a UAV system. In the requirement determination, a compromise is reached between the operational requirements and the SPEW specifications determined by the SWaP analysis. The system may be in an external pod or internally mounted distributed structure. Compatible internal designs will be preferred to prevent additional pod usage and maintain aerodynamics.

The SPEW system in Figure 1.8 shows that a self-protection system for a UAV will be composed of RF and IR warning and countermeasure systems. For low-maneuverable UAVs, such as using internal combustion engines and rotary engines, warning and countermeasure systems for laser-guided threats may be a suitable selection. There is a very vast difference between the SPEW systems of conventional airborne platforms and piloted UAVs. In both structures, pilots use displays and controls. For airborne platforms, the pilots are in the cockpit. However, the pilots are in the GCS for UAVs. For piloted UAVs, the data gathering and determining proper countermeasure processes occur in the SPEW system CMU. Although the SPEW system generates one or more options for appropriate countermeasures, the pilot is responsible for implementing one. The CMU system exchanges data with UAV flight and mission computers, and general coordination is conducted here. For this reason, the control and display data are exchanged between flight and mission control via a data link system.

For autonomous UAVs, SPEW system usage is a concept that needs to be worked on. In an autonomous structure, there will be no need to exchange data between the UAV flight and mission computer and GCS for a pilot's decision. The data may be exchanged for information or recording purposes only. In addition to gathering data and determining proper countermeasures, CMU is responsible for decision-making.

Precision decision-making requires fast data processing from bulky data, detecting threats, determining appropriate countermeasures, and finding the right decision for the situation. High-capacity computers and software are required for data analysis, threat classification and identification, optimization, decision-making and resource allocation.

SPEW system usage in piloted or autonomous UAVs is an innovation and a future concept. For this purpose, demonstrating the layout of SPEW systems into Class-II and III-type UAVs would be a conceptual projection. The operational requirements of a UAV system yield a projection for the suitable SPEW system's subunits and probable placements. Figure 1.9 shows a conceptual layout of an onboard SPEW system in a fighter UAV. The subunits are demonstrated from one sight; thus, a subsystem with a right-sided component has a counterpart on the left side. Also, except for antennas, the placement of subunits inside the aircraft would be suitable.

1.4   History of SPEW Systems

The history of the SPEW system is relatively new compared to that of EW. Historical information could be given from the beginning of SPEW systems. However, it is pertinent to summarize EW, which includes only essential milestones from its inception to the first appearance of the SPEW system.

There is no definite information about when and where the first EW application started. Strictly speaking, EW arose naturally due to the emergence and development of communications and radar theory. The first applications of EW were seen in the early 1900s.

The first recorded EW event occurred in 1904 in the Russo-Japanese War. It is assumed this was the first radio and wireless telegraph used in combat. Japan's Navy intercepted and analyzed Russian naval radio transmissions in the battles for intelligence. However, in April 1904, Russian wireless telegraphy stations were installed in the Port Arthur fortress, and Russian light cruisers successfully jammed wireless communication between a group of Japanese battleships. The spark-gap

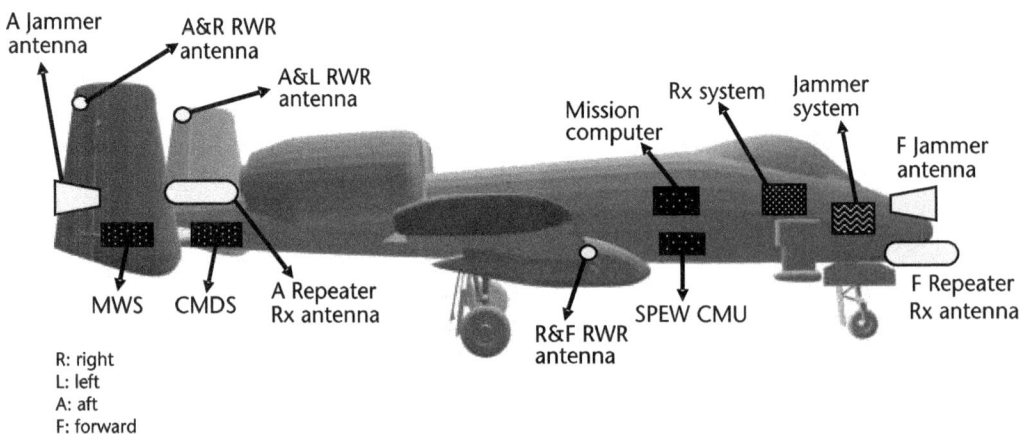

Figure 1.9 A conceptual SPEW system layout for a fighter UAV.

transmitters in the Russian stations generated senseless noise while the Japanese attempted to coordinate their efforts in bombing a Russian naval base.

Another recorded early EW event occurred in World War I. The Royal Navy's network picked up a spike in radio traffic from the German port of Wilhelmshaven. The navy commander Sir Henry Jackson focused on the radio signals, found the directions, and determined the movements of the German fleet in May 1916 [16]. The British fleet followed the German fleet according to the radio signal directions; however, some negations were experienced with this newly coined intelligence. The fleets came across the Danish peninsula of Jutland, and the largest sea battle of the war occurred. After a brutal but resultless battle, neither the German nor British fleets won a victory; however, this phenomenon passed into history as the first EW application.

Before World War II, *radar theory* was highly developed, and its applications began to be seen in the field of operations. During the war, radars were used in military applications. Radars' superior detection and guidance feature revealed the need to take countermeasures, and the first EW applications began to be seen in military operations.

This war, which emerged in the electromagnetic field, was called the *Battle of the Beam* by its characteristics. However, Germany's bombardment operations against England, the bombing operations of the Allies in Europe, the frequent change of electronic superiority between the parties, and the efforts to seize regional superiority with counterattacks were too intense to be handled with conventional methods. This war involving electronic systems resulted from the use of modern science and was called the War of the Wizards. During World War II, the vital importance of EW emerged. They were used in secret operations carried out in parallel with the war. EW applications during the war were also called crows or ravens, in addition to the battle of wizards.

ECM has been almost as rapid as the development of radars and communication systems. The first recorded electronic jamming event was the jamming of RF communication systems used by German tanks by British aircraft during the Libya campaign in November 1941. Compared to today's radios, radio sets on German tanks were primitive. Furthermore, they were only sometimes reliable due to atmospheric conditions, and networks could be overloaded. However, signaling in operation with tanks was undoubtedly a challenge with the communication available in 1941. The jamming equipment was also elementary compared to modern systems. Nevertheless, it worked, and the interruption of communication caused the tanks to become disorganized. The jamming event has been encouraging and has proven EW's operations successful.

Germany successfully applied RF technology to the military field and used radar-guided weapons. They inflicted heavy damage on the French coast and British ships through the English Channel. Germany's RF capabilities were not limited to these but were also used to jam radar systems. By jamming almost all British radars in the region, Germany passed the flagship *Scharnhorst* and a fleet through the Kaiser Wilhelm (Kiel) Channel from Brest to the North Sea. In this classic ECM operation, all British radars were jammed, and the British could not rely on any radar systems.

Although the movement of the German fleet was masked, the British learned valuable lessons from it. This operation proved the effectiveness of radar jamming. Furthermore, the Germans revealed their EW capabilities. This event pioneered the lessons learned with EW and resulted in *electronic countercountermeasures* (ECCM).

In 1942, a group of British Royal Air Force experts began working with US research personnel on a program called *radiation countermeasures*. This organization established a laboratory in Malvern/England in 1943 specifically to work on countermeasures and produced the first US/British-designed jammers. These systems were used in the B-17s, the first US aircraft deployed in communication and radar jamming operations.

England went from the Battles of Britain to the Battles of Berlin, from defensive to offensive. During the Battles of Berlin, Britain started to fight against German AI fighters rather than air defense weapons to break through the German defense. The Germans' high-sensitivity early warning radar system, Freya, detected the Allies' fighters and bombers during the daytime and in the dark. The detection was not only due to Freya's sensitivity but also because it was equipped with an IFF system, which had not been used before. Even though Freya's properties were very primitive compared to today's systems, its detection range could reach 110 km. For this reason, in the Battles of Berlin, attention was focused on the Freya early warning radars supported by the IFF feature and the ground radar named Wurzburg, which was quite precise compared to the period in which they were used and provided guidance to the AAA system.

The British aimed to learn about the capabilities of these radars to design the jammer that would neutralize the German air defense system. For this purpose, they captured the Wurzburg radar, which is directly connected to the firing system, with a commando operation. Research has shown that this radar has no ECCM other than its wide frequency band tunability. It was determined that the most effective countermeasure against this radar was chaff, consisting of small aluminum pieces with dipole properties. The British were unwilling to use chaff because they did not want Germany to find the working principle of the aluminum particles. Thus, Germans would have used chaff against them or produced a countermeasure. Chaff was used as an ECM for passive jamming or target cloaking. Another effect of chaff was the disruptions that it created in the radio communication used for information flow between radar operators and fighters during combat.

EW was first used as part of the general operational plan in the Normandy landings. To disguise their motives, the Allies made it appear that the activity was intense in Dover, a remote area from the landing, giving the impression that the landing would occur from the Calais region. This operation plan consists of three parts:

- *Part 1:* Increasing the radio communications around Dover to convince the Germans that their forces were concentrated in this area.
- *Part 2:* Jamming the radars around Calais to mislead that a landing would be made here.
- *Part 3:* Conceal the actual amphibious operation by jamming the radars around Normandy.

The planned operation was successfully implemented with the support of EW, and the German forces were prevented from taking the necessary precautions against the landing and bringing their strategic reserves to the region.

EW studies, which stagnated after World War II, revived during the Korean War. Most US equipment was used in World War II at the beginning of the war, and there were very few EW experts. North Korea had the RUS II radar, manufactured by the USSR, operating at 70 MHz. It has been noticed that, when this radar works, it causes the light on the landing gear of the US planes to turn on, thus providing an indirect warning that the radar has caught the aircraft. This coincidence is one of the fundamental events in the RWR systems used in the SPEW system. In the later phase of the war, North Korea used radars operating in the X-band (8–12 GHz) manufactured by China. This event led to the strengthening of the North Korean air defense system and to an increase in American losses. In the war that lasted for 3 years, the United States lost about 1,400 aircraft. A critical lesson from this war was that EW should become indispensable to air operations.

The SPEW system idea first emerged in the Vietnam War. North Vietnam built a powerful air defense system with the help of the USSR and the People's Republic of China. EW gained its importance today with the US Air Force and Navy Force operation against the SAM systems and radar-controlled AAA in North Vietnam in 1965. However, the ECM capability of the United States was only EB-66 *escort jammer* (EJ) aircraft. The ECM capabilities of EB-66s were effective against early warning radars, SAM, and AAA systems at the beginning of the operations. EB-66 aircraft were penetrating deep into North Vietnam with reconnaissance and strike forces. Over time, North Vietnamese defense systems engaged EB-66 aircraft as the primary target, forcing them to withdraw. This situation has led to the emergence of the self-protection concept.

Vietnam saw many conceptual EW developments, like SOJ and SEAD. The EB-66 aircraft was first used in the EJ role, and then they were used for SOJ. Further, some aircraft were dedicated to the "Wild Weasel" role, named after a project aimed at detecting and engaging missile threats. These aircraft used specialized equipment that enabled them to conduct the new mission role of SEAD. As the Vietnam War continued, more advanced and complicated EW systems, such as the EA 6B (Prowler) SOJ aircraft, were brought into service [17].

In the first phase of the SPEW concept, fighters started to use ECM pods that protect against radar-guided threats, so they no longer required an EJ with them, except in exceptional cases. Although they received the most attention, EW in Vietnam focused on more than radar threats. IR-guided missiles that were targeted using the large heat signature of aircraft were operationalized in Vietnam. These missiles motivated the development and fielding of a countermeasure in the form of flares [17]. The systems used in early SPEW applications were AN/APR-25 RWR, AN/ALQ-51, AN/ALQ-71 ECM pods, and AN/ALE-29 CMDS. The AN/APR-25 RWR's operating frequency is 2 to 12 GHz. The AN/ALQ-51 is a deceptive track breaker to counter S-band fire control radars. However, the operating frequency of the AN/ALQ-71 ECM pod is 1 to 8 GHz, and it can only be applied to noise jamming. The AN/ALE-29 CMDS has no automatic mode, and all three systems have a simple structure according to contemporary counterparts.

In the proceeding phases of the SPEW systems, the following progress and evolvements have occurred:

- SPEW systems have been used on different platforms, such as helicopters, vessels, cargo aircraft, and bombers. In addition to the pod structure, internally mounted, platform-specific systems have been developed.
- In the meantime, new detecting and countermeasure subsystems have been added to the SPEW systems for different platforms. These systems are active and passive MWSs, LWS, LCMS, DIRCM, smoke, disposable and towed decoys.
- Each new development has presented a more capable SPEW system. These developments have occurred in the frequency range, sensitivity, processing capacity, speed, and jamming power and techniques. In addition, their weight, volume, and power requirements have been reduced. In addition, high-capacity computers increase processing capabilities while shortening the processing time of CMUs.
- Additional improvements have occurred in the threat detection and handling capability. The first occurred in detecting and identifying simultaneous threat numbers in RF and IR warning systems. Furthermore, some improvements have taken place in the accuracy of the *direction of arrival* (DOA) and *time of arrival* (TOA). Another one has been in spoofing systems like the *digital radio frequency memory* (DRFM) units that came into play and changed the nature of the jamming. DRFM is a computer-controlled digital device used in radar jamming systems. SPJ systems employing DRFM can rapidly and accurately generate coherent jamming based on the memorized threat radar signal.

At this point, narrowing the scope and giving the technological developments a time sequence for only the RF part of the SPEW systems would be beneficial. It will be helpful to inspect the developments and the requirements for the RWR and SPJ systems separately.

RWR systems:

- Little was known about the threat at the beginning of the RWR journey. Moreover, the RWR design philosophy was based on gathering as much signal information as possible. Later, this philosophy changed as the sorting analysis and prioritization of lethal pulses were needed.
- The first RWR used in the operational area is AN/APR-25. It was used with the AN/APR-26 launch warning receiver, the pioneer of missile warning systems. During the Vietnam War, these two systems entered service in 1966 and were used in U-8, U-21, F-4, F-104C, F-100, and F-105D. Its operational frequency range is 2 to 12 GHz. The system only used crystal-video detection techniques in the threat bands. Due to its low technology, AN/APR-25 suffered from poor discrimination and a high falsealarm rate. Also, the system was not automatic. When these handicaps were combined, the system's unsuitability for single-seat aircraft such as the F-104C and F-105D was understood. However, the system had successfully served the Wild Weasels, a dedicated *electronic weapons officer* (EWO), and the F4Cs had a backseater. These crews tried

to sort out the dangerous signals from the spurious ones. Nevertheless, the impact of the AN/APR-25 system was optimistic since it was the first sample of a new concept and technology.

- The updated version of the AN/APR-25 system was the AN/APR-36 system, which used a superheterodyne analysis Rx. The system increased the reaction success with cooperation between the pilot and EWO. Furthermore, it provided an instant alert and maximum reaction time for evasive action. However, it was still not suitable for single-seat aircraft. The design philosophies in AN/APR-25 and AN/APR-36 systems were based on obtaining as much signal data as possible.

- As the higher technological SAM systems came onto the battlefields, a new system, AN/ALR-45, emerged in 1969. This RWR used *crystal video Rx* (CVR) and was the first digital system that incorporated hybrid microcircuits using digital logic and clock drivers. This system's RWR frequency range sets the standard levels at 2 to 18 GHz, and four cavity-backed spiral antennas feed four CVRs. Furthermore, the AN/ALR-45 system started discarding nonlethal threat information. This system had another important property: its MIL-STD-1553B bus-compatible display terminal. The prioritization of threats and emitter tagging became possible with computational power. It was used in F-8, F-14, A-4, RA-5C, A-6, EA-6B, A-7, F-100, F-4, F/A-18, and CF-104 platforms.

- In the mid-1970s, the first digital, software-controlled RWR, AN/ALR-46, was produced. It had a *tuned radio frequency* (TRF) receiver, and the technical specifications were similar to the AN/ALR-45. Based on the processing concept, this system created software routines so that probability statements on frequency, *pulse repetition interval* (PRI), *pulse width* (PW), modulation, and any other required signal properties could be constantly analyzed. Also, the computer would deinterleave pulse trains and be capable of squadron-level reprogramming using flight line equipment. The first computer designed explicitly for EW applications was developed during this period.

- The AN/ALR-67 and the AN/ALR-69 RWR systems were developed in the following period. Furthermore, their technical properties have continued to improve up to the present. The first AN/ALR-67 system was used with the third-generation fighters and has continuously been developed. The second and third versions are used with the fourth-generation fighters. However, the AN/ALR-69 RWR system is an improved variant of AN/ALR-46. It also has an improved, fully digital version called AN/ALR-69A. Furthermore, AN/ALR-74 is an improved version of this system. The first contribution of the systems occurs in the computational and signal processing capabilities. The other is in the extended operational frequency, 0.5 to 20 GHz. Also, four high-band antennas with 360° RF coverage, four wideband, and a narrowband superheterodyne Rx are the other contributions to the RWR technology.

- One RWR system with significant technical contributions produced after the 1990s is AN/ALR-91. The system's technological innovation can continuously collect and analyze radar signals over the 0.5 to 18.0-GHz frequency range. Prior and current systems use the band-sampling technique. Thus, they collect and analyze signal data of one band at a time. Three or four bands are usually required to cover the entire frequency range. As a result, detecting radar signals on other bands is impossible when band sampling is underway. The ALR-91 system does not require band-sampling, as all bands are collected simultaneously. This capability allows a short detection time for all tactical threat radars.

- The other AN/ALR-93 is a lightweight, high-sensitivity, C through J-band system designed to operate in dense, complex emitter environments with a high probability of intercept capability. It is an effective system to use against these modern threats, including lethal radars that operate in CW, wide pulse widths and high-duty cycle, *track-while-scan* (TWS) radars, high-duty cycle emitters capable of masking the detection of other radars, and agility in PRI and RF radars. The Rx configuration of the ALR-93 system contributes to its effective operation. Baseband Rx and a wide acquisition bandwidth *instantaneous frequency measurement* (IFM) Rx cover the entire frequency spectrum in continuous bands while maintaining a high probability of intercept. The system incorporates a superheterodyne Rx for increased sensitivity and selectivity with narrowband frequency search modes.

- The developments have occurred not only on airborne platforms but also on naval surface platforms. The AN/ALR-66 is a multipurpose *electronic support measure* (ESM) suite for aircraft and surface vessels. It has been successfully integrated into E-6A, P-3, SH-2F, SH-3H, A-7, and F-4 aircraft. However, The AN/ALR-66(V)6 is a variant available for shipboard applications.

- Today's requirements from the RWR systems can be summarized as follows:
  - Full range frequency coverage in 0.5 to 40 GHz.
  - Continuous 360° azimuth and maximum possible elevation RF coverage.
  - High signal processing capability and reprogrammability.
  - High sensitivity and AOA accuracy.
  - Real-time detection of all the threat signals, including LPI radars, in any operational environment.
  - Fully digital and compatible with MIL-STD-1553B and ARINC-429.
  - Handling CW and pulsed radar signals simultaneously.
  - Low volume, weight, and power requirement.

- Some other RWR types in service are given below:
  - ALR-400 (Spain): A new generation, wideband digital RWR system operates from 0.5 to 42 GHz (configurable). It is used in EF-18A/B, A400M, C-295, CH-47, Cougar, TIGER, NH90, and CH-53 platforms.
  - AN/APR-39 (United States): This digital RWR family has four versions designed for helicopter platforms.
  - BOW-21 (Sweden): This modular RWR was designed for the JAS-39 Gripen aircraft.
  - L-150 (Russia): This digital RWR, with an internally mounted system, works in the 1.2 to 18.0-GHz frequency range and covers 360° horizontal and 60° vertical RF coverage. It is used in Russian aircraft.
  - SkyGuardian 2000 (United Kingdom): This typical RWR system covers the most common radar frequencies from 2 to 18 GHz. It is used in EH-101 and Tornados.

SPJ systems:

Every SPJ system must have an Rx system. Otherwise, the system can only jam preset frequencies with limited power and may fall outside the operation of the SPJ concept. Furthermore, emissions may be dangerous in some aspects without detecting an RF threat and waste resources. For this reason, operating an SPJ system without an RWR system is not usual, and the development of the RWR systems is given above. The products of SPJ systems from primitive to advanced over time and the growth in requirements accordingly are summarized below.

- The first SPJ AN/ALQ-51 system was used in the US Navy in the last quarter of 1965, after shooting down a plane in a SAM attack in August 1965. In January 1967, the US Air Force used AN/ALQ-71 pods [18].
- AN/ALQ-51 Mod I was a deceptive track breaker to counter S-band fire control radars that employ pulse ranging, frequency modulating carrier wave pulse, and conical scan. AN/ALQ-51 was used with AN/ALQ-41, X-band track breaker. The system performed *range gate pull-off* (RGPO), frequency translation, inverse conical scan deception, and angular deception techniques. The upgraded version of AN/ALQ-51 Mod II was a S/E/F-band deception jammer and track breaker. AN/ALQ-51 family SPJ pods used on F-4, A-4, F-8E/J, A-3, A-5A, RA-5C, RF-101, and 147F platforms. In 1967, the AN/ALQ-51 systems were replaced by its upgraded system, AN/ALQ-100. The system provided azimuth, elevation, and range deception against fire control radars. The AN/ALQ-100 operated in the 1.8 to 8.0-GHz band for pulse radar and the 4.0 to 8.0-GHz band for continuous wave radars. The AN/ALQ-100 would simultaneously deceive radars in one or more frequency bands or modes of operation.
- The AN/ALQ-71 was an air-to-ground active electronic attack pod. Its frequency range was 1 to 8 GHz, emission power was 90W, and it had two channel noise jammers. The system could jam ground early warning radars or naval vessels. It was used on RF-101C, F-105F, A-7, B-52, B-57, EB-66, F-4, T-33, and F-16 platforms. Also, another SPJ pod used in the Vietnam War was AN/ALQ-72. It was an air-to-air active electronic attack pod to jam air or ground radars. Its frequency range was 9 to 20 GHz, emission power was 100W, and it had two channel noise jammers. Its emission power was 100W.
- A critical milestone in the SPJ technology was AN/ALQ-119. In 1972, it was first used in combat. It covered the 2 to 20-GHz frequency band. Also, it was one of the first jamming systems that could simultaneously perform the dual-mode noise/deception jammer. The AN/ALQ-119 and the AN/ALR-46 RWR system combination were called Compass Tie. The AN/ALQ-119 proved its success during the Vietnam War, the Yom Kippur War, and, to a lesser extent, during Operation Desert Storm. It was used on F-4, F-15, F-16, and A-10 aircraft.
- A pioneer usage for a multiple-false-target generator in the SPJ system technology was provided by AN/ALQ-122 SNOE (*Smart Noise Operation Equipment*). It is a long-range radar jammer carried on the B-52H. The system automatically searches, acquires, and tracks threat radar signals, generating a low-duty-cycle ECM program to deny range and azimuth [19].

- The AN/ALQ-131 was designed to provide advanced broadband coverage against various radar-guided weapons. The system is automatic, and the frequency range is 2 to 20 GHz. The technological contributions of the system have been broadband coverage and a modular self-protection structure. The ALQ-131 is used by F-16, A-10, F-4, and C-130 aircraft. A second version of the system has been realized by adding a new Rx and processor.
- The AN/ALQ-135 was one of the first internal or integrated countermeasure systems and incorporated miniaturized component technology. It was a part of the *tactical EW system* (TEWS) component for the F-15 fighters. It was installed in varying configurations in the F-15A, C, E, and S models and operated with the AN/ALR-56 RWR, the AN/ALQ-128 RWR, and the AN/ALE-45 CMDS systems in different TEWS types [20].
- Another internal countermeasure system was Zeus, an integrated defensive aids suite (DAS). This system was one of the pioneers in providing a complete defensive capability consisting of an RWR and a jammer. It could also interface with the host platform's other EW units.
- The AN/ALQ-162(V) was a compact, lightweight countermeasure set that provides self-protection against CW radar-guided threats for fixed- and rotary-wing aircraft. After the first version, many versions were developed. The main contribution of the versions has been the ability to protect against CW and pulse Doppler radar threats simultaneously. It can be installed inside a pod, pylon, or internal configurations. The AN/ALQ-162(V)6 is the latest variant featuring a microwave power module that provides more than double the previous model's output power.
- The first DRFM EW application was AN/ALQ-165 Airborne SPJ (ASPJ). The system is an automated modular reprogrammable active RF deception jammer designed to contribute to the electronic self-protection of the host tactical aircraft from various air-to-air and surface-to-air RF threats. The basic system consists of five *line-replaceable units* (LRUs), which include two receivers, two transmitters, and one processor. Each LRU is interchangeable among different aircraft. Additional transmitters with larger RCS can be installed on aircraft to increase the effective radiated power. AN/ALQ-165 started development in 1979 but was canceled in 1992 because of not operationally suitable. This cancelation was because the system did not meet the required criteria for mission reliability or built-in-test (BIT) effectiveness. The system replaced the AN/ALQ-214 system's technique generator.
- In parallel with radar-guided SAM, AAA, and AI systems developments, improved ECM systems have been developed for airborne platforms. These systems increase radiated power, reduce ECM response time, and improve reliability. These systems utilize a central programmable computer for data analysis and system control, with independent microprocessors to direct the RWR, display, jamming, and countermeasures dispensing functions. In parallel with these improvements, the receiving, sampling, and memory capabilities of the DRFM systems have developed. The pod-structured examples of these systems are AN/ALQ-184, Russian Khirbiny, and Israel's EL/L-8222SB. However, the improved internal mounted self-protection ECM system examples

are AN/ALQ-178, AN/ALQ-187, and *Suite of Integrated Radio Frequency Countermeasures* (SIRFC) AN/ALQ-211.

- Furthermore, some EW self-protection systems are more conceptual and project for the following generation system requirements. It will also be beneficial to give some of these systems. The AN/ALQ-214(V) *RF countermeasures* (RFCM) system is designed to provide a range of aircraft types with next-generation protection from RF threats. The system's primary application is the F/A-18E/F carrier-borne, multirole combat aircraft where it is teamed with an RWR, the AN/AAR-57 Common Missile Warning System (CMWS), and the *Advanced Strategic/Tactical Expendable* (ASTE) infrared (IR) decoy flare to create a total defensive aids package [20]. The AN/ALQ-254 Viper Shield is an integrated EW suite developed for the F-16C/D Block 70/72 multirole fighter aircraft. The system will be integrated with the AN/APG-83 *Scalable Agile Beam Radar* (SABR). This integration will deepen situational awareness by merging the threat picture of the radar with that generated by the AN/ALQ-254's ESM. Another SPEW system is ALL-in-SMALL, a next-generation unified suite for airborne self-protection. ALL-in-SMALL has the properties of combination edge technologies, modularity, open architecture, and *Multilevel Redundancy* (MLR) design and growth capabilities. These properties provide operational solutions and flexibilities for all airborne platform types, such as fighters, helicopters, mission aircraft, and transporters. As seen from the sample systems, the most crucial contribution of these systems arises from their additional ESM capabilities.
- Some sophisticated SPEW systems for advanced aircraft can be considered here. SPS-170 family of self-protection jammers associated with Su-34 and Su-35 aircraft. The L402 Himalayas is an ECM suite intended to protect the Sukhoi Su-57 fifth-generation aircraft of the Russian Air Force. The AN/ASQ-239 system is an electronic warfare suite to protect the F-35, a fifth-generation aircraft. The AN/ALR-94 EW suite provides advanced self-protection to the F-22, a fifth-generation fighter of the US armed forces, by detecting and defeating surface and airborne threats. The specific properties of advanced EW self-protection systems are that they provide fully integrated radar warning, targeting support and self-protection to detect and defeat surface and airborne threats.
- Also, naval SPEW systems were developed in parallel. The most prevalent system used for this purpose is the AN/SLQ-32(V) series EW system. The designs are the US Navy's standard radar threat detection, analysis, and jamming suites. The system was initially developed to replace the AN/WLR-1 receiver and AN/ULQ-6 radar jammer combination for surface ships. The AN/SLQ-32(V) is modular and employs lens-feed, multibeam reception antennas for all signals except those in the lowest-frequency bands. The antenna technology consists of an array of receivers fed through coaxial cables by a strip line, multibeam, and parallel-plate lens. This arrangement provides a set of separate, contiguous, high-gain outputs that benefit from the available aperture's total gain. Each array offers more than an octave of frequency coverage within the system. The AN/SLQ-32(V) has been developed in five distinct variants [20].

- Another radar jammer system for naval vessels is Scorpion. The system is designed to counter long-range search target acquisition and missile seeker radars operating in the 7.5 to 18-GHz band search and lock-on modes. The Scorpion system is lightweight and high-power and uses effective jamming techniques. It can be integrated with Racal's Cutlass and Type-242 ES systems and different types of ES equipment.
- APECS II is a modular, 0.5 to 18-GHz band ES and ECM system designed for various vessel types and intended to counter the most known radar threats. It is a fully automatic system that identifies and jams threat emitters using instantaneous frequency-measuring receivers and microprocessor-controlled jammers. The jamming array is a phased array assembly, and high-speed switching enables up to 16 threats to be jammed simultaneously. APECS II comprises four subsystems: a mast-mounted ES antenna assembly, a system processing equipment package, an operator's console, and port and starboard ECM transmitter units with internally mounted transmission arrays [20].
- Scorpius-N (ELL-8256SB) is a long-distance RF ES and ECM system for naval vessels. The system effectively intercepts, analyzes, locates, tracks, and jams various airborne, shipborne, and shore-based systems, including fire control radars, search radars, *airborne early warning* (AEW) sensors, SAR, and missile seekers. The main properties of the system are multibeam, multijamming techniques, high sensitivity, and high *effective radiated power* (ERP), sophisticated jamming techniques using DRFM, and the ability to track and intercept *low-probability-of-intercept* (LPI) radars.

1.5   General Structure of Self-Protection EW Systems

The primary purpose of this book is to focus on the SPJ systems. SPJ systems are subsystems of the SPEW system. Therefore, understanding the relationship between the SPJ and SPEW systems is essential for the remaining part of the book. For this purpose, the place of the SPJ in a generally structured SPEW is described here.

Section 1.1 mentions the subsystems of a general SPEW system, and Figure 1.1 presents the block diagram of an available SPEW system. In parallel with the given subsystems, all the subsystems of the SPEW systems may be classified by the operational principles, as shown in Table 1.3.

The CMU subsystem coordinates all the SPEW system functions. As stated before, displays and controls may be dedicated to each subsystem, class, or whole system. CMU takes the required information from the capable RF and EO sensors to generate a countermeasure decision. RFJ is an RF band subsystem, and the CMU makes its usage decisions according to information from the RWR subsystem. Similarly, when using DIRCM or flare, active or passive MWS senses a threat activation, and, simultaneously, RWR does not sense any threat signal. Furthermore, the LWS detects a threat signal when using LCMS or smoke.

For using RFJ or chaff against an RF-guided threat, CMU must take the information about the *emitter identification* (EID) and direction from the RWR subsystem. The threat signal is discriminated from the received signals of many different

Table 1.3   The Classifications of the SPEW Subsystems

RF Band Subsystems	• RWR • RFJ
Optical Band Subsystems	• Active MWS • Passive MWS • LWS • DIRCM • LCMS • Smoke
Towed Decoys	• Towed decoy acts as a target for the incoming missile • Towed decoy jammer
Expandable Decoys	• CMDS • *Expandable Active Decoy* (EAD)—act as a target • EAD—jammer
Central and Side Subsystems	• CMU • Displays • Controls • Data Bus • Power Supply

sources. In this process, previously defined threats are detected and identified, and the threat information with its emitter is obtained. Then their approximate locations with range and direction are achieved. After RWR sends this information to CMU, the RFJ system applies preprogrammed jammer techniques. In addition to RFJ, some jamming methods may require chaff, maneuver, or both.

The most critical point for obtaining a highly efficient self-protection jamming depends on the accuracy of the threat's information in the environment. Furthermore, it is not ignored that the situational awareness obtained from an RWR system can only be as accurate as the data and skill it has programmed through its *Mission Data Set* (MDS) or *Pre-Flight Message* (PFM). So the first step of the self-protection jamming efficiency is related to the intelligence about the systems in the operational field, well-trained EW staff, and a properly equipped *Electronic Warfare Support Center* (EWSC) that makes possible the PFM programming. After achieving a complete PFM, another critical issue, jamming techniques, is developed. To that end, the requirements are as follows:

- Detailed technical threat information in the probable operational region and a detailed analysis of the threats.
- Ultimate information about the RWR and RFJ systems, not only their technical capabilities but also their limits, pros, and cons, should be known.
- An in-depth information about the EW doctrine and EW tactics.

Developing jamming techniques depends on using intelligence, systems, and technical and military information. In addition, the efficiency of the SPJ will be as much as the capabilities of the jamming technique.

References

[1]   AAP-6 (2021), *NATO Glossary of Terms and Definitions (English and French)*, NATO Standardization Office (NSO), December 15, 2021.

[2]   Joint Publication 3-13.1 (2012), *Electronic Warfare, USA Joint Chiefs of Staff (CJCS)*, February 8, 2012.

[3]   Adamy, D. L., *EW 102: A Second Course in Electronic Warfare*, Norwood, MA: Artech House, 2004.

[4]   Welch, M., and M. Pywell, *Electronic Warfare Test and Evaluation*, NATO RTO AGARDograph 300 Flight Test Technique Series, Vol. 28, December 2012.

[5]   Zygmunt, M., and K. Kopczynski, "Laser Warning System as an Element of Optoelectronic Battlefield Surveillance," *Proc. SPIE 11442, Radioelectronic Systems Conference*, Jachranka, Poland, 2019, 1144202.

[6]   Cheng, C., and J. Tsui, *Introduction to Electronic Warfare from the First Jamming to Machine Learning Techniques*, Denmark: River Publishers Series in Signal, Image and Speech Processing, 2021.

[7]   Knott, E., J. Shaeffer, and M. Tuley, *Radar Cross Section*, 2nd ed., Raleigh, NC: SciTech Publishing, 2004.

[8]   Pakfiliz, A. G., "Increasing Self-Protection Jammer Efficiency Using Radar Cross Section Adaptation," *Computers & Electrical Engineering*, Vol. 98, 2022, p. 107635.

[9]   US Environmental Protection Agency, *Phase I Final Rule and Technical Development Document of Uniform National Discharge Standards (UNDS)*, Section 2: Vessels of the Armed Forces, April 1999.

[10] Norcutt, L. S. "Ship Self-Defense System Architecture," *Johns Hopkins APL Technical Digest*, Vol. 22, No. 4, 2001, pp. 536–546.

[11] Benaskeur, A., E. Bosse, and D. Blodgett, "Combat Resource Allocation Planning in Naval Engagements," *Defence R&D Canada—Valcartier*, Technical Report, DRDC Valcartier TR 2005–486, August 2007.

[12] Blom, J. D., *Unmanned Aerial Systems: A Historical Perspective*, Fort Leavenworth, KS: Combat Studies Institute Press, 2010.

[13] Austin, R., *Unmanned Aircraft Systems*, New York: John Wiley & Sons, 2010.

[14] Dalamagkidis, K., "Classification of UAVs," in *Handbook of Unmanned Aerial Vehicles*, K. P. Valavanis and G. J. Vachtsevanos, (eds.), New York: Springer, 2015, pp. 83–90.

[15] Fahlstrom, P. G., and T. J. Gleason, *Introduction to UAV Systems*, 4th ed., New York: John Wiley & Sons, 2012.

[16] Bray, H., "You Are Here: From the Compass to GPS," *The History and Future of How We Find Ourselves*, New York: Basic Books, 2014.

[17] Woldhuis, D., *The Path to 5th Generation Electronic Warfare*, Air Power Development Centre, F3-G, Department of Defense, Canberra, Australia, 2018.

[18] Dickson, J. R., *Electronic Warfare In Vietnam: Did We Learn Our Lesson?* Air War College Research Report, Maxwell Air Force Base, Alabama, Air University USAF, May 1987.

[19] Blake, B., (ed.), *International Electronic Countermeasures Handbook*, Norwood, MA: Artech House, 2004.

[20] Streetly, M., (ed.), *Jane's Radar and Electronic Warfare Systems 2002–2003*, Janes Information Group; 2002–2003 edition, June 2002.

CHAPTER 2

Principles of Radar Systems and Radar-Guided Threat Systems

The emergence of EW showed up with the development of radar systems, and their mutual requirements have derived their enhancements in parallel. Radars mainly consist of the backbone of the air defense systems. Moreover, primarily, middle-range and long-range weapon guidance systems generally depend on radar systems. The requirements and the developments of the EW platform protection systems are sourced from radar technology. Thus, considering SPEW systems without discussing radar principles and radar-guided threat systems would be deficient. SPEW systems detect the radar signals and use proper ECM techniques for the radar-based threat systems. For this purpose, more than obtaining technical specifications about radar systems will be required. In other words, one must have detailed information about the theoretical principles of radar systems. So a radar system inspection from different aspects would be appropriate before discussing the SPEW systems. During tracking and launching missiles, radar-guided weapons are the threat systems. However, it changes the role of a victim when being jammed.

Understanding the operational and technical principles of the radar system is essential for radar-guided threat systems. Although radar-guided threat systems use different guidance methods, they use the same principles regarding radar operation rules. For this purpose, giving sufficient information about radar theory and guidance methods would be appropriate. The initial part of this section informs about radar basics, operation modes, and classification of radars. Then radar equations are derived for monostatic and bistatic cases. Radar operations for pulsed, CW, and pulsed Doppler modes are inspected separately. The last part presents tracking radar theory and analysis of radar guidance.

2.1 Radar Basics

2.1.1 The RF Spectrum

The RF spectrum is part of the EM spectrum with frequencies from 0 Hz to 3 THz. *International Telecommunication Union* (ITU) defines the RF spectrum into 12 bands. Each band begins at a wavelength, which is a power of 10 (10^n) meters, with the corresponding frequency of $3 \times 10^{(8-n)}$ Hz. Only the bands used for radar purposes are considered here. Thus, the frequency band between *extremely low frequency* (ELF) and *medium frequency* (MF), which corresponds to a 3-Hz to 3-MHz band, is omitted. Moreover, *terahertz/tremendously high frequency* (THF) represents 300 GHz to 3 THz and is excluded. Table 2.1 shows the ITU RF bands related to radar operation.

In addition to ITU classification, radar frequency classification by the *Institute of Electrical and Electronics Engineers* (IEEE) has subdivided the RF band differently. These bands, also known as radar design bands, are given in Table 2.2.

The NATO frequency bands' nomenclature originated during World War II for military radar applications. In 2014, the European Union (EU), NATO, and the United States military agreed on a set of EU-NATO-US ECM frequency bands for EM frequencies used primarily for radar. This classification is known as radar-frequency bands, which are defined by NATO for ECM systems. Thus, the classification is given in Table 2.3 and is generally used when discussing radar and EW systems.

This book considers the IEEE-defined RF bands given in Table 2.2 for radar purposes unless stated otherwise. When radar and EW are dealt with together, Table 2.3 is also used in addition to Table 2.2.

2.1.2 Radar Concept

Radar determines objects' distances and angular positions relative to a specific location. This location may be the radar, another radar, or a regional operations center.

Table 2.1 Radar-Related RF Bands and Frequency Ranges Defined by the ITU

Band Designator	Origin of Designator	Frequency Range
HF	High Frequency	3–30 MHz
VHF	Very High Frequency	30–300 MHz
UHF	Ultra-High Frequency	0.3–3 GHz
SHF	Super High Frequency	3–30 GHz
EHF	Extremely High Frequency	30–300 GHz

Table 2.2 Radar RF Bands Defined by the IEEE

Band Designator	Origin of Designator	Frequency Range
HF	High Frequency	3–30 MHz
VHF	Very High Frequency	30–300 MHz
UHF	Ultra-High Frequency	0.3–1 GHz
L	Long Wave	1–2 GHz
S	Short Wave	2–4 GHz
C	Compromise Between S and X	4–8 GHz
X	Used in World War II for fire control (X)	8–12 GHz
Ku	Kurz-Under	12–18 GHz
K	Kurz (Short in German)	18–26 GHz
Ka	Kurz-Above	26–40 GHz
V	Very High-Frequency Band	40–75 GHz
W	W Follows V in the Alphabet	75–110 GHz

Table 2.3   ECM Frequency Bands Defined by NATO

Band Designator	Frequency Range
A-band	0–250 MHz
B-band	250–500 MHz
C-band	500 MHz–1 GHz
D-band	1–2 GHz
E-band	2–3 GHz
F-band	3–4 GHz
G-band	4–6 GHz
H-band	6–8 GHz
I-band	8–10 GHz
J-band	10–20 GHz
K-band	20–40 GHz
L-band	40–60 GHz
M-band	60–100 GHz

In a radar operation, RF energy is transmitted to the environment generally using a directed antenna. Then the radar receives the backscattered signals in its direction from the objects. The critical point is that only a tiny part of the energy returns to the radar Rx antenna. The backscattered signals are called echoes. Radar uses the return that it receives to find the direction and distance of the object.

Radars are used for many different purposes, and according to usage aims, frequencies vary in a wide range, as given in Tables 2.1 and 2.2. Radar frequency may be reduced to HF ranges for particular purposes like detection beyond the horizon. Low-frequency systems can be used for foliage penetration applications, as well as in ground-penetrating applications. Operating frequencies of military radar systems are occasionally L and Ku-bands or 1 to 18 GHz. Above 18 GHz, RF waves suffer severe weather and atmospheric attenuation. Therefore, radars utilizing these frequency bands are limited to short-range applications, such as police traffic radars, short-range terrain avoidance, and terrain following radars. In the automotive industry, very small radars, operating at 75–76 GHz, are used for parking assistants, blind spots, and brake assists. Some radar systems operating above 90 GHz are used as laboratory experimental or prototype systems.

The term target is used to describe almost any object to be detected in radar literature. The targets of radars are airborne platforms, vessels, land vehicles, land-based structures, a specific point on the Earth, and all kinds of precipitation and aerosols. However, objects not assumed as targets, such as buildings, terrain features, and sea waves, may also cause high signal power returns. It is appropriate to discuss some rules of RF propagation to understand the mechanism of the backscattered signals generated by objects.

Suppose waves of a transmitter continue to propagate in space unless a thwart exists in the propagation path. Far from their origin, waves will have spread out enough that they will appear to have the same amplitude everywhere on the plane perpendicular to its direction of travel, specifically in the near vicinity of the observer. These types of waves are called plane waves. They are idealized waves whose value, at any moment, is constant over a plane normal to the direction of propagation.

LOS propagation is a characteristic of *electromagnetic* (EM) radiation, meaning that waves travel directly from the source to the receiver. RF waves, a subband of the EM spectrum, travel in a straight line above a specific frequency. The rays or waves may be diffracted, refracted, reflected, or absorbed by the atmosphere and obstructions with material and generally cannot travel over the horizon or behind obstacles. At low frequency, below approximately 3 MHz, due to diffraction, RF waves can travel as ground waves, which follow the contour of the Earth in contrast to LOS propagation. Thus, low-frequency over-the-horizon (OTH) radars can make detection beyond the horizon.

Skywaves or ionospheric waves can travel long distances at MF, HF, and lower portions of VHF under certain conditions. Skywaves are refracted back to Earth by the ionosphere, thus giving radio transmissions in approximately 1 to 30-MHz range a potentially global reach. However, these effects are insignificant at frequencies above 30 MHz and in lower levels of the atmosphere. EM waves higher than 30 MHz will follow the shortest direct path between transmitting and receiving antennas, and this direct path is referred to as the LOS path of propagation. Thus, any obstruction between the Tx antenna and the Rx antenna above 30 MHz will block the signal, just like the light transmission.

Since military radar's operating frequencies are higher than 30 MHz, LOS operation principles are valid. LOS obstruction can be met when an obstacle exists between the radar and targets, as in Figure 2.1(a). If the output power is sufficient, radar can detect Target A; however, it will not see Target B due to the mountains. LOS obstruction also occurs in flat regions, such as over the sea, when targets are beyond the LOS horizon and their altitudes are not sufficiently high. In Figure 2.1(b), the height of Target A is high enough that radar can detect it. Target B's altitude is low and is beyond the LOS horizon; thus, the radar cannot see it.

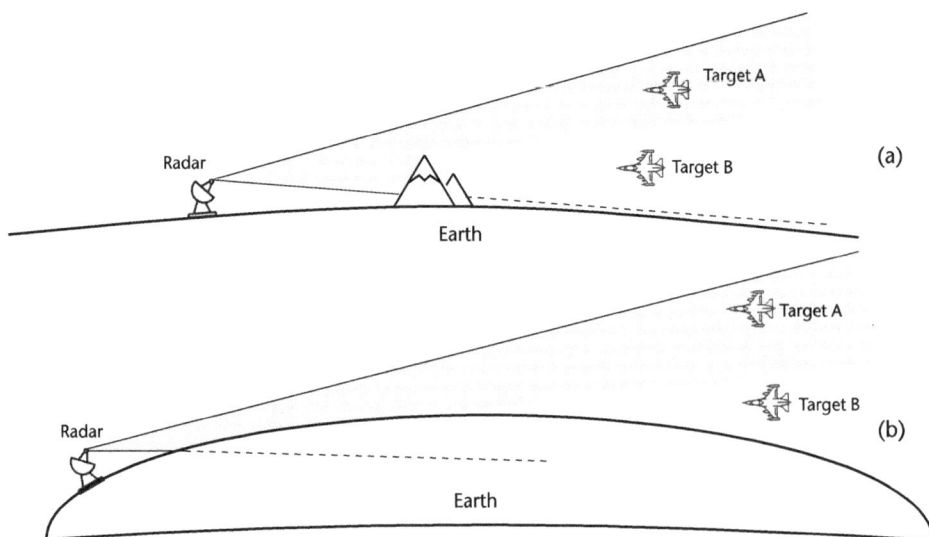

Figure 2.1 LOS obstruction for radar operation: (a) obstacle and (b) LOS horizon.

The LOS horizon can be calculated for km-m and NM-ft pairs by the rule of thumb as follows:

$$d(\text{km}) = 3.57\sqrt{h(\text{m})} \tag{2.1}$$

$$d(\text{NM}) = 1.23\sqrt{h(\text{ft})} \tag{2.2}$$

In the above equations, d shows the LOS horizon range according to an altitude, and h represents the radar antenna or target altitude. The maximum distance between the radar (R) and a target (T) for LOS propagation can be calculated as:

$$d_{R-T}(\text{km}) = 3.57\left(\sqrt{h_R(\text{m})} + \sqrt{h_T(\text{m})}\right) \tag{2.3}$$

$$d_{R-T}(\text{NM}) = 1.23\left(\sqrt{h_R(\text{ft})} + \sqrt{h_T(\text{ft})}\right) \tag{2.4}$$

At this point, a close but different concept, *RF line-of-sight* (RF-LOS), will be convenient to mention. The calculated visual LOS horizon differs from the RF wave propagation horizon or RF-LOS. Without a mitigating factor, the range of radar detection is limited by RF-LOS. This horizon goes beyond the calculated visual horizon due to direct radiation, reflected ground waves, and occasionally refraction. The visual and RF horizons are shown in Figure 2.2.

A good rule of thumb is achieved by multiplying the height of the radar system by 4/3. Thus, the effective or RF-LOS to the horizon is expressed as:

$$d_{\text{RF}}(\text{km}) = 3.57\sqrt{\frac{4}{3}h(\text{m})} \tag{2.5}$$

$$d_{\text{RF}}(\text{NM}) = 1.23\sqrt{\frac{4}{3}h(\text{ft})} \tag{2.6}$$

The same is true for target altitudes. So the maximum detection distance between the radar (R) and a target (T) for RF-LOS propagation can be calculated as:

$$\left(d_{R-T}\right)_{\text{RF}}(\text{km}) = 3.57\left(\sqrt{\frac{4}{3}h_R(\text{m})} + \sqrt{\frac{4}{3}h_T(\text{m})}\right) \tag{2.7}$$

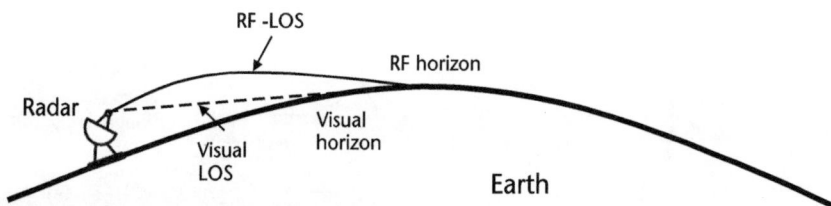

Figure 2.2   Visual and RF horizons.

$$\left(d_{R-T}\right)_{\mathrm{RF}}(\mathrm{NM}) = 1.23\left(\sqrt{\frac{4}{3}h_R\,(\mathrm{ft})} + \sqrt{\frac{4}{3}h_T\,(\mathrm{ft})}\right) \qquad (2.8)$$

2.1.3 RF Wave Propagation Characteristics

In practice, the propagation characteristics of RF waves vary substantially depending on the transmitted signal's frequency and strength. An Rx far from the Tx can sense the EM wave, thereby, the RF wave, with a reduced power while the distance increases. Various phenomena occur when an EM wave is incident on a surface. These phenomena depend upon the wave's polarization, the surface's geometry, material properties, and the surface's characteristics relative to the wavelength of the EM wave [1]. Before explaining these phenomena, their general schemes are depicted in Figure 2.3.

- *Reflection:* Depending on the media, the EM wave may bounce in another direction when it hits a smooth object more significantly than the wave. This bounce is called reflection. When a plane wave encounters a medium change, some waves may propagate or be reflected in the new medium. The part that enters the new medium is the refracted portion, and the other is the reflected portion. Reflection creates multipath, which can degrade the received signal's strength and quality and cause data corruption or cancel signals.
- *Refraction:* The propagation speed will change when the EM wave enters the new medium. The transmitted wave will change direction to match the incident and transmitted wave at the boundary. If the new medium has a higher refraction index, the propagation speed is lower, so the wavelength will become shorter. If the new medium has a lower index of refraction, the wavelength will become longer. The frequency must stay the same because of the boundary conditions. The phenomenon of changing the direction of

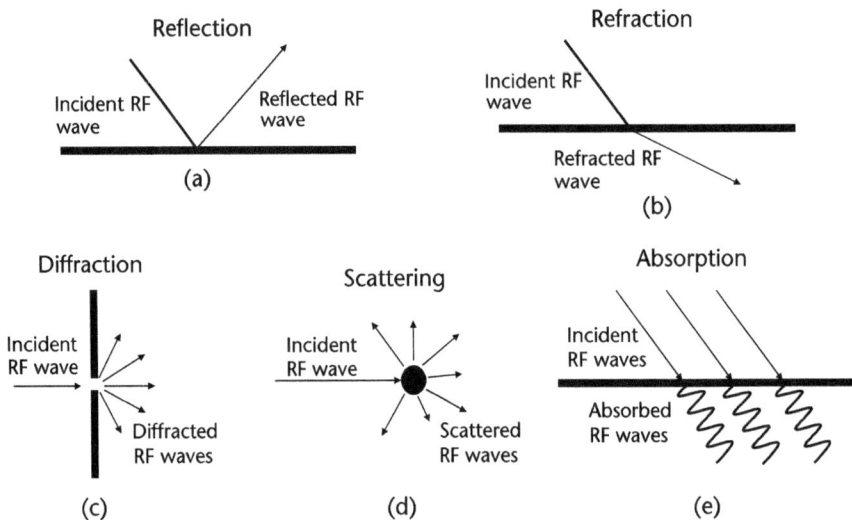

Figure 2.3 Various RF propagation phenomena for different surface properties.

propagation change is called refraction. The refraction may cause lower data rates, retransmissions, and reduced capacity.

- *Diffraction:* Transmitted waves may be received by an Rx even though a large object between Tx and Rx obstructs them. This phenomenon is called diffraction. Although a shadow zone is immediately behind the obstacle, the signal will diffract around it and start to fill the void. Diffraction becomes more prominent than usual when the obstacle becomes sharper, like a knife edge. For an RF signal, the definition of a knife edge depends upon the frequency and wavelength of the signal.

  To understand diffraction, one should consider Huygens' principle. This principle states that every point on a wavefront may be regarded as a source of secondary spherical wavelets that spread out in the forward direction at the speed of light. The new wavefront is the tangential surface to all these secondary wavelets.

  For low-frequency signals, a mountain ridge may provide a sufficiently sharp edge. A more rounded hill will not produce such a marked effect. Furthermore, low-frequency signals diffract more substantially than higher-frequency ones. Thus, signals on the long wave band provide coverage even in hilly or mountainous terrain, whereas signals at VHF and higher would not. As a result of diffraction, RF shadow can occur, causing dead coverage zones or receiving degraded signals.

- *Scattering:* Scattering occurs when an EM wave impinges upon a surface at an incident angle. In the far field, if the surface smoothness is electrically large compared to the wavelength, there is no scattering, and the only wave propagation from the surface may be the reflection. However, if the surface is rough and a primary reflected component is in the specular direction, the incident wave is scattered in many different directions, which is called scattering.

  The smoothness or roughness of a surface determines the degree of scattering. The surface roughness is related to the wavelength of the impinging wave. In the lower-frequency range, scattering is generally negligible since any protrusions on the surface are very small compared to the wavelength of the impinging wave.

  As a result of scattering, RF encounters a rough surface and is reflected in multiple directions. The RF signal dissipates into multiple reflected signals in many directions. Reflection in the incident wave direction will be only a tiny portion of the signal. So the received signal power in this direction will be reduced.

- *Absorption:* Whenever an EM wave is present in a material other than free space, there will be some loss of strength with distance due to ohmic losses. Different materials of the objects can cause absorption; precipitation and gases also cause absorption. Radar waves propagating through rain precipitation suffer a loss in signal power. This power loss is due to absorption and scattering from the rain droplets. Heaver rain rate will result in more absorption and scattering, thus leading to more power loss. Attenuation due to rain is also a function of frequency or radar wavelength [2].

- *Depolarization:* The scattered waves will have different polarization for many objects than the incident waves. This phenomenon is known as depolarization or cross-polarization. The effects of transmission and reflection depend upon the orientation of the incident wave's polarization relative to the plane of incidence. If the incident wave is circularly or elliptically polarized, its transmitted and reflected waves can alter.

Perfect reflectors reflect waves such that an incident wave with horizontal or vertical polarizations remains the same, but the phase is shifted 180°. Furthermore, a *right-hand circular* (RHC) incident wave becomes *left-hand circular* (LHC) when reflected from a perfect reflector, and vice versa. Thus, when a radar uses LHC waves for transmission, the receiving antenna needs to be RHC polarized to capture them by the Rx because they propagate in the opposite direction.

2.1.4 Radar Cross-Section

The behavior of the RF propagation and the alteration when an RF wave is incident on a surface are of great importance for radar operation. The basic operation principle of a radar system is composed of three steps:

- RF wave propagation from the radar Tx.
- RF wave encounters an object and scatters in different directions according to the object's shape and the RF signal's frequency.
- Radar Rx receives the return of the RF signal in its direction.

This three-step operation is depicted in Figure 2.4. In the figure, the solid lines represent the RF wave propagation from the radar Tx. When an RF wave reaches an object, scattering occurs in different directions. The scattering wave in the direction of the Rx composes the backscattered signal or return signal, which is depicted in the dashed line in Figure 2.4.

Figure 2.4(a) shows that Rx may be in the same location as Tx, called monostatic radar operation. However, Rx may be in a different location than Tx, which is called bistatic radar operation, given in Figure 2.4(b).

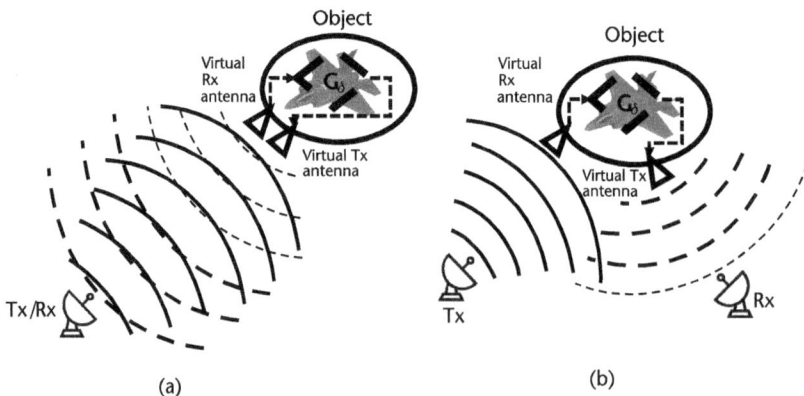

Figure 2.4 Radar operation: (a) for Tx and Rx in the same location and (b) for Tx and Rx in different locations.

The description of the radar operation process from another point of view is object-centric. When an object is exposed to an incident wave, it behaves like a repeater antenna. This process can be modeled with virtual Rx and Tx antennas [3]. At the beginning of the process, the object intercepts part of the incident wave, and an analogy can be established between the object's behaviors and the Rx antenna. After intercepting, the object reflects or reradiates the wave in different directions, like a Tx antenna. The reradiating pattern is similar to the erratic directional antenna pattern, and its shape is due to the shape and material of the object. The amount of RF power reflected from the central object in the spherical direction is RCS. To simplify the RCS analysis of an object, the spherical RCS can be split into two-dimensional (2-D) components: azimuth and elevation. RCS is an essential component in radar operation, and the reradiated RF signal strength in the Rx direction is used in calculations. Therefore, in Figures 2.4(a, b), the backscattered signal powers to the radar Rx are different and determined by the target's RCS.

The RCS of the target, usually represented by the symbol σ, is a function of its geometric cross-section (G_{CS}), reflectivity (R), and directivity (D) [4]. Using these three identities, RCS is represented as follows:

$$\sigma = G_{CS} \times R \times D \tag{2.9}$$

In the equation, G_{CS} is the object's size as viewed from the aspect of the radar Rx. R is the ratio of the power leaving the target versus the radar power that illuminates the target. D is the ratio of the power scattered in the direction of the radar Rx versus the amount of power for scattering omnidirectionally. The RCS of an object is the vector sum of the reflections from each part of it.

RCS behaves like a gain element in the radar signal transmission path, as shown in Figure 2.4. The power reflected from an object in the direction of radar Rx is equivalent to the reradiation of the power captured by an antenna with an RCS area. Thus, the figure's RCS gain (G_σ) in decibels is expressed as follows [4, 5]:

$$G_\sigma = -38.54 + 10\log_{10}\sigma + 20\log_{10}f \tag{2.10}$$

In the equation, when frequency (f) is in megahertz, and RCS (σ) is in m^2, the unit stabilization constant is taken as -38.54. Also, G_σ is defined as the ratio of the signal leaving the target in the radar Rx direction to the signal arriving at the target. Both signals are referenced to isotropic antennas in decibels.

2.2   Basic Radar Block Diagrams for Operation Modes

The primary radar block diagrams are suitable for increasing comprehension of radar operation principles. The essential components of a radar system are an antenna, Rx, Tx, and processor. Radar operations can be classified as pulsed, CW, and pulsed Doppler. More than a general radar block diagram would be required to include all the properties of different radar operation modes. Thus, it will be appropriate to treat different radar operation classes separately.

The fundamental forms of pulsed radar block diagrams for monostatic and bistatic radar operations are given in Figure 2.5. Even though most of the operational principles and the obtained information are similar, discussing them individually would be beneficial. Figure 2.5(a) gives the fundamental block diagram for monostatic pulsed radar operation. The sinusoidal signal with the radar's operating frequency is generated in the *local oscillator* (LO) or CW generator. Then, this sinusoidal signal is separated into pulses with the pulse modulator. Also, linear frequency modulation (LFM) is used in pulsed radars for pulse compression. This time, the pulse modulator in Figure 2.5(a, b) provides to form modulated pulse signals. Pulsed radars have a low-duty cycle and transmit narrow pulsed, high-power RF signals. Since pulses are transmitted only a small percentage of the time, the same antenna may be used for both Tx and Rx. This operation mode is called a monostatic radar. The modulator generates RF pulses that the transmitter will give to the environment at high power.

The duplexer lets the transmitted pulse be transmitted to the antenna, and the received pulse be sent to the Rx. The power of the transmitted pulse is much higher than the received pulse. Therefore, Rx is protected by turning it off while the pulse is transmitted. Rx sends the received pulses to the processor. If the system is for tracking, the data processor keeps the antenna on the target and uses the amplitude

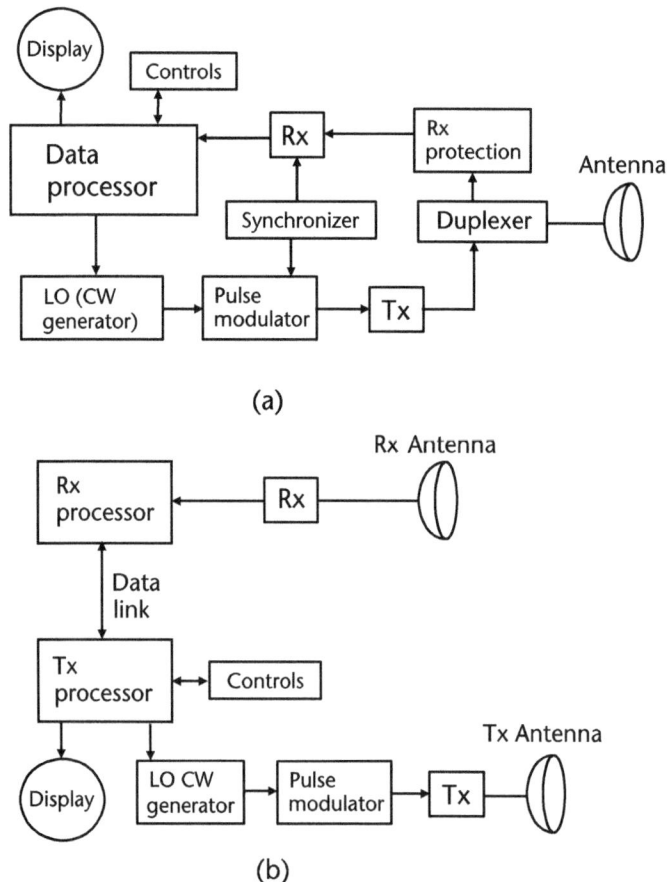

Figure 2.5 Pulsed radar operation block diagram: (a) for monostatic and (b) for bistatic.

of the received signal to perform tracking functions. It also conducts range tracking to focus the radar on a single target. Information about the target's location is displayed on the screen. Controls include operating modes and target selection.

Figure 2.5(b) shows the basic block diagram of the bistatic pulsed radar operation. In this operation mode, Tx and Rx are in different locations, and generally, a long distance exists between them. There is a constant stream of data and synchronization between Rx and Tx, even if they are not in the same location. In some tracking radar applications, such as *lobe on receive only* (LORO), the coordination between Rx and Tx operation becomes more complex. The data link enables transmission and can be wired or wireless, depending on the application. The most common bistatic radar application is semiactive RF-guided missile applications. The Tx is located on or near the launch platform, on land, at sea, or in the air. However, the Rx is on the missile, which is somewhere between the launch platform and the target. Tx and Rx antennas are different and in distinct places. Since Tx and Rx have discrete hardware, they also have separate processors. Generally, the control and the display parts occur in the Tx part of the bistatic radars.

The basic block diagram of a CW radar operation is given in Figure 2.6. Tx signal is always present in the CW radars, meaning it always has separate antennas for Tx and Rx. CW radars represent a family of unmodulated and modulated CW radars. The block diagram in Figure 2.5(b) covers the CW radar family with minor changes. The unmodulated CW radars operate with the Doppler principle to detect nonstationary targets. Thus, these types of CW radars are called *CW Doppler* (CWD) radars. They give accurate measurements of relative velocities but cannot obtain range information since unmodulated CW radars cannot measure distance. The CWD radars do not use the *frequency modulation* (FM) modulator in Figure 2.6 since they use the sinusoidal signal without modulation in their operating frequencies. Tx is modulated with FM in CW radars to perform the distance measurement. These radars are called *frequency-modulated continuous wave* (FMCW) radars.

The CW radars transmit RF continuously and are expected to receive a much weaker backscattered signal than the transmitted signal. For this reason, sufficient RF isolation must be provided between the two antennas so that the Tx signal does not saturate Rx. The CWD radar's Rx compares the frequency of the received signal with the transmitted frequency to determine the Doppler shift caused by the target's relative velocity. Thus, data flow from Tx to Rx is shown. However, in an FMCW radar, Rx takes information about transmitted frequency and modulation specifications from Tx.

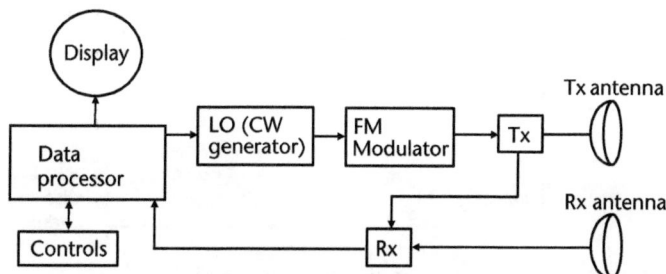

Figure 2.6   CW radar operation block diagram.

In pulsed radars, successive transmitted pulses are not coherent; thus, they cannot differentiate between foreground targets and background clutter. The foreground targets and background clutter pairs differ for various radar platforms. For land-based radar systems, airborne targets and mountain-based clutter may be defined. Furthermore, for airborne radars, pairs may be counted as lower-altitude airborne platforms and ground clutter or vessel and sea clutter for airborne radar systems. Clutter may be avoided by keeping the radar beam from striking the clutter source, as in early radars. However, this kind of measure limits the radars' operational capability. Another attempt to detect the target in background clutter is to determine the beat between the frequencies of the target echoes and the simultaneously received clutter. However, the clutter is spread over many frequencies. In addition to the returns of target echoes, beats due to clutter frequencies exist. Pulsed Doppler radars have been developed to overcome the problem of discriminating targets from background clutter [6].

Radar systems use pulsed Doppler operations to obtain range and velocity information. The *moving target indicator* (MTI) and *pulsed Doppler radar* (PDR) are used pulsed Doppler operation principles. They are used for finding and tracking a target in a high-clutter environment. The basic block diagram of a pulsed Doppler operation is given in Figure 2.7. It differs from a pulsed radar in two aspects: one is coherency, and the other is high *pulse repetition frequency* (PRF). Coherence, the radar's ability to accurately measure the received signal phase, is the relationship between the stages of transmitted and the received pulses. Cutting a radar's transmitted pulses from a CW, the RF phase of successive echoes from the same target will be coherent, enabling their Doppler frequency to be easily measured.

Low PRF waveforms can provide accurate, long, unambiguous range measurements but apply severe Doppler ambiguities. MTI radar operates with low PRF, and the limits of low PRF operation are assumed to be below 10 kHz. Medium PRF waveforms must resolve range and Doppler ambiguities; however, they provide

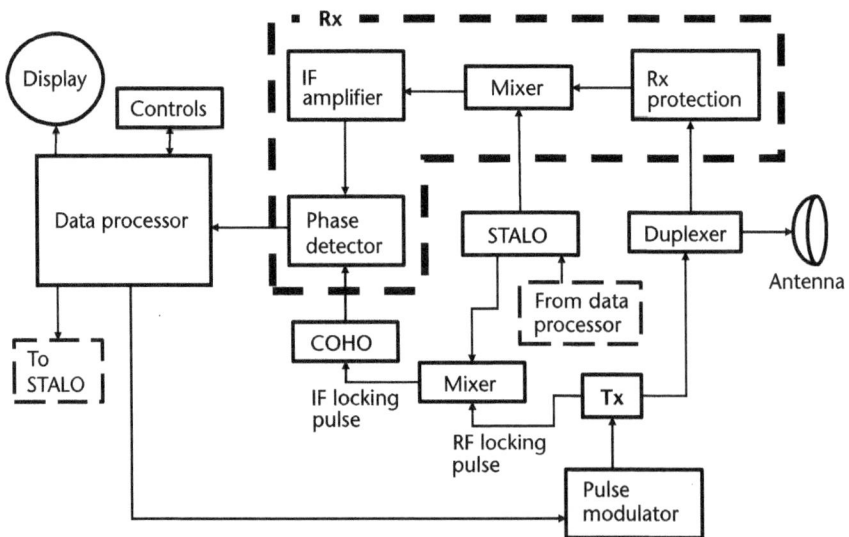

Figure 2.7 Pulsed Doppler operation block diagram.

adequate average transmitted power compared to low PRFs. The medium PRF operation limits are between 10 and 100 kHz. High PRF waveforms can provide superior average transmitted power and clutter rejection capabilities. However, high PRF waveforms are highly ambiguous in the range [2]. High PRF operation limits are assumed between 100 and 300 kHz. Pulsed Doppler radars use medium and high PRF waveforms.

Although the Rx and data processor structures change for MTI and pulsed Doppler radars, we can generalize both operations using the pulsed Doppler block diagram in Figure 2.7. Let us consider the pulsed Doppler operation block diagram. This time, the dashed line represents the Rx system, and its components are drawn separately inside the dashed line. The local oscillator must have a *stable local oscillator* (STALO). The RF echo signal is heterodyned with the STALO signal to produce the *intermediate frequency* (IF) signal as in a conventional superheterodyne Rx. The required coherent reference for pulsed Doppler radar is provided by the *coherent oscillator* (COHO). The COHO is a stable oscillator with the same frequency as the IF used in the Rx. A portion of the transmitted signal is mixed with the STALO output to produce an IF-locking pulse signal whose phase is directly related to the Tx. This IF pulse is applied to the COHO and causes the phase of the COHO oscillation to lock in step with the phase of the IF reference pulse. The phase of the COHO is then related to the phase of the transmitted pulse and may be used as the reference signal for echoes received from that particular transmitted pulse. In the subsequent transmission, another IF locking pulse is generated to relock the phase of the CW reference signal formed by the COHO until the next locking pulse comes along. Thus, the Rx can fulfill the Doppler frequency measurement, and the radar is said to be coherent upon receipt.

2.3   Radar Classification

Radars may be classified in many aspects. These classifications may be according to their functions, applications, and frequency bands. Also, some additive classifications of radars can be established for platforms, modulation, and scan patterns. However, the three categories mentioned above are sufficient for the requirements. This book focuses on military radars and, more specifically, radar-guided systems. We will first discuss the different classifications for radar systems without excluding any radar systems and then narrow the explanations to tracking radars. It would be appropriate to mention each classification separately.

The first classification is for radar functions. Figure 2.8 shows this classification, and as seen from the figure, the radar functions are separated mainly into two parts: primary and secondary radars.

The classification of radars in Figure 2.8 shows us the main radar operation in primary radars. However, secondary radar operation is also mentioned here. Primary radar sends high-frequency signals at certain intervals toward the sectors determined in the scanning area. The target reflects signals sent as pulses or CW, and the backscattered signals are received in the direction of the radar Rx. Received echo signals are further processed to obtain target information. The most crucial

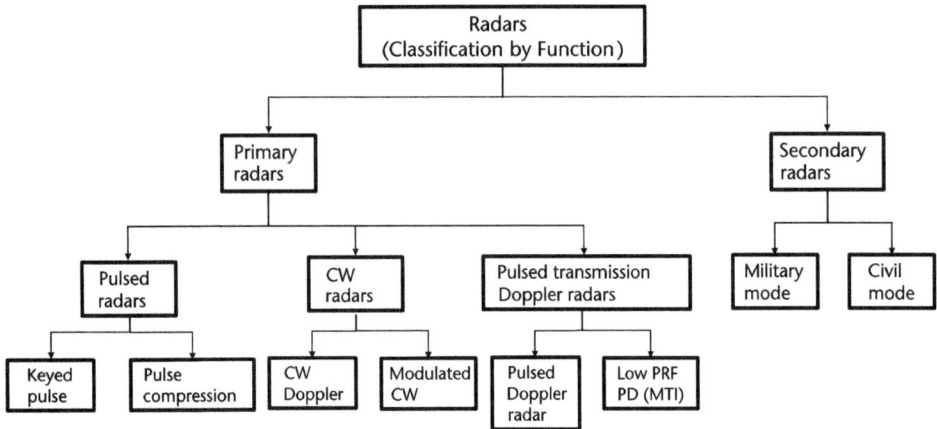

Figure 2.8 Radar classification based on functions.

disadvantage of primary radars is the drastic reduction in the received power concerning transmitted signal power.

Secondary radars work with a different principle than primary radars. In a secondary radar operation, each platform has its Tx-Rx pairs, which is a disadvantage according to primary radar operation. The ground unit, called the interrogator, transmits coded pulses toward the target. The transponder on the airborne platform receives and decodes the pulsed signal, prepares the suitable answer, and then sends the interrogated information back to the ground unit. The interrogator in the ground unit demodulates the answer.

Since secondary radar systems use a one-way signal, their path losses are lower than those of primary radars. Therefore, their Tx output powers can be lower than those of their primary radar counterparts. The secondary radar interrogates information like the aircraft's identity, position, and altitude.

Examples of secondary radars are identification friend or foe (IFF) for military purposes and secondary surveillance radars for civil aviation. Furthermore, *tactical air navigation* (TACAN) and *distance measurement equipment* (DME) are other kinds of secondary radar applications for military and civil applications, respectively. In IFF systems, secondary radars are always used with a primary radar system in military applications. However, TACAN systems are used solitarily to achieve direction and range according to a specific location.

Inspecting subordinates of the primary radar in further detail will be convenient.

- *Pulsed radars:* As stated before, a pulsed radar transmits high-power, high-frequency pulses to the environment in a directed manner. Then it waits for the echo of the transmitted signal for some time, called *pulse repetition interval* (PRI), before it transmits a new pulse. Within this period, it receives the pulsed signals reflected by reflective objects in their directions. The choice of PRI and PW decide the range and range resolution of the radar, respectively. The measured antenna position and the backscattered signal's arrival time can determine the target range and bearings. Pulsed radars utilize low PRF and are used for ranging where target velocity, or Doppler shift, has secondary importance.

Keyed (on/off) pulse radars use a very short signal transmission with turn-on and turn-off processes. Their frequencies are constant, and keyed pulse radars' PW and PRI (or PRF) characterize them. If its pulses repeat themselves constantly, this is a constant or fixed PRI pulsed radar. However, PRI may vary in each period, and it may be stagger or jitter.

Pulse compression or intrapulse modulation improves the range resolution of pulse radar without decreasing its *signal-to-noise ratio* (SNR). Range resolution for a given radar can be enhanced by shortening pulses. However, short pulses reduce the average transmitted power, resulting in a decrease in SNR. Since the average transmitted power directly affects the receiver SNR, increasing the PW while maintaining adequate range resolution is often desirable. This solution can be achieved by modulating the transmission pulse internally. When processing the return signal, a frequency comparison is conducted in the received echo, which makes it possible to localize the reflecting object within the pulse [2].

• *CW radars:* These radars transmit a high-frequency signal without interruption, and the reflected energy is received and processed simultaneously and continuously. Two antennas are used in CW radars to avoid interruption of the continuous radar energy emission, and they may be bistatic or monostatic. These radars must ensure excellent RF insulation between Tx and Rx antennas. The Rx does not necessarily have to be in the same location as the Tx. It has subordinate classes according to the modulation of the CW signal.

An unmodulated CW radar, or CWD radar, transmits a signal constant in frequency and amplitude. CW radar transmitting unmodulated power can measure the speed only by using the Doppler effect. It cannot calculate a range or difference between two reflecting objects. There will not be any change in the frequency of the received echo signal for stationary objects. However, the frequency of the echo signal is shifted by the Doppler frequency of an object moving at a radial velocity. CWD radars are specifically designed to measure this Doppler frequency.

The transmitted signal is constant in amplitude but modulated in frequency or phase in the modulated radars. The radars modulated in frequency are called FMCW radars. Modulation in these radars makes it possible to determine the distance according to the runtime measurement principle. The distance can then be determined by a given phase code's frequency shift or delay time. The advantage of these radar sets is that evaluation takes place without a time out for the reception, and thus, the measurement result is continuously available. This kind of radar is often used as a "radar altimeter" to measure the exact height during the landing procedure of an aircraft. It is also used as an early-warning (E/W) radar, wave radar, and proximity sensor. Doppler shift is not always required for detection when FM is used.

• *Pulsed transmission Doppler radars:* The pulsed transmission Doppler radars are categorized as MTI and PDR [7]. A PDR determines the range of a target using pulse-timing techniques and uses the Doppler effect of the returned signal to determine the target's velocity. It combines the features of pulse radars and CW radars by increasing the complexity of the electronics. Like pulsed radars, using time gating is a substantial property of PDR Rx. Time gating prevents Tx RF leakage from occurring on Rx. Thus, a single antenna can be used for Tx and Rx, which is not feasible for CW radar for high Tx/Rx isolation requirements. PDR uses high and medium PRF schemes to determine output power requirements.

MTI radars aim to discriminate targets against the clutter. Generally, MTI is a mode of operation rather than a radar type. If radar is used to detect the movable target, then the radar must receive only the return signal due to that movable target. However, the radar receives the echo signals of both stationary objects and movable targets.

Clutter primarily comprises unwanted stationary ground and sea reflections with limited relative motion concerning the radar. Hence, the radar should be designed in such a way that it takes into consideration only the echo signal due to the movable target, not the clutter. For this purpose, MTI radars use the Doppler effect principle to distinguish nonstationary targets from stationary objects. In the Doppler effect, the frequency of the received signal will increase if the target moves through the radar's direction and vice versa. The MTI radars use low PRF to avoid range ambiguities. Low PRF waveforms can provide accurate, long, unambiguous range measurements but cause severe Doppler ambiguities.

Another radar classification can be made according to the application, given in Figure 2.9. The subclasses of radar applications can be clustered into six groups. The first one is the surveillance radars, and many essential radar systems are considered in this group. These are long-range detection radars, especially part of the IADS systems. *Ground control intercept* (GCI), *airborne early warning* (AEW), *airborne ground surveillance* (AGS), and OTH radars are mentioned in early warning radar systems.

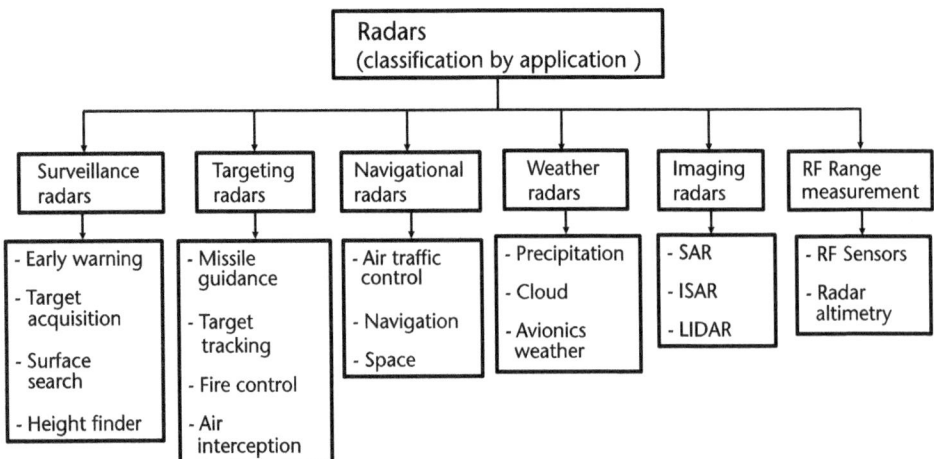

Figure 2.9 Radar classification based on applications.

The acquisition function of the SAM and AAA systems can be counted in target acquisition radars. Another critical subgroup is surface search radar, which is composed of surface search, coastal surveillance, and harbor surveillance radars. Most of the radar systems mentioned in the surveillance radars are military radars. Some subgroups of this group are in the scope of this book, like target acquisition radars.

The second subclass, the most important for us in this book context, is targeting radars. Missile guidance radars are composed of the tracking function of the AAM, ASM, SAM, and SSM radar systems. *Fire control* (FC) and AI radar groups indicate that many systems are used on various platforms. Both have search, acquisition, and target tracking modes. Additionally, AI radars have target illumination and missile guidance modes.

Even though the remaining four groups are not direct threats to the mother platforms of the self-protection systems, they are mentioned here. Navigational radars are standard on commercial ships and commercial aircraft. *Air traffic control* (ATC) radars use primary and secondary radars. The secondary radar usage in civil aviation is Mode-S and DME; however, the military counterpart of this usage is IFF. *Ground control approach* (GCA) radar and *precision approach radar* (PAR) systems are other navigation radar applications. Space tracking and space-based radar systems are considered in the space radar subgroup.

Weather radars locate precipitation, calculate its motion, and estimate its type. Modern weather radars are mostly pulse-Doppler radars, capable of detecting the motion of rain droplets in addition to the intensity of the precipitation. A *precipitation radar* is a weather radar operating in the S-band or C-band. It can measure the reflectivity of water droplets, which is proportional to the precipitation rate. Another kind of weather radar is cloud radar, which displays clouds' position and shape. Cloud radars use very high frequencies between K and W-bands. An avionics weather radar is usually an X-band multifunction radar for navigation and collision avoidance, combined with a weather channel as a precipitation radar. It is in the nose of the aircraft and operates in only one sector in the flight direction. An aircraft's weather radar displays the precipitation areas in real time as colored areas.

Imaging radar methods attempt to calculate a map-like image from the received information. Imaging radar sensors measure two dimensions of coordinates to create a map-like picture of the observed object or area. A *synthetic aperture radar* (SAR) is a coherent, active RF imaging method. A SAR is mounted on a moving platform such as an aircraft, a UAV, or a satellite. EM waves are transmitted sequentially, and the echoes are collected, digitized, and stored data for subsequent processing. Transmission and reception occur at different times and map to various positions. The well-ordered combination of the received signals builds a virtual aperture much longer than the physical antenna width. *Inverse synthetic aperture radar* (ISAR) operates with a similar operational principle, except that ISAR uses the target's movement to generate its radar image. ISAR is used in military applications to identify and target objects by their movement. *Light detection and ranging* (LIDAR) is a remote sensing method that uses light as a pulsed laser to measure ranges to the Earth. These light pulses, combined with the data recorded by the system, generate precise, three-dimensional (3-D) information for mapping the Earth and its surface characteristics. Low-profile altitude flying vehicles like helicopters and UAVs use LIDAR for obstacle detection and avoidance to navigate safely through environments.

Different types of radar sensors include millimeter-wave, CW Doppler, and FMCW radar sensors. The sensor that uses millimeter waves is known as a millimeter-wave radar sensor. Millimeter-wave radars are used in automobiles for collision avoidance. A CWD radar sensor uses Doppler to measure the object's speed at different distances. The FMCW radar sensors have high-range resolution and accuracy and are used for robust sensing of autonomous vehicles. A *radar altimeter* is an airborne system that measures the distance between the antenna and the ground directly below it. Conventional altimeters measure barometric altitude and give the aircraft's height above sea level. This property makes the radar altimeter a precious piece of equipment utilized as a flight aid and part of many aircraft's ground proximity warning systems.

The last classification of the radar systems is according to frequency bands [2]. This classification can be given as follows:

- *HF and VHF (3–300-MHz) radars:* These frequency bands are very busy, and frequency allocations are set with strict rules. Therefore, the bandwidth for military radar systems is limited and highly contested worldwide [2]. Early warning radars of extremely long-range or OTH radars are in this band. Since the angle determination and angular resolution accuracy depend on the wavelength ratio to antenna size, high accuracy is not expected from these radars.
- *UHF (0.3–1-GHz) radars:* Radars in 300 MHz to 1 GHz are used for long-range early warning radars, such as military early warning radar for the *medium extended air defense system* (MEADS), or as wind profilers in weather observation. These frequencies are damped only very slightly by weather phenomena and thus allow a long range. Relatively new methods, ultrawideband radars, transmit low pulse power in the UHF band and are mainly used for technical material investigation or partly in archaeology as *ground penetrating radar* (GPR).
- *L-band (1–2-GHz) radars:* The L-band radars are used in long-range military and air traffic control search and surveillance operations for up to 460 km (250 NM). They transmit high power, wide bandwidth, and intrapulse modulation pulses to achieve longer ranges. Their maximum detection range is limited by RF LOS, calculated by the earth's curvature, the radar's altitude, and targets.
- *S-band (2–4-GHz) radars:* The S-band radars are used for ground- and ship-based medium-range search and surveillance. Also, the *airborne warning and control systems* (AWACS) radars use this band. In civil aviation, *airport surveillance radars* (ASRs) detect and display the aircraft's position in the terminal area with a medium range of up to 110 km (60 NM). In the S-band, the atmospheric attenuation is higher than in the L-band. For this reason, radar systems in the S-band require a much higher pulse power to achieve long ranges.
- *C-band (4–8-GHz) radars:* Short- and medium-range military battlefield radars operate in this band. The antennas are small enough to quickly install and achieve high angular accuracies and resolution for weapon control. Radar systems' performance is severely affected by weather conditions, and they often

use circularly polarized antennas to overcome this problem. The C-band is the operational frequency band for most weather radar systems.

- *X-band (8–12-GHz) radars:* In this frequency range, radar systems are used in land-based, airborne, and sea-surface platforms for military purposes. Also, X-band radars are suitable for maritime and aviation navigational purposes in civil and military applications. These radars are used in various applications, such as airports, to control air traffic and long-range surveillance. The radars in this band have relatively small antennas, and sufficient angular accuracy can be achieved, which supports military use as airborne radar. These radars are also suitable for SAM system TTRs. Another significant use of AI radars for fighters is air-to-air engagements and dogfights. However, the sea surface X-band radar acquires, tracks, and discriminates the flight characteristics of ballistic missiles. X-band is also convenient for spaceborne or airborne SAR and ISAR applications for military electronic intelligence, civil geographic mapping, and monitoring.

- *Ku-band (12–18-GHz) radars:* Ku-band radars are used in missile guidance system applications. The small antenna size makes them very handy in low-weight applications. The primary users of the Ku-band are the military services for airborne and shipborne radars. Different military radars are used in this band as listed here [8]:

  - Airborne multimode Doppler navigation;
  - Airborne fire-control radars;
  - Airborne search and interception radars;
  - Airborne weather and navigation radars;
  - Airborne terrain following and terrain avoidance radars;
  - Battlefield weapons-locating radars;
  - Battlefield aircraft approach landing system radars;
  - Battlefield mobile;
  - Portable ground surveillance radars, weapons guidance, and control radars;
  - Shipborne weapons' fire-control radars and navigational radars.

  Some nonmilitary radar applications in the Ku-band include avionics, such as *microwave landing systems* (MLSs), search radars for atmospheric research studies, and airport *surface movement radars* (SMRs). Furthermore, satellites observe Earth in this band and operate spaceborne active sensors such as altimeters, scatterometers, and precipitation radars.

- *K-band and above (higher than 18-GHz) radars:* The achievable angular accuracies and range resolution are superior to other bands. In *air traffic management* (ATM) applications, these radars are often called SMRs. Vehicle speed detection radars exist in this band at about 24 GHz. Airborne navigational and mapping radars, weather avoidance, aircraft radar beacon systems, airborne terrain following, avoidance radars, and cloud measurement radars are used for military purposes in the 32–36-GHz band.

Also, some research and development studies for conceptual radars are conducted above 90 GHz. These radars are pulse-Doppler fire-control radars and airborne radiometer beacons. Synthetic vision radar research has been shown at around 94 GHz. Civil operations consist of ground-based CW pulse radars in the high-resolution profiling of clouds at these frequencies.

2.4　Radar Equations

A radar system operates with the principle of the body reflecting high-frequency RF energy to obtain a return or echo signal; reflecting or backscattering means RF scattering from an object in the direction of the radar Rx. How an RF wave is transmitted and received is chosen according to the radar system's intended use. Generally, one focused measurement, such as range, becomes prominent, and other information, such as velocity and elevation, may be considered a less important objective.

Determining radars' detection range for different targets is essential for radar operation planning, specifying technical requirements, and design process. Here, the radar range equation or, more commonly, radar equation is obtained. For this purpose, a two-step derivation is considered here. First, we get a one-way radar equation as the model of the transmitted radar power in the RWR Rx. Then, by adding the RCS of the target, two-way radar equations are obtained for monostatic and bistatic radar cases.

2.4.1　One-Way Radar Equation

The one-way radar equation is the first step of the radar equation derivation, and its outline is given in Figure 2.10.

The one-way radar equation aims to obtain the signal power at the target (P_R) due to the ERP of the radar Tx. The ERP is the multiplication of the Tx power and antenna gain. In the one-way radar equation derivation, RCS is not considered first, and it is assumed that there is an Rx at the target's distance.

In long distances, the uniform power density spherical surface appears flat to a receiving antenna, which is very small compared to the sphere's surface. Therefore, the far-field wavefront is considered planar, and the rays are approximately parallel. The following formulation gives the average power density (P_{av}) for plane waves in W/m^2.

$$P_{av} = \frac{E_{peak}^2}{2Z_0} = \frac{E_{rms}^2}{Z_0} \qquad \left\{ E_{rms} = \frac{E_{peak}}{\sqrt{2}} \right\} \qquad (2.11)$$

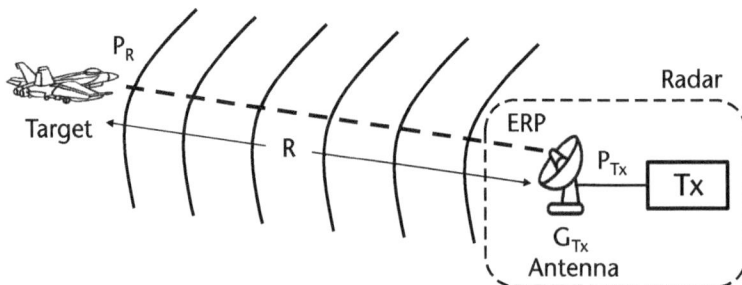

Figure 2.10　One-way radar equation outline.

In the above equation, E_{peak} is the peak value, E_{rms} is the *root mean square* (RMS) value of the far-field electric field in volts/meter, and Z_0 is the characteristic impedance of the medium in ohms. Z_0 can be obtained using the following equation:

$$Z_0 = \sqrt{\frac{\mu}{\varepsilon}} = \sqrt{\frac{\mu_0 \mu_r}{\varepsilon_0 \varepsilon_r}} = 377\sqrt{\frac{\mu_r}{\varepsilon_r}} \ \Omega, \ \mu_0 = 4\pi \times 10^{-7} \ \text{H/m}, \ \varepsilon_0 = 8.85 \times 10^{-12} \ \text{F/m} \quad (2.12)$$

The equation μ and μ_r represent the medium's absolute and relative magnetic permeability, respectively. μ_0 represents the permeability of free space. Also, ε and ε_r represent the medium's absolute and relative electrical permittivity, respectively. ε_0 shows the permittivity of free space. Air's relative permeability is 1, and its relative permittivity at sea level is approximately 1.0006, which can also be assumed to be 1 with acceptable precision. As a result, air's characteristic impedance converges to the free-space impedance, defined as $Z_0 = 377\Omega$ (or $120\pi \ \Omega$). Thus, the following equation relates the relationship between the magnitudes of power density and RMS electric field intensity.

$$P_D = \frac{E_{rms}^2}{Z_0} = \frac{E_{rms}^2}{377} \quad (2.13)$$

Power density (P_D) is defined as the amount of power per unit area (W/m²), and it is preferable to electric field intensity in calculations above 100-MHz frequencies [5]. Sometimes, Rx sensitivity is stated with a field intensity in µV/m rather than W or mW. In this case, the Rx power calculation is made using the following equation:

$$P_{Rx} = P_D A_c = \frac{E_{rms}^2 A_c}{Z_0} = \frac{E_{rms}^2}{480 \times \pi^2} \frac{c^2}{f^2} G \quad (2.14)$$

In (2.14), A_c is the physical aperture area or effective capture area of the Rx antenna, defined as:

$$A_c = \frac{G\lambda^2}{4\pi} = \frac{Gc^2}{4\pi f^2} \quad (2.15)$$

Also, c is the speed of light (3×10^8 m/s), f is the frequency in hertz, λ is the wavelength in meters, and G is the unitless Rx antenna gain. At this point, mentioning antenna gain and directivity would be helpful to shed light on some issues mentioned in future analyses. *Radiation intensity* (U) in a specific direction is the power radiated from an antenna per unit solid angle (W/unit solid angle). It is a far-field parameter, and the directivity (dimensionless) can be represented in terms of the radiation intensity as follows [9, 10]:

$$D = \frac{U}{U_0}, \ \left\{ U_0 = \frac{P_{rad}}{4\pi} \right\} \Rightarrow D = \frac{4\pi U}{P_{rad}} \quad (2.16)$$

In the equation, U_0 is the radiation intensity of an isotropic source, and P_{rad} is the total radiated power. If the direction is not specified, the direction of maximum radiation intensity is implied. The direction of the maximum radiation intensity is defined as the antenna directivity and expressed as:

$$D_{max} = \frac{4\pi U_{max}}{P_{rad}} \quad (2.17)$$

The directivity is independent of antenna losses and equal to the absolute gain in the same direction if the antenna is lossless. The antenna gain is a concept that is close to directivity, except for efficiency. The antenna gain considers the antenna's efficiency in addition to its directivity. Antenna gain is the ratio of the radiation intensity in one direction to the radiation intensity of an isotropic antenna for the same input power (P_{in}). The dimensionless gain is given as:

$$G = \frac{4\pi U}{P_{in}} \quad (2.18)$$

The antenna gain is considered the maximum radiation direction if no direction is specified.

$$G_{max} = \frac{4\pi U_{max}}{P_{in}} \quad (2.19)$$

Notice that the main difference between (2.17) and (2.18) is the powers used in the denominators. The relationship between radiated power (P_{rad}) and input power (P_{in}) is defined as [9]

$$P_{rad} = \eta P_{in} \quad (2.20)$$

In the above equation, η shows the dimensionless antenna radiation efficiency. When using the efficiency obtained in (2.20) for establishing a correlation between the antenna gain and directivity, the following equation is obtained:

$$G = \eta D \quad (2.21)$$

The gain and directivity are dimensionless using equations from (2.16) to (2.21). However, obtaining them in dB form is a common approach. The following equation pair transforms a linear number into its dB form and a dB into a linear number.

$$A(\mathrm{dB}) = 10 \times \log_{10}(A) \Leftrightarrow A = 10^{\frac{A\,(\mathrm{dB})}{10}} \quad (2.22)$$

When considering a dB number, we think of a ratio in the linear number domain. Expressing an absolute physical value in the dB domain may pose a problem. To that end, we utilize the ratio property to obtain the dB counterpart of the absolute physical value. For expressing a power value in the dB domain, the power level may be divided into 1W or 1 mW and then transformed into dB value,

$$P(\text{dBW}) = 10 \times \log_{10}\left(\frac{P(\text{W})}{1\,\text{W}}\right) \quad \text{or} \quad P(\text{dBm}) = 10 \times \log_{10}\left(\frac{P(\text{W})}{1\,\text{mW}}\right) \quad (2.23)$$

A similar dB domain unit is dBi, which is used for antennas. It represents the boresight gain of an antenna compared to an ideal isotropic antenna that emits energy equally in all directions. Another absolute dB unit used for antenna gain is dBd. Sometimes, dBd, referenced to a half-wave dipole, may be confused with dBi. One may assume that the referenced gain is dBd when using dB for antenna gain. The relationship between dBi and dBd is defined as

$$\text{dBi} = \text{dBd} + 2.15 \qquad (2.24)$$

Another absolute dB unit is dBm², which expresses an object's RCS according to the 1 m² ratio. The dB representations of the derived equations in linear form are also mentioned throughout the book.

After mentioning the relationship between electric field intensity and power density and the meaning of antenna gain and directivity in this context, let us define power density according to the Tx power. For this purpose, isotropic antennas are considered first. An isotropic antenna is used as a reference when calculating its directivity and gain. It is assumed that an isotropic antenna, which cannot be physically realized, radiates uniform power density in all directions.

The power density of the isotropic antenna at any distance R is the Tx power divided by the surface area of the sphere $(4\pi R^2)$, considering this distance as the radius. The surface area of the sphere increases with the square of the radius. So the power density decreases with the square of the sphere's radius, as shown here:

$$\left(P_D\right)_i = \frac{P_{\text{Tx}}}{4\pi R^2} \qquad (2.25)$$

The formula given in (2.25) is the classical *inverse square law* found in many other fields of physics. In (2.25), P_{Tx} represents the Tx power either in peak or average. Depending on the P_{Tx}, P_D may be obtained as peak or average power density.

Radars use directional antennas to direct most radiated power in a particular direction. So, while the power density is found, the antenna gain is also included in the calculations. In the Tx operation, the power density created by radar with G_{Tx} antenna gain at a certain R distance is the product of the antenna gain and the power density generated by an isotropic antenna at the same distance. In this case, the power density caused by the radar at a distance of R,

$$P_D = \left(P_D\right)_i G_{Tx} = \frac{P_{Tx}G_{Tx}}{4\pi R^2} = \frac{ERP}{4\pi R^2} \rightarrow \left\{ERP = P_{Tx}G_{Tx}\right\} \qquad (2.26)$$

As stated before, ERP is obtained by multiplying the Tx power and antenna gain in the linear domain. However, in the dB domain, ERP is obtained by summing them. P_{Tx} may be dBW or dBm, and G_{Tx} is dB, and their sum gives the result depending on the unit of P_{Tx}.

If we return to Figure 2.10 again, assuming there was an Rx antenna on the target, the derivations would be more explicit. An Rx antenna captures only a portion of the ERP power limited by the effective capture area (A_c), given in (2.15). The power at the antenna terminal is the effective capture area of the Rx antenna multiplied by the power density.

The capture area is constant for identical Rx_1 and Rx_2 antennas, no matter how far from Tx. As shown in Figure 2.11, area A_1 and area A_2 are equal. Suppose the distance between Rx_2 and radar is m times longer than between Rx_1 and radar. According to (2.26), the power density on Rx_2 decreases by m^2 times compared to Rx_1. This phenomenon is formulated as follows.

$$\left.\begin{array}{l} \left(P_D\right)_{Rx_1} = \dfrac{P_{Tx}G_{Tx}}{4\pi R_{A_1}} \\[2ex] \left(P_D\right)_{Rx_2} = \dfrac{P_{Tx}G_{Tx}}{4\pi R_{A_2}} = \dfrac{P_{Tx}G_{Tx}}{4\pi\left(mR_{A_1}\right)} \end{array}\right\} \rightarrow \left(P_D\right)_{Rx_2} = \frac{\left(P_D\right)_{Rx_1}}{m^2} \qquad (2.27)$$

Now, a relation for Rx power can be written by combining (2.15) with (2.26),

$$P_{Rx} = P_D A_c = \frac{P_{Tx}G_{Tx}G_{Rx}\lambda^2}{\left(4\pi R\right)^2} = \frac{P_{Tx}G_{Tx}G_{Rx}c^2}{\left(4\pi Rf\right)^2} \qquad (2.28)$$

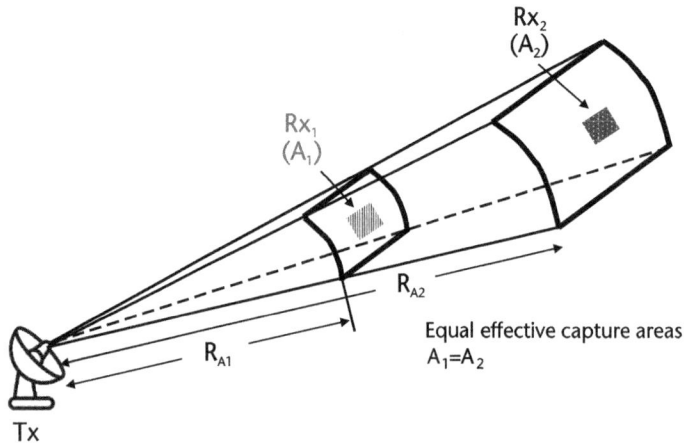

Figure 2.11 Effective capture areas for identical Rx.

The equation in (2.28) is the one-way radar equation for calculating the signal power at the Rx input on the target. In an ordinary two-way radar equation, no Rx antenna is considered on the target, and the required power is not the signal power at the Rx input. However, this operation is an example of receiving a radar signal by the RWR, ED, or ELINT Rx, and this equation can be applied to all LOS cases. Also, the equation in (2.28) can be divided into physically meaningful parts. The Rx power comprises Tx ERP, Rx antenna gain, and the one-way space (or path) loss (L_S). With this point of view, the one-way radar equation can be written as follows:

$$P_{Rx} = \frac{P_{Tx}G_{Tx}G_{Rx}c^2}{(4\pi R f)^2} = (ERP)_{Tx} \, G_{Rx} L_S \tag{2.29}$$

where $ERP = P_{Tx}G_{Tx}$ and $L_S = (c^2)/(4\pi R f)^2$.

2.4.2 Two-Way Radar Equation for Monostatic Operations

The two-way radar equation will be derived using the results obtained in the one-way radar equation. The derivations will be made for monostatic radars where the Tx and Rx of the radar are in the same location. The two-way radar equation, more commonly known as the radar equation, is calculated in two successive stages for monostatic radars. The first stage refers to approaching the target, including the RCS effect, and the second stage refers to returning from the target to the radar, as shown in Figure 2.12.

The signal power calculation produced by the Tx ERP at the target includes a one-way path loss. As seen in Figure 2.12, one-way space losses are the same for approaching a target and in the return path since the ranges are R for both. Including the target's RCS in the calculation, the backscattered signal strength (P_R) from the target to the radar Rx is obtained as follows:

$$P_R = \frac{P_{Tx}G_{Tx}G_\sigma \lambda^2}{(4\pi R)^2} \rightarrow G_\sigma = \frac{4\pi\sigma}{\lambda^2} \tag{2.30}$$

This power level (P_R), backscattered through the radar Rx direction at distance R, forms the first stage of the radar equation. The difference between (2.30)

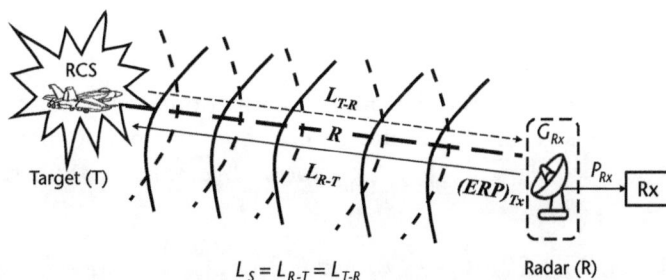

Figure 2.12 Two-way radar equation outline for monostatic radar operation.

and (2.29) is using G_σ instead of G_{Rx}. Furthermore, when comparing (2.30) and (2.26), G_σ is included in the calculations. Similar to an Rx antenna, a target cuts off part of the RF signal, but unlike the Rx, it scatters the incident signal in many directions, including the radar's direction. The amount of power backscattered by the target's RCS (σ) in the radar Rx's direction determines the amount of power returned to the radar. The RCS represents the target's size as seen by the radar, and the RCS is a property of the target with the areal dimension. RCS magnitude is not the same as the physical area. However, the power reflected in the radar direction for a target is equivalent to the power reradiated by an antenna in the RCS area. Therefore, the RCS replaces the Rx antenna's effective capture area (A_c) in the one-way radar equation. In this equation, G_σ is the RCS gain and depends on the target's frequency in m^2.

In the second stage of the process, the signal backscattered power from the target and returned to the Rx is calculated. This stage refers to a one-way loss calculation, and its mathematical expression is given in (2.31).

$$P_{Rx} = \frac{P_R G_{Rx} \lambda^2}{(4\pi R)^2} \rightarrow \left\{ P_R = \frac{P_{Tx} G_{Tx} \sigma}{4\pi R^2} \quad \text{and} \quad \lambda^2 = \frac{c^2}{f^2} \right\} \tag{2.31}$$

Combining (2.30) and (2.31) equations obtained from the two stages, the two-way radar equation of monostatic radars is derived as follows.

$$P_{Rx} = P_{Tx} G_{Tx} G_{Rx} \left(\frac{\sigma c^2}{(4\pi)^3 R^4 f^2} \right) \tag{2.32}$$

As mentioned earlier, P_{Tx} can be either the peak or the mean value, and correspondingly, the corresponding value of P_{Rx} is obtained. However, in practice and calculations, the peak value is mainly preferred. When we use (2.32), it should be noted that the units are kept the same.

Generally, monostatic radars use the same antenna for Tx and Rx purposes. Due to the duality principle, the antenna gains are equal ($G_{Rx} = G_{Tx} = G$) in Rx and Tx operations in this case [9]. Thus, the two-way radar equation obtained in (2.32) can be rewritten as:

$$P_{Rx} = P_{Tx} G^2 \left(\frac{\sigma c^2}{(4\pi)^3 R^4 f^2} \right) = \frac{P_{Tx} G^2 \sigma c^2}{(4\pi)^3 R^4 f^2} \tag{2.33}$$

The above equation is the most used equation for monostatic radars. In the radar equations, it is generally not directional, but the mean RCS is used to calculate the average echoed signal [11]. The dB form of (2.33) can be written as follows:

$$10\log P_{Rx} = 10\log P_{Tx} + 20\log G + 10\log \sigma + 20\log\left(\frac{c}{f}\right) - 40\log R - 30\log(4\pi) \tag{2.34}$$

After that, we will neglect to write the base 10 for simplicity and use it as $10\log(A)$. The equation in dB form obtained can be simplified by making certain classifications [5].

$$10\log P_{\mathrm{Rx}} = 10\log P_{\mathrm{Tx}} + 20\log G + G_\sigma - 2L_{\mathrm{S}} \qquad (2.35)$$

The RSC gain G_σ is defined below:

$$G_\sigma\,(\mathrm{dB}) = 10\log\left[\frac{4\pi \times \sigma \times f^2}{c^2}\right] = 21.46 + 20\log f + 10\log\sigma \qquad (2.36)$$

Here, 21.46 is the compensation coefficient calculated from known values when the frequency is GHz, and the RCS is m^2. In (2.10), the same equation is given with a -38.54 coefficient for the frequency is megahertz, and the RCS is m^2. The one-way space loss L_{S} is shown here:

$$L_{\mathrm{S}}\,(\mathrm{dB}) = 20\log\left[\frac{4\pi f R}{c}\right] = 92.45 + 20\log(f \times R) \qquad (2.37)$$

In (2.37), 92.45 is the compensation coefficient calculated from known values when the frequency is gigahertz and the range is kilometers.

The two-way radar equation for monostatic radars is also often called the radar range equation. The radar range equation is a two-way radar equation adjusted to find the maximum range. The maximum radar range (R_{max}) refers to the distance limit at which the target cannot be detected and accurately processed. This limit is found at a distance where the received echo signal equals the Rx sensitivity (S_{min}).

$$R_{\mathrm{max}} = \left[\frac{P_{\mathrm{Tx}}\,G^2 c^2 \sigma}{(4\pi)^3 f^2 S_{\mathrm{min}}}\right]^{\frac{1}{4}} \qquad (2.38)$$

The dB form of the above equation is written as follows:

$$40\log R_{\mathrm{max}} = 10\log P_{\mathrm{Tx}} + 20\log G + 10\log\sigma - 10\log S_{\mathrm{min}} \\ -20\log f - 30\log 4\pi + 20\log c \qquad (2.39)$$

By collecting the constants in a group and using specific units to write (2.39) in a more straightforward form, we can obtain:

$$S_{\mathrm{min}} = -103.43 + P_{\mathrm{Tx}} + 2G - 20\log f - 40\log R + 10\log(\sigma) \qquad (2.40)$$

where f = operating frequency (in MHz); R = range between the radar and the target (in km); and σ = RCS of the target (in m^2).

The -103.43 dB is a compensation constant obtained from the known constants. Furthermore, when selecting the antenna gain (G) in dB and the Tx power (P_{Tx}) in dBm, the Rx sensitivity of the radar is obtained in dBm.

The Rx sensitivity calculations are explained elaborately in Chapter 4. However, here, the mathematical definition of the sensitivity is given, and the radar equation given in (2.38) will be rearranged. The Rx sensitivity is defined as

$$S_{min} = \left(kT_0B\right) \times (F) \times \left(\frac{S}{N}\right)_{min} \tag{2.41}$$

The equation shows that the sensitivity comprises three components separated by parentheses. The first parentheses term represents the noise floor or thermal noise level, k is the Boltzmann constant (1.38×10^{-23} Joule/K), T is the effective system noise temperature in degrees, and B is the radar Rx operating bandwidth. The second term is the noise figure of the radar Rx, and the third term shows the minimum *signal-to-noise ratio* (S/N) required to discriminate the sought signal components from the unwanted signals with a given probability of detection. The following equation is obtained when we substitute the sensitivity definition given in (2.41) in the radar equation (2.38).

$$R_{max} = \left[\frac{P_{Tx}\ G^2c^2\sigma}{\left(4\pi\right)^3 f^2\left(kT_0B\right)(F)(S/N)_{min}}\right]^{\frac{1}{4}} \tag{2.42}$$

Occasionally, finding $(S/N)_{min}$ value according to the maximum detection range is in question. Thus, the equation in (2.42) can be written in the following form. Furthermore, considering the radar losses (L) to reduce the SNR would be appropriate.

$$\left(\frac{S}{N}\right)_{min} = \frac{P_{Tx}G^2c^2\sigma}{\left(4\pi\right)^3 f^2\left(kT_0B\right)(F)LR_{max}^4} \tag{2.43}$$

2.4.3 Two-Way Radar Equation for Bistatic Operations

In bistatic operation, the radars' Tx and Rx are in different locations and use separate antennas. Figure 2.13 represents the physical state of the radar's Tx and Rx for bistatic operation.

The most common bistatic radar application is semiactive guided missile applications. The operation method of the bistatic radar is given by an air-to-air semiactive guided missile example, as in Figure 2.13. The Tx, which broadcasts the RF guidance signal, is on the missile launch platform. The launch platform can be a land, sea, or air platform. However, Rx is on the missile somewhere between the launch platform and the target. The Tx and Rx antennas are separate in distinct locations, and the gains of the Rx and Tx antennas are often different. In addition, since the distance between the target and the radar differs from the distance between the

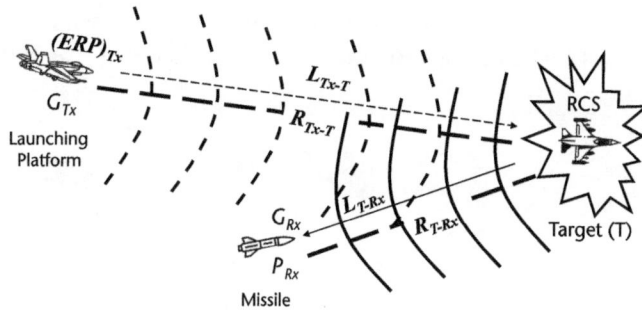

Figure 2.13 Two-way radar equation outline for bistatic radar operation.

target and the missile, the free space losses between the radar Tx-target and the target-missile Rx are various.

The following two components are calculated separately and combined to obtain the bistatic radar equation. Although these are similar to the components obtained in monostatic radars, they have two crucial differences. First, since Tx and Rx are in different places, the distance between Tx-target ($R_{Tx\text{-}T}$) and between target-Rx ($R_{T\text{-}Rx}$) are different. Second, since the Rx and Tx antennas are different, the G_{Rx} and G_{Tx} gains are different. These factors also make up the difference between the monostatic and bistatic radar equations.

In the first stage of the process, the signal power created by Tx on the target is calculated. This one-way space loss calculation uses $R_{Tx\text{-}T}$, the distance between Tx and the target. After including the RCS in this calculation, the reflected signal strength from the target to the Rx is obtained. The power from Tx to target $P_{Tx\text{-}T}$ is as follows.

$$\left.\begin{array}{c} P_{Tx-T} = \dfrac{P_{Tx}\ G_{Tx}G_\sigma\lambda^2}{\left(4\pi R_{Tx-T}\right)^2} \\[12pt] G_\sigma = \dfrac{4\pi\sigma}{\lambda^2} \end{array}\right\} \Rightarrow P_{Tx-T} = \dfrac{P_{Tx}G_{Tx}\sigma}{4\pi\left(R_{Tx-T}\right)^2} \tag{2.44}$$

The RCS gain used in bistatic radars differs from that used in monostatic radars. In monostatic radars, the RCS gain is the same in Tx and Rx. However, the RCS gain in bistatic radars is different due to the unlike aspect angles of Tx and Rx, and the value of the RCS gain in the Rx direction is taken into the calculations.

In the second stage of the process, the signal's power reflected from the target and returned to Rx is calculated. This process is a one-way space loss calculation for the distance between the target and Rx ($R_{T\text{-}Rx}$). The Rx antenna gain is considered G_{Rx}, which differs from the Tx antenna gain. In the following equation, $P_{T\text{-}Rx}$, which is the signal strength reflected from the target to the radar Rx, is calculated.

$$P_{T-Rx} = \dfrac{P_{Tx-T}G_{Rx}\lambda^2}{\left(4\pi R_{T-Rx}\right)^2} \tag{2.45}$$

Combining the equations formed by the one-way calculation stages given in (2.44) and (2.45), the bistatic radar equation is written as follows.

$$P_{Rx} = P_{Tx}G_{Tx}G_{Rx}\left[\frac{\sigma c^2}{(4\pi)^3 f^2 R_{Tx-T}^2 R_{T-Rx}^2}\right] \tag{2.46}$$

Using the peak power of Tx in the above equation, the peak power of Rx is obtained. The equation can be written in dB form as:

$$10\log P_{Rx} = 10\log P_{Tx} + 10\log\left(G_{Tx} \times G_{Rx}\right) + 10\log\sigma - 20\log f + \ldots$$
$$\ldots + 20\log c - 30\log 4\pi - 20\log\left(R_{Tx-T} \times R_{T-Rx}\right) \tag{2.47}$$

Known values such as the speed of light and 4π are clustered under certain constants for specific units to simplify the resulting logarithmic equation in terms of computation and can be written as follows:

$$10\log P_{Rx} = 10\log P_{Tx} + 10\log\left(G_{Tx} \times G_{Rx}\right) + G_\sigma - L_{S(Tx)} - L_{S(Rx)} \tag{2.48}$$

In the equation, $L_{S(Tx)}$ is the one-way space loss from Tx to the target, $L_{S(Rx)}$ is the one-way space loss from the target to Rx, and G_σ is the target's RCS gain factor in the Rx direction. When the frequency is taken as gigahertz, distance in kilometers, and RCS m², these values are calculated as follows.

$$L_{S(Tx)} = 92.45 + 20\log\left(f \times R_{Tx-T}\right), \quad L_{S(Rx)} = 92.45 + 20\log\left(f \times R_{T-Rx}\right)$$
$$G_\sigma = 21.46 + 20\log f + 10\log\sigma_{Rx} \tag{2.49}$$

Also, the following equation is obtained for $(S/N)_{min}$ when we substitute the sensitivity definition given in (2.41) in the bistatic radar equation (2.46).

$$\left(\frac{S}{N}\right)_{min} = \frac{P_{Tx}G_{Tx}G_{Rx}\sigma c^2}{(4\pi)^3 f^2 \left(kT_0 B\right)(F) R_{Tx-T}^2 R_{(T-Rx)max}^2} \tag{2.50}$$

In the equation, $(S/N)_{min}$ is obtained for the maximum range between the target and the Rx. Furthermore, this analysis can be conducted for the maximum range between Tx and the target.

2.5 Radar Operations

In general, radar systems can be operated using three different signal types. The radar signal types are pulsed, CW, and pulsed Doppler. From a signal point of view, pulsed and CW signals may be considered separate groups, and PDR may be a combination. However, taking the three signal groups into account and inspecting them is a convenient classification in operation and functionality.

2.5.1 Pulsed Radar Operation

In the pulsed radar operation, a train of pulsed waveforms is used. Pulsed radars operate with constant-frequency pulse or LFM pulse signals. The pulsed radar block diagrams for monostatic and bistatic radar operations are given in Figure 2.4. Even though the operational and functional differences are apparent for pulsed and PDRs, there is a substantial difference between their PRF rates. Low PRF radars are used for ranging but not the target velocity. Because of this, the nature of a low PRF radar is range-unambiguous and Doppler ambiguous. Medium PRF radars have both range and Doppler ambiguities. However, high PRF radars have only ambiguity for the radial velocities [7].

The detection range is obtained using (2.44) for monostatic and (2.48) for bistatic radar operations. Detecting and measuring a target's distance is a pulsed radar's main aim. However, pulsed radars do not have an interest in determining the radial velocity of a target. Thus, low PRFs are primarily used for pulsed radars. Before inspecting the signal structure of pulsed radar signals, giving some definitions and relationships about the radar pulses would be appropriate. Figure 2.14, demonstrating the transmitted and received pulses, is given to elaborate on the concepts in pulsed radars.

A pulsed radar transmits and receives a train of pulses. Each pulse is composed of a windowed sinusoidal signal in the operating frequency. The window width is the *pulse width* (PW), the starting point of the window is the leading edge, and the ending point is the trailing edge. In monostatic radars, when they use the same antenna, the Rx is closed during the emission of the Tx. This duration is PW (τ), and the magnitude of the rectangular envelope is peak power (P_t). After transmitting the pulses, Rx receives the backscattered signals with much less power in monostatic radars. This situation is illustrated with some exaggeration in Figure 2.14. Also,

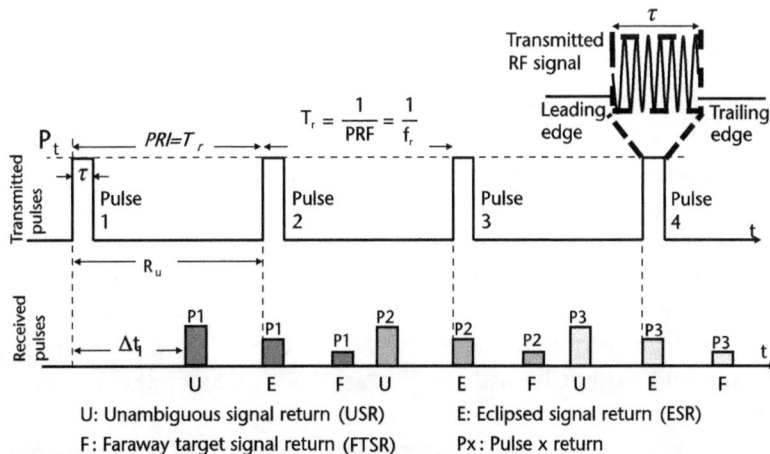

Figure 2.14 Pulsed radar transmitted and returned signals.

receiving signal powers depends on the distance between the radar and the target. Rx is shut down during transmission, and a death time arises according to the PW (τ). Thus, an undetectable distance, calculated as follows, exists.

$$R_\tau = \frac{c \times \tau}{2}$$

(2.51)

The time between the pulses is the *pulse repetition interval* (PRI), and the reciprocal of PRI is the *pulse repetition frequency* (PRF). PRI may be constant or change from pulse to pulse according to the modulation type. In the given figure, the interval between the pulses is assumed to be steady. The range corresponding to the two-way time delay PRI duration is known as the unambiguous radar range, R_u. If a radar transmits only one pulse in range measurements, all intervals can be measured without uncertainty. However, when it sends a second pulse, the unambiguous range measurement has the limit of the PRI time before it becomes ambiguous. An unambiguous range is used for radar system design considerations. For two reasons, the unambiguous range becomes essential for extracting information from the receiving pulses. The first is when more than one signal is available for the correct information, making the valid value uncertain. Second, some conditions do not allow accurate information to be extracted. The unambiguous range definition is as follows:

$$R_u = \frac{c \times T_r}{2} = \frac{c}{2 \times f_r}$$

(2.52)

The radar-to-target distance is calculated as in (2.53) in the time interval between Pulses 1 and 2. This distance is in the limits of the unambiguous range and is called *unambiguous signal return* (USR). Also, this is valid for the subsequent pulse intervals.

$$R_1 = \frac{c \times \Delta t_1}{2}$$

(2.53)

In Figure 2.14, the *eclipsed signal return* (ESR) exists when a reflected signal from a pulse appears simultaneously as the next pulse is transmitted, and the signal cannot be detected. Eclipsing is the intersection of a return signal and a transmitted pulse phenomenon. The *faraway target signal return* (FTSR) appears after sending the next pulse, and this signal is ambiguous. Since the distances of ESR targets are farther than the targets of USRs according to the radar, the return powers of ESR signals are lower than those of USR signals. The FTSR signals have the lowest power level since the target source of the FTSR signal is the farthest from the radar.

The transmitted and received signal scheme for a monostatic pulsed radar in Figure 2.14 cannot be used for bistatic radars. Since the Tx and Rx are in different locations, mentioning an echo return is impossible for bistatic radar operation. Protection is not used for Rx since there will not be a requirement to shut down the

Rx during Tx transmission. Thus, mentioning death time, USR, ESR, and FTSR times is meaningless for bistatic radars.

Another important term in radar operation is the average transmitted power. It can be obtained using peak power (P_{Tx}) and the duty ratio or duty cycle (D). The duty cycle is defined as the ratio of the PW to the PRI: $D = \tau/T_r$.

$$P_{av} = P_{Tx} \times D = P_{Tx} \times \frac{\tau}{T_r} \qquad (2.54)$$

The range resolution (ΔR) is a radar metric that describes its ability to detect targets close to each other as distinct objects [2]. In unmodulated pulse radars, the range resolution relationship is as follows: the radar frequency bandwidth (B) equals $1/\tau$.

$$\Delta R = \frac{c \times \tau}{2} = \frac{c}{2 \times B} \qquad (2.55)$$

Minimizing the pulse width can improve range resolution. However, this process reduces the average transmitted power and increases the operating bandwidth. This problem can be solved using modulated pulses or pulse compression techniques without reducing average power and increasing bandwidth. This time, the magnitude of B depends on the PW and the modulation bandwidth, and equality $B = 1/\tau$ spoils. Thus, the range resolution is calculated using $\Delta R = c/(2B)$.

The typical signal forms encountered in radar systems are the narrow bandpass signals. They have a limited angular-frequency bandwidth of 2W centered about a carrier angular frequency $\pm\omega_0$. If the signal bandwidth is B and f_0 is very large compared to B, then the signal $x(t)$ is referred to as a narrow bandpass signal. A narrow bandpass signal can be written in terms of the angular frequency and the ordinary frequency (or frequency) of the carrier signal as [2, 12]

$$x(t) = a(t)\cos\left(\omega_0 t + \phi(t)\right) = a(t)\cos\left(2\pi f_0 t + \phi(t)\right) \qquad (2.56)$$

$a(t)$ is the amplitude modulation or natural envelope, f_0 is the frequency defined as $f_0 = \omega_0/2\pi$ and $\phi(t)$ is the instantaneous phase of the narrow bandpass signal. Also, the narrow bandpass is expressed as:

$$x(t) = a_c(t)\cos 2\pi f_0 t - a_s(t)\sin 2\pi f_0 t \qquad (2.57)$$

The in-phase component $a_c(t)$ and the quadrature component $a_s(t)$ in the above equation are baseband signals bounded by W and defined as follows:

$$a_c(t) = a(t)\cos\phi(t)$$
$$a_s(t) = a(t)\sin\phi(t) \qquad (2.58)$$

In light of these statements, the complex envelope of the bandpass signal $x(t)$ is written as:

$$s(t) = a_c(t) + ja_s(t) \tag{2.59}$$

Also, it is essential to establish the relationship between the bandpass signal and its complex envelope. For this purpose, we consider the Hilbert transform of the bandpass signal. The Hilbert transform operator and its Fourier transform are defined in the following equation.

$$h(t) = \frac{1}{\pi t} \underset{F^{-1}}{\overset{F}{\rightleftharpoons}} H(\omega) = \exp\left(-j\frac{\pi}{2}\right)\mathrm{sgn}(\omega) = -j\,\mathrm{sgn}(\omega) \tag{2.60}$$

The Hilbert transform is computed as the convolution between the signals $x(t)$ and the Hilbert transform operator $1/(\pi t)$. The $\mathrm{sgn}(\omega)$ function is defined in (2.61).

$$\mathrm{sgn}(\omega) = \begin{cases} 1 \; ; & \omega > 0 \\ 0 \; ; & \omega = 0 \\ -1; & \omega < 0 \end{cases} \tag{2.61}$$

The definition of the Hilbert transform of a bandpass signal $x(t)$ and its Fourier transform is as follows. In the below equation, $\otimes$ represents convolution:

$$H\{x(t)\} = \hat{x}(t) = x(t) \otimes h(t) = x(t) \otimes \frac{1}{\pi t}$$
$$F\{\hat{x}(t)\} = \hat{X}(\omega) = X(\omega)H(\omega) = -j\,\mathrm{sgn}(\omega)X(\omega) \tag{2.62}$$

As the equation shows, the Hilbert transform results in a phase shift of $\pi/2$ on the $x(t)$ spectra. So the Hilbert transform of a narrow bandpass signal given in (2.57) is obtained as:

$$\hat{x}(t) = a_c(t)\sin 2\pi f_0 t + a_s(t)\cos 2\pi f_0 t \tag{2.63}$$

Defining the analytic signal $\psi(t)$ is beneficial at this point. The analytic signal of the bandpass signal $x(t)$ is achieved by eliminating the negative frequency contents of $X(\omega)$.

$$\Psi(\omega) = \begin{cases} 2X(\omega); & \omega > 0 \\ X(\omega); & \omega = 0 \\ 0; & \omega < 0 \end{cases} \rightarrow \Psi(\omega) = X(\omega)(1 + \mathrm{sgn}(\omega)) \tag{2.64}$$

We obtain the following analytic signal when we take the inverse Fourier transform of $\Psi(\omega)$. This signal is referred to as the pre-envelope of $x(t)$ since the complex envelope of $x(t)$ is obtained by taking the modulus of $\psi(t)$ [2].

$$\psi(t) = F^{-1}\{\Psi(\omega)\} = x(t) + j\hat{x}(t) \qquad (2.65)$$

After considering the Hilbert transform and pre-envelope of the bandpass signal $x(t)$, we can now establish the relationship between the bandpass signal and its complex envelope.

$$x(t) = \text{Re}\{s(t)\exp(j2\pi f_0 t)\} \qquad (2.66)$$

Considering (2.57), (2.63), and (2.65) together and using them with (2.66), we obtain the following result:

$$s(t)\exp(j2\pi f_0 t) = x(t) + j\hat{x}(t) = \psi(t) \qquad (2.67)$$

Now we can continue the analysis of the pulse radar signal with magnitude $1/\sqrt{\tau_0}$ and width τ_0,

$$x(t) = \frac{1}{\sqrt{\tau_0}}\text{Rect}\left(\frac{t}{\tau_0}\right)\cos(2\pi f_0 t), \quad \text{where Rect}\left(\frac{t}{\tau_0}\right) = \begin{cases} 1, & -\frac{\tau_0}{2} \le t \le \frac{\tau_0}{2} \\ 0, & \text{otherwise} \end{cases} \qquad (2.68)$$

The complex envelope of the above constant frequency, or unmodulated, pulse radar signal is defined as follows [12]:

$$s(t) = \frac{1}{\sqrt{\tau_0}} \times \text{Rect}\left(\frac{t}{\tau_0}\right) \qquad (2.69)$$

Figure 2.15 shows the relationship between the pulse radar signal $x(t)$ and its corresponding complex envelope $s(t)$. The pulse signal is processed in an I/Q detector to obtain the complex envelope, and the I (in-phase) and Q (quadrature) components are discriminated.

Defining the pulse radar signal's *ambiguity function* (AF) at this point would be appropriate. For this purpose, a short discussion on the matched filter would be suitable. The range resolution for simple pulses that do not use any modulation, the pulse width is strictly related to the range resolution in (2.55) as $\Delta R = c\tau/2$. In the presence of modulated pulses with an instantaneous bandwidth B, the range resolution is calculated by $\Delta R = c/(2B)$. This expression can be motivated by the matched filter whose output in time is reminiscent of the autocorrelation function of the transmitted waveform. The instantaneous bandwidth of the transmitted

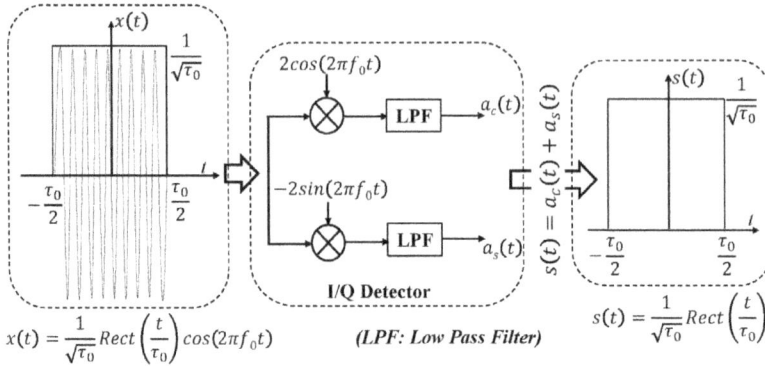

Figure 2.15 Discrimination of in-phase and quadrature components.

waveform represents an important parameter that determines the minimum distance at which the radar can resolve two point-like scatterers.

Another aspect that should be considered is the frequency domain behavior of the radar waveform. If a target moves, the reflected echoes incorporate an additional frequency modulation due to the Doppler component. As a result, the magnitude spectrum of the received echoes will be shifted according to the Doppler frequency value. If this displacement is of the order of the instantaneous bandwidth, the matched filter will cut off most of the received energy.

Analyzing the matched filter behavior in both time and frequency domains leverages the definition of the AF, a powerful analytical tool for waveform design. It characterizes the behavior of a waveform along the range and Doppler dimensions in terms of resolution, sidelobe behavior, ambiguities, and phenomena such as range-Doppler coupling. The effect of a Doppler mismatch can be appreciated from the ambiguity function by varying the Doppler frequency. The peak values decrease when the Doppler mismatch increases. Two strategies can be pursued to mitigate this loss. The first strategy uses a Doppler filter bank where each filter is matched to a specific Doppler frequency of interest. In contrast, the second approach exploits a suitably shaped waveform whose magnitude spectrum is such that the instantaneous bandwidth is much greater than the frequency displacements of interest, leading to quasi-constant peaks within the tolerance region. In both cases, a frequency mismatch always exists that will cause a reduction in the observed peaks and, hence, in the system's detection capabilities. In addition, the peak can no longer be centered at the same instant as in a perfect Doppler match, but it can move as the Doppler mismatch increases due to range-Doppler coupling [13].

The AF represents the time response of a filter matched to a given finite energy signal when the signal is received with a time delay (or range) τ and a Doppler shift ϕ relative to the nominal values (zeros) expected by the filter. Constant values of time delay Δt and constant values of Doppler frequency shift f_d are used in this book. However, as a function variable, ϕ is used for the Doppler shift, and τ is used for the time delay. Let $s(t)$ be the transmitted waveform, and then the complex ambiguity function (CAF) can be defined as the cross-correlation between $s(t)$ $\exp(j2\pi\phi t)$ and $s(t)$.

$$\chi(\tau,\phi) = \int_{-\infty}^{\infty} s(t)s^*(t+\tau)\exp(j2\pi\phi t)\,dt \qquad (2.70)$$

Notice that the above function represents the output of a matched filter when a frequency mismatch ϕ occurs between the incoming signal (due to target radial velocity) and the matched filter. The unique characteristic of the matched filter is that it produces the maximum achievable instantaneous SNR at its output when a signal plus noise (when Gaussian noise is assumed) exists in its input.

To get the maximum range and Doppler resolution, the square of the modulus of this function should be minimized at $\tau \neq 0$ and $\phi \neq 0$. With these minimizations, the output of the matched filter provides the maximum instantaneous SNR and exhibits the most achievable range and Doppler resolutions. The matched filter is the optimal linear filter for maximizing the SNR in the presence of additive stochastic noise. One can maximize the SNR at the receiver by using a matched filter for a deterministic signal in white Gaussian noise. The matched filter is a time-reversed and conjugated version of the signal. The matched filter is shifted to be causal, so convolving the unknown signal with a conjugated time-reversed version of the delayed signal is obtained. The modulus square of (2.70) is referred to as the *ambiguity function* (AF) [2].

$$\left|\chi(\tau,\phi)\right|^2 = \left|\int_{-\infty}^{\infty} s(t)s^*(t+\tau)\exp(j2\pi\phi t)\,dt\right|^2 \qquad (2.71)$$

A positive ϕ implies a target moving toward the radar. Positive τ means a target farther from the radar than the reference ($\tau = 0$) position. The radar AF is a 2-D function of the Doppler shift and the range. Since Doppler shift and range are canonically conjugate variables, an uncertain relationship exists between them. It is not possible to measure both these quantities with infinite accuracy.

The AF is central to echolocation as it relates signal characteristics to the system and parameter estimation performance. Estimation performance as measured by the Cramer-Rao bound to be proportional to the resolution of the AF, which depends upon the characteristics of the transmitted signal. When the parameters are the usual delay and Doppler parameters associated with radar, the parameter space is often referred to as the phase space because the parameters are directly related to range (delay) and velocity (Doppler). For the monostatic line-of-sight geometry, range and time delay are described by $r \approx c\tau/2$ for the narrowband representation. However, the relationship between the velocity and Doppler shift is $v \approx \phi c/2f_0$ for the wideband representation. The accurate AF represents a surface. AFs arise naturally from the maximum likelihood criteria for detection and estimation in a Gaussian environment. It maps a time function with one dimension into another 2-D function composed of time delay and frequency [14].

The AF measures the imperfections of the matched filter, often known as the filter side lobes. Taking the square root of both sides of the AF equation in (2.71) results in the following equation [12]:

$$\left|\chi(\tau,\phi)\right| = \left|\int_{-\infty}^{\infty} s(t)s^*(t+\tau)\exp(j2\pi\phi t)\,dt\right| \tag{2.72}$$

This equation turns the analysis for different signals into a more manageable form. Also, in this form of AF, the sidelobes are not suppressed as much as in (2.71). In this book, the AF is referred to as (2.72). So the AF of the unmodulated pulse radar signal given in (2.69) is obtained as

$$\left|\chi(\tau,\phi)\right| = \begin{cases} \left|\left(1-\dfrac{|\tau|}{\tau_0}\right)\dfrac{\sin\left[\pi\tau_0\phi\left(1-|\tau|/\tau_0\right)\right]}{\pi\tau_0\phi\left(1-|\tau|/\tau_0\right)}\right|, & |\tau| \le \tau_0 \\ 0, & \text{elsewhere} \end{cases} \tag{2.73}$$

If we set the Doppler shift variable to zero in the AF, we get the variation in the cut along the delay axis,

$$\left|\chi(\tau,0)\right| = \begin{cases} 1-\dfrac{|\tau|}{\tau_0}, & |\tau| \le \tau_0 \\ 0, & \text{elsewhere} \end{cases} \tag{2.74}$$

However, when we set the time delay variable to zero, we obtain the variation in the cut along the Doppler axis,

$$\left|\chi(0,\phi)\right| = \left|\frac{\sin\pi\tau_0\phi}{\pi\tau_0\phi}\right| \tag{2.75}$$

Using the obtained results for an unmodulated pulse, we can extend the analysis for the coherent train of identical unmodulated pulses, which is defined in mathematical form as follows,

$$x(t) = \sum_{n=0}^{N-1} \frac{1}{\sqrt{\tau_0}}\text{Rect}\left(\frac{t-nT_r}{\tau_0}\right)\cos(2\pi f_0 t) \tag{2.76}$$

In the equation, T_r is the PRI. The complex envelope of the coherent train of identical unmodulated pulses is described by

$$s(t) = \frac{1}{\sqrt{N}}\sum_{n=0}^{N-1}\frac{1}{\sqrt{\tau_0}}\text{Rect}\left(\frac{t-(n-1)T_r}{\tau_0}\right) \tag{2.77}$$

The AF of a coherent pulse train is defined as follows for the practical case of $T < T_r/2$ [12].

$$|\chi(\tau,\phi)| = \begin{cases} \dfrac{1}{N} \displaystyle\sum_{p=-(N-1)}^{N-1} \left\{ |\chi_T(\tau,\phi)| \left| \dfrac{\sin\left[\pi\phi\left(N-|p|/T_r\right)\right]}{\sin\left(\pi\phi T_r\right)} \right| \right\}, & |\tau| \leq NT_r \\ 0 & \text{elsewhere} \end{cases} \qquad (2.78)$$

where $|\chi_T(\tau,\phi)|$ is the AF of an individual pulse, which is given by (2.73).

For pulsed radars, using unmodulated pulses is one of two cases. For this purpose, it would be appropriate to consider the LFM pulse radar signal. The range resolution for unmodulated pulse radars is given in (2.55) according to the pulse width and bandwidth. Narrowing down the pulses improves range resolution for an unmodulated pulse radar. However, short pulses decrease the average transmitted power and the SNR. In unmodulated pulse radar operation, pulse compression is not mentioned, and the time-bandwidth product equals 1.

Since the average transmitted power is desirable to increase the pulse width while maintaining adequate range resolution, these properties are obtained using pulse compression techniques and the matched filter receiver. Using LFM-modulated pulse signals is a prevalent technique for increasing SNR without reducing range resolution. This process compresses matched filter output by a factor obtained from the time-bandwidth product. Thus, long pulses and high SNR can be achieved by using LFM. An LFM-modulated pulse signal is shown in Figure 2.16.

The LFM pulse radar signal with magnitude $1/\sqrt{\tau_0}$ and width τ_0 is defined as follows:

$$x(t) = \frac{1}{\sqrt{\tau_0}} \text{Rect}\left(\frac{t}{\tau_0}\right) \cos\left(2\pi f_0 t + \frac{\pi B}{\tau_0} t^2\right) \qquad (2.79)$$

where B is the LFM sweep frequency band, and f_0 is the operating frequency. The complex envelope of the LFM pulse signal is obtained by using (2.59) as follows:

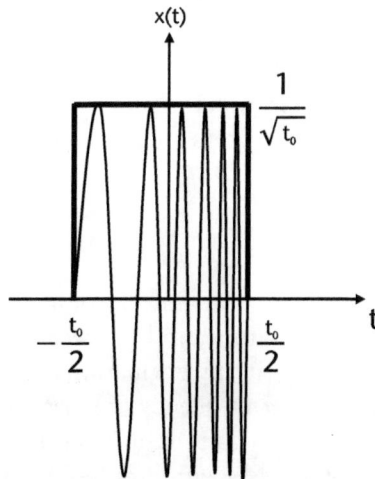

Figure 2.16 LFM modulated pulse signal.

$$s(t) = \frac{1}{\sqrt{\tau_0}} \times \text{Rect}\left(\frac{t}{\tau_0}\right) \exp\left(j\frac{\pi B}{\tau_0}t^2\right) \qquad (2.80)$$

Because of the variation in the frequency, defining the instantaneous frequency of the LFM pulse signal would be beneficial. The instantaneous frequency $f_i(t)$ is calculated by differentiating the argument of the sinusoidal term of the LFM pulse signal given in (2.79),

$$f_i(t) = \frac{1}{2\pi}\frac{d}{dt}\left(2\pi f_0 t + \frac{\pi B}{\tau_0}t^2\right) = f_0 + \frac{B}{\tau_0}t \qquad (2.81)$$

The AF of an LFM pulse signal is obtained by using (2.72) as follows:

$$|\chi(\tau,\phi)| = \begin{cases} \left(1 - \frac{|\tau|}{\tau_0}\right)\left|\dfrac{\sin\left[\pi\tau_0\left(\phi \pm B(\tau/\tau_0)\right)\left(1 - |\tau|/\tau_0\right)\right]}{\pi\tau_0\left(\phi \pm B(\tau/\tau_0)\right)\left(1 - |\tau|/\tau_0\right)}\right|, & |\tau| \le \tau_0 \\ 0 & \text{elsewhere} \end{cases} \qquad (2.82)$$

2.5.2 Continuous-Wave Radar Operation

CW radars are categorized into short-range and small-range radars and may use unmodulated or modulated waveforms. Two antennas are used in CW radars for continuous radar energy emission. These antennas are for Tx and Rx purposes, and RF insulation is required if used in the same location. The unmodulated CW radars are called CW Doppler (CWD) radars and are used to measure velocity. They are inappropriate for range measurement since there must be time information for determining the target range from transmitting and receiving signals. Thus, modulated CW radars extract the target range by comparing the transmitting and receiving signals. The general block diagrams for CW radar operations are given in Figure 2.6.

Compared with pulsed radar signals, a CW signal is more difficult for an adversary Rx to detect. This is because the CW signal has a very narrow bandwidth, and there would be no detection if a filter were not used to match the signal. Thus, CW waveforms are generally classified as low probability of intercept (LPI) waveforms [12].

2.5.2.1 CW Doppler Radars

We first consider the CWD radars into account. CWD radar transmits an unmodulated continuous frequency tone, and the received echo signal is processed to estimate a target's radial velocity by evaluating the phase change. This phase change is due to a shift in Doppler frequency (f_d) arising from the reflection of a moving target. However, CW radars cannot obtain the range information. A coarse range estimation can be obtained by either pulse-Doppler operation or by transmitting two distinct frequency tones referred to as *frequency shift keying* (FSK) [15].

The signal processing block diagram of a CWD radar system is given in Figure 2.17. A CWD radar system's Doppler frequency measurement resolution depends on the bandwidth of the *individual narrowband filters* (NBFs). The individual NBF bandwidth is to be as narrow as possible to achieve higher accuracy Doppler measurements and minimize the amount of noise power. The Doppler filter bank is occasionally implemented using a *fast Fourier transform* (FFT) of size N. Suppose the individual NBF bandwidth or the FFT bin is Δf. In this case, the effective radar Doppler bandwidth is $N\Delta f/2$. The 1/2 factor is used to add the effects of both negative and positive Doppler frequencies [2].

As an FFT implements the NBF bank, only finite-length data sets can be processed simultaneously. The length of the sets is commonly referred to as the dwell interval (T_D). The reciprocal of the dwell interval gives the frequency resolution or the bandwidth of the individual NBFs (Δf). After that, the frequency resolution is selected, and accordingly, the size of the NBF bank is calculated as follows:

$$N = \frac{2B}{\Delta f} = 2B \times T_D, \quad \left\{\frac{1}{T_D} = \Delta f\right\} \tag{2.83}$$

where B is the maximum resolvable frequency bandwidth by the FFT. Given such information, we can extend the pulsed radar equation for CW radars. First, we can consider the radar equation for monostatic pulsed radar (2.43) and substitute peak power P_t with the average power P_{av} defined in (2.54) as follows:

$$\left(\frac{S}{N}\right)_{min} = \frac{P_t G^2 c^2 \sigma}{(4\pi)^3 f^2 (kT_0 B)(F)LR^4_{max}} = \frac{P_{av}T_r G^2 c^2 \sigma}{(4\pi)^3 f^2 (kT_0)(F)LR^4_{max}} \tag{2.84}$$

For obtaining the CW radar equation from the above equation, the power over the dwell interval P_{CW} is substituted for the average transmitted power, and the PRI (T_r) is replaced by T_D. CW radars use different antennas for transmitting and

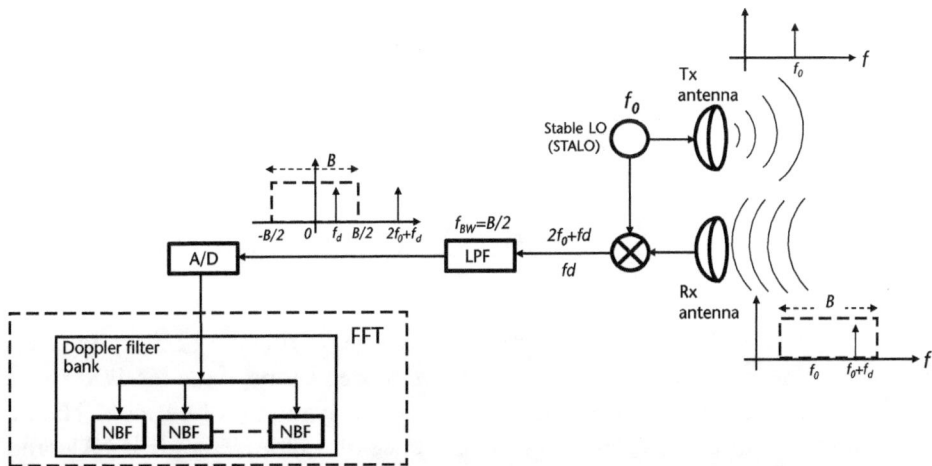

Figure 2.17   CW radar signal processing block diagram.

receiving, so gains must be taken separately for Rx (G_{Rx}) and Tx (G_{Tx}). As a final point for the CW radar equation, adding a loss term L_{win} associated with the type of window used in computing the FFT would be appropriate. Thus, the CW radar equation can be written as [2]:

$$\left(\frac{S}{N}\right)_{\min} = \frac{P_{CW} T_D G_{Tx} G_{Rx} c^2 \sigma}{(4\pi)^3 f^2 (kT_0)(F) L R_{\max}^4 L_{win}} \tag{2.85}$$

As given in (2.88), the Doppler resolution is inversely proportional to the duration of the signal coherently processed by the receiver or the dwell interval (T_D). As mentioned before, this duration cannot be infinitely long. A periodic waveform is assumed to modulate the carrier signal with period T_m. Thus, Rx is matched with N integer periods, in other words, with the NT_n duration signal. The tool for analyzing the response of such an Rx is the *periodic ambiguity function* (PAF). The PAF is used in CW signals in a role like that the regular AF serves for finite energy signals. The CW signal constructed from N coherent periods is called a reference signal $x_N(t)$ and is defined as follows [12]:

$$x_N(t) = \mathrm{Re}\left\{s_N(t)\exp\left(j2\pi f_0 t\right)\right\} \tag{2.86}$$

The complex envelope of the reference signal limited with an N period is defined in the following form:

$$s_N(t) = \frac{1}{\sqrt{N}}\sum_{n=1}^{N} s_n\left[t - (n-1)T_m\right] \tag{2.87}$$

Let us assume that all the N periods are identical, and in this context, all the sequential periods are equal to the first period $s_1(t)$. The received signal (at zero relative delay and zero Doppler) is not limited to N periods and will be described as an infinite periodic signal defined as follows

$$x(t) = \mathrm{Re}\left\{s(t)\exp\left(j2\pi f_0 t\right)\right\} \tag{2.88}$$

The complex envelope of the received signal given above is

$$s(t) = \sum_{n=-\infty}^{\infty} s_1\left[t - (n-1)T_m\right] \tag{2.89}$$

Figure 2.18 shows the complex envelopes defined in (2.87) and (2.89).

The matched Rx correlates the complex envelope of the received periodic signal $s(t)$ with a reference periodic signal $s_N(t)$ of identical N periods. The normalized response of this processor in the presence of a Doppler shift is PAF, and periodic AF is defined as:

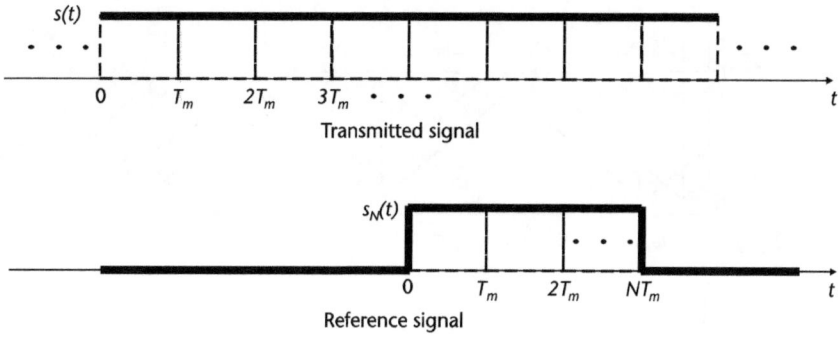

Figure 2.18   Complex envelopes of transmitted and reference CW radar signals.

$$\left|\chi(\tau,\phi)\right| = \left|\frac{1}{NT_m} \int_0^{NT_m} s(t)s^*(t+\tau)\exp(j2\pi\phi t)dt\right| \qquad (2.90)$$

2.5.2.2   Modulated CW Radars

One cannot obtain range information using CWD radars with one frequency. For this purpose, CW radars use modulated waveforms to measure range and Doppler information. Even though there are various types of modulated CW radar signals, we will limit our analysis to LFM radar signals here since LFM is prevalent in most modern radar systems.

The LFM waveform cannot continually increase or decrease in practical CW radars in one direction. Thus, the modulation's periodicity is usually utilized. Figure 2.19 shows symmetric triangular LFM waveforms for stationary targets (Figure 2.19(a)) and moving targets (Figure 2.19(b)). In Figure 2.19, solid lines show the transmitted signal with frequency f_t, and dashed lines represent the received signal with frequency f_r. Starting the inspection for stationary targets would be appropriate. The time delay Δt in Figure 2.19(a) is used for range calculation for stationary targets as follows:

$$\Delta t = \frac{2R}{c} \rightarrow R = \frac{\Delta t \times c}{2} \qquad (2.91)$$

The period of a chirp, T_r, is composed of up-chirp and down-chirp portions. The time interval between the starting point of a chirp and the peak point is rising or decaying time t_p. This time duration can also be obtained from the peak point to the end point of the chirp. It can be concluded that rising and decaying times are equal, and t_p is half of the chirp period. The modulating frequency f_m is the reciprocal of the chirp period, $1/T_r$. The slope of the up-chirp or down-chirp portion shows the frequency change rate, f', and is calculated using the peak frequency deviation Δf and rising (or decaying) time.

$$f' = \frac{\Delta f}{t_p} = \frac{\Delta f}{T_r/2} = \frac{2\Delta f}{1/f_m} = 2f_m\Delta f \qquad (2.92)$$

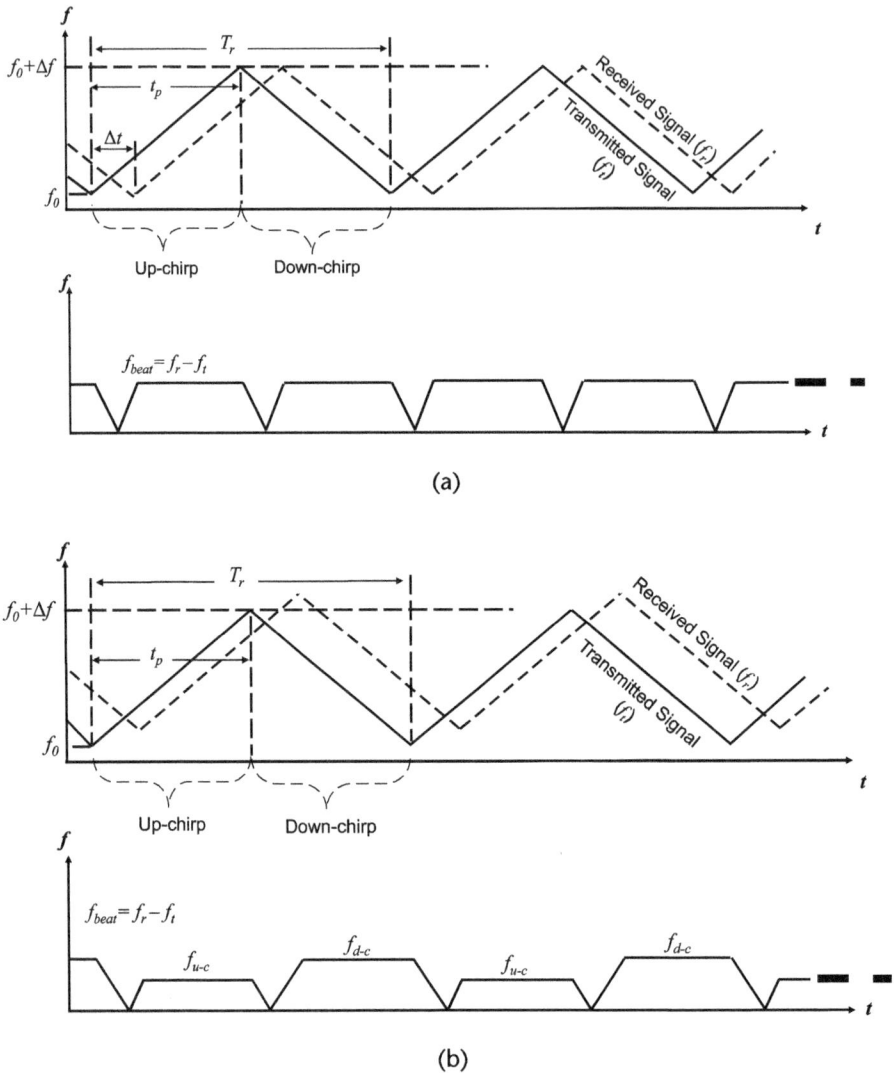

Figure 2.19 Symmetric triangular LFM waveforms and corresponding beat frequencies: (a) for a stationary target, and (b) for a moving target.

The beat frequency, f_{beat}, is the difference between the received signal frequency and the transmitted signal frequency. In stationary targets, beat frequency variation according to time is periodic. In this case, the beat frequency can be defined in terms of time delay and frequency change rate,

$$f_{beat} = \Delta t \times f' = \left(\frac{2R}{c}\right) \times \left(2f_m \Delta f\right) = \frac{4R f_m \Delta f}{c} \tag{2.93}$$

Figure 2.19(b) demonstrates the moving target case. In this case, the periodicity of the beat frequency corrupts, and the equation given in (2.93) cannot be used for obtaining beat frequency. This time, the only method to find the beat frequency is to find the difference between the received and transmitted signals' frequency. The beat frequency comprises two components: up-chirp frequency, $f_{u\text{-}c}$, and down-chirp frequency, $f_{d\text{-}c}$. The up-chirp and down-chirp frequencies may change according to the target's position and velocity variations in the elapsing time. The up-chirp frequency is defined as [2]:

$$f_{u\text{-}c} = \frac{2R}{c}f' - \frac{2v_T}{\lambda} \tag{2.94}$$

As seen from the right side of the equation, the up-chirp frequency is composed of two components. The first component $(2Rf'/c)$ is due to the range delay, and the second one $(2v_T/\lambda)$ emerges from the Doppler frequency. In (2.94), v_T represents the radial velocity of the target. However, the down-chirp frequency is obtained as follows:

$$f_{d\text{-}c} = \frac{2R}{c}f' + \frac{2v_T}{\lambda} \tag{2.95}$$

The difference between the up-chirp and down-chirp frequencies gives the target's radial velocity. Besides, the sum of the up-chirp and down-chirp frequencies results in the range. The equations of the target's radial velocity and range are

$$v_T = \frac{\lambda}{4}\left(f_{d\text{-}c} - f_{u\text{-}c}\right), \quad R = \frac{c}{4f'}\left(f_{u\text{-}c} + f_{d\text{-}c}\right) \tag{2.96}$$

A symmetric triangular FMCW signal's complex envelope for a single period is defined as [16]:

$$s_1(t) = \begin{cases} \exp\left(j\dfrac{2\pi\Delta f}{T_r}t^2\right) & 0 < t \leq \dfrac{T_r}{2} \\[3mm] \exp\left(-j\dfrac{2\pi\Delta f}{T_r}(t - T_r)^2\right) & \dfrac{T_r}{2} < t \leq T_r \end{cases} \tag{2.97}$$

Following the above equation, the symmetric triangular FMCW complex signal envelope is as follows:

$$s(t) = \sum_{n=-\infty}^{\infty} s_1\left[t - nT_m\right] \tag{2.98}$$

The PAF is the generalization of the periodic autocorrelation function to the nonzero Doppler shift case, which describes a correlation receiver's response to a continuous signal modulated by a periodic waveform. Following the conventional definition of the AF for signals with finite energy, the *single-period ambiguity function* (SPAF) of a symmetric triangular FMCW waveform is

$$\chi_{T_r}(\tau,\phi) = \frac{1}{T_r}\int_0^{T_r} s(t)s^*(t+\tau)\exp(j2\pi\phi t)\,dt \tag{2.99}$$

To complete the analysis, the relationship between the SPAF and PAF is established as follows. Thus, the PAF of the symmetric triangular FMCW signal is obtained as follows [17].

$$\left|\chi_{NT_r}(\tau,\phi)\right| = \left|\chi_{T_r}(\tau,\phi)\right|\left|\frac{\sin(\pi\phi NT_r)}{N\sin(\pi\phi T_r)}\right| \tag{2.100}$$

2.5.3 Pulsed Doppler Operation

This section will mention MTI and PDR systems since they combine pulsed and CW radar operations. The general block diagram for pulsed Doppler operations is given in Figure 2.7. A PDR system and a pulsed radar system with MTI operation are similar. Both methods try to obtain both range and velocity information. The difference between them exists in the unambiguous velocity and range information. This difference shows itself for the PDRs extracting unambiguous velocity within finite limits at the cost of unambiguous range information. However, the MTI radars have the property of extracting unambiguous range information within finite limits at the expense of ambiguous velocity information.

Most MTI radars do not measure the Doppler frequency if the only objective is to separate moving and stationary targets. However, the PDRs are designed to extract Doppler information to find and track targets in a high current environment. An essential feature of a PDR system is its capability to detect the presence of a frequency rather than catching a signal above an amplitude threshold. This feature makes it an inherently automatic device that can discriminate signals at nearly noise levels.

As stated before, different radar types are required for various PRF schemes, such as low, medium, and high. MTI radars use low PRF waveforms. Low PRF waveforms provide sorting clutter from targets based on range. They can suppress sidelobe detections at short ranges and reduce dynamic range requirements. However, low PRF radars are excessively ambiguous in Doppler, resulting in multiple blind speeds and the inability to measure radial target velocity.

Furthermore, low PRF radars have poor ground-moving target rejection. High PRF waveforms can provide considerable average transmitted power and perfect clutter rejection capabilities. However, high PRF waveforms are excessively ambiguous in range. Radar systems using medium and high PRFs are often called PDR [2, 7].

2.5.3.1 MTI Radars

The signal processing techniques are applied to pulsed-Doppler signals to enhance the SNR detection performance, discriminate between actual targets and interference, clutter, or jamming, and extract the required information about targets, such as range and velocity. The primary role of the MTI is to provide rejection of *main beam clutter* (MBC) and *sidelobe clutter* (SLC). MBC and SLC are demonstrated in Figure 2.20 for the single pulse and low PRF case. The main beam is seen to intersect the hill/mountain edge ahead of the radar and gives rise to a strong MBC. The sidelobes intersect the ground or hills/mountains around the radar and give rise to SLC, which is considerably spread in range and Doppler.

In many respects, the clutter situation for a surface-based radar simplifies the airborne case. Some surface-based radars may be on moving platforms and mounted at some considerable height above the cluttered surface. Radars mounted onboard ships may be moving at the mast height above the sea surface and can be treated the same way as airborne radars. The exception is that the platform velocity and altitude are much more modest. However, ground-based air surveillance radars are typically stationary and at ground level. In such cases, all the SLC and MBC fall at zero Doppler [18].

The received power from clutter is obtained using the following equation:

$$P_c = \frac{P_{\mathrm{Tx}} G^2 \lambda^2 (\sigma_c)}{(4\pi)^3 R^4}, \quad \{\sigma_c = \sigma_{\mathrm{MBC}} + \sigma_{\mathrm{SLC}}\} \tag{2.101}$$

In the equation, σ_{MBC} is the main beam clutter RCS, and σ_{SLC} is the sidelobe clutter RCS, given in Figure 2.20.

The clutter mainly comprises unwanted stationary ground reflections with limited relative motion concerning the radar. Therefore, its power spectral density concentrates around $f = 0$. The overall clutter spread is not always zero, exhibiting some Doppler frequency spread. The clutter power spectrum can be written as the sum of stationary and random (due to frequency spreading) components [2].

$$S_c(f) = \frac{P_c}{T_r \sigma_f \sqrt{2\pi}} \sum_{k=-\infty}^{\infty} \exp\left(-\frac{(f - k/T)^2}{2\sigma_f^2}\right), \quad \{\sigma_f^2 = \sigma_v^2 + \sigma_s^2 + \sigma_w^2\} \tag{2.102}$$

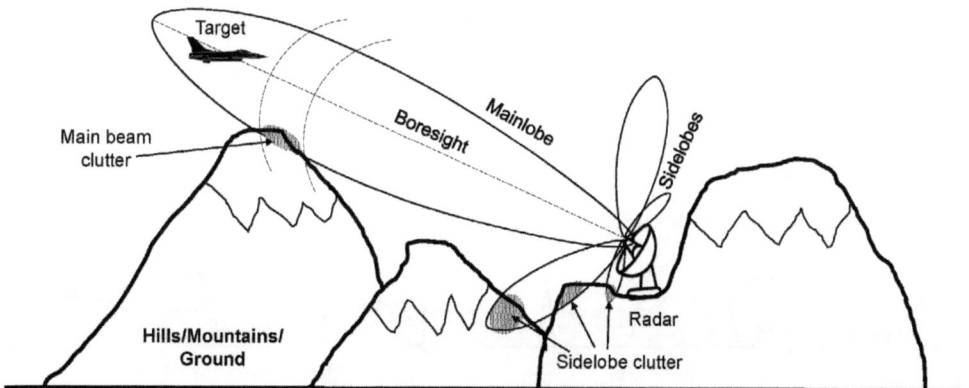

Figure 2.20 Main beam clutter and sidelobe clutter for a ground base radar.

The equation shows that the clutter power spectrum has a Gaussian distribution. In (2.102), T_r is the PRI, P_c is the clutter power, and σ_f is the clutter spectral spreading parameter. The spectral spreading parameter's components are: σ_v is the clutter spread due to platform motion, σ_s is the antenna scan rate, and σ_w the clutter spread due to wind.

The equation states that the clutter PSD is periodic, and its period equals f_r (1/T_r). Furthermore, the clutter PSD extends about each multiple integer of the PRF. Since $\sigma_f \ll f_r$ is always valid, the spread is considerably small. Considering these assumptions, Figure 2.21 depicts (2.102).

One can represent a target as a point reflector moving relative to element clutter scatterers. The target's radial motion causes its return to be shifted in frequency concerning the clutter scatterers, and the return signal spectrum centered around the Tx's carrier frequency. The target's power spectral density is determined by an antenna scanning motion, which induces spectral spreading onto the target. The target and ground clutter spectra are shown in Figure 2.22 [19].

When we look at the Rx spectrum, clutter is concentrated around DC and multiple integers of the radar PRF. Since most clutter power is focused on the zero-frequency band, filtering the receiver output around DC eliminates or suppresses clutter in CW radars. Pulsed radar systems may utilize special filters that distinguish between low-velocity or stationary targets and moving ones. This filter class is known as the MTI. The MTI systems do not attempt to measure the Doppler shift, and hence velocity, of target returns; the goal is to admit returns with a Doppler shift and reject those without one. This Doppler discrimination is accomplished using an MTI clutter cancellation filter or delay line cancelers in the baseband.

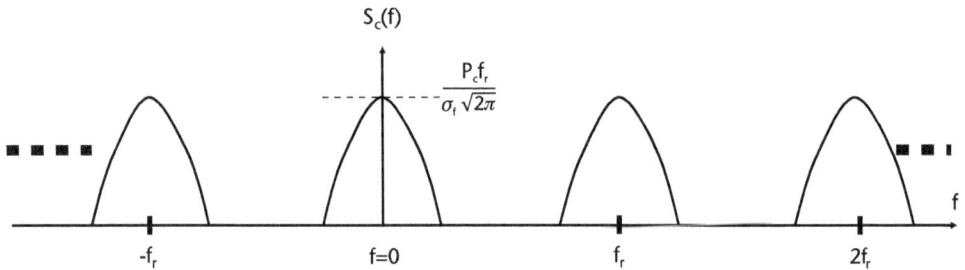

Figure 2.21 Power spectral density of the typical clutter.

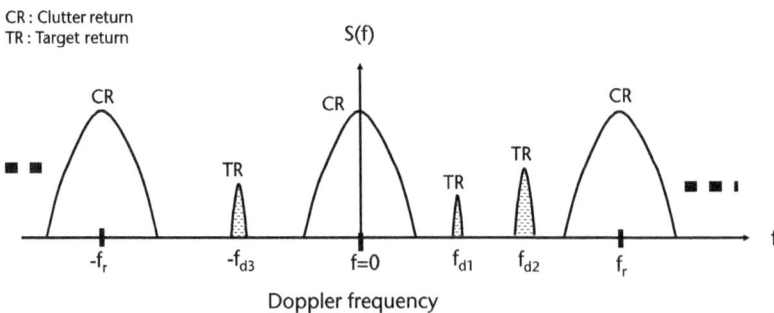

Figure 2.22 Power spectral density of target and clutter.

An MTI filter must have a deep stopband at DC and integer multiples of the PRF to suppress clutter returns. Figure 2.23(a) shows the spectrum of an MTI filter response and an MTI filter input, which consist of target and clutter returns. Furthermore, Figure 2.23(b) gives the power spectral density of the MTI filter output.

As shown in Figure 2.23, an adaptive detection threshold level is applied to the MTI filter output. In practical applications, surface clutter frequently specifies the limit on target detection. Since the clutter varies considerably over the radar's range/velocity detection space, so would the false alarm rate if a fixed threshold level were used. Furthermore, the clutter levels and statistics depend highly on parameters such as altitude, scan angle, and surface type. Maintaining a constant false alarm rate would be desirable by employing a variable threshold level appropriate to the local clutter levels and statistics. Such a detection mechanism is called a *constant false alarm rate* (CFAR) detector based on an adaptive threshold.

MTI clutter cancellation filters severely attenuate targets with Doppler frequencies equal to nf_r. Since Doppler is proportional to the target's radial velocity ($f_d = 2v/\lambda$), target speeds that produce Doppler frequencies equal to integer multiples of PRF are known as blind speeds, defined as follows:

$$v_{\text{blind}} = \frac{n\lambda f_r}{2}, \quad n \geq 0 \tag{2.103}$$

In Figure 2.23, the target return on the spectrum with $f_{d3} = f_r$ Doppler frequency is at blind speed and cannot produce a signal for detection in the MTI filter output. In MTI systems, it is generally assumed that the interpulse period was more significant than the required unambiguous range. If the resulting blind speeds present a severe limitation, a staggered or high PRF system is used.

The return echoes from moving targets vary in amplitude; however, pulses from stationary targets are constant. A clutter cancellation filter is applied to the

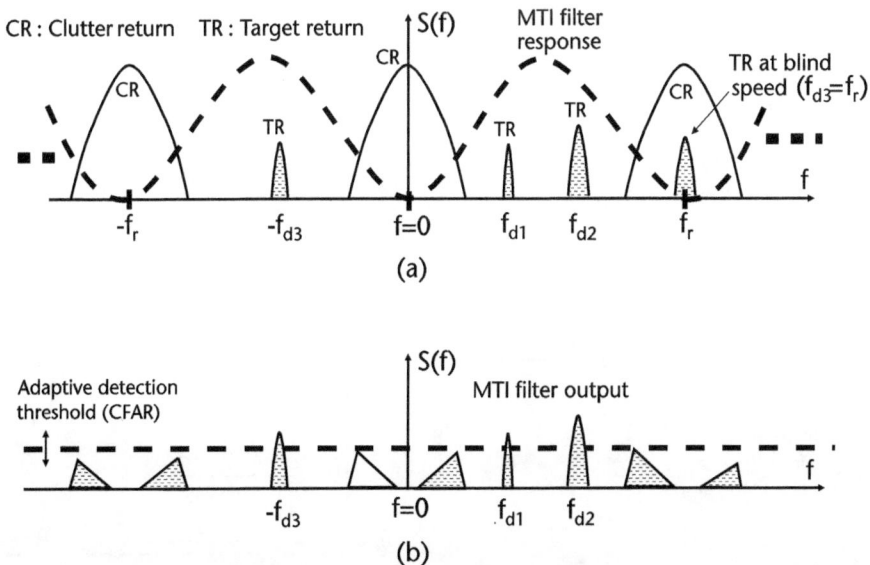

Figure 2.23 MTI filter operation.

output of the phase detector. The simplest form of the cancellation filter is single delay canceler. Single stands for a single delay line. The delay is equal to the radar PRI, T_r. The structure of a single delay line canceler in a general MTI radar block diagram is shown in Figure 2.24. The coherent reference signal with the operating frequency, f_0, is supplied by the *coherent oscillator* (COHO). The COHO is a stable oscillator whose input signal is the same as the IF in the Rx part. The *local oscillator* (LO) has a *stable LO* (STALO).

The reference signal and the output signal from the IF amplifier are the inputs of the phase detector, and the output of the phase detector is the response of a filter matched to zero Doppler. The canceler's impulse response is $h(t)$. The output $y(t)$ equals the convolution between the impulse response and the input $v(t)$. When the input is taken as a unit impulse function, we will obtain the single delay line canceler's impulse response. The output signal and impulse response of the canceler are as follows,

$$y(t) = v(t) - v(t - T_r) \xrightarrow{v(t) = \delta(t)} h(t) = \delta(t) - \delta(t - T_r) \quad (2.104)$$

We obtain its frequency response by taking the Fourier transform of the single-delay line canceler.

$$FT\{h(t)\} = H(\omega) = 1 - e^{-j\omega T_r}, \quad \{\omega = 2\pi f\} \quad (2.105)$$

Furthermore, in the z-domain, the response of the canceler is defined as,

$$z\{h(t)\} = H(z) = 1 - z^{-1} \quad (2.106)$$

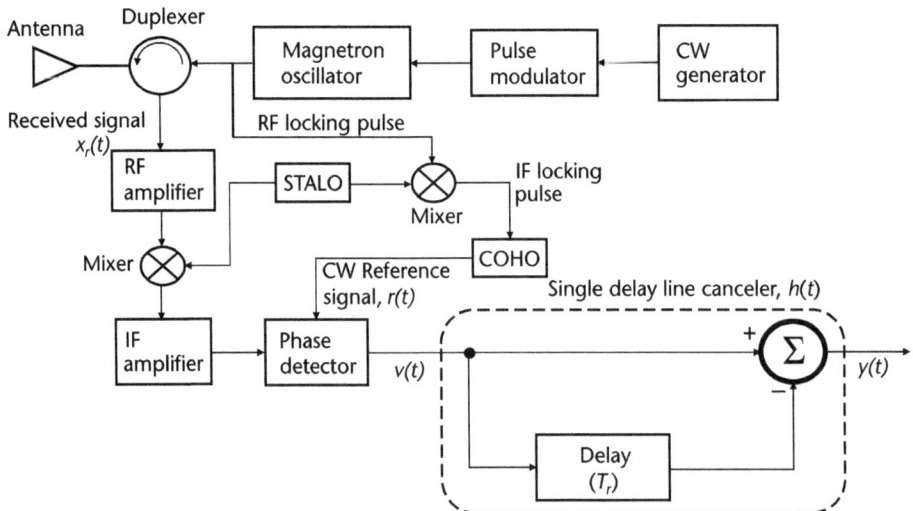

Figure 2.24 Block diagram of MTI radar with single delay-line canceler.

The frequency response of a single-delay line canceler only sometimes has as broad a clutter-rejection null as desired in the vicinity of DC. The clutter rejection notches can be widened using two single-delay line cancelers in a cascade. The basic configuration of a double-delay line canceler is shown in Figure 2.25.

The pulse response of the double-line canceler is given in (2.107). The frequency response of the double-line canceler is obtained by taking the Fourier transform of the pulse response. Its frequency response is simply the square power of the response of a single canceler. The double-line canceler's frequency response in the z-domain is also given in (2.107).

$$\left. h(t) = \delta(t) - 2\delta(t - T_r) + \delta(t - 2T_r) \right\} \begin{array}{l} \xrightarrow{\; FT \;} H(\omega) = 1 - 2e^{-j\omega t} + e^{-j\omega t} \\[2ex] \xrightarrow{\; z \;} H(z) = 1 - 2z^{-1} + z^{-2} \end{array} \qquad (2.107)$$

The normalized responses of single-line and double-line cancelers are shown in Figure 2.26.

The double canceler has a deeper notch and flatter passband response than a single canceler. Thus, for MTI operations, the double canceler frequency response is better than the single canceler. More complex delay-line cancelers can be produced by including further stages of the single delay-line canceler in the series.

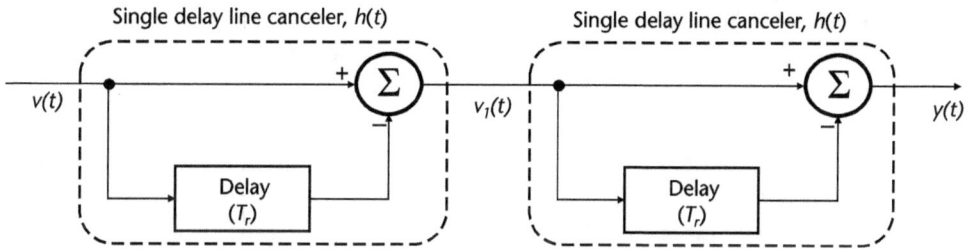

Figure 2.25   Block diagram of double delay-line canceler.

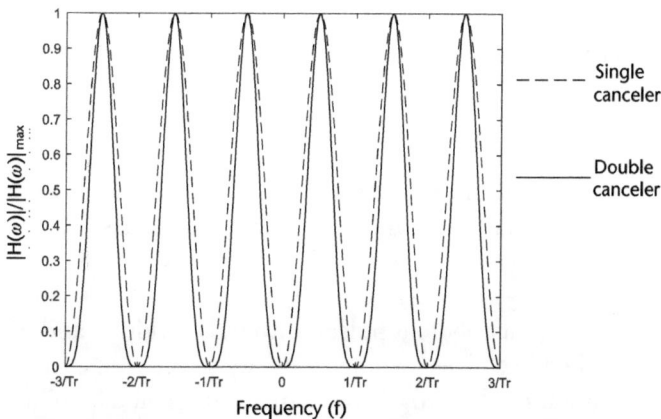

Figure 2.26   Frequency responses of a single canceler and a double canceler.

2.5.3.2   PDRs

Finding out the shortcomings of low PRF would be beneficial in understanding why we require high PRF waveform systems. Classical pulsed radars without Doppler sensing use low PRF waveforms. These radars are occasionally used for search and surveillance duties, and low PRF is used to obtain the maximum possible unambiguous range. The maximum detection range of the radar for all target and clutter returns must be, at most, maximum R_u to avoid range ambiguity. Thus, all the returns are first-trace echoes, and their range may be determined from a simple range delay timing method. Low PRF waveforms have low-duty cycles and employ many range cells matched to the range resolution of the radar distributed throughout the receiving period. The target range is easily obtained from low PRF systems.

All SLC and MBC have zero or near-zero Doppler frequency shifts for ground-based radars and may be rejected using an MTI radar clutter cancellation filter. Airborne systems that use platform motion compensation can correct MBC to a center Doppler frequency of 0 Hz (DC). The DC signal can then be rejected using an MTI clutter filter. However, SLC is spread over a considerable Doppler band, depending on platform velocity, and cannot be filtered out since targets of interest would also be eliminated.

The baseband output of a coherent pulsed radar is a sampled form of the Doppler signal. The PRF sets the sampling rate to avoid aliasing ambiguities, which must be at least twice the expected Doppler frequency. Low PRF radars seldom employ PRFs that are sufficiently high to avoid such ambiguities. Low PRF leads to ambiguous repetition in the frequency domain; hence, velocity ambiguities arise. The spectral spread of SLC from a fast-moving airborne platform causes ambiguous repetition of the SLC in the frequency domain. This phenomenon compounds the clutter problem since MBC and its ambiguous repetition occupy much of the Doppler band. SLC plus multiple ambiguous repeats of SLC occupy the whole Doppler band. High clutter levels exist across the Doppler band for a high-velocity airborne platform with low PRF. So the detection performance is severely reduced. If MBC were filtered away, it would result in overwide bands of blind velocities and the rejection of a high proportion of the targets of interest. Given their high degree of Doppler ambiguity and clutter-related problems, particularly those associated with MBC, low PRF waveforms are not prevailing for most airborne pulse Doppler radar applications. The solution to these problems is to increase the PRF so that Doppler and velocity ambiguities are avoided [18].

A high PRF wave means that waveforms with a PRF high enough to avoid velocity ambiguity. So according to the application, this rate is within the limits of high and sometimes medium PRF. The radar systems utilizing high and medium PRFs are commonly called PDR systems. The PDRs use range-gated Doppler filter banks but do not use delay line cancelers as in MTI radars. Furthermore, PDRs operate at a higher duty cycle than MTIs. Commonly, MTI radars are more prevalent than PDRs, but PDR systems are generally more capable of reducing clutter.

PDRs transmit coherent pulses with pulse width τ and PRI T_r to detect moving targets masked in clutter by range-velocity selectivity. Using a coherent pulse train, a PDR combines pulse radar's range-measurement capability with CW radar's frequency discrimination capability. The pulses are samples of a single unmodulated sine wave. The spectrum consists of lines with spacing equal to the repetition frequency for a fixed repetition rate. When a moving object reflects a coherent pulse train, the spectrum lines are Doppler frequency shifted an amount proportional to the object's radial velocity. When several objects with different velocities are present, the resultant echo is a superposition of a corresponding number of pulse trains, each with its Doppler frequency shift.

Figure 2.27 shows a typical pulsed radar block diagram. The coherent transmission and reception parts are classic; the received signal is reduced to IF, and the IF signal is amplified. The range gates are used to output the IF amplifier. Range gates are filters that open and close at time intervals corresponding to the detection range. A range gate selects only those pulse trains coincident in time with the pulse train echoes from the target. A narrowband filter following the range gate often selects only a single spectral line corresponding to a particular Doppler shift, thus attenuating all those trains that pass the range gate but do not have the proper Doppler shift [20].

The width of open and close time intervals corresponds to the required range resolution. The radar Rx is often implemented as a series of contiguous range gates in time. The width of each gate is achieved through pulse compression. The clutter rejection can be accomplished using MTI or other clutter rejection techniques. The NBF bank is typically implemented using an FFT, where the bandwidth of the individual filters corresponds to the FFT frequency resolution [2].

Doppler radar signal processing aims to divide the antenna's 3-dB beamwidth intersection with the ground into resolution cells. These cells form a range-velocity resolution map, illustrated in Figure 2.28. The axes of the resolution map are range and velocity (Doppler frequency). The range resolution, ΔR, is accomplished in real time using range gating and pulse compression. Doppler resolution is obtained from the coherent processing interval.

Using the matched filter is essential for obtaining high-range resolution. Clutter rejection is performed on each range bin. Then all samples from one dwell within

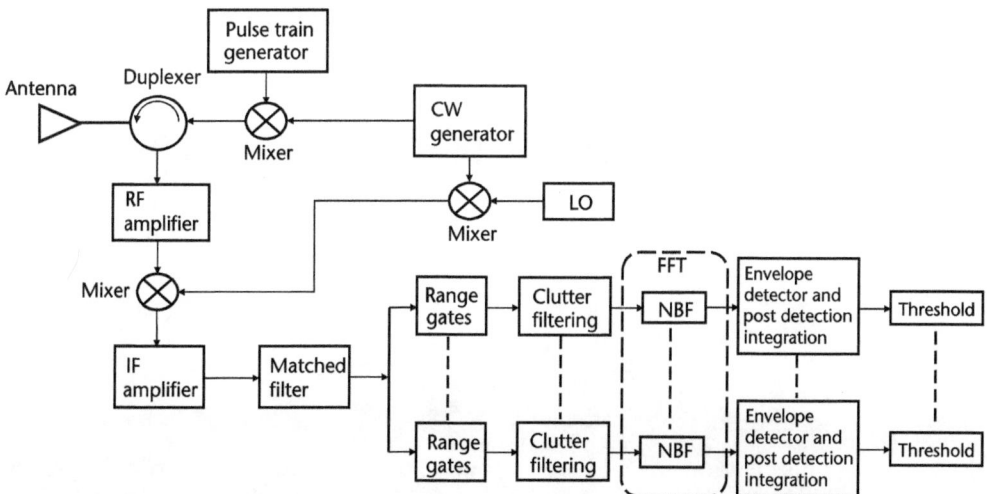

Figure 2.27   Block diagram of pulsed Doppler radar.

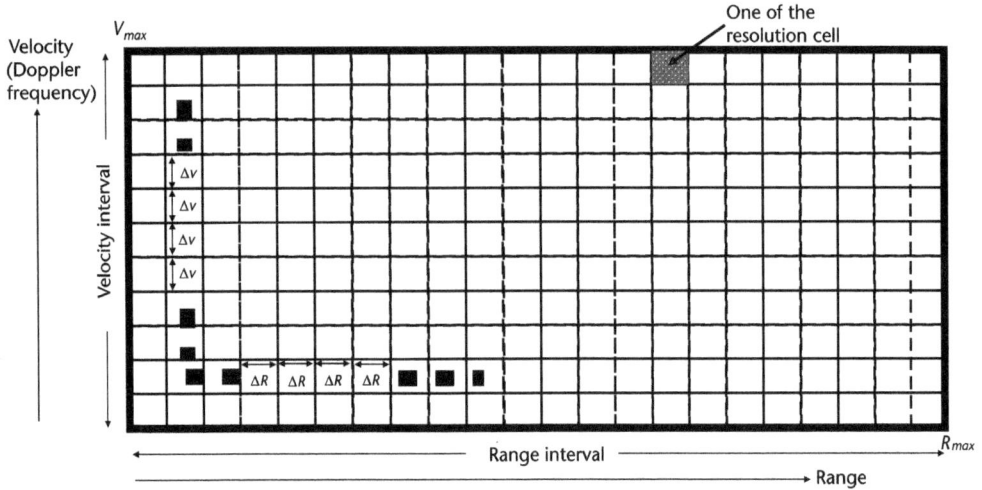

Figure 2.28 Range-velocity resolution map.

each range bin are processed using an FFT to resolve the return signals in Doppler and the corresponding radial velocities of the targets. A peak in each resolution cell corresponds to a specific target detection at that range and velocity.

The transmitted signal's carrier frequency of a coherent pulse Doppler radar is f_0, the signal's PRF is f_r, and the pulse width is τ. Each spectral component reflected from a moving target undergoes a Doppler frequency shift defined as:

$$f_d = \frac{2v_T}{\lambda} = \frac{2v_T f_0}{c} \tag{2.108}$$

where v_T is the radial velocity of the target and λ is the wavelength. As seen from the equation, the higher the carrier frequency, the greater the magnitude of the Doppler frequency for a given target velocity. Higher carrier frequencies increase the sensitivity of the Doppler shift on target velocity. This definition has several secondary effects on the selection of PRF. The first is a greater spread of Doppler shifts requiring higher PRF values to avoid velocity ambiguity. The second is the increased sensitivity of the Doppler shift on target velocity at higher carrier frequencies, which concerns the relationships between velocity resolution and the integration time. The Doppler resolution is the reciprocal of the integration time (t_i) or the coherent processing interval, and the velocity resolution is as follows [18]:

$$\Delta v_T = \frac{\lambda \Delta f_d}{2} \xrightarrow{\Delta f_d = \frac{1}{t_i}} \Delta v_T = \frac{\lambda}{2 \times t_i} \tag{2.109}$$

This relationship illustrates that velocity resolution is inversely proportional to the carrier frequency. A higher carrier frequency achieves better velocity resolution for a fixed integration time. For the FFT processing of the Doppler signals in the radar baseband, the Doppler resolution is given by the PRF divided by the FFT point size (N), that is,

$$\Delta f_d = \frac{f_r}{N} \tag{2.110}$$

Multiple combinations of PRF and N exist for a fixed value of the Doppler frequency resolution and the integration time. Higher values of PRF are likely to result in more significant numbers of transmitted pulses in a given integration time duration, so the process requires larger FFT sizes. Lower values of PRF tend to result in smaller FFT sizes for a given integration time duration. However, if the integration time is the same in both cases, the Doppler resolution will be the same.

Blindness occurs when an echo is received during the transmitted pulse since the receiver is isolated during the transmitted pulse. The received echo completely overlaps the transmitted pulse, resulting in a condition of total eclipse and, hence, blind ranges at

$$R_b = \frac{c \times T_r \times n}{2} = \frac{c \times n}{2 \times f_r}, \quad \{n = 0,1,2,...\} \tag{2.111}$$

2.6 Tracking Radars

Radars that form a track for one or more targets and generate position and speed information for these targets are called tracking radars. Before tracking, radar detects targets and finds their range, angular position, and speed according to their measurement accuracy. The requirements for the accuracy of a tracking radar's range, angle, and velocity measurements are better than those for a search radar.

The radar systems used for the same duties (i.e., searching, acquisition, tracking) utilize parametrically similar operating modes. For this reason, using frequency is compulsory rather than a selection. Thus, it would be beneficial to mention the radar operating frequencies; however, there are flexible boundaries between classifications. Long-range surveillance radars operate in the L-band (1–2 GHz), and moderate-range surveillance radars use the S-band (2–4 GHz). Long-range tracking radars utilize the C-band (4–8 GHz); however, some long-range tracking radars use the upper portion of the S-band. Finally, medium-range and short-range tracking radars use the X-band (8–12 GHz), and some very precision tracking radars operate in the Ku-band (12–18 GHz). Thus, the tracking radar operating frequency band is 2–18 GHz.

When a tracking radar detects echo signals, the target potential of the echo is examined, and if the object is considered a target, a track is initiated. After the initiation process, the track continuation or track maintenance phase is conducted. The main aim of track maintenance is to improve the sensor accuracy and obtain estimates of variables not measured by the sensor. Commonly used track maintenance algorithms for a tracking radar without measurement origin uncertainty are *Kalman filter* (KF), α-β-(γ) filter, *extended KF* (EKF), and *multimodel* (MM) approaches like *interacting MM* (IMM). The selection of the filters' type and specifications depends on the target maneuvering conditions. Measurement origins are specific when target echo is measured in a clutter-free environment.

A tracking radar uses nearest neighbor standard KF and *probabilistic data association filter* (PDAF) and *interacting multimodel PDAF* (IMMPDAF) for tracking maintenance of a single target in a cluttered environment. However, for multiple targets in clutter, a tracking radar uses *joint PDAF* (JPDAF), *multiple hypothesis tracking* (MHT), and *probabilistic multihypothesis tracking* (PMHT) algorithms.

Probabilistic-based algorithms, such as PDAF, MHT, and PMHT, perform a decision process to discriminate targets from clutter. Thus, an association is made between the related echoes and the targets. This way, the process aims to make an accurate association and exclude the clutter. The tracking maintenance process seeks to provide precise estimation information about the target range, azimuth, and elevation. Weapon systems use this information for precision targeting of missile and gun systems. The continuation phase lasts for a predetermined condition. This condition is commonly adjusted according to sequential detections. The tracking is terminated if no target detection exists in several sequential measuring steps.

The principle of tracking radar is to direct the antenna's boresight (or central axis) on the target by using the target's estimated range, azimuth, and elevation information from the tracking algorithms. For this purpose, the tracking radars use the error signal to adjust the antenna's boresight. The error signal emerges from the difference between the target and boresight directions. The tracking radar aims to overlap the target direction with the boresight of the radar's antenna or converge the angular error to zero. For this reason, the estimated position of the target obtained from the tracking algorithm must be precise and real-time.

The types of tracking radars differ based on efforts to reduce tracking errors. The following sections explain the different types of tracking radar.

2.6.1 Single-Target Tracking Radar

Tracking radars are primarily used in military and some civilian applications. In the military, tracking radars are engaged for missile guidance and fire control. However, radar systems in the commercial field are used for tracking incoming and departing airplanes or controlling airport traffic. *Single-target tracking* (STT) radars continually engage one target at a high data rate. An STT radar measures the location of a target according to the used coordinate system. It provides data that may be used to determine the target trajectory and to predict its future position. These radars provide continuous position data on a single target. Typical STT radars use a pencil beam (the 3-dB beamwidth angle is between 1° and 2°) to receive echoes and use the target's angle, range, and Doppler information for track maintenance. However, the STT radar's pencil beam brings about a significant problem. STT radars depend on a separate wide-beam angle radar for target initiation, such as search and surveillance.

Tracking can be conducted with range/velocity information and angle information. Also, the used information divides the tracking techniques into two groups. First, we will mention the range and velocity gating and then explain the angle tracking. Afterward, we will elaborately describe different application types for angle tracking.

The tracking, carried out using the range and velocity information, is applied to the gate-forming principle. The process of continuously estimating the range of a moving target is known as range tracking. Since the range of a moving target changes with time, the range tracker must be constantly adjusted to keep the target locked in range. This tracking technique can be obtained by forming a range gate technique. A *range gate* is a signal processing technique that selects signals within a given time. The range gate allows signals to pass through only within the specified time. The most common configuration for range-gate tracker implementations is a split-gate tracker or an early-gate and late-gate tracker.

The radar echo signals from all targets in the radar antenna beam are applied to the input of the early, late, and on-target gates. The early gate passes only the first half of the tracked target echo signal. The late gate starts at the end of the early gate and only passes the target echo's second half. The on-target gate passes the complete tracked target echo. The output of the on-target echo gate is not required in the range tracker itself; this gated target echo signal is often used in many radar configurations for subsequent signal processing and display [21].

A range-gate tracker tracks the centroid of the echo signal corresponding to the desired target. The early and late samples are equal if a target echo is located precisely at the center of a range gate. Figure 2.29 shows the concept of split gate monitoring for the equality of early and late samples.

The early and late-gate outputs are applied to the integrators. The late-gate integrator is identical to the early-gate integrator. However, the semiconductor switch used in the early-gate integrator is on during the off period of the semiconductor switch in the late gate. A voltage occurs at the late-gate integrator output while the switch is on, which is also valid for the early-gate integrator. This voltage is proportional to the target echo signal energy.

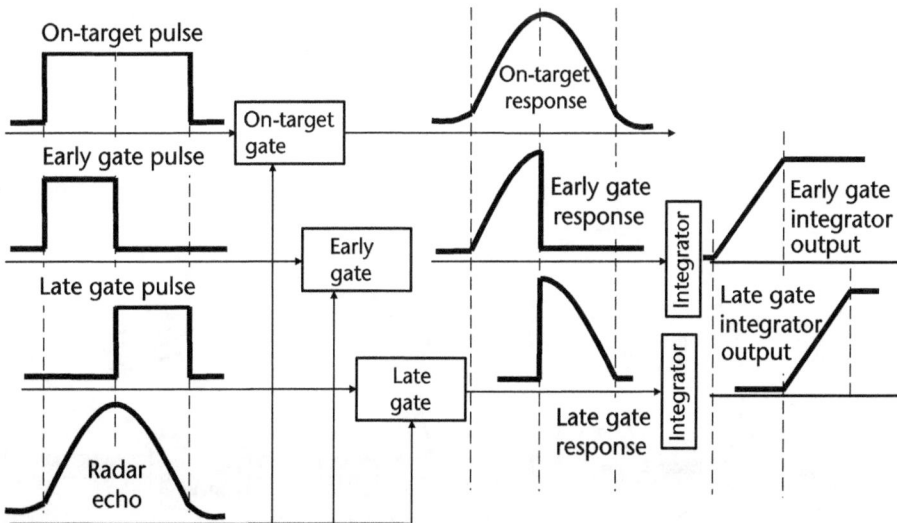

Figure 2.29 Signals samples of the range-gate (split-gate).

The voltage outputs from the early-gate and late-gate integrators are applied to a difference amplifier. If the target echo energy during the on-period of the early gate is equal to that of the late gate, the two integrators will have equal output voltages. So the difference amplifier input voltages will be equal, and the output will be 0V. This condition exists when the range tracker correctly follows the target return's centroid. However, the difference amplifier output will be nonzero if the partition between the early and late gates is not centered on the target echo signal.

Similarly, estimating the velocity of a moving target is known as velocity tracking. For this purpose, a similar gating process, like range gating, is applied. However, this time, the Doppler frequency of the target return is used. For this purpose, the narrowband filter bank, shown in Figure 2.17, is used.

Tracking radars continuously measure the target's angular position in the azimuth and elevation coordinates. The accuracy of early-generation angle tracking radars depends on the size of the beamwidth. However, most modern radar systems achieve excellent angular measurements by utilizing different tracking techniques, as given in the following sections. Before discussing more sophisticated tracking techniques, mentioning a primary method called sequential lobing will be beneficial.

2.6.1.1   Lobe-Switching Radar

Lobe switching is also called sequential lobing or sequential switching. A lobe-switching radar has a tracking accuracy limited by the pencil beamwidth. The lobe switching technique is straightforward to implement. In this technique, the pencil beam of the radar antenna must be symmetrical. The symmetrical means the radar antenna's azimuth and elevation beamwidths are equal. The beam is switched alternately between two positions. The difference between the strengths of the return signals for the two positions shows the angular displacement between the target and the switching axis. The switching axis, called the tracking axis, is commonly adjusted as the antenna's boresight.

In Figure 2.30, the operational principle of lobe switching is given. The figure provides two targets, and the beam returns according to the targets' position. Target

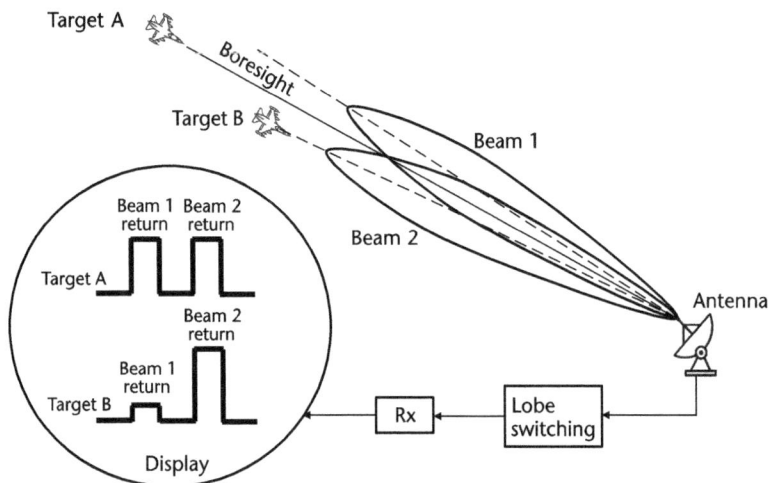

Figure 2.30   Operational principle of lobe switching.

A is on the tracking axis; the return signal strengths of beams 1 and 2 are the same, and the difference is zero. Target B is out of the boresight tracking axis and is close to the central axis of beam 2. For this reason, the return signal strength is higher than beam 1, and a nonzero error signal is produced.

The sign of the difference shows the required antenna moving direction to match the target direction with the boresight. When the switching axis is in the direction of the target, the return signal strengths for the two beams are equal. Lobe switching aims to zero the return signal difference and keep the target in the boresight axis.

Lobe switching is a well-known tracking technique, and the targets can detect these radars using RWRs. Furthermore, targets can use inverse gain jamming to break the tracking lock of the lobe-switching radars.

2.6.1.2 Conical Scan Radar

Conical scan radar systems send a signal with a beam slightly offset from the antenna's boresight and then rotate the beam to form a circle around the centerline. The working method of conical scanning radar is shown in Figure 2.31. A target focused on the boresight direction is illuminated by a small, constant portion of each rotating beam. In other words, if the target is in the boresight (or tracking axis) direction, an echo signal of the same power intensity is received in each rotating beam. Target A produces constantly received echo signal intensity, and it is assumed on the boresight.

If the target is not on the boresight, the return echo intensity increases when the beam reaches the target direction and weakens for beams in other directions. In this case, the powers of the return echoes form a sinusoidal envelope. The difference between this sinusoidal envelope's maximum and minimum values increases as the target moves away from the centerline. As shown in Figure 2.31, target C is further from the centerline than target B. Therefore, the difference (P_{dif}) between the maximum and minimum echo powers reflected from target C is more significant than that from target B.

In conical scan radars, a range gate is set to select the echo signal of interest and reject all others. To handle the extensive dynamic range, *automatic gain control*

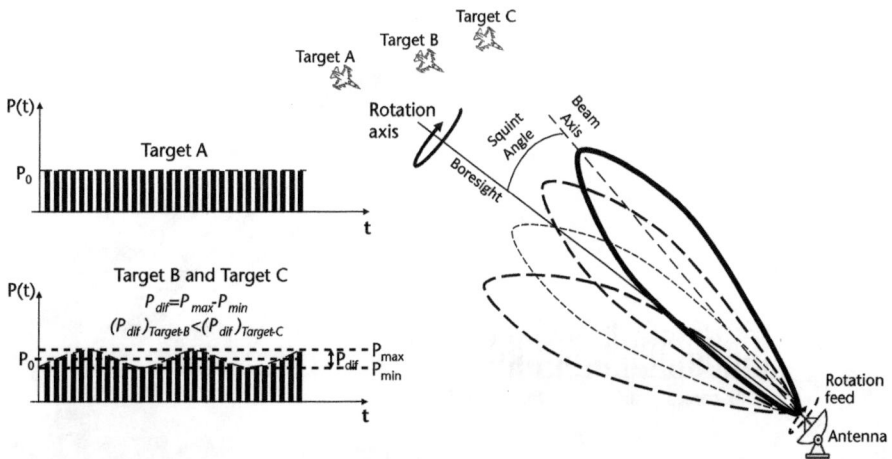

Figure 2.31 Operational principle of conical scan radars.

(AGC) keeps the echo signal constant [11]. This difference in return echo signal intensity disappears when the antenna is rotated in the appropriate direction so that the target is in the boresight direction. Each beam receives the same intensity echo signal, aligning the boresight in the direction of the target.

2.6.1.3 Conical Scan on Receive Only Radar

Conical scan operation forms a central tracking area with a much smaller effective beamwidth than the rotating radar beam. These radars present an exact tracking solution for missile guidance systems. However, the beam scanning pattern of the conical scan tracking method can be detected by the target's RWR, as in lobe-switching. Thus, the self-protection system can apply inverse gain jamming to break the tracking lock. An unnoticeable ECCM method for the target's RWR is used to overcome this weakness of conical scan radars. This technique is called conical scan on receive only (COSRO).

As seen in Figure 2.32, COSRO radars consist of two antennas. The Rx antenna provides the same scanning pattern as the conical scan radars and receives the signals reflected from the target. This scan aims to rotate the Rx antenna beam to keep the target in the Rx boresight (or tracking axis) direction. The Rx and the Tx antennas' boresights are adjusted for overlapping. The amplitudes of the echo signals form an amplitude modulation in the Rx. Taking the target to the Rx boresight is accomplished by using the amplitude-modulated signal resulting from the conical scan of the Rx antenna. Suppose the target is not in the Rx boresight. In that case, the amplitude levels of the echo signals coming to the radar Rx antenna vary according to the instantaneous position of the beam. If the target moves away from the scanning axis, the ripple increases, and when it gets closer, the ripple decreases. The amplitude levels of the reflection signals take a constant value when the target comes to the Rx boresight. According to the amplitude-modulated signal, Rx and Tx antennas' boresights are moved with slight changes in the target direction.

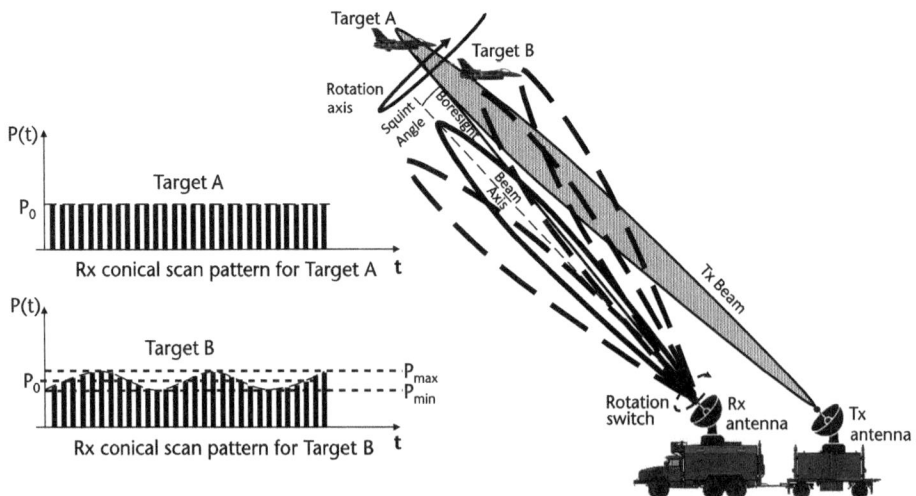

Figure 2.32 Operational principle of COSRO radars.

The COSRO radar's Rx beam has a similar shape to a transmitted beam, being rotated near the radar boresight axis. A target's RWR would detect the COSRO radar as a steady, nonscanning, pulse radar beam. Thus, the beam provides only basic radar parameters such as frequency, PRF, and pulse duration and no information for determining the beam scanning used in inverse gain jamming.

2.6.1.4 Lobe on Receive Only Radar

Tracking radars, used for estimating the future location of a target, perform scanning by looking at areas where the target is not present but is likely to go. When a tracking radar performs this scan with a single beam, even a simple ECM application such as chaff clouds will cause the entire radar screen to be covered with clutter echo. Furthermore, in the single-beam method, the RWR system on the target platform can obtain indications to detect the lock before the radar locks on the target. The impact of chaff clouds and being detected by RWR systems are also valid for the lobe-switching method, which uses two switching lobes sequentially. Furthermore, lobe-switching radars can apply a deceptive technique called inverse gain jamming by self-protection jammers to break the tracking lock.

Single-beam tracking and lobe-switching radars can develop an ECCM technique against deception using lobe on receive only (LORO). This ECCM technique, passive lobing or silent lobing, is used to eliminate the chaff effect and against inverse gain jamming. LORO is also used for not being detected by RWRs. The LORO method overcomes these problems by scanning only with Rx antennas and having the Tx antenna in a separate location from the Rx. The operational process of the LORO radars is given in Figure 2.33.

In LORO operation, the radar's Rx conveys the obtained target location data to the Tx. The Tx antenna emits several narrow beam pulses along the direction of the target. This working method of Tx emits narrow beam pulses, which is similar to the lobe-switching radar working method. The difference between these

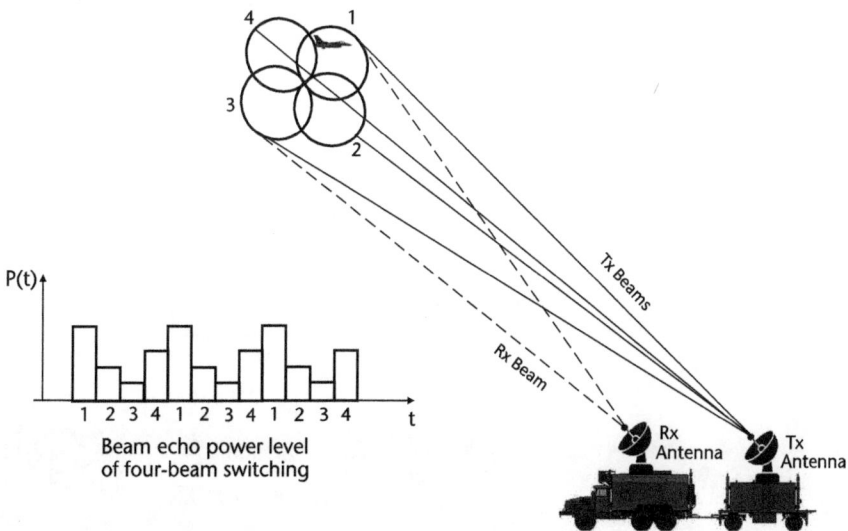

Figure 2.33 Operational principle of LORO radars.

two methods is in the antennas of Rx and Tx. While LORO systems use different antennas for Tx and Rx, lobe-switching radars use the same one for Tx and Rx.

The LORO operation given in Figure 2.33 is for a four-beam pulse or, in general terms, four-beam switching. These pulses are used in the acquisition mode of the target tracking radars (TTR). In other words, in acquisition mode, the narrow beam TTR scans the broad position segment determined by the surveillance radar.

2.6.1.5 Monopulse Radar

Tracking radars often choose the monopulse method as it is highly accurate and not easy to deceive. Monopulse radars determine a tracking solution based on a single pulse rather than the lobe switching or the conical scan. For this reason, the tracking data rate is higher and potentially more accurate than the others. Another advantage is that the tracking is based on the simultaneous reception of the target return in all channels. Adapting to echo variations for the elapsed time in monostatic radars is more satisfying than in other techniques.

Monopulse radar system is mainly used for target angle measurement and tracking. Monopulse radars are similar in general structure to conical scan radar systems. However, the beam does not rotate around a center and is not oriented only in one direction at a time. The target's angular position or direction information is determined by comparing the amplitudes of the pulse signals from the two or four simultaneous beams. The term monopulse comes from this system's ability to detect the target's angular position from just one impact.

However, in practice, the target's angular position is obtained from multiple pulses, which do not fit the system's name. Using multiple pulses increases the target detection probability and improves angle measurement accuracy. Figure 2.34 shows the return amplitude measurements taken from the beams of a monopulse radar operating with four simultaneous pulses.

Monopulse radars may use amplitude comparison, phase comparison, and a mixture of both to perform the tracking task. Amplitude comparison monopulse tracking is similar to the lobe-switching technique in that four squinted beams are required to measure the target's angular position. The difference is that the four

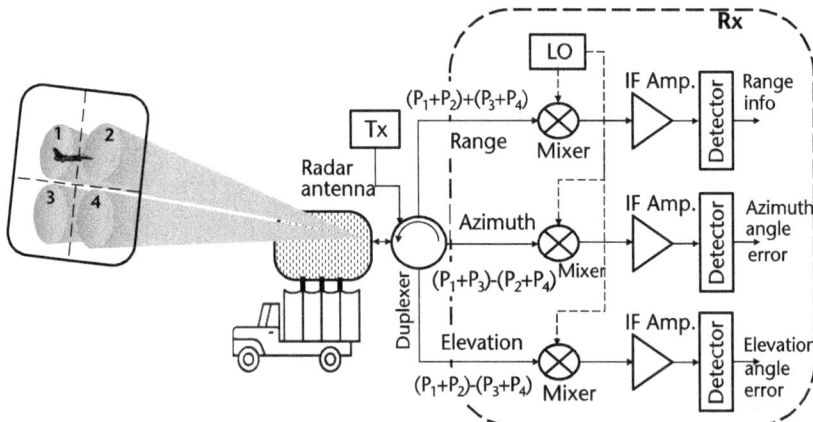

Figure 2.34 Operational principle of monopulse radars.

beams are generated simultaneously rather than sequentially. For this purpose, a special antenna feed is utilized to produce the four beams using a single pulse. Thus, the monopulse tracking is more accurate and is not susceptible to AM jamming and inverse gain jamming. Monopulse tracking radars can employ antenna reflectors and phased array antennas [2].

Figure 2.34 shows a typical monopulse antenna pattern and operational principle. The four beams 1, 2, 3, and 4 represent the four conical scan beam positions. Amplitude comparison monopulse processing requires that the four signals have the same phase and different amplitudes. It is helpful to look at the Rx stage of the monopulse radar to understand the range and angular error calculation with amplitude monopulse processing.

When the reflected signals are received, they are amplified separately and put in a different comparison process. This process is accomplished in the Rx part of the monopulse radar given in Figure 2.34 and can be described as follows:

- The top line of the duplexer output is used to compute the target range. The target range is derived by adding the signal powers from all four beams $((P_1 + P_2) + (P_3 + P_4))$. The output of the summation is then passed to the range detection circuit, which provides the range information. After obtaining the range information, a leading-edge or split-gate tracking loop is used.
- The middle line of the duplexer output is used to obtain the azimuth error. The signal powers from beam 1 and beam 3 are added, and the signals from beam 2 and beam 4 are added. The sums of these values are subtracted from each other, and the difference $((P_1 + P_3) - (P_2 + P_4))$ is then passed to the angle error detection circuit. This circuit provides the tracking angle error in azimuth. After obtaining an azimuth angle error, the radar system moves the position of the beams to equalize the power levels between the two pairs of sums. So the monopulse radar performs tracking in azimuth.
- Target elevation tracking error is derived using the bottom line of the duplexer output. The signal powers from beam 1 and beam 2 are added, and the signals from beam 3 and beam 4 are added. The sums of these values are subtracted from each other, and the result $((P_1 + P_2) - (P_3 + P_4))$ is sent to the angle error detection circuit to obtain the tracking angle error in elevation. The elevation error angle is used to correct the target's elevation position.

Phase comparison monopulse radars use a similar operational principle given in Figure 2.34. The main difference is that the four signals produced in amplitude comparison monopulse have similar phases but different amplitudes; however, in phase comparison monopulse, the signals have the same amplitude and different phases. Phase comparison monopulse tracking radars use two separate (or four independent) antennas or a minimum of a two-element array antenna for each azimuth and elevation coordinate and illuminate the same volume in space. Instead of being squinted like in an amplitude comparison monopulse system, the beams are kept parallel in a phase comparison monopulse system. Since the beams are parallel, if the target is in the boresight direction, the echo signal arrives at the two antennas simultaneously and has the same phase. Conversely, if the target is off the boresight, then the echo signal arrives at one antenna later than the other.

In the amplitude and phase comparison method, monopulse radars use a combination of amplitude and phase comparison for target tracking. The applied monopulse technique determines the nature of the information in the received signal before any processing. This means that the choice of a particular monopulse approach will also determine the structure of the radar antenna system.

2.6.2 Automatic Detection and Tracking

Since the radar's earliest use, operators have detected and tracked targets using visual inputs from *plan position indicators* (PPIs) and A scopes. Although operators can perform these tasks accurately, they are easily overloaded and quickly become exhausted. Many radars have *automatic detection and tracking* (ADT) systems to overcome this handicap [22]. An ADT system is a computer-based radar data processor with automated radar target detection, tracking, and correlation capabilities.

Target detection and tracking from radar observations are difficult because of false alarms and the simultaneous presence of multiple targets with a probability of detection P_D is less than unity. At each scan, the radar provides position measurements in range and azimuth according to the radar position. These measurements are used for radar ADT systems to perform detection and tracking. A generic block diagram of an ADT system for the tracking radars is shown in Figure 2.35.

Sensing a target in background echoes, atmospheric noise, or noise generated in the radar Rx is the detection process. Noise always accompanies the target's echo signal in each stage of the radar Rx, in which most of the noise is generated. However, noise is not the only clutter source that increases the false alarm rate. In the operational fields, different clutter sources like land shapes, plant cover, sea waves, rain, and birds cause excessive false alarms for a fixed-threshold system. The radar detection threshold is adjusted according to the Rx noise, clutter, and interference. This adjustment is accomplished by *constant false alarm rate* (CFAR) processing methods, which aim to control the number of false alarms in a changing and unknown environment.

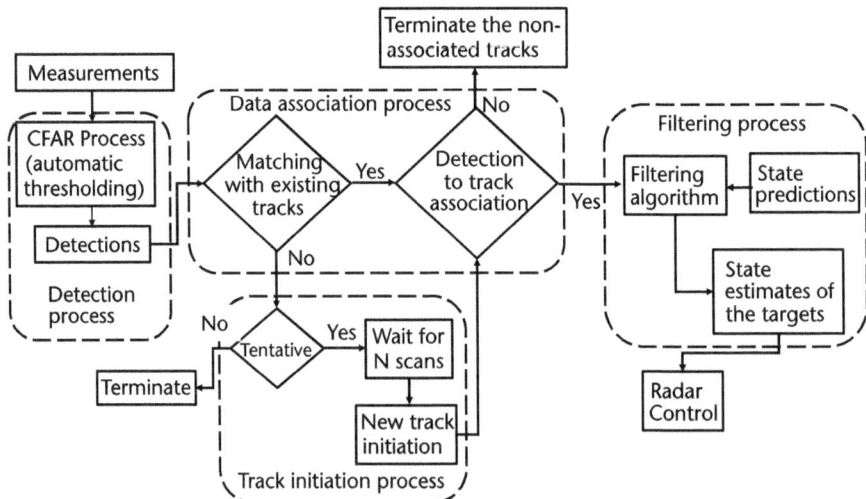

Figure 2.35 Conceptual principle of automatic detection and tracking process.

A conventional Rx system uses a straightforward thresholding method to identify target detections even when thermal noise is interference. The assumed situation is the noise-limited detection case in which only white Gaussian noise is present at the Rx input. Narrowband filtering within the Rx converts the noise voltage envelope to a Rayleigh *probability density function* (PDF). The summation of the noise voltage envelope and target signals is then compared with a fixed threshold voltage to decide whether a target is present. A noise spike could also exceed the threshold and trigger a false alarm. The decision for target presence is made in each range cell, velocity cell, and beam position or angular cell. The *false alarm rate* (FAR) is a function of the relative levels of the threshold voltage and noise voltage envelope in each range, velocity, and angle cell. This relationship can be written as follows [18]:

$$T_{FA} = \frac{1}{FAR} = \frac{1}{B_N}\exp\left(\frac{v_T^2}{2v_N^2}\right) \tag{2.112}$$

T_{FA} is the mean time between false alarms, v_T is the threshold voltage, v_N is the root mean square (RMS) noise voltage, and B_N is the noise bandwidth. *FAR* is a function of the probability of false alarms (P_{FA}) and the bandwidth in which the noise is detected, the noise bandwidth, B_N.

$$FAR = P_{FA}B_N \tag{2.113}$$

Furthermore, the probability of detection (P_D) for a given SNR may be deduced, assuming additive Gaussian noise.

$$SNR = \frac{\log P_{FA}}{\log P_D} - 1 \tag{2.114}$$

The radar detection performance can be described entirely for the known, fixed values of v_T and v_N. The ($v_T^2/2v_N^2$) term in (2.112) is known as the *threshold-to-noise ratio* (TNR) and is a power ratio. The false alarm rate is susceptible to TNR. The threshold level is scaled from the noise level as follows:

$$v_T^2 = k_1\left(2v_N^2\right) \tag{2.115}$$

where k_1 is a multiplier that sets the TNR; the choice of the k_1 multiplier, therefore, sets P_{FA} and P_D. A fixed threshold voltage is perfect when the noise statistics, such as PDF or mean levels, remain fixed in the overall range, velocity, and angular cells. This situation will likely differ for noise jamming and other noise-like forms of interference and clutter. Noise jamming tends to vary over different reception angles, and clutter varies considerably over range and velocity axes; it is also angle-dependent. Furthermore, clutter statistics may exhibit spatial variations, such as a ramp transition from a region of the calm sea through regions of progressively rougher seas or the abrupt transition from land to sea clutter encountered along a coastline. In such circumstances, a fixed threshold would lead to considerable variations in P_{FA} (and P_D) with range, velocity, and angle. Should a threshold voltage set based on the worst-case conditions be used, it would undoubtedly be too high most of the time, leading to degraded P_D over much of the detection space of the radar.

If the TNR, $k_1 = v_T^2/2v_N^2$, can be held constant, then the FAR is also constant. Thus, if v_N were to change, the threshold, v_T, must vary in sympathy with it. This variation yields a CFAR function but only in a minimal sense because here it is assumed that the input is thermal noise in which only its RMS level varies; all other statistical quantities, notably its PDF, remain the same. Nevertheless, this does establish the critical principle of requiring an adaptive threshold. Adapting the threshold to the local noise, interference, or clutter statistics is necessary to obtain a proper CFAR function. For each cell in which a "target present" decision is made, a CFAR circuit must decide whether the voltage in that test cell exceeds the voltages in neighboring cells by some reasonable margin.

The CFAR detection methods can be inspected in two categories: the parametric and the nonparametric. The classification of CFAR methods depends on the pre-information about the distribution of background clutter and noise data. For the parametric CFAR processing techniques, the cell averaging CFAR, the greatest-of CFAR, the ordered statistic CFAR, and the trimmed mean CFAR methods can be counted. In the operational field, the clutter distribution severely mismatches the assumed one. Thus, the performance of a parametric CFAR detector would decrease substantially. However, the nonparametric or distribution-free detector can maintain a CFAR despite changes in the underlying distributions of the observed data. This inherent advantage of the nonparametric CFAR detectors makes them find wide applications in actual implementations. The most used nonparametric CFAR procedures are rank quantization and rank-sum detectors [23].

The tracking process consists of track initiation, data association, and filtering. In this context, the *track initiation* is the detection of new targets in the surveillance area. The *data association* is the assignment of measurements obtained in a cluttered environment to the established tracks. Furthermore, *filtering* can be defined as updating the targets' state using measurements and a model of the target motion [24]. It would be appropriate to deal with the track initiation, data association, and filtering separately.

- *Track initiation:* This stage is the first step of target tracking. Correct track initiation will effectively reduce the computational burden. By false track initiation, the remaining part of the target tracking will also be wrong, and the target will be lost.

  Conventional track initiation algorithms of radars use some predetermined sequential rules for deciding a measurement to be a target. For this purpose, they generally utilize geometrical relationships such as range, azimuth/elevation angle, and velocity among the measurements that must be combined. The measurement combination is a sequence of measurements obtained from consecutive radar scans. When sufficient conditions are met among the geometrical relationships of the sequential measurements, it is regarded as a tentative target. If the relationships continue to provide the predetermined conditions, then a decision for target initiation is made.

The number of measurement combinations in low-clutter environments is relatively low, and conventional algorithms' performance is probably adequate. Nevertheless, in higher-clutter environments, the number of combinations can increase exponentially, and the probability of success of traditional algorithms decreases rapidly. Initiating false tracks requires extreme computational power for data association and running tracking filters. As a result, the performances of automatic detection and tracking systems in heavy, cluttered environments may degrade considerably.

Conventional track initiation algorithms are categorized as sequential and batch methods [25]. Sequential methods produce the combination of measurements sequentially using the measurements that exceed a threshold for each radar scan. If measurement combinations are maintained during a time window of an initiation process, these combinations are determined as tentative tracks and following confirmed tracks. The well-known M/N logic widely used in the radar field rules the promotion from tentative to verified track.

This algorithm is based on the sliding window process, where N is the number of scans and M is the detections out of N scans in the acceptance gate. M and N are integer parameters selected as $1 \leq M < N$. If the number of detections over N consecutive scans is at least M, then the tentative track is confirmed. Otherwise, the track is terminated. The choice of M and N must be related to the target detection probability P_D and the false alarm probability P_{FA}. Reasonable values should satisfy the following constraints [24].

$$A_{mean}P_{FA} \leq \frac{M}{N} \leq P_D \qquad (2.116)$$

where A_{mean} is the mean area of the validation region. A continued track is terminated whenever it goes through a predetermined L consecutive missed detections.

The sequential methods have a simple structure and low computational complexity. However, the probability of false track initiation can be increased in high-clutter environments since the number of measurements may exceed a threshold that can be handled for each radar scan. In the case of batch methods, every measurement from some period of radar scans is stored. Then batch data of every possible measurement combination are processed simultaneously. These combinations are determined as targets if the measurement combination provides a predetermined threshold.

Conventional methods use deterministic approaches to process measurement information; score-based algorithms use probabilistic approaches. These state-estimation-based algorithms integrate track initiation, data association, and tracking filter stages into one. Usually, signal intensity, Doppler, and position information represent the state of measurements. Track scores are calculated based on the error covariance matrix between measurement and prediction, and the initiation of targets to track is determined according to these scores. Since the probabilistic approach for targets, as well as the estimation and prediction of the state, are used, these algorithms are more accurate and suppress false tracks effectively. However, they require more computational resources because all stages of the tracking system are integrated. So using separate algorithms for each stage in high-clutter environments is a common approach [26].

- *Data association:* Data association is essential to increasing the performance of radar tracking systems. The data association task identifies whether a measurement belongs to the target or is a false alarm. Furthermore, target measurements must be associated with specific targets for multiple targets. Data association aims to assign a given set of measurements to a set of tracks.

  An important distinction for the data association techniques can be hard and soft decisions. The hard assignment is where the data association approach decides which measurement is due to a particular target and assigns that measurement to the target. The target state estimate is updated, assuming that the assigned measurement is correct for this target. The soft assignment is where multiple measurements are assigned to each target with a certain probability. Rather than choose a single measurement to update the track, the track is updated using many possible assignments, and the collection of updated states is combined using the assignment probabilities, usually in a Bayesian framework. Another difference in data association techniques is the usage of single-scan and multiscan. The single-scan techniques consider the measurements only at the present scan. However, the multiscan methods envision measurements collected over a given number of past scans to correct possible previous association errors.

  Different algorithms have been developed to solve the data association problem. Two simple solutions are the *strongest neighbor filter* (SNF) and the *nearest neighbor filter* (NNF). SNF uses the signal with the highest intensity among the validated measurements for track updates and discards the others. In the NNF, the measurement closest to the predicted measurement is used. These simple techniques give successful results for detectable targets with low maneuvers. However, their performance fails for high maneuvering targets and increases false alarm rates. Thus, using only one measurement among the received ones and discarding the others is ineffective. An alternative approach is using all validated measurements with different probabilities, known as probabilistic data association (PDA). The standard PDA and its numerous improved versions have been developed to track a single target in clutter [27].

  Multiple target situations complicate data association, and the techniques used for single-target cases must be enhanced. The Joint Probabilistic Data Association (JPDA) algorithm tracks multiple targets by evaluating the measurement-to-track association probabilities and combining them to find the state estimate. For tracking high maneuvering targets in the presence of clutters, the PDA-based IMM estimator gives remarkable results. Also, its performance is comparable to that of the multiple hypothesis tracking (MHT) algorithm. MHT makes associations in a deterministic sense and exhaustively enumerates all possible associations as hypotheses. The hypothesis number increases with time, and the MHT algorithm requires additional pruning algorithms. The MHT algorithm is computationally exponential both in memory and time. A probabilistic MHT (PMHT) is proposed in which the associations are assumed statistically independent random variables to reduce

the computational load. Thus, there is no requirement for an exhaustive enu-
meration of associations as in the MHT algorithm [28].

- *Filtering:* After solving the data association problem, the next task is filtering in the target tracking process. Using measurements and a target motion
model, the filtering task estimates kinematic variables, such as the target's
positions and velocities. The primary tool used for filtering is the KF whenever
the model is linear or its linearization around the current estimate, known
as the EKF when the model is nonlinear. However, a single-model filter is
inadequate for tracking targets with fast maneuvering capabilities.

To this end, *multiple model* (MM) filters have been proposed to provide
greater flexibility in modeling different behaviors of the target. MM algorithms
use a bank of filters based on a specific model tailored to target behavior.
There exists a large variety of MM algorithms, such as *static multiple model*
(SMM), *generalized pseudo-Bayesian-MM* (GPB-MM), IMM, and *variable
structure IMM* (VS-IMM) [24].

2.6.3 Track-While-Scan Radar

In various operations, radars conduct the search process while tracking a target.
The most crucial property of *track-while-scan* (TWS) radar systems is to fulfill
multitasking and multitarget tracking. The TWS radars perform target tracking and
searching simultaneously. In this method, the radar operates directly as a search
radar and searches for targets in a substantial sector. When a target is detected,
the radar starts tracking that target. More than one target can be checked using
this method. Radars may use the automatic detection and tracking principle or the
monopulse principle in this operating method. The operational principle of the TWS
radar systems for automatic detection and tracking methods is shown in Figure 2.36.

TWS radars effectively use resource allocation. TWS radars allocate some of the
Tx power for scanning and some for tracking. The signal processing applications

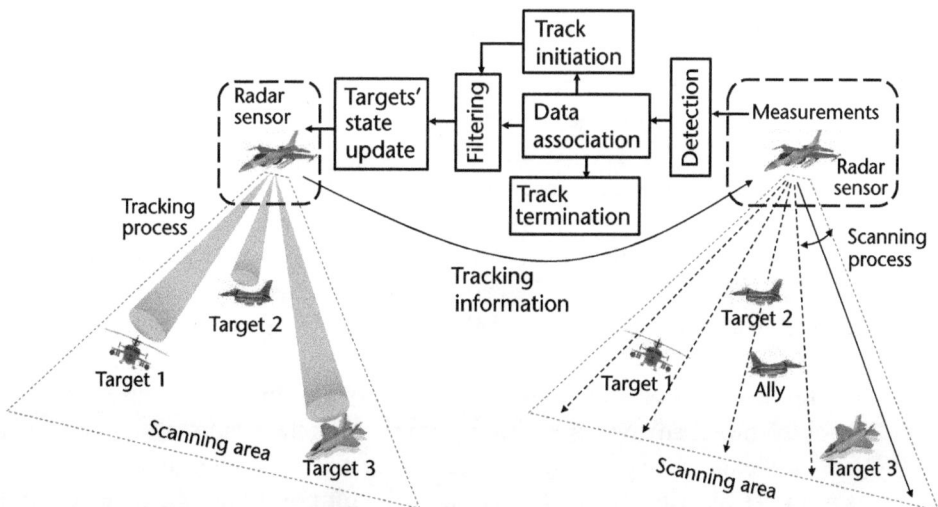

Figure 2.36 Operational principle of TWS radar systems.

used for this purpose are detection, data association, filtering, and tracking. The TWS is a hybrid operating method of conventional early warning, target tracking, and fire control radars. If we draw a general framework, a TWS radar can conduct the following processes:

- Constantly scans and detects new targets in the environment.
- Manages target tracking process for multitarget tracking.
- Constitutes a general picture of the airspace.
- Ensures the continuity of situational awareness.

TWS radars can change the scan regime and regions according to the detection status of the previous scan. In addition, the radar antenna can generate additional beams for precisely known targets during scanning. Although the radar does not keep a continuous beam on the target as in a conventional tracking method, it can allocate enough energy for missile guidance. Using a phased array antenna in this system provides simultaneous multibeam forming and directing on the target without changing the antenna boresight aspect. Thus, the phase array antennas form a beam on the target for the desired time without changing its direction in its fixed position.

2.7   Radar Guidance

Guidance systems aim to direct the missile or ammunition to hit the target. Missiles can be guided in the guidance process; however, ammunition cannot. It is usual to use the hit point calculations fulfilled before the firing instant for ammunitions, and making route correction cannot be possible for them. Guidance systems use different portions of the EM spectrum, such as RF and EO (i.e., IR, visual band, and ultraviolet (UV)). According to the guidance, different sensors and methods are used in the guidance systems. Some guidance systems use semiactive methods, which require a Tx in the launching platform and Rx on the missile.

However, in active methods, Tx and Rx are on the missile. Furthermore, for passive methods, only Rx is on the missile and uses the RF or EO transmissions of the victim platform. The guidance process lasts from firing to the missile's moment of impact. The missile guidance systems perform some or all the following procedures depending on the situation:

- The measurement process in which measurement information such as the position and velocity of the target is obtained.
- Defining the guidance route is done by providing feedback about the target.
- The deviations from the target hit point cause the route correction process.
- Rerouting process that recalculates the guidance line when necessary.

Radar-guided systems guide the missile or ammunition using RF waves reflected from the target surface or emitted from the target. A radar system is used in radar-guided weapons to identify and track targets. Since guidance of the ammunition is not possible after firing, detection of the hit point must be defined before firing. However, RF-guided missiles can be directed to the recalculated hit point for the changing situation after firing. The guidance systems on the launching platforms emit an RF transmission to use the RF reflection from the target, and an Rx system must be on the missile.

Similarly, the missile's Rx system is sufficient to identify and track victim RF emissions. However, if a missile conducts the transmission and the receiving of the reflection, this time, a whole radar system, with Rx and Tx, must be on the missile. Different radar guidance methods are described in the following sections.

2.7.1   Active Guidance

In active guided systems, the missiles find and track targets via their emissions, which means they have Rx, Tx, and the necessary electronics. The active guidance method is generally used in antiship missiles or ASM, SAM, and AAM for terminal homing. In the operational method of active guided missile systems, the target is first selected, and then the missile is fired. While the missile moves towards the target, it turns on its radar, detects the target, and steers to hit it. Missiles using active guidance have some crucial advantages as follows:

- The launching platform does not have to stay in the lethal area and can leave after firing the missile.
- As the missile-to-target range decreases, tracking can be fulfilled with higher accuracy.
- Jamming the missile gets tricky when the missile more closely reaches the target range.
- A fundamental guidance principle, *fire-and-forget*, can be obtained.

Although active guide missile systems have some vital advantages, they have disadvantages that cannot be ruled out. These disadvantages can be listed as follows:

- Active-guided missiles must contain an entire radar transceiver and electronics. Even with today's miniaturization technology, it is challenging to fit all required hardware into a missile without increasing its size and weight.
- Although producing these missiles is very expensive, they are single-shot, and the sophisticated electronics within the missile are destroyed upon impact.
- Detecting these kinds of missiles is highly possible for the modern RWR systems of the targets; since they are approaching the target, their transmission is sensed with increasing power. Early detection gives the targets time to take evasive action and deploy countermeasures. However, considering the accuracy of the active guidance homing method, interception avoidance is only possible if the target has a high maneuvering capability, like fighters.

The active guidance method has some crucial advantages and disadvantages that must be addressed. Active-guided missiles carry a self-contained radar system and are expensive. Thus, a trade-off between target importance and cost is essential in the operation planning phase. Therefore, the targets for active-guided missile systems should be carefully selected, and mission planning should be carried out meticulously.

2.7.2 Semiactive Guidance

Semiactive guidance includes illuminating or designating the target with transmission from a source other than the missile. These systems have only Rx in the missile's hardware. The Tx is in the launch platform, which can be airborne, shipborne, or land-based. After the missile is fired, it receives the signals reflected from the target. Like active guidance, semiactive guidance is generally used for terminal homing. Since no Tx exists on the missile, its weight is reduced, which means the range of the missile would increase. Thus, passive missiles always have a more excellent range than active ones.

Semiactive radar guidance is the most common type for medium-range and long-range AAM and SAM systems. The missile is only a passive detector of a radar signal provided by a launch platform as it reflects off the target. Semiactive missile systems generally use the bistatic continuous-wave radar principle. AAM systems that perform guidance with RF are a widely used bi-static radar application (i.e., Tx and Rx are in separate locations). Figure 2.37 shows a generic semi-active guidance method of the AAM system.

As shown in Figure 2.37, the Tx on the launching platform illuminates the target with the incident RF wave. The Rx on the semiactive guided missile receives the reflected RF wave to track the target. This is a standard bistatic radar application since the Tx and Rx are in different locations. The Tx and Rx aspects differ according to the target, so the target's RCS values differ for each. Also, the ranges between the Tx to target and the Rx to Rx vary.

The semiactive guidance systems determine the closing velocity using the flight path geometry shown in Figure 2.37. The Doppler frequency of the closing velocity is used to set the frequency location of the received CW signal. The offset angle of the missile antenna boresight is placed after the missile seeker acquires the target. The missile's Rx operates as a monopulse radar Rx that produces angle error measurements for position fixing. The flight path is controlled by producing navigation input to the steering system using angle errors produced by the antenna. Position fixing using the angle error information steers the missile's body to hold the target near the antenna's boresight. The offset angle is determined by flight dynamics using missile speed, target speed, and the distance between the missile and the target.

In the semiactive guidance method, the Tx must illuminate the target until the hitting instant. Thus, the boresight of the Tx antenna must be kept in the target direction. Furthermore, a *line of sight* (LOS) must be provided between the launching platform's Tx, the target, and the missile's Rx. For this reason, semiactive RF

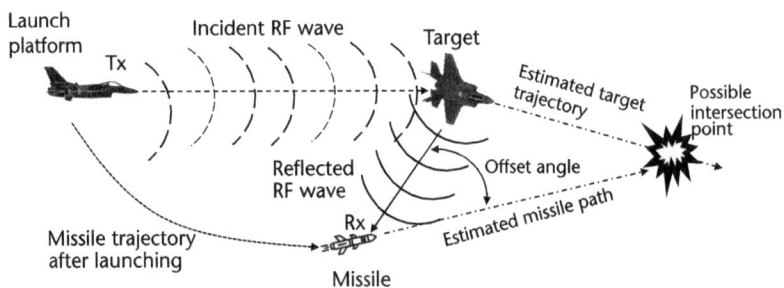

Figure 2.37 Semiactive guidance method for a generic AAM.

guidance requires the launching platform carrying the Tx to be at a particular location throughout the engagement. The most critical disadvantage of the semiactive guidance method is that the launch platform has to remain in the theater throughout the tracking process. However, missiles only have Rx and not Tx equipment, which can be counted as the advantage of semi-active guidance due to the structural simplicity, extended operational range, and hardware costs.

2.7.3   Command Guidance

Command guidance is when a ground-based fire control station, a naval platform, or an airborne platform tracks a target using a radar system and sends steering information for reaching an intersection point to a missile via data link signals. The guidance command system tracks the targets and the missiles by radar. It determines the position and velocity of the target and the missile and calculates whether their paths will intersect.

The command guidance method is generally used in radar-guided SAM systems. Figure 2.38 shows a general command guidance structure. Additionally, radar-guided AAA systems principally use command guidance. AAA systems calculate flight path and collision point, fire ammunitions at appropriate azimuth and elevation angles, and adjust the detonation time for the aircraft's predicted position. If the missile passes close to the target, its proximity or contact fuse will detonate the warhead, or the guidance system can estimate when the missile will pass near the target and send a detonation signal.

There are different operational methods for command guidance systems. These methods can be listed as follows:

- Some systems have dedicated communication data links between the command center and the missile.
- In another method, the tracking radar sends coded pulses to the missile as guidance commands.

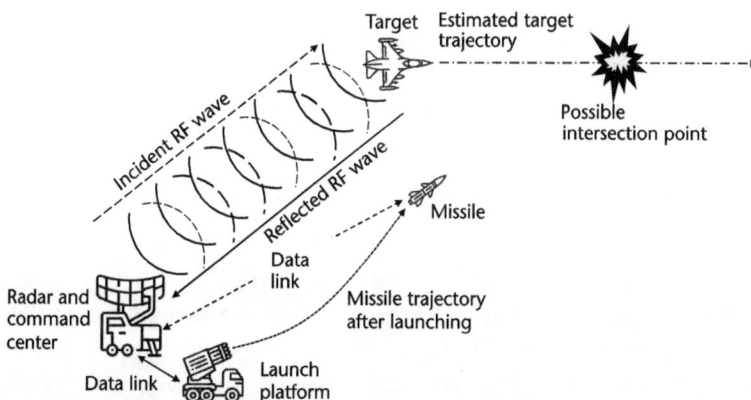

Figure 2.38   Command guidance method for a generic SAM site.

- Another method is known as *beam riding*. The missile senses the radar beam points at the target and uses the beam information to conduct an automatic correction process.
- A two-way data link is used between the command center and the missile as a different method. The command center sends the guidance information to the missile; however, the missile sends position information to the command center. These signals make it easier for the command center to track the missile.
- In some systems, a dedicated radar antenna is used to track the missile in addition to the primary tracking radar antenna.

2.7.4 Passive Guidance

If a missile system requires transmission for guidance, the target's RWR systems probably detect the threat and use an ECM for avoidance. Therefore, using the RF transmission of the target for guidance may be a suitable method. Passive radar guidance systems do not emit energy and do not receive commands from an external source. They locked onto the RF emission coming from the target itself. This emission may emerge from the target's search or tracking mode radar transmissions or avionic facilities.

Passive-guided weapon systems do not emit signals, so there is only one signal path from the target to the weapon. Furthermore, like active guidance, passive-guided weapons use the fire-and-forget method. Therefore, the launch platform may leave the area or hide after the gun is fired. Missiles that direct the target's radiation are called *antiradiation missile* (ARM) systems.

If the target performs noise jamming, it also starts making a distinctive broadcast. When the guided missile system pursues this emission, it may become an anti-radiation missile directed at the target's jamming radiation. Especially if the noise jamming is too powerful to allow the missiles to find and track the target normally, they can home in directly on sources of radar jamming. This mode of passive-guided missile operation is called *home-on-jam* (HOJ). The HOJ is an ECCM technique used for hard-kill purposes.

2.8 Radar-Guided Threat Systems

Up to this point, explanations for different radar operations, the principles of tracking radars, and the radar guidance methods are given. In this section, the types of radar-guided threat systems and their using platforms are mentioned.

2.8.1 Types of Radar-Guided Threat Systems

The principal categories of radar-guided threat systems are air-to-air missile (AAM), surface-to-air missile (SAM), air-to-surface missile (ASM), and anti-aircraft artillery (AAA) systems. The threats in the first three categories use missiles, which can be guided after launching. However, the AAA-type threats use ammunition that they cannot guide after firing. The categories may also be classified differently, such as target type (i.e., antiship, antiaircraft, or antitank). The following items explain the radar-guided threat systems according to the principal categories.

AAM: AAMs are weapon systems that fire missiles from an airborne platform, such as a fighter, cargo aircraft, helicopter, or UAV, to destroy another airborne platform. As guided weapons, AAM systems have changed the doctrine of aerial combat. These high-tech weapons, capable of destroying fast and maneuverable jet fighters and other air platforms at ranges of more than 100 km, determine who gets the air superiority.

AAMs are typically powered by one or more rocket motors, usually solid-fueled but sometimes liquid-fueled. Ramjet engines have also recently been used in AAM systems. In SAM systems, ramjet rocket motors have been used since the late 1950s, extending the range to 700 km. Using the ramjet rockets in AAM systems, the range of the missiles has recently extended up to 200 km. An example of the AAM application with ramjet is the Meteor (by MBDA), an active radar-guided beyond-visual-range. Also, with the research and development studies, this range may extend up to 500 km soon.

Radar guidance is typically used for medium-range or long-range missiles, where the IR signature of the target would be too faint for an IR detector to track. For the radar-guided AAM systems, the short-range can be defined as up to 10 km, the medium range can be defined as 10 to 25 km, and the long-range can be defined as higher than 25 km. These ranges are not strictly separated and may change according to operational requirements and the usage of different guidance methods, such as IR, video, and UV.

Short-range AAMs emphasize agility rather than range; thus, they have extreme maneuverability and high-speed capabilities. They can be fired at targets within visual range, sometimes called dogfight missiles. Most short-range AAM systems use IR guidance for their effectiveness in the visible range. Furthermore, IR-guided systems are not affected by jammers and stealth technology. Medium-range AAMs have larger warheads and longer ranges, which can be fired at targets beyond visual range. The AIM-120 AMRAAM (Advanced Medium-Range Air-to-Air Missile) is one of the most widely mature examples of a medium-range, radar-guided AAM system. The other examples are Matra-530, Marta-Mica, Rafael Derby, PL-11 and 12, F-80, AAM-4, Vympel R-27 and R-77, and Skyflash.

Long-range AAMs are the most advanced systems in navigation, rocket, and guidance technologies. They have massive warheads, very high speeds, and an enormous range. With today's technology level, they can destroy their targets from 200 km away, but this range is expected to extend shortly. Some examples of long-range AAM in operation are MBDA Meteor, Astra, K-100, PL-15, Bisnovat R-40, Vympel R-33 and R-37, Novator KS-172, and AIM-54 Phoenix.

AAM systems generally use four significant types of radar-guided missiles: active, semiactive, command (especially beam riding), and passive.

SAM: SAM systems are designed to be launched from the ground or sea surface to destroy airborne platforms or other missiles. SAM systems introduced a new concept for critical facilities and regional air defense, replacing most other antiaircraft weapons. All radar-guided SAM systems, from the smallest to the largest, generally include *identification friend or foe* (IFF) systems to help identify the target before engagement.

Short-range air defense systems (SHORAD) generally use EO bands, such as IR, video, and UV. Furthermore, smaller missiles, like MANPADs, generally use IR for guidance. The most significant advantage of these systems in the operational area is their fire-and-forget property. However, SAM systems do not use EO guidance methods beyond the visual range or for the medium and long ranges.

Medium-range and long-range range SAM systems use radar for early detection and guidance. Early SAM systems mainly used command guidance; after the 1960s, the semiactive guidance concept became more prevalent. The most vital advantage of the semiactive guidance method in SAM systems is using the advantage of sufficient space and power by being on the ground. Thus, most of the guidance and data link equipment can be located on the ground, and the electronic hardware on the missile can be reduced. Furthermore, radars can achieve more power for RF transmission and use additive hardware for a more sensitive Rx structure. The same situation is valid for larger vessels where space is not a problem. SQ-9, S-200, S-300, S-400, Aspide, Sky Bow TK-1, Thunderbird, Sea Dart, Patriot PAC-2, RIM-2 Terrier, RIM-7 Sea Sparrow, RIM-8 Talos, SA-6 Gainful, and SA-8 Gecko systems are examples of semiactive-guided SAM systems. Also, some systems use inertial guidance in midcourse and semiactive guidance in the terminal phase, such as Masurca, RIM-161 Standard Missile 3, and RIM-66 Standard MR systems.

Active radar guidance is rarely employed as the only guidance method of a SAM system. Generally, the active radar guidance method is used in the terminal phase, and for the mid-course phase, command guidance or inertial guidance usage is preferred. Akash and RIM-8 Talos systems are examples of command guidance in mid-course and active guidance in the terminal phase. CAMM, Aster, Quick Reaction Surface-to-Air Missile (QRSAM), Vertical Launch-Short Range Surface to Air Missile (VL-SRSAM), Sky Sword-II, MICA EM, and RIM-174 ERAM (SM-6) systems use the inertial guidance in mid-course and active guidance in the terminal phase. Sometimes, a dual-mode missile seeker, a passive IR seeker, and an active radar seeker are used. Examples of dual missile seeker systems are Arrow and Spyder.

Solid-fueled, liquid-fueled, or ramjet engines typically power SAMs. Since the late 1950s, they have used ramjet rocket motors. Some examples of ramjet motors using SAM systems are RIM-8 Talos, CIM-10 Bomarc, SA-6 Gainful, RIM-50 Typhon LR, RIM-8 Talos, Akash, and Bloodhound.

Long-range SAMs, such as the S-400 and S-500, may extend up to 500 km and up to 700 km, like the CIM-10 Bomarc. Long-range missiles are generally heavier; therefore, their operational concept is less mobile or fixed. Modern long-range weapons, including the MIM-104 Patriot and S-300 systems, offer relatively good mobility, but their effective ranges are up to 150 km. Medium-range SAMs are vehicle-mounted systems that can fire from mobile systems.

Medium-range designs, like the Rapier and 2K12 Kub, are designed to be highly mobile with very fast locked and loaded times. Many of these designs were mounted on armored vehicles, allowing them to keep pace with mobile operations in a conventional war. Developments in onboard maneuverability in medium-range SAM systems have also been made, changing their focus to intercept tactical ballistic missiles at low altitudes. This concept was considered for most medium-range radar-guided SAM systems during development. Thus, they have been used for air raids, cruise missiles, and tactical ballistic missiles. These systems include Terminal High Altitude Area Defense (THAAD), Aster 30 Block 1/Block NT, Sky Bow III, David's Sling, CAMM-MR, S-300, S-400, Triumph, and MIM-104 Patriot. Additionally, most sea-skimming missiles are classified in medium-range radar-guided SAM systems, such as Sea Dart Mod 1, Sea Wolf, Barak 1 and 8, HQ-10, Umkhonto-IR Block I, RIM-7 Sea Sparrow, RIM-116 Rolling Airframe Missile, and RIM-174 Standard ERAM (SM6).

AAA: Air defense aims to detect hostile aircraft and destroy them. The active components of any air defense system are missiles and ammunition. Missiles can be guided after firing; however, ammunition cannot. When ammunition is used for air defense, projectiles must be aimed at the predicted position of the target at the time the shot reaches it. For this purpose, considering the speed and direction of both the target and the projectile are essential. As in radar-guided SAM systems, AAA systems also include IFF systems for discriminating the ally units from hostile targets before engagement.

AAA systems are used from the ground or shipboard against aerial attack. The first pioneer AAA system development began in 1910 when the airplane became an effective weapon. Initially, sensors were optical and acoustic devices until the 1930s. Soon after, radar systems superseded them and have been used since then. Although missiles have dominated the air defense systems, AAA systems still operate, especially in short ranges for regional defense and installation protection.

The radars of AAA systems detect and track targets and calculate the probable point of impact. Generally, these systems are integrated fire-control systems composed of target acquisition radar, target tracking radar, IFF system, direct-view optics, and fire-control computer. Some examples of AAA system radars are Hot Shot, PPRU-1 Dog Ear, Flycatcher, SON-50 Flap Wheel, AN/MPQ-64 Sentinel, X-TAR-3D, and RPK-2 Gun Dish.

The warships use a principle similar to AAA, radar-guided anti-aircraft weapons. Vessels typically have machine guns or cannons, which can be deadly to low-flying aircraft if connected to a radar-guided fire-control system. In naval operation principles, these weapons are close-in weapon systems utilized to defend military watercraft automatically against incoming threats such as aircraft, missiles, and small boats. Some examples of these systems are Phalanx CWIS (or Sea-Wiz), Panstir-M, AK-630, DARDO, and Goalkeeper.

The widespread use of UAVs brings back AAA systems to most modern armies. These relatively cheap, low, slow, and small targets are ideal for autocannons. Engaging these targets with expensive missiles is unsustainable in the long run. Therefore, AAAs are a much more economical way to counter UAVs. Some examples of AAA systems still in operation are Skyranger 30 and 35, Maneuver-Short Range Air Defense (M-SHORAD), Flakpanzer Gepard, ZSU-23-4 Shilka, Skynex, Skyshield, and Pantsir.

ASM: The ASM systems launch missiles from military aircraft at land or sea targets. They are like guided glide bombs but are regarded as missiles and usually contain some propulsion system. The most common propulsion systems for ASMs are rocket motors and jet engines. These also tend to correspond to the range of the missiles. Additionally, some ASMs are powered by ramjets, giving them both long range and high speed.

One of the significant advantages of ASMs for aircraft to ground or sea targets is the standoff distance they provide. This operation method allows them to launch the weapons outside the intense air defenses around the target's location. Most ASMs are fire-and-forget to take the most advantage of the standoff distance; they allow the launching platform to move away after launch. Some missiles have enough range to be launched over the horizon. These missiles utilize laser, IR, optical, inertial, or satellite guidance signals in the mid-course. The type of guidance depends on the kind of target. In the terminal phase, passive radar or active radar homing may be used against ships; however, these methods are less effective against multiple, small, fast-moving land targets. Several ASM systems use passive radar homing solely as the guidance method.

Some ASMs use radar command guidance, where the launch platform must be inside the operation field up to detonation. Also, some ASMs use *manual commands to the line of sight* (MCLOS) via a data link. Generally, missiles or glide bombs are steered with joysticks and use optic guidance. In this case, the launch platform will remain inside the theater until the impact ends.

This book covers ASMs that use radar guidance in any phase. Some ASMs are suitable for surface-to-surface usage but are discussed as ASM systems. Classified examples of ASM systems according to their mid-course and terminal phases are:

- The example systems for the inertial and satellite guidance in the mid-course and active or passive radar homing in the terminal phase ASMs are ASM-3, YJ-83, C-801, YJ-83, Exocet, AS.34 Kormoran, RBS 15, PJ-10 BrahMos, Ruram, HF-2 Hsiung Feng II, AGM-88 Harm E/G, AGM 84 Harpoon Block-II, Kh-15 Raduga, Kh-22, Kh-31, Kh-32, Kh-59 Ovod, KS-1 Komet, and Kh-35.
- The systems that use passive radar homing are Martel AS.37, Armat, Alarm (Air Launched Anti-Radiation Missile), Sea Eagle, AGM-88 Harm, and AGM-114L Longbow Hellfire.
- The ASM systems that use radar command guidance are AS.15TT, AGM-12 Bullpup, and Kh-23 Grom.

2.8.2 Radar-Guided Threat System Types for Different Platforms

Radar-guided threat or weapon systems may launch from ground-based, shipborne, and airborne platforms. They may vary according to the platform's properties, such as the size of the warheads, range of the missile, guidance methods, operational principles, detection range, tracking methods, and targets. The launch platforms' speed and maneuvering capabilities also affect the operational performance of the radars and missiles. In practice, the mechanical vibration due to platform maneuvers causes oscillation, affecting the Rx operation of radars.

Land-based, shipboard, and airborne platforms may use various versions of the same threat systems, and sometimes different platforms dispose of the same weapon systems. The operation and target properties define the situation's required *platform and weapon system pair* (PWSP). Furthermore, *intelligence, surveillance, and reconnaissance* (ISR) information about the targets' location, defense capabilities, and vulnerabilities impact the PWSP selection.

Another critical determiner for selecting PWSP is the types of operations, such as defense and attack. The C2 and *command, control, communications, computers, and intelligence* (C4I) are essential for properly using the weapon systems. C2 is exercising authority and direction by a properly designated commander over assigned and attached forces in accomplishing the mission. However, the C4I system links passive defense, active defense, and attack operations to provide a timely assessment of the threat. The assumption for the platforms mentioned here is that they operate within a C4I framework, which takes the operational orders from a C2 structure.

Radar-guided weapons in defense operations protect selected assets and forces from attack by destroying air assault forces and tactical and strategic missiles. In the defense position, land-based systems, which may be a SAM or an AAA site, use radar-guided weapons against tactical and strategic missiles and assault airborne platforms. These platforms may be fighters, bombers, transport aircraft, helicopters, and UAVs.

In the defensive role, a shipborne platform must protect maritime transport, bases and ports, and sensitive and critical naval power. Furthermore, individually executed missions, defending a vessel against threats and sometimes ensuring regional air superiority are parts of their defensive duties. A maritime ship uses radar-guided weapon systems against tactical and strategic missiles and airborne platforms. These missiles may have sea-skimming properties, and SAM and anti-aircraft weapons with machine guns or cannons must have the specifications to avoid them.

The airborne platforms that use radar-guided weapons for defensive purposes are generally fighters, and their role is *air interception* (AI). This operation is conducted as *air combat maneuvering* (ACM) or dogfighting and relies on offensive and defensive basic fighter maneuvering to gain an advantage over an aerial opponent. Aircraft utilize AAM systems for these purposes. In most countries' air force concepts, UAVs will soon take over AI duty. So one should expect conceptual and structural changes in the AAM systems to adapt them for UAVs. The threats for AI aircraft are air attack packs, which consist of fighters, fighters, bombers, transport aircraft, helicopters, and UAVs. The air attack may be a harassing attack with a small group of aircraft.

However, an all-out air attack may be conducted with an air assault force composed of many packs with different air vehicles. Also, *close air support* (CAS) supports firepower in defensive operations to destroy, disrupt, suppress, fix, harass, neutralize, or delay enemy targets as an element of joint fire support. Radar-guided ASM systems are effective against naval vessels, RF-emitting land-based military assets, and facilities.

The attack operations of radar-guided weapons can be defined as offensive actions intended to destroy and disrupt enemy assets, units, and facilities. Radar-guided land-based systems are SAM or AAA sites and conduct defensive operations. For this reason, land-based systems are not considered for attack purposes.

The offensive duties of the naval ships are to destroy the enemy forces by combat. Within this scope, the targets of the vessels may be the enemy's naval ships, aircraft, helicopters, UAVs, coasts, ports, industrial facilities, and maritime transport. As discussed, inertial guidance during mid-course can enhance radar-guided weapons' effectiveness. Furthermore, their accuracy and target portfolio may be increased using a dual-mode missile seeker, such as an IR seeker with a radar seeker. The probable targets of the vessels with radar-guided weapons are the enemy's naval ships, aircraft, helicopters, UAVs, maritime transport, and RF-emitting facilities. Furthermore, using dual-mode weapon systems may include the coasts, ports, and industrial facilities into targets of shipboard radar-guided weapon systems.

The airborne platforms have the most effective, versatile, and flexible attacking capabilities. The airborne platforms' attacking operations are ACM or dogfight, air strike, strategic bombing, and CAS. These operations are generally carried out by military aircraft such as bombers and strike fighters. However, they can also include other aircraft types, such as helicopters, transport planes, and UAVs. The radar-guided weapon systems occasionally participate in ACM, airstrike, and air support operations. ACM is the tactical art of moving, turning, and positioning a fighter aircraft to reach a position where it will attack another aircraft. In ACM, the airborne platform uses radar-guided AAMs in middle and long ranges. Air strikes are commonly delivered from fighters, bombers, ground attack aircraft, and attack helicopters. The official definition includes all sorts of targets, including enemy air targets. Still in widespread use, the term is usually narrowed to a tactical attack on a ground or naval objective. It is also commonly referred to as an air raid. For ships and RF-emitting land-based facilities, aircraft use most probably radar-guided ASMs.

CAS is defined as air action by aircraft against hostile targets near friendly forces that require detailed integration of each air mission with the fire and movement of those forces. CAS provides supporting firepower in offensive and defensive operations. In CAS operations, aircraft may also use radar-guided AAMs for air targets, ASMs for vessels, and RF-emitting land-based targets.

2.9   Problems

Problem 2.1: A reconnaissance aircraft searches for a specific fighter, whether in a region or not. For this purpose, the reconnaissance aircraft uses the ELINT Rx and searches for the fighter's radar system. The radar system is operated at 10.2 GHz; the peak output power is 15 kW, and the antenna gain is 10 dB. The ELINT Rx sensitivity is −80 dBm, and its antenna gain is 15 dB. The region is the open sea, and the distance between the two aircraft is 400 km. The flight altitude of the fighter is 200 ft; however, the height of the reconnaissance aircraft is 15,000 ft. Can the ELINT Rx detect the fighter for these conditions?

Problem 2.2: You are part of a threat programming team for a self-protection system used in fighters. In the meantime, your team leader gives you a duty about a threat system that you must immediately add to the threat list. Until this time, no intelligence has been obtained about the system, and after doing a quick survey, you reach some information from the open-internet environment as follows:

- It is a C-band (4 to 8 GHz) radar-guided SAM system.
- Its engagement range is 4 to 25 km.
- It is a pulsed radar and has a constant PRI.

After finding this information, you must decide on approximate technical details to accomplish threat programming duty. For this purpose, you calculate the tracking radar's ERP, peak, and average powers. Your approximation for the radar antenna gain is 30 dB, and the Rx's sensitivity is −80 dBm. The maximum probable RCS of the fighter aircraft that use the self-protection system is 12 m^2.

(*Note:* Find the radar system's worst-case ERP, peak, and average power values in the C-band to determine the self-protection jammer requirements. Also, assume that 25 km corresponds to an unambiguous range limit and 4 km corresponds to radar transmitting time.)

Problem 2.3: An MTI radar operates at 5.4 GHz with a PRI of 10 kHz. Find the first four blind speeds of this radar.

Problem 2.4: An LFM pulse signal $x(t)$ is given as follows:

$$x(t) = \text{Rect}(100t)\cos\left(3.14 \times 10^{10}t + 15.7 \times 10^8 t^2\right)$$

Obtain the complex envelope of the above signal.

Problem 2.5: Write the ambiguity function (AF) for a single unmodulated pulse signal with a pulse width of 5 μs and an operating frequency of 3 GHz. Also, write MATLAB code to draw the AF of the single unmodulated pulse signal in 3-D mesh and contour formats.

Problem 2.6: Write the AF for a coherent, identical, unmodulated pulse train signal with a pulse width of 5 ms, three pulses in the train, a PRI value of 15 ms, and an operating frequency of 3 GHz. Also, write MATLAB code to draw the AF of the coherent identical unmodulated pulse train signal in 3-D mesh and contour format.

Problem 2.7: An unmodulated pulse signal and an LFM signal are defined as follows:

(a)
$$x(t) = \frac{1}{\sqrt{\tau_0}}\text{Rect}\left(\frac{t}{\tau_0}\right)\cos(2\pi f_0 t), \quad \text{where} \begin{cases} \tau_0 = 1 \ \mu s \\ f_0 = 1 \ \text{GHz} \end{cases}$$

(b)
$$x(t) = \frac{1}{\sqrt{\tau_0}}\text{Rect}\left(\frac{t}{\tau_0}\right)\cos\left(2\pi f_0 t + \frac{\pi B}{\tau_0}t^2\right), \quad \text{where} \begin{cases} \tau_0 = 10 \ \mu s \\ f_0 = 5.4 \ \text{GHz} \\ B = 2 \ \text{MHz} \end{cases}$$

For the signals, obtain the analytic signals and instantaneous frequencies.

Problem 2.8: An FMCW radar's wavelength is 1 cm, and the frequency sweep is 800 kHz. Let the rising or decaying time be $t_p = 40$ ms.

(a) Calculate the radial velocity of a target if its mean Doppler shift is 20 kHz.

(b) Compute the beat frequency during up-chirp and down-chirp portions of the slope corresponding to the target at $R = 200$ km.

Problem 2.9: Consider a pulsed Doppler radar with a PRF of 150 kHz and an operating frequency of 9.4 GHz. Find the required minimum FFT point size in the baseband processor to distinguish the 1-m/s velocity difference between two or more moving target returns.

Problem 2.10: A semiactive radar-guided missile is launched from a fighter aircraft to a chopper. The Tx is on the fighter and is 4 km from the target. The Tx specs are:

- Operating frequency: 9 GHz;
- Output power: 10 kW;
- Antenna gain: 20 dB.

The Rx is on the missile, and its antenna gain is 15 dB. If the distance between Rx and the target is 200m and the target's RCS in the Rx direction is 12 m², calculate the power at Rx.

Solutions

Solution 2.1: First, the dB values are turned into linear values,

$$G_{Tx(dB)} = 10 \text{ dB} \rightarrow G_{Tx} = 10$$
$$S_{min(dB)} = -80 \text{ dBm} = -80 \text{ dBm} - 30 \text{ dB} = -110 \text{ dB} \rightarrow S_{min} = 10 \text{ pW}$$
$$G_{Rx(dB)} = 15 \text{ dB} \rightarrow G_{Rx} = 31.62$$

To obtain detection range, we may use a one-way radar equation or (2.30),

$$P_{Rx} = \frac{P_{Tx}G_{Tx}G_{Rx}c^2}{(4\pi R f)^2} \rightarrow P_{Rx}$$

$$= \frac{(15 \times 10^3) \times (10) \times (31.62) \times (3 \times 10^8)^2}{(4\pi \times 400,000 \times 10.2 \times 10^9)^2} \approx 0.162 \text{ nW}$$

and

$$P_{Rx(dBW)} = 10\log(0.162 \times 10^{-9})$$
$$= -97.89 \text{ dBW} \rightarrow P_{Rx(dBm)} = -67.89 \text{ dBm}$$

Since $P_{Rx} > S_{min}$ (−67.89 dBm > −80 dBm), the fighter's radar signal can be detected if we do not consider the Earth's curvature. This time, we will find the maximum detection distance between the ELINT Rx and the fighter's radar Tx for RF-LOS propagation. For this purpose, we can take the 4/3 Earth model into account, given in (2.8); the effective or RF-LOS to the horizon is obtained as:

$$\left(d_{\text{R-T}}\right)_{\text{RF}}\ (\text{NM}) = 1.23\left(\sqrt{\frac{4}{3}h_R\ (\text{ft})} + \sqrt{\frac{4}{3}h_T\ (\text{ft})}\right)$$

$$\left(d_{\text{R-T}}\right)_{\text{RF}}\ (\text{NM}) = 1.23\left(\sqrt{\frac{4}{3}15000} + \sqrt{\frac{4}{3}200}\right) = 194.03\ \text{NM}$$

$$\left(d_{\text{R-T}}\right)_{\text{RF}} = 194.03\ \text{NM} \times 1.852 \approx 359.34\ \text{km}$$

Thus, the distance between the ELINT aircraft and the fighter (400 km) is higher than the LOS range (359.34 km) for these flight altitudes; the ELINT Rx cannot detect the fighter's radar signal. The ELINT aircraft must increase its altitude or come close to the fighter for detection.

Solution 2.2: Rx is shut down during transmission, and a death time arises according to the PW (τ). Thus, using undetectable distance, PW can be calculated by using (2.51) as follows:

$$R_\tau = \frac{c \times \tau}{2}, \quad 4\ \text{km} = \frac{3 \times 10^5 \times \tau}{2} \Rightarrow \tau = 26.7\ \mu s$$

Using the unambiguous range definition (2.53), the PRI is calculated as follows:

$$R_u = \frac{c \times T_r}{2} = \frac{c}{2 \times f_r}, \quad R_{\text{Tx}} = 25\ \text{km} = \frac{3 \times 10^5 \times T_r}{2} \Rightarrow T_r = 167\ \mu s$$

and the duty cycle is obtained by using (2.52)

$$D = \frac{\tau}{T_r}, \quad D = \frac{\tau}{T} = \frac{26.7 \times 10^{-6}}{167 \times 10^{-6}} \approx 0.16$$

The relationship between the maximum range and sensitivity of the two-way radar equation is given in (2.38). Thus,

$$R_{\text{max}} = \left[\frac{P_{\text{Tx}}G^2 c^2 \sigma}{(4\pi)^3 f^2 S_{\text{min}}}\right]^{\frac{1}{4}} \rightarrow S_{\text{min}} = \frac{P_{\text{Tx}}G^2 \lambda^2 \sigma}{(4\pi)^3 R_{\text{max}}^4}$$

For substituting the dB values in the above equation, all the dB values must transform to linear form,

$$G = 30\ \text{dB} \rightarrow G = 1{,}000$$

$$S_{\text{min}} = -80\ \text{dBm} \rightarrow S = 10 \times 10^{-12}\text{W} = 10\ \text{pW}$$

$$\lambda = \frac{c}{f} = \frac{3 \times 10^8}{8 \times 10^9} = 0.0375\text{m}$$

Peak transmitted power is calculated using the above equation.

$$P_{Tx} = \frac{S_{min}(4\pi)^3 R_{max}^4}{G^2\lambda^2\sigma} = \frac{10^{-11} \times (1981.4) \times (25,000)^4}{(1,000)^2(0.0375)^2(12)} = 458,657.4W$$

So the ERP can be found as follows:

$$ERP = 458,657.4 \times 1,000 = 458.657 \text{ MW} \rightarrow ERP = 86.6 \text{ dBW}$$

Using (2.59), the average power is obtained as

$$P_{av} = P_{Tx} \times D = 458,657.4 \times 0.16 = 73.385 \text{ kW}$$

Solution 2.3: The given values are

$$f = 5.4 \text{ GHz} \qquad f_r = 10 \text{ kHz}$$

First, we calculate the wavelength of the operating frequency,

$$\lambda = \frac{c}{f} = \frac{3 \times 10^8}{5.4 \times 10^9} = 0.056\text{m} = 5.6 \text{ cm}$$

Now, we can obtain the first blind speed by using (2.103)

$$v_{blind} = \frac{n\lambda f_r}{2}, \quad n \geq 0$$

$$v_{blind(1)} = \frac{\lambda f_r}{2} = \frac{0.056 \times 10 \times 10^3}{2} = 280 \text{ m/s}$$

The remaining three sequential blind speeds are found as follows:

$$v_{blind(2)} = \frac{2\lambda f_r}{2} = 560 \text{ m/s}$$

$$v_{blind(3)} = \frac{3\lambda f_r}{2} = 840 \text{ m/s}$$

$$v_{blind(4)} = \frac{4\lambda f_r}{2} = 1,120 \text{ m/s}$$

Solution 2.4: Figure 2.39 shows the discrimination of in-phase and quadrature components.

The LFM pulse signal is a bandpass signal. First, the bandpass signal can be written in terms of the quadrature components and the operating frequency of the carrier signal.

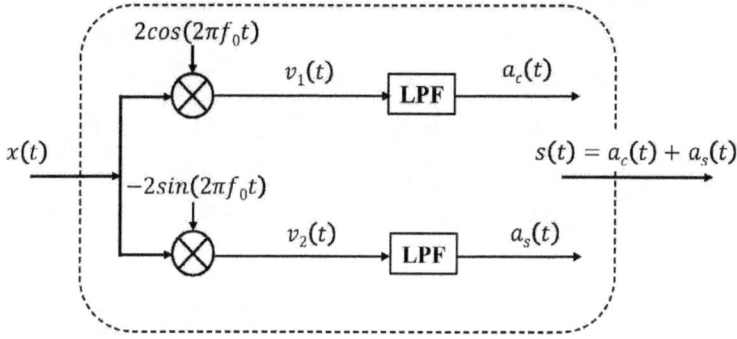

Figure 2.39 Discrimination of in-phase and quadrature components.

$$x(t) = \text{Rect}(100t)\cos\left(3.14 \times 10^{10}t + 15.7 \times 10^{8}t^{2}\right)$$
$$x(t) = a_c(t)\cos 2\pi f_0 t - a_s(t)\sin 2\pi f_0 t$$

At the output of the mixer, the following signals are defined in terms of the band-pass signal and the quadrature components.

$$v_1(t) = x(t) \times 2\cos\left(2\pi f_0 t\right)$$
$$v_1(t) = 2a_c(t)\left(\cos 2\pi f_0 t\right)^2 - 2a_s(t)\cos\left(2\pi f_0 t\right)\sin\left(2\pi f_0 t\right)$$

$$v_2(t) = x(t) \times (-2)\sin\left(2\pi f_0 t\right)$$
$$v_2(t) = -2a_c(t)\cos\left(2\pi f_0 t\right)\sin\left(2\pi f_0 t\right) + 2a_s(t)\left(\sin 2\pi f_0 t\right)^2$$

So the output of the mixer can be obtained as

$$v_1(t) = \left\{\text{Rect}(100t)\cos\left(3.14 \times 10^{10}t + 15.7 \times 10^{8}t^{2}\right)\right\} \times 2\cos\left(3.14 \times 10^{10}t\right)$$
$$v_1(t) = \text{Rect}(100t)\cos\left(15.7 \times 10^{8}t^{2}\right) + \text{Rect}(100t)\cos\left(6.28 \times 10^{10}t + 15.7 \times 10^{8}t^{2}\right)$$

$$v_2(t) = \left\{\text{Rect}(100t)\cos\left(3.14 \times 10^{10}t + 15.7 \times 10^{8}t^{2}\right)\right\} \times (-2)\sin\left(3.14 \times 10^{10}t\right)$$
$$v_2(t) = \text{Rect}(100t)\sin\left(15.7 \times 10^{8}t^{2}\right) - \text{Rect}(100t)\sin\left(6.28 \times 10^{10}t + 15.7 \times 10^{8}t^{2}\right)$$

The following are obtained from the outputs of the LPFs.

$$a_c(t) = \text{Rect}(100t)\cos\left(15.7 \times 10^{8}t^{2}\right)$$
$$a_s(t) = \text{Rect}(100t)\sin\left(15.7 \times 10^{8}t^{2}\right)$$

Finally, the complex envelope is obtained as follows.

$$s(t) = a_c(t) + ja_s(t)$$
$$s(t) = \text{Rect}(100t)\left(\cos\left(15.7 \times 10^8 t^2\right) + j\sin\left(15.7 \times 10^8 t^2\right)\right)$$

Solution 2.5: For this purpose, the equation given in (2.73) is used,

$$|\chi(\tau,\phi)| = \left|\left(1 - \frac{|\tau|}{\tau_0}\right)\frac{\sin\left[\pi\tau_0\phi\left(1 - |\tau|/\tau_0\right)\right]}{\pi\tau_0\phi\left(1 - |\tau|/\tau_0\right)}\right| \quad |\tau| \le \tau_0$$

In the question, pulse width (τ_0) is given as 5 µs, so the AF can be written as follows:

$$|\chi(\tau,\phi)| = \left|\left(1 - \frac{|\tau|}{\left(5 \times 10^{-6}\right)}\right)\frac{\sin\left[\left(5 \times 10^{-6}\right)\pi\phi\left(1 - |\tau|/\left(5 \times 10^{-6}\right)\right)\right]}{\left(5 \times 10^{-6}\right)\pi\phi\left(1 - |\tau|/\left(5 \times 10^{-6}\right)\right)}\right|$$
$$|\tau| \le 5 \times 10^{-6}$$

Figure 2.40 provides the 3-D mesh format and contour format drawings of the AF for the single unmodulated pulse signal in MATLAB.

The MATLAB code for drawing the 3-D mesh and contour plots of Problem 2.5 is as follows:

```
% 3-dimensional mesh format and contour format drawings of the
% AF for the single unmodulated pulse signal
clear, clc, close all;
% pw == pulse-width in seconds
pw = 5e-6;
```

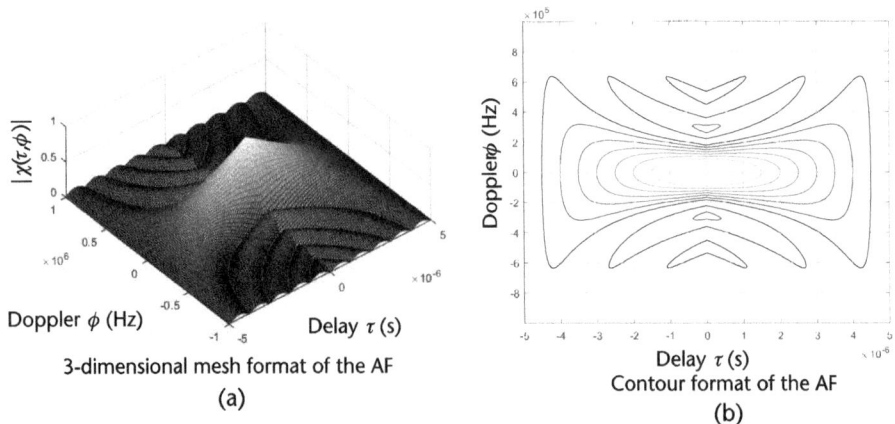

3-dimensional mesh format of the AF

(a)

Contour format of the AF

(b)

Figure 2.40 AF for the single unmodulated pulse signal given in Problem 2.5.

```
k = 0;
ts = 2*pw/200;
for tau = -pw:ts:pw
k = k + 1;
j = 0;
% Doppler frequency limit fd1
fd1=5;
fd = linspace(-fd1/pw,fd1/pw,201);
val1 = 1. - abs(tau) / pw;
val2 = pi * pw .* (1.0 - abs(tau) / pw) .* fd;
x(:,k) = (abs( val1 .* sin(val2)./(val2)));
end
taux = linspace(-pw,pw, size(x,1));
fd1=5;
fdy = linspace(-fd1/pw,fd1/pw, size(x,1));
mesh(taux,fdy,x);
xlabel ('Delay \tau (s)');
ylabel ('Doppler \phi (Hz)');
zlabel ('|\chi(\tau,\phi)|')
figure(2)
contour(taux,fdy,x);
xlabel ('Delay \tau (s)');
ylabel ('Doppler \phi (Hz)'); grid
```

Solution 2.6: For this purpose, the equation given in (2.78) is used,

$$|\chi(\tau,\phi)| = \frac{1}{N} \sum_{p=-N-1}^{N-1} \left\{ |\chi_T(\tau,\phi)| \left| \frac{\sin\left[\pi\phi\left(N - |p|/T_r\right)\right]}{\sin\left(\pi\phi T_r\right)} \right| \right\} \quad |\tau| \le NT_r$$

and

$$|\chi_T(\tau,\phi)| = \left| \left(1 - \frac{|\tau|}{\tau_0}\right) \frac{\sin\left[\pi\tau_0\phi\left(1 - |\tau|/\tau_0\right)\right]}{\pi\tau_0\phi\left(1 - |\tau|/\tau_0\right)} \right| \quad |\tau| \le \tau_0$$

The following is given in the question: pulse width $\tau_0 = 5$ ms, the number of pulses in the train $N = 3$, and the PRI $= 15$ ms.

So the AF can be written as follows:

$$|\chi(\tau,\phi)| = \frac{1}{3} \sum_{n=-2}^{2} \left\{ |\chi_T(\tau,\phi)| \left| \frac{\sin\left[\pi\phi\left(3 - |p|/(0.015)\right)\right]}{\sin(\pi\phi \times 0.015)} \right| \right\} \quad |\tau| \le 0.045$$

where

$$|\chi_T(\tau,\phi)| = \left| \left(1 - \frac{|\tau|}{\left(5 \times 10^{-3}\right)}\right) \frac{\sin\left[\pi\left(5 \times 10^{-3}\right)\phi\left(1 - |\tau|/\left(5 \times 10^{-3}\right)\right)\right]}{\pi\left(5 \times 10^{-3}\right)\phi\left(1 - |\tau|/\tau_0\right)\left(5 \times 10^{-3}\right)} \right| \quad |\tau| \le \left(5 \times 10^{-3}\right)$$

Figure 2.41 gives the 3-D mesh format and contour format drawings of the AF for the coherent identical unmodulated pulse train signal in MATLAB.

The MATLAB code for drawing the 3-D mesh and contour plots of Figure 2.41 is as follows:

```
% 3-dimensional mesh format and contour format drawings of the
% AF for the coherent unmodulated pulse train signal
clear,clc,close all;
% pw1 == pulse-width in seconds
pw= 5e-3;
% Pulse repetition interval
pri = 15e-3;
% Number of pulses in the train
n = 3;
bw = 1/pw;
q = -(n-1):1:n-1;
ts = 0:1e-5:pri;
[K, L] = meshgrid(q, ts);
K = reshape(K, 1, length(q)*length(ts));
L = reshape(L, 1, length(q)*length(ts));
tau = (-pw * ones(1,length(L))) + L ;
fd = -bw:(2/7.5):bw;
%keyboard
[T, F] = meshgrid(tau, fd);
K = repmat(K, length(fd), 1);
L = repmat(L, length(fd), 1);
N = n * ones(size(T));
s1 = 1.0-(abs(T))/pw;
s2 = pi*pw*F.*s1;
s3 = abs(s1.*sin(s2)./(s2));
```

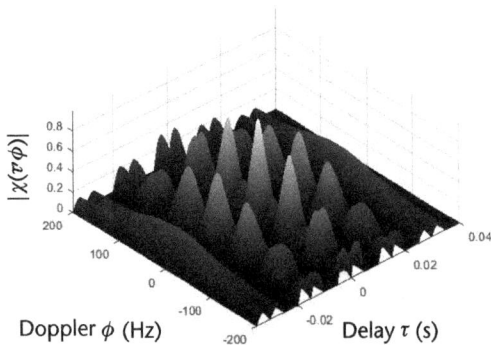

3-dimensional mesh format of the AF

(a)

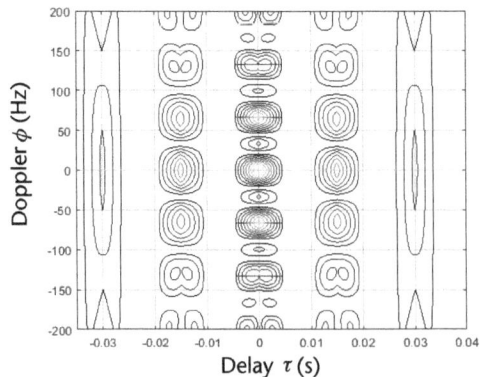

Contour format of the AF

(b)

Figure 2.41   AF for the coherent identical unmodulated pulse train given in Problem 2.6.

```
s4 = abs(sin(pi*F.*(N-abs(K))*pri)./sin(pi*F*pri));
x = s3.*s4./N;
[rows, cols] = size(x);
x = reshape(x, 1, rows*cols);
T = reshape(T, 1, rows*cols);
indx = abs(T) > pw;
x(indx) = 0.0;
x = reshape(x, rows, cols);
%3-D mesh plot (time, doppler, magnitude AF)
figure(1)
time = linspace(-(n-1)*pri-pw, n*pri-pw, size(x,2));
doppler = linspace(-1/pw, 1/pw, size(x,1));
mesh(time, doppler, x); %shading interp;
xlabel ('Delay \tau (s)');
ylabel ('Doppler \phi (Hz)');
zlabel ('|\chi(\tau,\phi)|')
axis tight;
%Contour plot (time, doppler, magnitude AF)
figure(2)
contour(time, doppler, (x));
xlabel ('Delay \tau (s)');
ylabel ('Doppler \phi (Hz)');
grid;
axis tight;
```

Solution 2.7:

(a) First of all, we can define the signal in numerical form as follows:

$$\text{Rect}\left(\frac{t}{\tau_0}\right) = \begin{cases} 1, & -\frac{\tau_0}{2} \le t \le \frac{\tau_0}{2} \\ 0, & \text{otherwise} \end{cases} \rightarrow \text{Rect}\left(10^6 t\right) = \begin{cases} 1, & -5 \times 10^{-7} \le t \le 5 \times 10^{-7} \\ 0, & \text{otherwise} \end{cases}$$

$$x(t) = \frac{1}{\sqrt{\tau_0}} \text{Rect}\left(\frac{t}{\tau_0}\right)\cos\left(2\pi f_0 t\right) \rightarrow x(t) = \left(10^3\right)\text{Rect}\left(10^6 t\right)\cos\left(2\pi \times 10^9 t\right)$$

Now, we can obtain the quadrature components using the discriminator given in Figure 2.39.

$$v_1(t) = x(t) \times 2\cos\left(2\pi f_0 t\right)$$
$$v_2(t) = x(t) \times (-2)\sin\left(2\pi f_0 t\right)$$

So the outputs of the mixers can be obtained as

$$v_1(t) = \left\{(10^3)\mathrm{Rect}(10^6 t)\cos(6.28 \times 10^9 t)\right\} \times 2\cos(6.28 \times 10^9 t)$$
$$v_1(t) = (10^3)\mathrm{Rect}(10^6 t)\left\{\cos(0) + \cos(12.56 \times 10^9 t)\right\}$$

$$v_2(t) = \left\{(10^3)\mathrm{Rect}(10^6 t)\cos(6.28 \times 10^9 t)\right\} \times (-2)\sin(6.28 \times 10^9 t)$$
$$v_2(t) = (10^3)\mathrm{Rect}(10^6 t)\left\{\sin(0) - \sin(12.56 \times 10^9 t)\right\}$$

At the outputs of the LPFs, we obtain the quadrature components as follows:

$$a_c(t) = \mathrm{Rect}(10^6 t) \qquad a_s(t) = 0$$

The analytic signal is defined in (2.65),

$$s(t)\exp(j2\pi f_0 t) = x(t) + j\hat{x}(t) = \psi(t)$$

and for obtaining an analytic signal, we use (2.59) for achieving a complex envelope,

$$s(t) = a_c(t) + ja_s(t) \rightarrow s(t) = \mathrm{Rect}(10^6 t)$$

So the analytic signal is

$$\psi(t) = \mathrm{Rect}(10^6 t)\exp\left(j(6.28)(10^9)t\right)$$

The instantaneous frequency is obtained using (2.81), but the signal is an unmodulated signal $B = 0$. Thus

$$f_i(t) = \frac{1}{2\pi}\frac{d}{dt}\left(2\pi f_0 t + \frac{\pi B}{\tau_0}t^2\right) \xrightarrow{\;B=0\;} f_i(t) = f_0 = 1 \text{ GHz}$$

(b) This time, we obtain the quadrature components and the complex envelope of the following signal,

$$x(t) = \frac{1}{\sqrt{\tau_0}}\mathrm{Rect}\left(\frac{t}{\tau_0}\right)\cos\left(2\pi f_0 t + \frac{\pi B}{\tau_0}t^2\right), \quad \text{where} \begin{cases} \tau_0 = 10 \ \mu s \\ f_0 = 5.4 \text{ GHz} \\ B = 2 \text{ MHz} \end{cases}$$

$$x(t) = (316.2)\mathrm{Rect}(10^5 t)\cos\left((3.4 \times 10^{10})t + (6.28 \times 10^{11})t^2\right)$$

The outputs of the mixers are obtained as follows:

$$v_1(t) = \left\{(316.2)\text{Rect}(10^5 t)\cos\left((3.4 \times 10^{10})t + (6.28 \times 10^{11})t^2\right)\right\} \times 2\cos(3.4 \times 10^{10}t)$$

$$v_1(t) = (316.2)\text{Rect}(10^5 t)\cos(6.28 \times 10^{11}t^2) + (316.2)\text{Rect}(10^5 t)\cos(6.8 \times 10^{10}t + 6.28 \times 10^{11}t^2)$$

$$v_2(t) = \left\{(316.2)\text{Rect}(10^5 t)\cos\left((3.4 \times 10^{10})t + (6.28 \times 10^{11})t^2\right)\right\} \times (-2)\sin(3.4 \times 10^{10}t)$$

$$v_2(t) = (316.2)\text{Rect}(10^5 t)\sin(6.28 \times 10^{11}t^2) - (316.2)\text{Rect}(10^5 t)\sin(6.8 \times 10^{10}t + 6.28 \times 10^{11}t^2)$$

The following are obtained from the outputs of the LPFs:

$$a_c(t) = (316.2)\text{Rect}(10^5 t)\cos(6.28 \times 10^{11}t^2)$$

$$a_s(t) = (316.2)\text{Rect}(10^5 t)\sin(6.28 \times 10^{11}t^2)$$

Finally, the complex envelope is obtained as follows:

$$s(t) = a_c(t) + ja_s(t)$$

$$s(t) = (316.2)\text{Rect}(10^5 t)\left(\cos(6.28 \times 10^{11}t^2) + j\sin(6.28 \times 10^{11}t^2)\right)$$

So the analytic signal is

$$\psi(t) = s(t)\exp(j2\pi f_0 t)$$

$$\psi(t) = (316.2)\text{Rect}(10^5 t)\left(\cos(6.28 \times 10^{11}t^2) + j\sin(6.28 \times 10^{11}t^2)\right)\exp(j(3.4)10^{10}t)$$

The instantaneous frequency is obtained using (2.81),

$$f_i(t) = \frac{1}{2\pi}\frac{d}{dt}\left(2\pi f_0 t + \frac{\pi B}{\tau_0}t^2\right) \rightarrow f_i(t) = f_0 + \frac{B}{\tau_0}t = \left(5.4 \times 10^9 + 2 \times 10^{11}t\right) \text{ Hz}$$

Solution 2.8: We can apply to Figure 2.19 to solve this problem,

    (a) We can calculate the target's radial velocity using the following equation, which is given in (2.108),

$$v_T = \frac{f_d c}{2f_0} = \frac{f_d \lambda}{2}, \quad \text{where} \begin{cases} f_d = 20 \text{ kHz} \\ \lambda = 1 \text{ cm} \end{cases}$$

$$v_T = \frac{(20 \times 10^3) \times (0.01)}{2} = 100 \text{ m/s}$$

(b) The slope of the up-chirp or down-chirp portion in Figure 2.19 shows the frequency change rate, f', calculated using the peak frequency deviation Δf and rising (or decaying) time. For this purpose, we can use (2.92),

$$f' = \frac{\Delta f}{t_p}, \quad \text{where} \begin{cases} \Delta f = 800 \text{ kHz} \\ t_p = 40 \text{ ms} \end{cases}$$

$$f' = \frac{800 \times 10^3}{40 \times 10^{-3}} = 20 \times 10^6 \text{ Hz/s}$$

We use (2.94) and (2.95) to compute the beat frequency during the up-chirp and down-chirp portions of the slope. Thus, the beat frequency for the up-chirp amount is

$$f_{\text{u-c}} = \frac{2R}{c} f' - \frac{2v_T}{\lambda}, \quad \text{where} \begin{cases} R = 200 \text{ km} \\ v_T = 100 \text{ m/s} \\ f' = 20 \times 10^6 \text{ Hz/s} \end{cases}$$

$$f_{\text{u-c}} = \frac{2 \times 200 \times 10^3}{3 \times 10^8} 20 \times 10^6 - \frac{2 \times 100}{0.01} = 6{,}666.67 \text{ Hz} \simeq 6.7 \text{ kHz}$$

The beat frequency for the down-chirp portion is obtained as follows.

$$f_{\text{d-c}} = \frac{2R}{c} f' + \frac{2v_T}{\lambda}$$

$$f_{\text{d-c}} = \frac{2 \times 200 \times 10^3}{3 \times 10^8} 20 \times 10^6 + \frac{2 \times 100}{0.01} = 46666.7 \text{ Hz} \simeq 46.7 \text{ kHz}$$

Solution 2.9: The velocity resolution is 1 m/s. Using (2.109), we can calculate the required Doppler resolution,

$$\Delta v_T = \frac{\lambda \Delta f_d}{2} \rightarrow \Delta f_d = \frac{2\Delta v_T}{\lambda} = \frac{2\Delta v_T f_0}{c}$$

$$\Delta f_d = \frac{2(1)(9.4 \times 10^9)}{3 \times 10^8} = 62.7 \text{ Hz}$$

Now we can obtain the minimum FFT size by using (2.110),

$$\Delta f_d = \frac{f_r}{N} \rightarrow N = \frac{f_r}{\Delta f_d} = \frac{150 \times 10^3}{62.7} \simeq 2{,}393$$

For better implementation of the FFT, N is extended to the next power of 2 by zero padding. Thus, the total number of samples for some positive integer n is

$$N_{FFT} = 2^n \geq N$$

$$N_{FFT} = 2^{11} = 4{,}096 \geq 2{,}393$$

Solution 2.10: First, we can make the dB-linear value transformation as follows:

$$G_{Tx(dB)} = 20 \text{ dB} \rightarrow G_{Tx} = 10^2 = 100$$

$$G_{Rx(dB)} = 15 \text{ dB} \rightarrow G_{Rx} = 10^{1.5} = 31.62$$

and the wavelength of the operating frequency is calculated as follows:

$$\lambda = \frac{c}{f} = \frac{3 \times 10^8}{9 \times 10^9} = 0.033 \text{m}$$

This situation defines a bistatic radar application. The power at the Rx on the missile can be obtained by using (2.46)

$$P_{Rx} = \frac{P_{Tx}\, G_{Tx} G_{Rx} \sigma c^2}{(4\pi)^3 f^2 R_{Tx-T}^2 R_{T-Rx}^2} = \frac{P_{Tx}\, G_{Tx} G_{Rx} \sigma \lambda^2}{(4\pi)^3 R_{Tx-T}^2 R_{T-Rx}^2}$$

$$P_{Rx} = \frac{(10{,}000)(100)(31.62)(12)(0.033)^2}{(1981.4)(4{,}000)^2 (200)^2} = 3.26 \times 10^{-10} \text{W}$$

$$P_{Rx} = -94.86 \text{ dBW} = 64.86 \text{ dBm}$$

References

[1]   Seybold, J. S., *Introduction to RF Propagation*, New York: John Wiley & Sons, 2005.

[2]   Mahafza, B. R., *Radar Systems Analysis and Design Using MATLAB*, 3rd ed., Boca Raton, FL: CRC Press Taylor & Francis Group, 2013.

[3]   Pakfiliz, A. G., "Increasing Self-Protection Jammer Efficiency Using Radar Cross Section Adaptation," *Computers & Electrical Engineering*, Vol. 98, 2022, p. 107635.

[4]   Adamy, D. L., *EW 102: A Second Course in Electronic Warfare*, Norwood, MA: Artech House, 2004.

[5]   Avionics Department, *Electronic Warfare and Radar Systems Engineering Handbook*, Point Mugu, CA: Naval Air Warfare Center, 2013.

[6]   Stimson, G. W., *Introduction to Airborne Radar*, 2nd ed., Mendham, NJ: Scitech Publishing, 1998.

[7]   Stralka, J. P., and W. G. Fedarko, "Pulsed Doppler Radar," in *Radar Handbook*, M. I. Skolnik, (ed.), 3rd ed., New York: McGraw-Hill, 2008, pp. 4.1–4.48.

[8]   Camacho, J. P., "Federal Radar Spectrum Requirements," US Dept. of Commerce, National Telecommunications and Information Administration, 2000.

[9]   Balanis, C. A., *Antenna Theory Analysis and Design*, 4th ed., New York: John Wiley & Sons, 2016.

[10]   International Electrotechnical Commission (IEC), IEC 60050—International Electrotechnical Vocabulary, 712-02-42, 1992.

[11]   Meikle, H., *Modern Radar Systems*, Norwood, MA: Artech House, 2001.

[12]   Levanon, N., and E. Mozeson, *Radar Signals*, New York: John Wiley & Sons, 2004.

[13]   Hao, C., et al., *Advances in Adaptive Radar Detection and Range Estimation*, New York: Springer, 2022.

[14]   Ricker, D. W., *Echo Signal Processing*, Boston, MA: Kluwer Academic Publishers, 2003.

[15]   Santra, A., and S. Hazra, *Deep Learning Applications of Short-Range Radars*, Norwood, MA: Artech House, 2020.

[16]   Liu, J., X. Zhou, and Y. Zhu, "Ambiguity Function of Symmetric Triangular Frequency Modulated Continuous Waveform," *IOP Conference Series: Materials Science and Engineering.* Vol. 768, No. 5, IOP Publishing, 2020.

[17]   Levanon, N., and A. Freedom, "Periodic Ambiguity Function of CW Signals with Perfect Periodic Autocorrelation," *IEEE Transactions on Aerospace and Electronic Systems*, Vol. 28, No. 2, 1992, pp. 387–395.

[18]   Alabaster, C., *Pulse Doppler Radar: Principles, Technology, Applications*, Edison, NJ: Scitech Publishing, 2012.

[19]   Schleher, D. C., *MTI and Pulsed Doppler Radar*, Norwood, MA: Artech House, 1991.

[20]   Cohen, M. N., et al., *Radar Design Principles: Signal Processing and the Environment*, 2nd ed., Mendham, NJ: Scitech Publishing, 1999.

[21]   Bruder, J. A., "Range Tracking," in *Principles of Modern Radar*, J. L. Eaves and E. K. Reedy, (eds.), New York: Chapman & Hall, 1987, pp. 541–566.

[22]   Trunk, G. V., "Automatic Detection, Tracking, and Sensor Integration," in *Radar Handbook*, 2nd ed., M. I. Skolnik, (ed.), New York: McGraw-Hill, 1990, pp. 8.1–8.23.

[23]   Meng, X., "Performance Evaluation of RQ Non-Parametric CFAR Detector in Multiple Target and Nonuniform Clutter," *IET Radar, Sonar & Navigation*, Vol. 14, No. 3, 2019, pp. 415–424.

[24]   Benavoli, A., et al., "Knowledge-Based Radar Tracking," in *Knowledge-Based Radar Detection, Tracking, and Classification*, F. Gini and M. Rangaswamy, (eds.), New York: John Wiley & Sons, 2008, pp. 167–196.

[25]   Blackman, S. S., *Multiple-Target Tracking with Radar Applications*, Dedham, MA: Artech House, 1986.

[26]   Lee, G., et al., "Probabilistic Track Initiation Algorithm Using Radar Velocity Information in Heavy Clutter Environments," *2018 15th European Radar Conference (EuRAD)*, 2018, pp. 277–280.

[27]   Kirubarajan, T., and Y. Bar-Shalom, "Target Tracking Using Probabilistic Data Association-Based Techniques with Applications to Sonar, Radar, and EO Sensors," in Chapter 8 in *Handbook of Multisensor Data Fusion*, D. L. Hall and J. Llinas, (eds.), Boca Raton, FL: CRC Press, 2001.

[28]   Pakfiliz, A. G., "A New Method for Surface-to-Air Video Detection and Tracking of Airborne Vehicles," *International Journal of Pattern Recognition and Artificial Intelligence*, Vol. 36, No. 1, 2022, pp. 415–424.

CHAPTER 3

Jamming of Radar-Guided Systems

Weapon systems primarily rely on radar and electro-optical (EO) guidance methods. EO guidance, encompassing infrared (IR), optical (or visible light), and ultraviolet (UV) methods, has witnessed a remarkable and rapid evolution in recent years. Laser-guided weapons, for instance, often utilize IR bands, with their operating wavelength spanning the EO spectrum. Each guidance method, whether radar or EO, presents a unique set of advantages and disadvantages. Understanding these characteristics is crucial for making informed decisions in weapon system development and defense technology research. While radar systems boast a more extended history in warfare, the rapid evolution of EO systems is now making substantial contributions, sparking intrigue and excitement in the field.

3.1 EO Guidance Systems and Countermeasures Against Them

Optical guidance stands as one of the most mature and extensively studied technologies. The research focuses on object detection and tracking in the visual spectrum, or image processing, which has been a prominent area for many years and has led to significant advancements. Notably, the use of machine learning and deep learning algorithms has revolutionized the performance of object detection and tracking capabilities over the past two decades. In addition to the technological developments of optical-guided systems, their passive nature, which does not emit any transmission, is a significant advantage. This passive nature, combined with machine learning and deep learning algorithms, not only advances the capabilities of optical-guided systems but also instills confidence in their reliability and effectiveness, reassuring our audience of their potential. However, optical guidance requires daylight and direct sight between the sensor and targets. This direct sight, including cloud or foggy environments, encompasses a broader meaning than LOS. The visual range, which sets the operating range limit, is another drawback for optical-guided weapon systems. Using the smoke generator and camouflage are the main countermeasures against optical-guided threats.

IR-guided systems have similar advantages to optical guidance; however, they do not need daylight and may also operate in the dark. They are passive systems and do not make emissions. However, they require direct sight, and clouds or foggy environments negatively affect their performance. In IR guidance, targets' IR emission, accumulated in some parts of their body, is a very discriminative sign for detection and tracking. This makes the process easier than optic guidance, which generally detects and tracks depending on the target shape. Furthermore, this discriminative

131

sign increases their operating range to the optical-guided systems up to a level. The countermeasures for IR-guided threats may include IR decoys (or flares), DIRCMs, IR suppressors, and smoke generators. Along with new scanning techniques, multispectral detectors have also been incorporated. Multispectral IR provides target identification under conditions where one wavelength cannot identify the target. They could be UV and IR or short and medium-wave IR.

Defining the UV environment is difficult since it cannot be seen. It is composed of Sun radiation; therefore, little energy is present outside daylight. It is also strongly absorbed by the atmosphere. However, due to the scattering of the UV radiation from the Sun, which is not strongly absorbed, the sky appears to have relatively uniform radiation. Thus, when the sky is observed in the UV, there is often no intensity variation whether the sky has clouds. This ideal background can easily discriminate against aircraft and other objects [1].

Like optical and IR-guided systems, UV-guided systems require direct sight between the sensor and targets. The wavelengths associated with UV are well-defined, ranging from 10 to 400 nm. However, the regions that are useful for detecting aircraft are much tighter. The definitions of the bands are UVA 315–400 nm, UVB 280–315 nm, UVC 100–280 nm, and far UV 10–100 nm.

The wavebands, including the UVC and the far UV, produce little background radiation since the atmosphere strongly absorbs these wavelengths. The region of these wavebands is called solar blind, as there is no background radiation either day or night from the Sun. This region discriminates a hot object from the rest of a scene, such as missile plumes and flares. Thus, the solar blind region is suitable for missile warning systems. However, missile seekers aim to detect and track an airborne target. They use UVA and UVB, which have short wavelengths, to discriminate airborne targets from the background. Furthermore, using these wavebands excludes flares and cannot be a countermeasure for UV guidance. The smoke formulation is an essential countermeasure against UV-guided threats.

The IR, optical, and UV guidance methods are passive and do not emit radiation. However, laser-guided weapons, operated in the IR and optical spectrum, are generally semiactive systems that emit radiation. Like IR and optical bands, the systems require direct sight. This time, direct sight must be provided between the designator and target and for the return path between the missile Rx and target. However, laser-guided systems may operate in daylight and dark, unlike optical bands. The smoke generator and laser countermeasure system are the main countermeasures against laser-guided threats.

3.2   Radar Guidance Systems and Countermeasures Against Them

Radar-guided weapon systems are active systems that emit radiation, except passive-guided systems. This property may be considered a weakness for them since they reveal themselves. However, radar-guided systems have significant advantages that make them indispensable. These advantages are listed here:

- Their effective range cannot be reached using other guidance methods.
- Their accuracy in far distances is unique.

- Daylight or darkness does not affect their performance.
- Precipitation, clouds, and foggy environments may prevent their operational performance from improving. Radars' technical specifications, such as operating frequencies and polarizations, determine the magnitude of these effects.
- They require LOS for operation, not direct sight, such as the EO-guided systems.

The physics limitations dictate radar technology's superiority over EO technology, especially at long ranges. Using EO-guidance methods for short-range AAM and short-range air defense (SHORAD) systems seems more effective in terms of detection and tracking. However, radar-guided systems are superior to EO-guided counterparts in the middle-range and long-range AAM and air defense systems. For this reason, to think that radar-guided systems will continue to be increasingly used in the future is a prediction with merit.

Radar guidance is associated with radar technology and is continuously updated. Technology, ECCM properties, and enhanced target ranges are the new challenging trends in improving radar systems. The research and development (R&D) studies for radar systems are significant; these studies are anticipated to increase growth.

The radar ECM techniques may be classified as hard-kill and soft-kill at the highest level. The hard-kill techniques include using antiradiation missiles (ARM) and directed energy weapons (DEW). For assuming a hard-kill method to be an ECM, the mother platform must be under threat by a radar-guided weapon. Using an ARM weapon when the mother platform is not the primary target, such as for SEAD operation purposes, might not be counted as an ECM technique but an EA method. Furthermore, using an HOJ missile is considered an ECCM technique rather than ECM.

The soft-kill techniques include passive and active methods. The classification's main difference is whether to radiate emission by the target platform. The passive ECM techniques reduce RCS and use chaff. However, active methods cover jamming and using decoys. Active and passive ECM methods constitute a significant part of EW's R&D studies. In this part, a discussion about the jamming principles and jamming types is given. Additionally, one can find the derivation of the jamming equations for different situations. Also, the readers can obtain information about jamming operations' technical and tactical principles in a mutual requirement frame.

3.3 Radar-Jamming Principles

Operation types dictate and determine the jamming principles. An air attack or air strike operation aims to destroy the predetermined targets. During these missions, the attacking airborne platforms, such as fighters, bombers, helicopters, or UAVs, are the targets of the enemy defense's radar-guided weapons. It is like being hunted on the hunt. When an aircraft in the striking force uses jamming for self-protection or to protect its wingman and squadron, it is considered ECM jamming. Thus, the jammer types are called *self-protection jammers* if they defend themselves or *escort jammers* (EJ) if they protect their wingman and squadrons. The jamming types used in SPJ and EJ may be in the form of noise or deception.

However, some jamming operations are conducted to protect ally forces from a different location or to help to obtain air superiority, like SEAD operations. This time, the jammer platform does not aim to protect itself or the accompanies in the nearfield region; it seeks to attack electronically. These types of operations are called stand-off jamming or stand-in jamming.

ECM has ceased in the new EW concept, and EA has been used instead. However, most scientific studies still use ECM instead of countermeasure actions that cover jamming. In the EA concept, SOJ and EJ systems are classified as offensive EA activities. The activities that can be under the offensive EA are:

- Jamming enemy radar or electronic command and control systems.
- SOJ and SIJ systems' operations are part of offensive EA. Also, according to its operational aims, the function of EJ systems is classified as a part of offensive EA activities. This point of view is likely actual because, in most scenarios, a threat does not directly aim at an EJ platform. The similarity between SOJ and EJ systems is functional as they maintain different platforms. In contrast, the difference between them shows itself in jamming calculations because the EJ operational geometry is similar to the SPJ.
- Using antiradiation missiles to suppress enemy air defenses.
- Using electronic deception techniques to confuse enemy intelligence, surveillance, and reconnaissance systems.
- Using directed-energy weapons to neutralize an enemy's equipment or capability [2].

According to the viewpoint, SPJ activities may be covered under *electronic protection* (EP) or defensive EA. SPJ may be considered part of EP since this jamming concept aims to protect the mother platform. However, jamming is a countermeasure that accommodates attack action. For this reason, it may be envisioned in the defensive part of EA. The focus of this book is SPJ, and it is assumed that both EP and defensive EA cover it. As stated in Chapter 1, the EP involves actions to ensure effective, friendly use of the electromagnetic spectrum despite the enemy's use of electromagnetic energy. EP may be active or passive measures, and SPJ may be considered active EP measures. In contrast, defensive EA involves the usage of the electromagnetic spectrum to protect personnel, facilities, capabilities, and equipment. The activities that can be under the defensive EA are listed as follows:

- SPJ and, as stated above, according to the jamming calculations, EJ systems;
- Use of expendables such as chaff and flares;
- Using expendable jammers and towed decoys;
- DIRCM systems and counter-radio-controlled improvised explosive device systems.

This book's viewpoint covers old and new concepts to avoid misunderstandings. In this context, SPJ has assumed an element of ECM according to the old concept, which is considered a subunit of EP and defensive EA concerning the new concept. In any form of jamming, including radar jamming, the jammer platform emits radiation, which reveals itself. Therefore, the first principle of radar jamming is not to use jamming unless necessary.

The jamming principles change according to the scenario, the aims, and the operation type. When an air attack operation is conducted, a SEAD operation is probably carried out to neutralize enemy IADS structure. For this purpose, all kinds of jammer systems, including SOJ, SIJ, EJ, and SPJ, are in operation. Thus, both offensive and defensive EA methods are used together. It must be noted that the main aim of an operation cannot be EW.

The EW capabilities and assets are force multipliers, not weapon systems. Thus, they must obey the general synchronization of the operation. Therefore, the second radar jamming principle is that EW operational plans must be compatible and synchronous with the general operational plan.

To use a radar jammer in any theater, sufficient technical and operational information about the threat systems must be known. Gathering intelligence is a vital requirement for effective jamming. An ELINT operation must be conducted to accumulate the technical and locational intelligence for the threat radar systems. Then a detailed threat analysis is undertaken to extract the required technical data and radars' locations. Furthermore, a COMINT operation may support the ELINT operation in obtaining a probable C2 structure. The obtained information determines the required jamming scenarios and the jammer packages. The third radar jamming principle is to get intelligence data for the operational field's radar threats and make a threat analysis before the operation.

After the analysis study, the required technical and operational information is extracted. This information is then used to program jammer systems. Different jammer systems have various programming requirements. SOJ systems aim to suppress enemy air defense, and, for this purpose, they jam a variety of threat radars; most are *early warning* (E/W). The jamming technique in the SOJ system does not have to apply a specific jamming technique to each threat system, but it must cover all of them. Also, the ESM system on the SOJ platform does not have to detect each active radar in the field. Thus, rough clustering for the radar signals may be sufficient for ESM purposes.

However, the situation for the SOJ systems is not valid for SIJ systems. Since SIJ platforms aim to jam one or two radar threats, they must define the active threats and apply specific jamming. SPJ systems aim to protect the mother platform from radar-guided weapons. So the threats must be determined precisely, and RWR is programmed to discriminate the active threats' emitters. Studying the effectiveness of jamming techniques before an operation is also crucial. An unignorable issue for this study is determining the effectiveness of each jamming technique separately and together. Sometimes, using jamming techniques simultaneously for double or triple-engaged threats may reduce the jamming effectiveness. The fourth radar jamming principle is the jamming scenario for different jammer systems, which must be studied before being applied to the operational field.

In a dense environment, detecting active emitters may be problematic for each jammer system. The RWR of an SPJ system must be very precise and detect each emitter in the environment. However, the ESM Rx of an SOJ system does not need such accurate emitter discrimination, and a clustering method may be sufficient for it. Also, the operational precision of RWR or ESM systems determines the efficiency of the jamming source. Thus, applying detection and jamming methods to jammer systems is essential. For this purpose, a preprogramming development center requires experienced technical support personnel. The fifth radar jamming principle establishes a preprogramming center for effectively operating the jammer systems.

The final principle is about the hardware. A suitable jammer system must be on a proper platform to provide jamming support to an operation. For this purpose, operational research must be conducted for SOJ and SIJ platforms and systems. However, the fighters are the main factors for an air force, and their entities depend on the operational requirements, technological and economic factors, political preferences, and logistic structure of a country. For this reason, fighters are definite, and suitable SPJ and EJ systems are selected for these platforms. Thus, the sixth radar jamming principle is obtaining proper equipment and platforms for different jamming requirements.

3.4   Radar Jamming Types

Jamming is defined in [3] as "the deliberate radiation, reradiation or reflection of electromagnetic energy with the object of impairing the effectiveness of hostile electronic devices, equipment or systems." Jamming application against radar systems is called radar jamming. The radar jammer systems may be classified as active, passive, or on-board and off-board jammers. The active jammers emit radiation, such as SOJ, SIJ, EJ, SPJ, and active decoy systems; these systems may be named electronic radar jammers. However, passive jammers, such as chaff and corner reflectors, do not radiate. Corner reflectors are metal prisms with many sides that have a similar effect to jamming with chaff. They reflect radar signals to their sources to confuse the hostile radars and remain undetected.

Another classification is about on-platform or off-platform jammers. On-platform jammers are SOJ, SIJ, EJ, and SPJ. In contrast, the systems launched from the platform, such as chaff and active decoy systems, are off-platform jammers. In this classification, the usage of corner reflectors may vary. When used for jamming, they are launched and counted as off-platform. However, when they are used for deception, especially in naval or land-based decoys, they are used as on-platform. Furthermore, in maritime navigation, they are placed on bridge abutments, buoys, ships, and lifeboats to ensure that corner reflectors show up strongly on radar screens.

Here, we focus on electronic radar jammers or simply radar jammers. We will also discuss the radar jamming types only in this class. These jammers emit radiation, such as SOJ, SIJ, EJ, SPJ, and active decoy systems. The radar jamming types applied by these systems are considered here. In this context, the radar jamming types are separated into two main categories: noise and deceptive jamming. These jamming types are explained respectively in the following sections.

3.4.1   Noise Jamming

Noise jamming aims to decrease the SNR of the enemies' radar Rxs by increasing the noise level. Noise jammers produce noise signals in the required frequency by modulating an RF carrier signal with noise, as given in Figure 3.1. The targeted radar frequency is adjusted in the *local oscillator* (LO), and the noise generator produces noise. This noise may be white Gaussian or colored, such as pink, gray, blue, or brown. *White Gaussian noise* (WGN), or simply *white noise* (WN), is a noise signal whose power spectrum is flat. In a different definition, WN has almost constant integrated power at various frequency bands of the same bandwidth. WN is made of nearly all the frequencies and will have constant power at all these frequencies; hence, it is also analogous to white light, emitting all the frequencies in the same proportion.

Colored noise will have different integrated power at various frequency bands of the same bandwidth. It will have a diverse power spectrum depending on whether it is pink, gray, blue, or brown. Based on this, power concentration varies at different frequencies. Thus, noise generators aim to generate WN, the most effective noise for reducing SNR in the radars' Rx. Since the noise in the return echo signal in a radar's Rx is WN, applying the WN-modulated jammer signal produces the maximum decreasing effect on the SNR.

After obtaining the noise-modulated RF signal, it is filtered and amplified with a bandpass filter. Knowing the radar Rx bandwidth is crucial for increasing the jamming efficiency since there is a direct proportion between the bandwidth and power of the jamming signal. Then, generally using a directed antenna, the noise signal is transmitted to the threat radar or radars. Noise jammers rely on high power levels to saturate the radar Rx or worsen its sensitivity by decreasing its SNR. At the end of the noise jamming process, the return echoes remain below the noise level. Thus, the noise jammers ensure that the range, azimuth, and altitude information of the targets cannot be detected in the threat radars' Rx.

The noise-jamming techniques can be classified according to the jammer signal bandwidth. These techniques are spot noise with a narrow bandwidth, barrage noise with an extensive bandwidth, and sweep spot or sweep modulation noise. Furthermore, the noise waveforms can be CW or pulsed. Noise jamming can be used by defensive EA (SPJ) and offensive EA (SOJ, SIJ, and EJ) to deny targeting information to the enemy radar or radars. The forms of noise jamming techniques are classified as follows.

- *Spot noise jamming (SNJ):* This method occurs when a jammer focuses all its power on a single frequency. This property of SNJ renders the technique

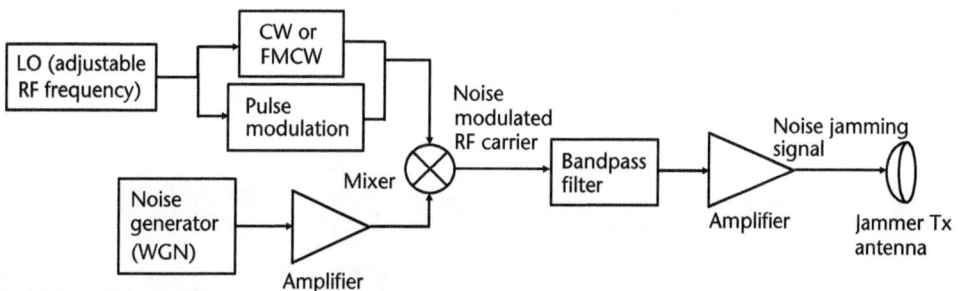

Figure 3.1 A generic noise jammer block diagram.

inoperative against frequency-agile radars. When it is necessary to jam radars with a range of frequencies, SNJ would also be ineffective.

The primary advantage of spot jamming is its power density. Radar Rx can be countered at longer ranges than a barrage jammer of equal output power. However, knowing the victim radar operating frequency and bandwidth is vital for the SNJ technique.

The echo signal received by the radar's Rx is corrupted with WN and defined as follows,

$$x_{Rx}(t) = Ax(t - t_0) + n_i(t), \quad n_i(t) \sim N(0, \eta_0) \tag{3.1}$$

where A is the return signal's magnitude, $x(t)$ is the transmitted signal, t_0 is an unknown time delay proportional to the target range, and noise, n_i, is independent and identically distributed and drawn from a zero-mean normal distribution (WN) with variance η_0. The noise generation process is assumed to be ideal, using the block diagram in Figure 3.2 at the radar operating frequency. The LO produces a signal in the radar's operating frequency, modulated with pulse trains, remaining intact for CW or FM modulated for FMCW. A similar signal with the victim radar, $z(t)$, is obtained at this point. Then the signal is modulated with WN, whose variance is η_0,

$$z_i(t) = z(t - t_0) \times n_i(t), \quad n_i(t) \sim N(0, \eta_0) \tag{3.2}$$

After obtaining the noise-modulated signal, it is applied to a bandpass filter, and then the signal at the filter output is amplified. The bandpass filter is adjustable; knowing the bandwidth of the radar's Rx is essential for optimum adjustment.

Figure 3.2 shows an example of an SNJ and radar echo signals in the radar's Rx. The radar's operating frequency is assumed to be 2 GHz, PRF is 10-kHz PRF with unmodulated pulsed signal, and Rx bandwidth is 200 MHz. The SNJ signal

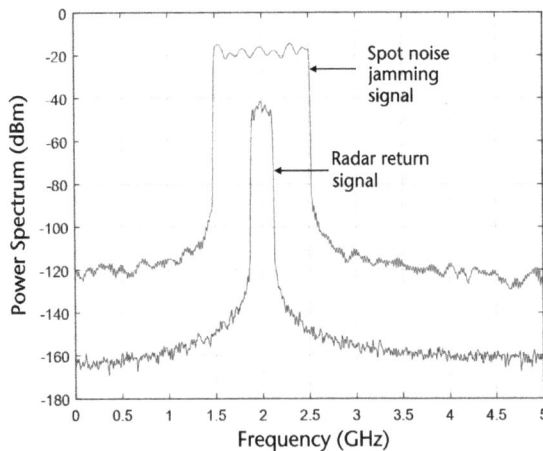

Figure 3.2 Spot noise-jamming signal effect on a radar Rx.

also uses the same carrier frequency as the radar signal. First, it is modulated with the same pulse train, then with the WN. The bandwidth of the jamming signal is 1 GHz. If the jammer's bandwidth decreases, the power level will increase.

- *Barrage noise jamming (BNJ):* Noise-jamming techniques are usually used with different concepts at SOJ, SIJ EJ, and SPJ systems. Especially in its operational principle, SOJ systems jam more than one radar system, generally a whole EADS or a part of EADS. Thus, SOJ systems must jam more than one carrier frequency and probably have different bandwidths and modulations. However, some radar systems use ECCM techniques for LPI and antijamming, such as frequency hopping, frequency chirp, and *direct sequence spread spectrum* (DSSS). This time, the jammer has to increase the bandwidth to obtain the required jamming effectiveness.

Another problem arises from using wide bandwidth for jamming. Noise jamming depends on power density to be effective. Power density is a function of the frequency bandwidth of the jamming signal. A jammer can concentrate energy in a narrow band if it covers a narrow frequency range. If it covers a wide frequency range, the energy is spread over that range. Barrage jamming is a jamming technique where high power is sacrificed for the continuous coverage of several radar frequencies.

Figure 3.3 shows an example of barrage noise jamming against three different radars. The radars' operating frequencies are 1, 2, and 3 GHz. The corresponding PRFs with unmodulated pulsed signals are 10, 20, and 40 kHz. Rx bandwidth is 200 MHz. The SNJ signal uses a 2-GHz carrier frequency; first, it is modulated with the 10-kHz pulse train and then with the WN. The bandwidth of the jammer is 3 GHz. Increasing the frequency bandwidth of the jamming signal causes a decrease in the power level. Thus, the jamming will be ineffective.

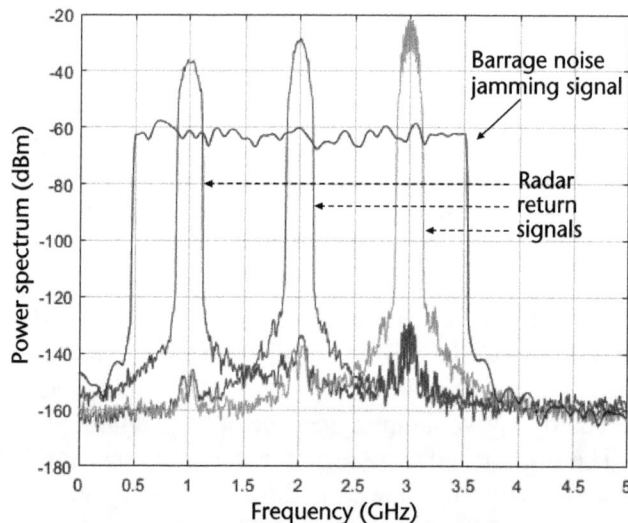

Figure 3.3   Barrage noise-jamming signal effect on three radar Rx.

- *Sweep-spot noise jamming (SSNJ):* Sometimes jamming multiradars effectively with a limited power source jammer is required. This requirement may be provided by shifting a jammer's full power from one frequency to another. This shifting motion, or sweeping, jams multiple frequencies quickly and is called SSNJ. The jammer focuses all its power on a single frequency, although not all the victim radars in different frequencies can be jammed simultaneously.

Figure 3.4 demonstrates an example of SSNJ against the three radars given in the barrage jamming. Furthermore, the jammer properties are the same as in the SNJ example. In each step of the figure, the return signal and the jamming signal effect on the radars' Rx are shown. In this application, the SSNJ signal uses the same carrier frequency modulated with the same pulse train of the related radar signal. Then the resulting pulse trains are modulated with the WN. The bandwidth of the jamming signal is 1 GHz in each case.

In the first step of Figure 3.4, the jammer concentrates all its power on the first radar and jams only this radar. In the second step, the jammer changes its power focus from the first radar to the second radar; this time, only the second radar is jammed. Moreover, in the third step, the jammer is directed to the third radar, which jams only it. Three factors determine SSNJ's effectiveness: the applied power in the spot, the bandwidth, or frequency range; the spot covers; and the sweep rate. When we assume that the power and the bandwidth are sufficient for effective jamming, the last one becomes vital for a general jamming performance. Sweep rate involves frequency-changing speed and dwelling time on a frequency. For efficient sweep planning, the technical specifications and the locations of the victim radars must be known, and the jammer structure must be suitable for the planned operation.

3.4.2 Deceptive Jamming

The effectiveness of the noise-jamming techniques depends on the ratio of the power density of the jamming signal to the return power in the radar's Rx. This ratio is called the *jamming-to-signal ratio* (J/S). The structure of the noise jammers is usually elementary, high-power systems. They can be effectively utilized in SOJ, SIJ, EJ, and SPJ systems. The high-power jamming signal usage is impossible for some applications due to the jammer platform's space, weight, and power requirements.

However, in some applications, multiple radars in different frequencies must be jammed simultaneously. Even a noise jammer at the highest power cannot help obtain sufficient jamming effectiveness. In this situation, using deceptive jamming is a good option. A deception jammer requires significantly less power than a noise-jamming system. The deception jammer obtains this advantage by utilizing a waveform identical to the waveform the radar's Rx is specifically designed to process.

The objective of deceptive jamming against radar is to mask the actual target by injecting suitably modified replicas of the fundamental signal into the victim system. The deceptive jamming can be applied against both search and tracking radars. The primary advantage of this form of jamming is that the victim radar absorbs all the jamming power, and the radar's processing gain is either partially or entirely negated. Since the deceptive jamming signal penetrates the radar's signal processor, it can be considered a direct attack on this radar function [4].

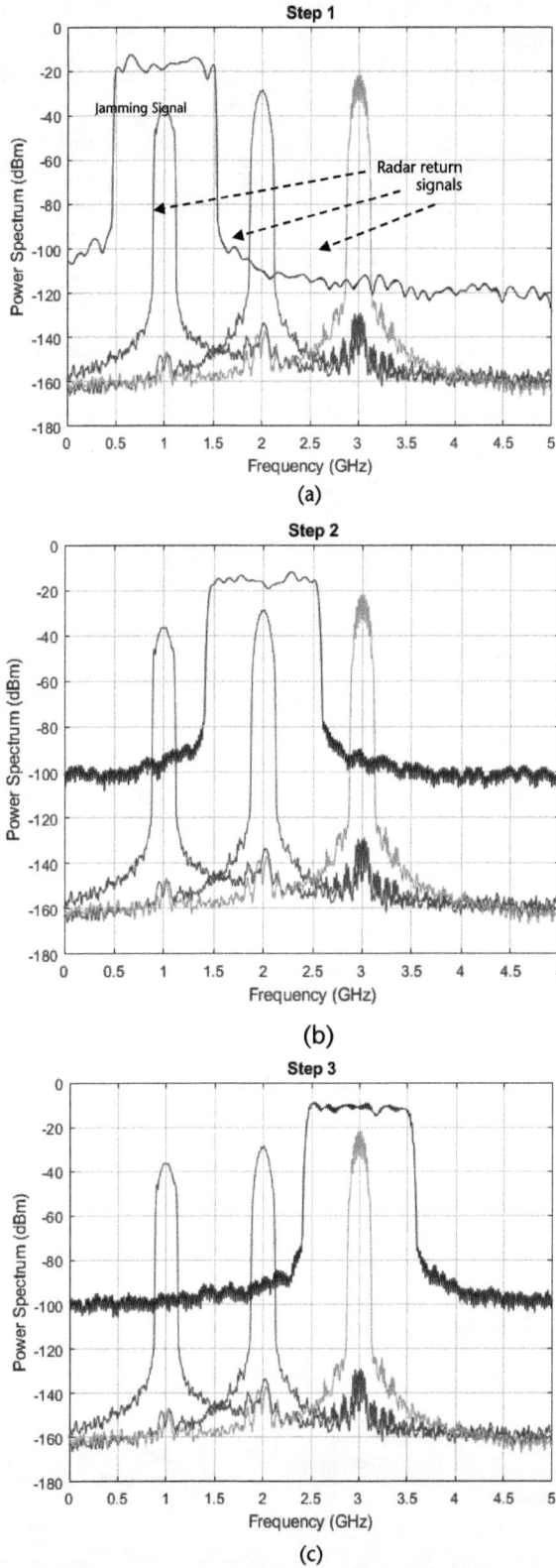

Figure 3.4   Sweep spot noise-jamming signal effect on three radar Rx.

Deception involves transmitting a signal as similar as possible to the source. These techniques are coherent and based on the reception and retransmission of the signal coming from the radar with additional modulations superimposed. The resultant waveform is coherent and naturally matched with the radar receiver, which cannot discriminate it from the natural echo. A radar is said to be coherent-on-receive if it records the phases of all transmitted pulses in its memory. The receiver phase reference is usually the most recent transmitted pulse phase [5].

Although deception jammers require less power than noise jammers, they are more complex. This complexity results in using highly dense digital integrated circuits, which provide the vast computing power necessary to implement this type of jammer. The deception jammers are more effective than noise jammers against modern radars that employ coherent integration techniques such as the pulsed Doppler and pulse compression. The reason is that radars using coherent integration techniques have a 30–60-dB processing gain against noise and attenuate the noise jammer signal by that amount while accepting any target-like return without attenuation.

The deceptive jamming waveform should be coherent and naturally matched with the radar's Rx, which will not be able to discriminate it from the natural echo. The deception jammer's transmitted power must be sufficient to overwhelm the radar target echo and deceive the radar's Rx to synchronize it with the false Tx with no symptom of the replacement at the waveform level. Figure 3.5 shows a block diagram of a general deceptive jamming system.

In the scheme given in Figure 3.5, the deceptive jamming system is on a platform that wants to be protected. The incident radar signal is received by the jammer's Rx antenna and sent to the preamplifier. The output of the preamplifier is directed to the memory component. The memory stores the victim radar's signal characteristics and passes these parameters to the control circuitry for processing. At the same time, a delay related to the preprogrammed jamming technique is added to the signal before amplifying and sending it back to the victim radar.

This process must be conducted almost instantaneously for every signal jammed. Any delay in the memory loop diminishes the effectiveness of the deception technique.

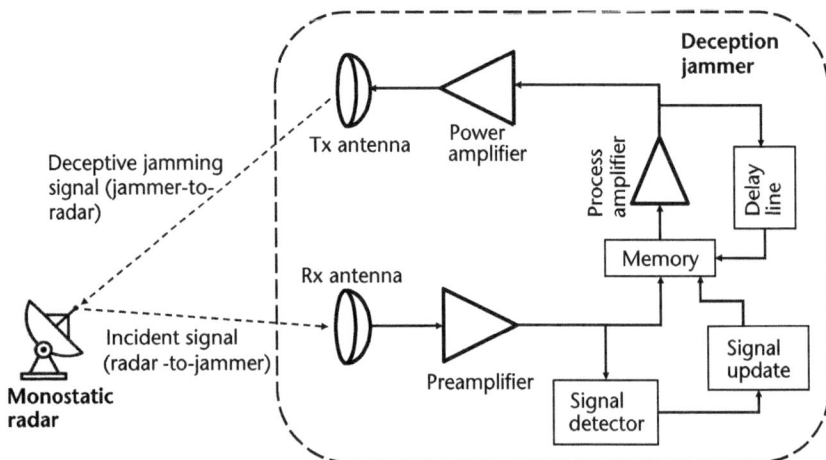

Figure 3.5 A general deceptive jamming system's block diagram.

Using *digital RF memory* (DRFM) reduces the time delay and enhances deception jammer effectiveness. Parallel processing also occurs during this period, which is called signal update. At this stage, the output of the preamplifier is checked to control the signal situation in the environment. If the signal related to the radar being jammed ceases to be received by the Rx, the memory section stops producing a jamming signal replica.

A deception jammer system employed in a self-protection role is designed to protect the mother platform from lethal radar-guided threats. Deceptive jamming systems must be programmed with detailed and exact signal parameters for each radar-guided threat to be effective.

Radar noise jamming can be employed with deception jamming techniques to maximize the impact of jamming on victim radars. The deception is mainly a coherent technique whose waveform is generated by a DRFM. Opening a separate bet and discussing DRFM, a critical usage in deceptive jamming, would be appropriate.

3.4.2.1   DRFM

The DRFM principle has been used since the mid-1970s. In general terms, a DRFM aims to capture different radar and modulation signals and copy these signals exactly. With appropriate delay and processing, DRFM can achieve a variety of jamming signals and jam multiple targets simultaneously. DRFM can deal with conventional noncoherent pulse radar and single-pulse radar signals, as well as pulse compression, pulse Doppler radar, and frequency-agile radar signals.

The essential task of a DRFM is to input an RF signal converted to a frequency low enough to be sampled by a high-speed *analog-to-digital converter* (ADC). The sampled signal is stored in a high-speed memory and can be retrieved and converted to the original signal using a *digital-to-analog converter* (DAC). A simplification of the DRFM function is to see it as a variable delay line for RF signals. DRFM is commonly used in many EW applications to generate false radar echoes. For example, by changing the delay in the transmission of pulses, the jammer can alter the range the radar detects and create false targets. There are two primary forms of DRFM architectures: wideband and narrowband structures. The wideband structure covers the whole band as determined by the DRFM bandwidth and stores signals occurring anywhere in the band. However, the multiple narrowband structure contains many narrowband DRFMs, each of which stores individual threat radar signatures. The wideband DRFM's structure is given in Figure 3.6.

The minimum system of DRFM is constructed with a downconversion module, a digital delay module, and an upconversion module. The incoming RF signal is downconverted into a general IF signal. Then a second adjustable LO is used for threat radar-related specific frequency levels in the operational band of DRFM. The ADC samples the IF signal and stores the sampled signal in memory. The controller is responsible for the memory's read-write control and sending the stored sample data to the DAC. The stored samples can be manipulated in amplitude, frequency, and phase to generate false target signals. The memory controller module adds the desired delay and Doppler shifts to simulate the false target range and velocity. The reconstructed digital signal is then upconverted back to RF and retransmitted. Then the output of the DAC is upconverted to the original RF in two stages to reconstruct

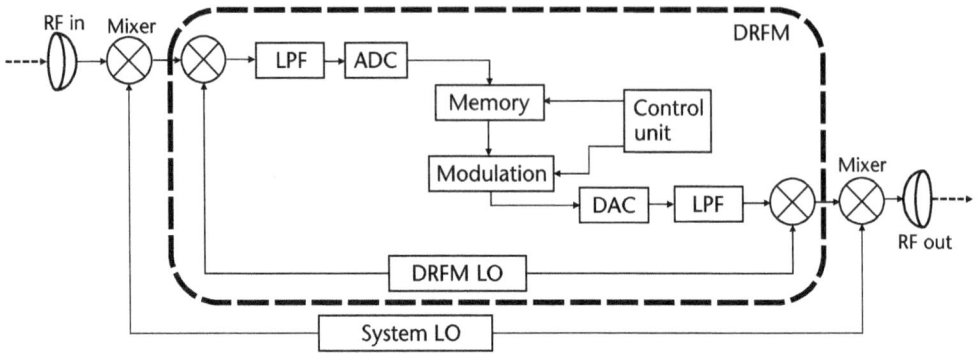

Figure 3.6 Block diagram of a generic wideband DRFM structure.

the original signal. Thus, coherent jamming is achieved when the same frequency is adopted in the downconversion and upconversion modules.

The wideband DRFM requires a state-of-the-art ADC that operates at high speed and instantaneously covers all the frequency bands. The narrowband DRFM must only cover the widest instantaneous radar bandwidth involved, about several hundred megahertz, and uses multiple DRFM structures containing relatively low-speed ADCs and memory devices. The multiple narrowband DRFM structure is given in Figure 3.7.

The functions of DRFM can be separated into range and velocity deception. A delay process achieves the range deception. The received radar signal is delayed, modulated, and amplified to distance deceive the pulse radar. The delay module is implemented by using a memory block. The received input signal is stored in the memory, and the time delays are periodically injected. This modified signal is amplified and retransmitted much stronger than the return signal. Increasing these time delays, the radar gate will detect an increase in the range and move off to a false target.

A Doppler process fulfills velocity deception. Doppler radars use the frequency difference between transmitted and received signals to determine the speed of the target aircraft. When the frequency of a received signal differs from the transmitted

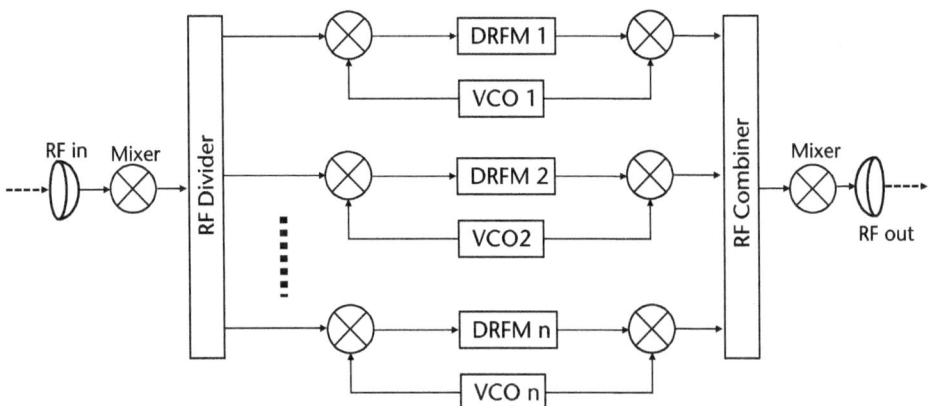

Figure 3.7 Block diagram of a generic narrowband DRFM structure.

signal, the Doppler effect will occur. The radar speed tracking system can easily capture the target based on the Doppler shift. A Doppler modulation process is conducted to deceive the radar regarding velocity.

Figure 3.8 shows the block diagram of the DRFM signal generation process and gives information about producing the range and velocity deception signals [6]. The block diagram represents both wideband and narrowband DRFM structures. If DRFM is narrowband, a *voltage-controlled oscillator* (VCO) is used. However, in the case of wideband DRFM, LO is used.

The signal transmitted from the radar and the resulting signal generated after t_1 at the DRFM input is defined as:

$$S_{Tx} = A\cos\left(2\pi f_0 t + \phi\right) \xrightarrow{t_1} S_{RFin} = A_1 \cos\left(2\pi f_0\left(t - t_1\right) + \phi\right) \qquad (3.3)$$

In the above equation, the transmitted signal magnitude A is much larger than the received magnitude A_1 ($A \gg A_1$) because of the one-way propagation loss. Furthermore, f_0 is the radar operating frequency, and ϕ is the phase component. The VCO/LO signal (S_{LO}) definition and the S_{LO} signal at the time DRFM receives the radar signal are defined as follows:

$$S_{LO} = \cos\left(2\pi f_{LO} t + \phi_1\right) \xrightarrow{t_1} S_{RFin} = \cos\left(2\pi f_{LO}\left(t - t_1\right) + \phi_1\right) \qquad (3.4)$$

where f_{LO} is the VCO/LO frequency, and the ϕ_1 is the phase component of the VCO/LO signal. In the input of the DRFM, the downconverted mixer-1 output is applied. The inputs of the mixer-1 are the RF input signal given in (3.3) and the VCO/LO signal given in (3.4). Thus, the downconverted signal obtained in the DRFM input is defined as:

$$S_{DRFMin} = A_1 \cos\left(2\pi\left(f_0 - f_{LO}\right)\left(t - t_1\right) + \left(\phi - \phi_1\right)\right) \qquad (3.5)$$

The store and forward time or delay time of DRFM is Δt, and the transmitted signal or RF output signal is defined as follows:

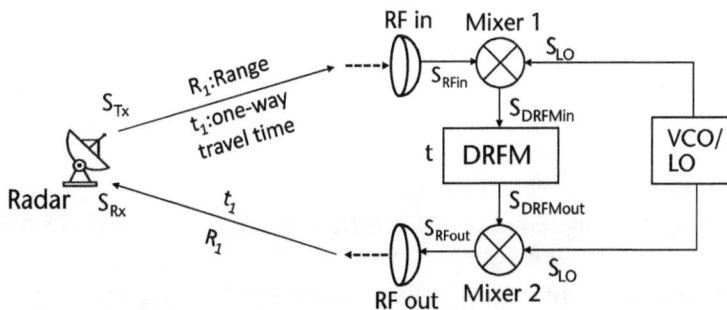

Figure 3.8 DRFM signal generation process block diagram.

$$S_{\text{DRFMin}} = A_2 \cos\left(2\pi f_0\left(t - t_1 - \Delta t\right) + \phi\right) \qquad (3.6)$$

Thus, the upconverted and downconverted frequencies remain the same, and the coherency between the input and output signals is provided. The DRFM systems may delay and amplify the signal to the required levels. So the signal magnitude A_1 is subject to a gain, and the output A_2 magnitude, which is higher than the input, is obtained. After obtaining the signal defined in (3.6), it is sent back to the radar, and the signal received by the radar is as follows:

$$S_{\text{Rx}} = A_3 \cos\left(2\pi f_0\left(t - 2t_1 - \Delta t\right) + \phi\right) \qquad (3.7)$$

In the above equation, an additive one-way propagation loss enters the process. Since the distance between the radar and the DRFM's mother platform does not change, the RF signal travel time will be t_1 again. The propagation losses will also drastically reduce the signal magnitude.

It is seen from (3.7) that the signal received by radar is the delayed transmit signal. If a Doppler shift is added, there will be a Doppler phase effect in the signal received by radar, but the impact between the adjacent pulses is the same, so the coherence is retained. A Doppler reference clock for synchronization generates a Doppler shift deception. Based on the system reference clock, a Doppler shift is added to the upconverted signal. Then the echo wave signal contains an additive Doppler shift with the true Doppler shift, which can confuse radar.

The *automatic gain control* (AGC) of the radar Rx changes its gain when detecting a cover pulse on the range gate. Thus, power level management is essential for DRFM systems. There must be a relationship between the received power signal and the transmit signal in such a way as to increase the efficiency of deceptive jamming, so the controllability of the power must be considered.

3.5 Radar Jamming Equations

The derivation of the radar jamming equations depends on the geometric position of the radar Rx and Tx, the jammer, and the protected platform relative to each other. Radar operation may be monostatic or bistatic, so the Rx and Tx may be at the same or different locations. However, the jammer may be on the protected platform, or the jammer and the protected platform are at various locations.

Also, the jamming may be noise or deceptive purposes. In any case, the power level that the jammer generates at the radar's Rx will be calculated similarly. However, the aims are different: the noise-jamming concept tries to increase the J/S at the radar's Rx, and the deceptive jamming intends to produce a similar effect with the target return echo at the radar's Rx.

The jamming equations for different situations are derived separately. In the derivations, obtaining J/S at the radars' Rx is aimed, so the calculations for noise and deceptive jamming equations are considered the same. However, there is a difference in the power management of the noise and deception-jamming purposes.

Different noise and deceptive jamming applications will be used for each scenario. Four different jamming scenarios use different equations in principle. All four cases are shown in Figure 3.9.

3.5.1 Case 1

In this case, the jammer is on the platform under threat, and the threat radar is a monostatic radar, which may be an E/W or command-guided SAM system. This case is shown in Figure 3.9(a). Since the jammer is on the target platform, the jammer is in the main lobe of the radar. Thus, jamming is applied from the main lobe. The radar's Rx and Tx are in the same location, and the range between the radar and target or the jammer, $R_{R\text{-}T}$, steers the calculation process.

For deriving the J/S, the target return signal strength (S) at the radar's Rx and the jamming power (J) at the radar's Rx are calculated separately. First, the target return signal at the radar's Rx is written by using (2.33) and adding the radar losses to the process as follows:

$$P_{Rx} = S = \frac{P_{Tx}G^2\sigma c^2}{(4\pi)^3 R^4 f^2 L_R}, \quad G_{Tx} = G_{Rx} = G \tag{3.8}$$

where P_{Tx} is the peak radar Tx power, G is the radar antenna gain for Tx and Rx operation, σ is the RCS of the target, c is the speed of light, R is the range between radar and the target, f is the operating frequency of the radar, and L_R is the total Rx and Tx radar losses.

The jammer signal in the radar's Rx is obtained by using a one-way radar equation given in (2.29):

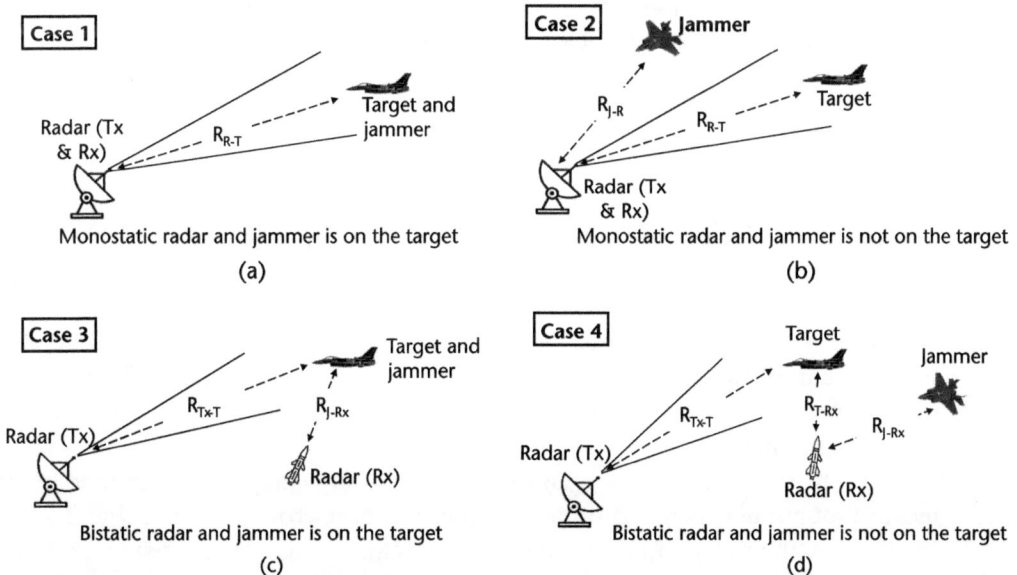

Figure 3.9 Different radar jamming scenarios.

$$P_{\text{Rx}(J)} = J = \frac{P_{\text{J}}G_{\text{Ja}}Gc^2B_{\text{R}}}{\left(4\pi Rf\right)^2 B_{\text{J}}L_{\text{J}}L_{\text{pol}}} \tag{3.9}$$

P_{J} is the jammer power, G_{Ja} is the jammer antenna gain, L_{J} is the jammer losses, and L_{pol} is the antenna polarization losses due to mismatching between jammer and radar antennas. B_{R} and B_{J} are bandwidths of the radar Rx and jammer, respectively. In modern jammer systems on the target platform, that is, self-protection jammer systems, the systems adjust $B_{\text{R}}/B_{\text{J}}$ to unity. Thus, $B_{\text{R}}/B_{\text{J}}$ is not considered in this case. Dividing (3.9) by (3.8), the J/S ratio is found as follows:

$$\frac{J}{S} = \frac{P_{\text{J}}G_{\text{Ja}}Gc^2\left(4\pi\right)^3 R^4 f^2}{\left(4\pi Rf\right)^2 P_{\text{Tx}}G^2\sigma c^2 L} = \frac{P_{\text{J}}G_{\text{Ja}}4\pi R^2}{P_{\text{Tx}}G\sigma L} \quad \left\{L = \frac{L_{\text{J}}L_{\text{pol}}}{L_{\text{R}}}\right\} \tag{3.10}$$

L is the total losses that enter the process, consisting of jammer losses (L_{J}), radar losses (L_{R}), and antenna polarization losses (L_{pol}). Note that the frequency of the J/S ratio obtained in (3.10) is not a part of the equation. The frequency does not affect the J/S ratio since the jammer frequency bandwidth in this operation scheme is expected to cover all the radar operating bandwidth with spot noise jamming. The J/S can be written in the logarithmic form as follows:

$$\left(\frac{J}{S}\right)_{\text{dB}} = \left(P_{\text{J}}\right)_{\text{dBm}} - \left(P_{\text{Tx}}\right)_{\text{dBm}} + 10\log G_{\text{Ja}} - 10\log G$$
$$-10\log\sigma + 10.99 + 20\log R - 10\log L \tag{3.11}$$

where $10\log_{10}(4\pi) = 10.99$. The jammer and radar Tx powers are used in dBm, defined in (3.11). For simplicity, after obtaining dBW by using standard logarithmic transformation of linear power value, adding 30 dB will also give dBm, such as

$$\left(P\right)_{\text{dBW}} = 10\log(P)\ \text{dBW} \rightarrow \left(P\right)_{\text{dBm}} = \left(P\right)_{\text{dBW}} + 30\ \text{dB} \tag{3.12}$$

Figure 3.10 shows the variation of the J/S ratio with the change in the distance between the target and the radar for Case 1 as a semi-logarithmic graph.

In Figure 3.10, J is the jamming signal power level in the radar Rx, and S is the power level in Rx of the radar signal reflected from the target. The slope of the S curve is 40 dB/decade, while the slope of the J line is 20 dB/decade. The intersection point of the J and S curves refers to the point on the radar's Rx where the jammer power equals the power of the target echo signal ($J = S$) and is called the crossover point.

The crossover point or range is the boundary of the ineffective jamming, which means that at radar-target distances farther from the crossover, jamming is active. In contrast, the radar can detect the target at closer radar-target distances despite the jamming signal. The crossover range (R_{CO}) can be calculated by substituting $J/S = 1$ in (3.10) as follows:

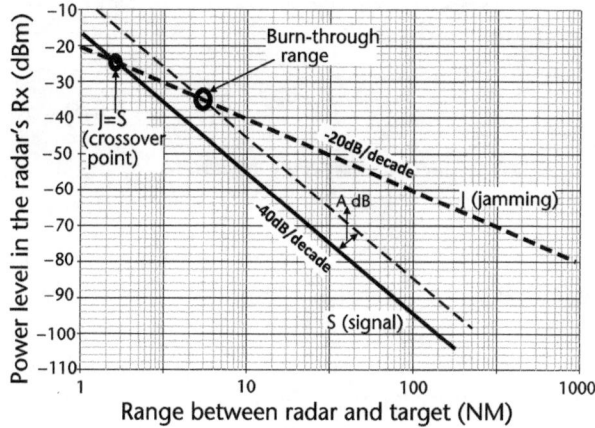

Figure 3.10   A generic J and S variance for the Case 1 jamming scenario.

$$\frac{J}{S} = 1 = \frac{P_J G_{Ja} 4\pi R_{CO}^2}{P_{Tx} G \sigma L} \rightarrow R_{CO} = \sqrt{\frac{P_{Tx} G \sigma L}{P_J G_{Ja} 4\pi}} \tag{3.13}$$

The crossover range equation can be written in a logarithmic form as:

$$20\log R_{CO} = \left(P_{Tx}\right)_{dBm} - \left(P_J\right)_{dBm} + 10\log G_{Tx} + 10\log \sigma$$
$$+ 10\log L - 10\log G_{Ja} - 10.99 \tag{3.14}$$

The crossover range in Figure 3.10, where $J = S$, can be regarded as the detection distance under jamming, but this is often not the case. For jamming to be effective, at $J/S > 0$ dB, the signal must not be at a level that can be distinguished from the noise floor level generated by the jamming. However, the situation may differ in radars where the Rx thermal noise level and noise figure level are deficient; in other words, the sensitivity is very high.

Although the jammer generates $J/S > 0$ dB at Rx, the signal can be discriminated from the increased noise floor due to jamming. The detection limit under jamming is determined not from the signal (S) curve but using the curve obtained by multiplying the S curve by a constant. This operation is defined as addition S (dBm) with a fixed number of dB in the logarithmic plane. Since the S curve is increased linearly due to this multiplication, the new curve is parallel to the S curve. This newly formed curve is shown with a dashed line in the figure, and the increment is A dB in logarithmic form.

Burn-through range (R_{BT}) is the distance the radar can detect the target under jamming. The burn-through range is shown in Figure 3.10. When viewed from the radar side, the minimum sufficient impact for detection under jamming, the minimum effect of jamming is calculated as follows:

$$\frac{J_{min\_eff}}{S} = 10^{\left(\frac{A}{10}\right)} \rightarrow R_{BT} = \sqrt{\frac{P_{Tx} G \sigma L J_{min\_eff}}{P_J G_{Ja} 4\pi S}} \tag{3.15}$$

In the above equation, $A = 10\log(J_{\text{min\_eff}}/S)$. Also, the logarithmic form of the above equation is written as

$$20\log R_{\text{BT}} = \left(P_{\text{Tx}}\right)_{\text{dBm}} - \left(P_{\text{J}}\right)_{\text{dBm}} + 10\log G + 10\log \sigma$$
$$+ 10\log L - 10\log G_{\text{Ja}} + A - 10.99 \tag{3.16}$$

Equations (3.10) and (3.11) can be used for noise and part of deceptive jamming. Now we can discuss deceptive jamming. In the general form, a deceptive jamming repeater has the structure of Figure 3.5. In this context, three power amplifiers are defined; however, it is not compulsory to use three separate amplifiers. Thus, a total jamming process amplifier might be described as if they are incorporated. In deception jammers, the jamming signal is not produced by the jammers. They capture the incident radar signal, amplify it, and retransmit it. However, the amplification is insufficient for closing ranges to radar, which are saturated. In this case, the jammer platform utilizes noise jamming rather than deception. The scheme is demonstrated in Figure 3.11.

For Case 1, high-power jamming, especially from long distances, creates a very high signal in radars compared to the reflection of their signals from the target. Experienced radar operators can understand that these high signals are jamming signals. Thus, the jamming activity loses importance as the false target is exposed. These explanations apply to deception jamming but not directly to noise jamming. The jamming power created in the radar Rx of the linear and saturated jamming methods is used by jammer systems on the targets for deceptive jamming. The power levels for the target reflected signal and the deceptive jamming signal at the radar Rx are shown in Figure 3.12 as a semi-logarithmic graph. Jammer systems using linear-constant gain and saturated power jamming processes are called repeater jammers.

As seen in Figure 3.12, the linear constant gain region of repeater jamming occurs only at far distances from the radar. The saturated power region is rapidly reached as the target approaches the radar.

If the jammer system on the target platform sends the received radar signal back by increasing the power level, both J and S are subject to two-way losses. The signal is not newly produced by the jammer but a radar signal amplified by the linear

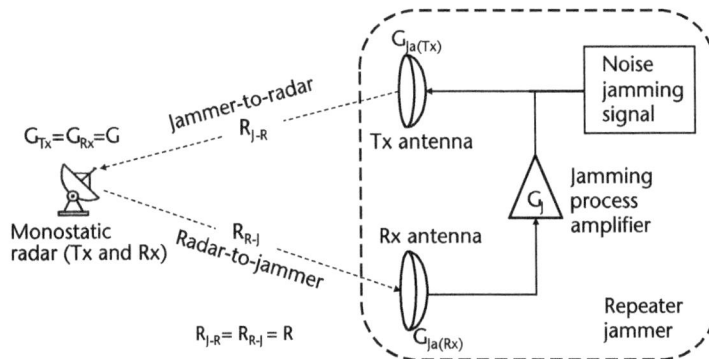

Figure 3.11 Deceptive jamming usage with noise jamming for the Case 1 jamming scenario.

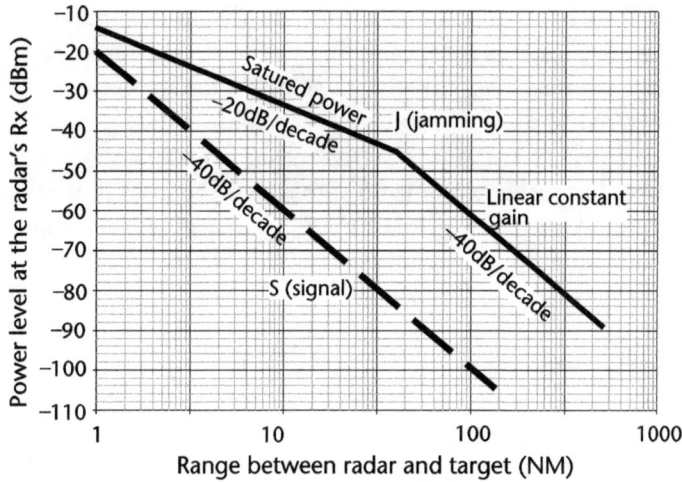

Figure 3.12   A generic *J* and *S* variance for repeater jamming.

constant gain. Figure 3.12 shows that the variation of the reflected echo and the linear constant gain jamming signal powers have a −40-dB/decade slope.

The saturated power *J/S* is calculated as in (3.10). For the linear constant gain part of Figure 3.12, *J/S* can be calculated by considering Figure 3.11. The signal (*S*) is calculated as in (3.8), and the jamming equation is obtained as follows:

$$J = \frac{P_{Tx}GG_{Ja(Rx)}c^2}{\left(4\pi R_{R\text{-}J}f\right)^2} \frac{G_J G_{Ja(Tx)}G\,c^2}{\left(4\pi R_{J\text{-}R}f\right)^2 L_J L_{pol}} \rightarrow \left\{R_{R\text{-}J} = R_{J\text{-}R} = R\right\} \tag{3.17}$$

In the above equation, *G* is the radar antenna gain in Rx and Tx. G_J is the jammer processing gain, and $G_{Ja(Rx)}$ and $G_{Ja(Tx)}$ are the jammer antenna gains in Rx and Tx, respectively. Since the jammer is on the target platform, the distance between the radar and jammer will not change in receiving and transmitting. The *J/S* can be written as follows:

$$\frac{J}{S} = \frac{G_{Ja(Rx)}G_J G_{Ja(Tx)}c^2}{4\pi\sigma f^2 L} , \quad \left\{L = \frac{L_J L_{pol}}{L_R}\right\} \tag{3.18}$$

where *L* is the total losses that enter the process, consisting of jammer losses (L_J), radar losses (L_R), and antenna polarization losses (L_{pol}).

3.5.2   Case 2

In this case, the jammer is not on the platform under threat, and the threat radar is a monostatic radar, which may be an E/W or command-guided SAM system. This case is shown in Figure 3.9(b). Since the jammer is on a platform other than the target platform, the jammer may not be in the main lobe of the radar. The radar's

Rx and Tx are in the same location, but the target and the jammer are at different locations. For this case, the ranges between the radar and target (R_{R-T}) and the radar and jammer (R_{R-J}) involve the calculation process.

For deriving the J/S, two different situations regarding jamming direction are considered. These situations are jamming from the main lobe and jamming from a sidelobe. First, the target echo signal and the jamming power at the radar's Rx are calculated for main lobe jamming. The target return signal at the radar's Rx is obtained by adapting the equation in (3.8) for the situation,

$$S = \frac{P_{Tx}G_{Tx}G_{Rx}\sigma c^2}{(4\pi)^3 R_{R-T}^4 f^2 L_R} \tag{3.19}$$

G_{Tx} and G_{Rx} are the radar antenna gain for Tx and Rx, respectively. The jammer signal in the radar's Rx is

$$J = \frac{P_J G_{Ja} G_{Rx} c^2 B_R}{\left(4\pi R_{J-R} f\right)^2 L_J L_{pol} B_J} \tag{3.20}$$

P_J is the jammer power, and G_{Ja} is the jammer antenna gain. B_R and B_J are bandwidths of radar Rx and jammer, respectively. Dividing (3.20) by (3.19), the J/S ratio is found as follows,

$$\frac{J}{S} = \frac{P_J G_{Ja} 4\pi R_{R-T}^4 B_R}{P_{Tx} G_{Tx} \sigma R_{J-R}^2 B_J L}, \quad \left\{ L = \frac{L_J L_{pol}}{L_R} \right\} \tag{3.21}$$

The logarithmic form of (3.21) is obtained as follows,

$$\left(\frac{J}{S}\right)_{dB} = \left(P_J\right)_{dBm} - \left(P_{Tx}\right)_{dBm} - BF_{dB} + 10\log G_{Ja} - 10\log G_{Tx} + 10.99 - \dots$$
$$\dots - 10\log\sigma - 10\log L + 40\log R_{R-T} - 20\log R_{J-R} \tag{3.22}$$

In the above equation, BF is the bandwidth reduction factor [7] and is defined as

$$BF_{dB} = 10\log\left[\frac{B_J}{B_R}\right] \rightarrow P_{J-N} = P_J - BF_{dB} \quad \left\{B_J \geq B_R\right\} \tag{3.23}$$

where P_{J-N} is the usable jammer power in dBm. The crossover range can be calculated by substituting $J/S = 1$ in (3.21) as follows:

$$\frac{J}{S} = 1 = \frac{P_J G_{Ja} 4\pi \left(R_{R-T}\right)_{CO}^4 B_R}{P_{Tx} G_{Tx} \sigma R_{J-R}^2 B_J L} \rightarrow \left(R_{R-T}\right)_{CO} = \left(\frac{P_{Tx} G_{Tx} \sigma R_{J-R}^2 L B_J}{P_J G_{Ja} 4\pi B_R}\right)^{\frac{1}{4}} \tag{3.24}$$

For the burn-through range calculation, J/S is substituted with $J_{\text{min\_eff}}/S$ in (3.24), as follows:

$$\left(R_{R-T}\right)_{\text{BT}} = \left(\frac{P_{\text{Tx}}G_{\text{Tx}}\sigma R_{\text{J-R}}^2 LB_J J_{\text{min\_eff}}}{P_J G_{\text{Ja}} 4\pi B_R S}\right)^{\frac{1}{4}} \tag{3.25}$$

For the sidelobe jamming, the target return signal is calculated as given in (3.19), and the jamming power at the radar's Rx is calculated as follows:

$$J = \frac{P_J G_{\text{Ja}} G_{\text{Rx(SL)}} c^2 B_R}{\left(4\pi R_{\text{J-R}}f\right)^2 L_J L_{\text{pol}} B_J} \tag{3.26}$$

In (3.26), $G_{\text{Rx(SL)}}$ is the radar Rx antenna gain for the side lobe in the jammer direction. The directed antennas have different sidelobes. The side lobes align in pairs according to boresight, and the related side lobe, according to the jammer direction, is calculated. Dividing (3.26) by (3.19) finds the J/S ratio for sidelobe jamming.

$$\frac{J}{S} = \frac{P_J G_{\text{Ja}} G_{\text{Rx(SL)}} 4\pi R_{\text{R-T}}^4 B_R}{P_{\text{Tx}}\ G_{\text{Tx}} G_{\text{Rx}} \sigma R_{\text{J-R}}^2 LB_J}, \quad \left\{L = \frac{L_J L_{\text{pol}}}{L_R}\right\} \tag{3.27}$$

In (3.27), G_{Rx} shows the main lobe gain. Since the gain of an antenna is the main lobe or the maximum gain, we do not add any additive indices to represent an antenna's main lobe gain. The logarithmic form of (3.27) is written as follows:

$$\begin{aligned}
\left(\frac{J}{S}\right)_{\text{dB}} = &\left(P_J\right)_{\text{dB}} - \left(P_{\text{Tx}}\right)_{\text{dB}} - BF_{\text{dB}} + 10\log G_{\text{Ja}} + 10\log G_{\text{Rx(SL)}} \\
&- 10\log G_{\text{Tx}} - 10\log G_{\text{Rx}} - 10\log\sigma + 10.99 \\
&- 10\log L + 40\log R_{\text{R-T}} - 20\log R_{\text{J-R}}
\end{aligned} \tag{3.28}$$

The crossover range can be calculated for side lobe jamming by substituting $J/S = 1$ in (3.27) as follows:

$$\left(R_{\text{R-T}}\right)_{\text{CO}} = \left(\frac{P_{\text{Tx}}G_{\text{Tx}}G_{\text{Rx}}\sigma R_{\text{J-R}}^2 LB_J}{P_J G_{\text{Ja}} G_{\text{Rx(SL)}} 4\pi B_R}\right)^{\frac{1}{4}} \tag{3.29}$$

For the burn-through range calculation, J/S is substituted with $J_{\text{min\_eff}}/S$ in (3.29), and the following is obtained:

$$\left(R_{\text{R-T}}\right)_{\text{BT}} = \left(\frac{P_{\text{Tx}}G_{\text{Tx}}G_{\text{Rx}}\sigma R_{\text{J-R}}^2 LB_J J_{\text{min\_eff}}}{P_J G_{\text{Ja}} G_{\text{Rx(SL)}} 4\pi B_R S}\right)^{\frac{1}{4}} \tag{3.30}$$

3.5.3 Case 3

In this case, the jammer is on the platform under threat, and the threat radar is a bistatic radar, which may be a semiactive guided SAM or AAM system. This case is shown in Figure 3.9(c). The radar's Tx and Rx are in different locations; thus, the distances between the radar's Tx and the target (or jammer), $R_{Tx\text{-}T}$, and the radar's Rx and the target, $R_{J\text{-}Rx}$, are different. Since the jammer is on the target platform, the jammer will be in the main lobe of the radar's Rx.

For deriving the J/S, the target return signal strength (S) at the radar's Rx is given as:

$$S = \frac{P_{Tx} G_{Tx} G_{Rx} \sigma c^2}{(4\pi)^3 R_{Tx\text{-}T}^2 R_{J-Rx}^2 f^2 L_R} \tag{3.31}$$

and the jamming power at the radar's Rx is calculated as follows:

$$J = \frac{P_J G_{Ja} G_{Rx} c^2 B_R}{\left(4\pi R_{J-Rx} f\right)^2 B_J L_J L_{pol}} \xrightarrow{B_R/B_J=1} J = \frac{P_J G_{Ja} G_{Rx} c^2}{\left(4\pi R_{J-Rx} f\right)^2 L_J L_{pol}} \tag{3.32}$$

In Case 3, jammer systems are on the target platform, as in Case 1, the systems adjust B_R/B_J to unity. Thus, the J/S ratio is found as:

$$\frac{J}{S} = \frac{P_J G_{Ja} 4\pi R_{Tx-T}^2}{P_{Tx} G_{Tx} \sigma L}, \quad \left\{ L = \frac{L_J L_{pol}}{L_R} \right\} \tag{3.33}$$

The RCS (σ) value is considered for the Rx boresight direction. The logarithmic form of the J/S ratio is written as follows:

$$\left(\frac{J}{S}\right)_{dB} = \left(P_J\right)_{dBm} - \left(P_{Tx}\right)_{dBm} + 10\log G_{Ja} - 10\log G_{Tx} + \dots$$
$$\dots 10.99 - 10\log L + 20\log R_{Tx-T} - 10\log\sigma \tag{3.34}$$

Figure 3.13 shows the variation of the J/S ratio with the change in the distance between the target (or jammer) and the bistatic radar Tx for Case 3 as a semilogarithmic graph.

Semiactive radar-guided threat systems with Tx on the launch platform and Rx on the missile are suitable examples for bistatic radar applications. Semiactive radar-guided missiles approach the target much faster than the launch platform comes close to the target. For this reason, the target's distance to Tx varies much less than the distance between the missile and the target. Figure 3.13 shows the radar's Rx power due to the jamming and the echo signal from the target as a function of the distance between the radar Tx and the target.

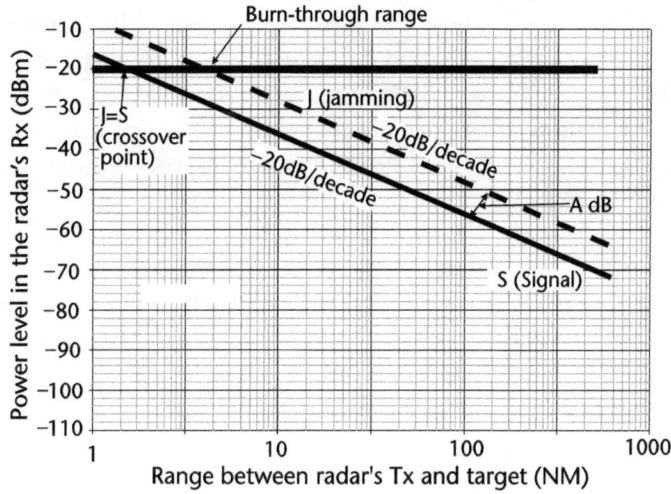

Figure 3.13 A generic J and S variance for Case 3 jamming scenario.

The diagram in Figure 3.13 for bistatic radar jamming differs from Figure 3.10 for monostatic radar jamming. This difference shows itself, especially on slopes. The slope of the echo signal curve changed from −40 dB/decade to −20 dB/decade, and the jamming curve changed from −20 dB/decade to a constant value. Also, the target's RCS is the same for the monostatic radar situation, even though the Tx and Rx antennae are different since they are in the same location. For the bistatic radar case, the Tx and Rx antennas are in other areas; thus, their aspect angles towards the target will probably change, and the RCS gets different values for each. However, for Case 3, given in Figure 3.9(c), the RCS can get the same value when the boresights of the missile and the launch platform face the target at the same angle.

The crossover range can be calculated in Case 3 by substituting $J/S = 1$ in (3.33). The interesting point for this case is that the crossover range calculation is made for the range between the radar Tx and the target (or jammer) as follows:

$$\frac{J}{S} = 1 = \frac{P_J G_{Ja} 4\pi \left(R_{Tx-T}\right)_{CO}^2}{P_{Tx} G_{Tx} \sigma L} \rightarrow \left(R_{Tx-T}\right)_{CO} = \sqrt{\frac{P_{Tx} G_{Tx} \sigma L}{P_J G_{Ja} 4\pi}} \tag{3.35}$$

For burn-through range calculation, J/S is substituted with J_{mineff}/S in (3.35), as follows:

$$\left(R_{Tx-T}\right)_{BT} = \sqrt{\frac{P_{Tx} G_{Tx} \sigma L J_{min\_eff}}{P_J G_{Ja} 4\pi S}} \tag{3.36}$$

3.5.4 Case 4

In this case, the jammer is not on the platform under threat, and the threat radar is bistatic. Case 4 is demonstrated in Figure 3.9(d). The radar's Tx and Rx are in different locations. Thus, the distances between the radar's Tx and the target, $R_{\text{Tx-T}}$, and the radar's Rx and the target, $R_{\text{T-Rx}}$, are different. For this case, which is different from Case 3, the jammer is not on the target and is in a different location. The distance between the jammer and the radar Rx is defined as $R_{\text{J-Rx}}$. So the jammer may or may not be in the main lobe of the radar's Rx, according to the geometric sequence.

For deriving the J/S, two different situations regarding jamming direction are considered. These situations are jamming from the main lobe and jamming from a side lobe. First, the target echo signal and the jamming power at the radar's Rx are calculated for main lobe jamming. The target return signal at the radar's Rx is obtained by using the equation in (3.31), and the jamming power at the radar's Rx is calculated as follows:

$$J = \frac{P_J G_{\text{Ja}} G_{\text{Rx}} c^2 B_{\text{R}}}{\left(4\pi R_{J-\text{Rx}} f\right)^2 L_J L_{\text{pol}} B_J} \tag{3.37}$$

Dividing (3.37) by (3.31), the J/S ratio for main lobe jamming is found as follows:

$$\frac{J}{S} = \frac{P_J G_{\text{Ja}} 4\pi R_{\text{Tx}-T}^2 B_{\text{R}}}{P_{\text{Tx}} G_{\text{Tx}} \sigma L B_J}, \quad \left\{ L = \frac{L_J L_{\text{pol}}}{L_{\text{R}}} \right\} \tag{3.38}$$

The logarithmic form of (3.38) is as follows:

$$\left(\frac{J}{S}\right)_{\text{dB}} = \left(P_J\right)_{\text{dB}} - \left(P_{\text{Tx}}\right)_{\text{dB}} - BF_{\text{dB}} + 10\log G_{\text{Ja}} - 10\log G_{\text{Tx}} + \dots$$
$$\dots 10.99 - 10\log \sigma - 10\log L + 20\log R_{\text{Tx}-T} \tag{3.39}$$

The crossover range can be calculated for main lobe jamming by substituting $J/S = 1$ in (3.38),

$$\left(R_{\text{Tx}-T}\right)_{\text{CO}} = \sqrt{\frac{P_{\text{Tx}} G_{\text{Tx}} \sigma L B_J}{P_J G_{\text{Ja}} 4\pi B_{\text{R}}}} \tag{3.40}$$

and for the burn-through range, J/S is replaced with J_{mineff}/S in (3.38) as follows:

$$\left(R_{\text{Tx}-T}\right)_{\text{BT}} = \sqrt{\frac{P_{\text{Tx}} G_{\text{Tx}} \sigma L B_J J_{\text{min\_eff}}}{P_J G_{\text{Ja}} 4\pi B_{\text{R}} S}} \tag{3.41}$$

Finally, the target echo signal and the jamming power at the radar's Rx are calculated for sidelobe jamming. The target echo signal at the radar's Rx is the same as the main lobe jamming case and uses the equation in (3.31). The jamming power at the radar's Rx is

$$J = \frac{P_J G_{Ja} G_{Rx(SL)} c^2 B_R}{\left(4\pi R_{J-Rx} f\right)^2 L_J L_{pol} B_J} \tag{3.42}$$

As stated, $G_{Rx(SL)}$ is the radar Rx antenna gain for the sidelobe in the jammer direction. Dividing (3.42) by (3.31), the J/S ratio is found for sidelobe jamming.

$$\frac{J}{S} = \frac{P_J G_{Ja} G_{Rx(SL)} 4\pi R_{Tx-T}^2 B_R}{P_{Tx} G_{Tx} G_{Rx} \sigma L B_J}, \quad \left\{ L = \frac{L_J L_{pol}}{L_R} \right\} \tag{3.43}$$

The logarithmic form of (3.43) is written as

$$\left(\frac{J}{S}\right)_{dB} = \left(P_J\right)_{dB} - \left(P_{Tx}\right)_{dB} - BF_{dB} + 10\log G_{Ja} + 10\log G_{Rx(SL)} - 10\log G_{Tx} - \ldots$$
$$\ldots 10\log G_{Rx} + 10.99 - 10\log \sigma - 10\log L + 20\log R_{Tx-T}$$

$$\tag{3.44}$$

The crossover and the burn-through ranges for sidelobe jamming are defined as follows:

$$\left(R_{Tx-T}\right)_{CO} = \sqrt{\frac{P_{Tx} G_{Tx} G_{Rx} \sigma L B_J}{P_J G_{Ja} G_{Rx(SL)} 4\pi B_R}}$$
$$\left(R_{Tx-T}\right)_{BT} = \sqrt{\frac{P_{Tx} G_{Tx} G_{Rx} \sigma L B_J J_{min\_eff}}{P_J G_{Ja} G_{Rx(SL)} 4\pi B_R S}} \tag{3.45}$$

3.6 Jamming Tactics

Jamming actions may be classified mainly as self-protection and support jamming. In self-protection jamming, the SPJ is on the mother platform, which is under threat. However, in support jamming, the jammer and the protected platforms are different. *Stand-in-jammer* (SIJ), SOJ, and EJ are the components of support jamming.

However, classifying the jamming tactics according to geometry is more common. In this context, they are classified into three main groups: self-protection, stand-off, and stand-in jamming. Since the jamming geometry is similar for SPJ and EJ, they are considered under SPJ. The jamming tactics are inspected in the following topics.

3.6.1 Self-Protection Jamming

In practical applications, deception jamming is generally used in SPJ systems, and noise jamming is used in *support jammer* (SJ) systems. As the name suggests, SPJ is a jamming system that protects the mother platform. SPJ systems may also be called self-screening, defensive EA, or deception ECM.

SPJ systems aim to protect the mother platform from the destructive effects of RF-guided threat systems. These platforms include fighters, helicopters, transport aircraft, UAVs, and vessels. As stated in Chapter 1, SPJ is a sub-unit of SPEW systems or suites. These systems include RWR, SPJ, CMDS, MWR, IRCM, LWR, and IR suppressor. The RF parts of the SPEW systems are the RWR, SPJ, and chaff dispenser of the CMDS. Also, for IR-guided and passive radar guidance systems, MWR systems are used as the only detection device. With this aspect, MWS is a common subsystem for both RF and IR sections. Utilizing the RF subsystems together in the coordination of a central computer protects the mother platform from radar-guided missiles.

While the SPEW system protects the mother platform from radar-guided missiles, it uses the SPJ system, coordinates the chaff dispenser and determines appropriate maneuvering directions in a coordinated and protection-enhancing manner. While the system automatically performs some operations, such as jamming and chaff dispensing, it does not complete the maneuvering process automatically and leaves the decision to the pilot or operator. SPJ uses the EID information from RWR and the threat-appropriate jamming techniques in its library while performing jamming activities.

The equations obtained in Cases 1 and 3, given in Section 3.5, are the mathematical models that can be used for self-protection jamming. The equations can be utilized both for deception and noise jamming. Case 1 represents command-guided systems; however, Case 3 is for semiactive guided threat systems. These equations can also be applied to EJ systems. Generally, SPJ systems use deceptive jamming, and EJ prefers noise jamming. The threat platforms may be land-based, airborne, or shipborne. The conceptual usage of airborne SPJ and EJ systems against different threat platforms is shown in Figure 3.14.

As seen in Figure 3.14, SPJ systems protect mother platforms against different radar-guided systems. Indeed, the threats are not limited to the guided systems given in the figure. SPJ systems may be used against any radar guidance method in case of being detected. SPJ can attack the RF part of the threat radar Rx by producing false targets or screening the platform echo. Furthermore, it may attack the signal processing part of the Rx by getting away from the range or velocity gates using different methods. In older-generation SPJ systems, the jammer beams are wide and cover the semiactive guided systems' missiles. The beamwidth can be adjusted, especially for the new-generation SPEW systems using phased array antennas. Thus, the missiles of the semiactive guidance threats may remain out of the jammer beam. For this reason, obtaining reliable intelligence about the threat systems is very important for the success of the jamming process.

However, EJ systems have more capable jammer and RWR systems, but they do not carry ammunition or missiles. They can move with the air mission package and protect them from radar-guided threats. The protection may be for multiple

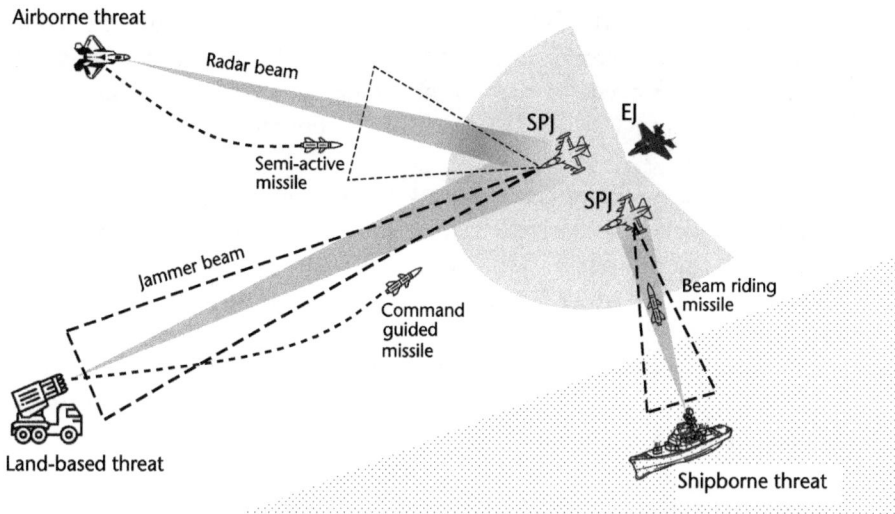

Figure 3.14 Conceptual usage of airborne SPJ and EJ systems.

threats, such as in Figure 3.14, or against a single threat. When EJ is used against various threats, noise jamming is probable. However, noise and deceptive jamming are possible when it is used against a single threat. It can be concluded that both SPJ and EJ apply to jam from the main beam of the threat radar systems.

3.6.2 Stand-Off Jamming

The stand-off jamming units remain outside the effective range of enemy weapons or lethal borders and are an element of SEAD operation. A SEAD operation aims to neutralize, destroy, or temporarily degrade surface-based enemy air defenses by destructive or disruptive means [8]. The SOJ systems are a part of EW's support of SEAD. They provide screening for the air strike, attack, and assault units that penetrate enemy defenses. Usually, the SOJ systems jam radar systems and may jam the C2 of the *enemy air defense system* (EADS).

Figure 3.15 shows the SOJ system protecting a task group against an EADS system modeled by a SAM radar system with five *early warning* (E/W) radars and three radar-guided missile firing capabilities. E/W and SAM systems within the EADS structure are connected to a C3 center with a data link. SOJ protection is not conducted solely; it may be supported by additive hard-kill and soft-kill countermeasures. Hard-kill methods may be ARM or *high-speed antiradiation missiles* (HARM), and soft-kill methods use EW countermeasures in other platforms. Thus, SOJ is a part of SEAD operation, and, in this operational principle, the aircraft in the air strike task group have their own SPJ. Furthermore, an EJ may protect the task group as well.

The SOJ system performs its task without entering the enemy missile range. The border, which is the ability of the EADS to destroy the enemy air strike or attack pack physically, is called the lethal border. In this case, the SOJ system operates outside the destructive border of the EADS. However, the offensive or primary task group protected by the SOJ system performs its duties within the lethal boundary of the enemy air defense. The disadvantage of the SOJ geometry is that crossover or burn-through occurs earlier on the attack units because the jammer must remain at a very long range while the attack units approach the enemy defense assets to a concise range.

The equations obtained in Cases 2 and 4, given in Section 3.5, are the mathematical models that can be used for stand-off jamming. Case 2 represents E/W and command-guided systems. However, Case 3 is for semiactive guided threat systems. According to the situation, SOJ systems generally use noise jamming, such as spot, barrage, or sweep-spot. Also, the jamming may be from the main beam or side lobes.

The SOJ system affects the different types of radars that make up the EADS. EADS is an integrated system with elements that can fully control the area they protect. These elements are listed as follows:

- E/W radars systems;
- SAM systems;
- Different types of radars, such as airborne E/W radars, coastal surveillance radars, meteorological radars;
- C3 center;
- C3 data transmission systems provide data and communication flow between radars and the C3 center.

The detection range of E/W radars is longer than the search and tracking mode of the SAM system radar, so the coverage areas of these radars form the EADS detection border. It enables radar systems to create a typical coverage area by data transfer within the C3 structure. Thus, targets detected by any radar in the EADS are also seen by all other radars. In real applications, the regional air picture is created in the C3 center, and for this purpose, all radars transfer the data that they have obtained to this center.

The SAM system radar operates in two or three different modes. They change the specifications in each mode, such as frequency, PRF, peak power, and modulation. One is the search mode, and the other is the tracking and launch modes. In some systems, these modes are combined or separate. The operational ranges of the tracking and launch modes of the SAM radars define the lethal border of EADS.

The SOJ system jamming the E/W radars suppresses the EADS's detection range, and the detection boundary is withdrawn into the interior of the enemy's defense area. This situation is illustrated in Figure 3.15. However, the SOJ system's jamming of the SAM radar prevents the air strike task group from detecting the SAM radars at the regular detection distance. As a result, the SOJ system aims to pull back the track initiation distance, which causes the tracking accuracy to deteriorate.

3.6.3 Stand-In Jamming

Another EW that supports the SEAD operation unit is SIJ. The SIJ systems are used against EADS radars that cannot be suppressed by soft-kill means, which are SOJ, EJ, and SPJ units, and hard-kill means, which are ARM and HARM units, in an integrated SEAD operation.

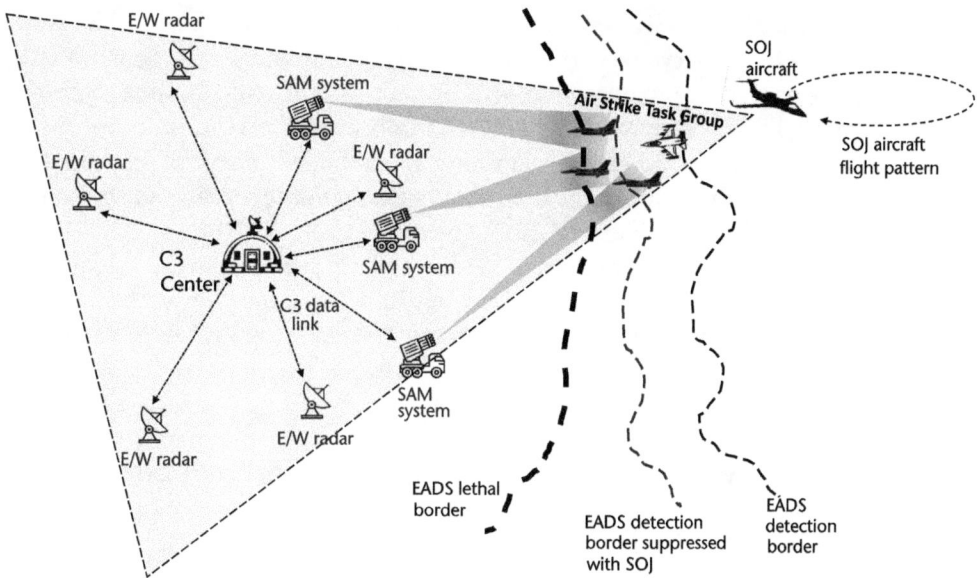

Figure 3.15 Conceptual usage of SOJ system.

The number of radars that need to be suppressed using the SIJ is usually tiny, and a separate SIJ system is planned for each of these radars. The required power level is very low since the SIJ applies at a close range. Furthermore, the aircraft's RCS must be small to pass the lethal border of the EADS and approach the target radar. For this reason, SIJ systems are generally used with UAV platforms.

SIJ can be used against E/W and SAM systems for different purposes. In the planning phase of a SEAD operation, using hard-kill and soft-kill units and the path planning of the primary mission pack is considered to accomplish the task with minimum losses. Most of the EADS units are suppressed with SOJ systems, and the remaining units are planned individually. Appropriate hard-kill methods, such as ARM or HARM, and soft-kill methods, such as SIJ, are dedicated to these individual radar systems.

The EADS's detection boundary can only be withdrawn if all the E/W radar systems are jammed effectively. When the SOJ aircraft cannot jam some E/W radars, they are suppressed with SIJ systems to reduce the detection range of the EADS. The related E/W radars are neutralized with SIJ systems to reduce the detection range of the EADS. For this purpose, the SIJ system aims to jam the victim E/W radar from the main beam. However, this action requires preplanning for the threat radar location and probable operational path of the offensive ally pack. Since E/W radars continuously scan the area, SIJ systems must be placed appropriately to jam the radars from the main beams. For this purpose, SIJ systems are located between the threat radars and the mission aircraft.

If a SAM system has a locational advantage in being jammed by EJ and SPJ systems or is out of the impact of the SOJ aircraft, this situation must be determined in the planning phase. Thus, a complete SEAD operation requires detailed and exact planning. In this case, using an ARM or HARM to destroy the SAM site or utilizing an SIJ to degrade it temporarily should be added to the EW plans. If using soft kill is selected, an SIJ system will be dedicated to jamming the SAM system. This action will not serve to withdraw the detection boundary of the EADS for general SEAD operation purposes but will protect the primary mission pack from local threats. The SIJ system jams the SAM radar from the same sector of the protected primary task group. The SIJ jamming is conducted from the victim SAM radar's main beam. Thus, in both cases, SIJ systems jam the victim radars from the main beams, as shown in Figure 3.16.

The mathematical models used for stand-off jamming are also valid for SIJ. The differences are not in the equations but in the distance between the victim radar and the jammer, and the jamming is accomplished from the main beam rather than the side lobes.

3.7 Jamming Methods of Radar-Guided Missile Systems

Chapter 2 defined radar guidance techniques, such as active, semiactive, command, and passive. We will not mention these methods here again. This part describes jamming methods for radar-guided missiles and related mathematical models.

3.7.1 Jamming Methods for Active Guidance

Active guided systems have their own Tx and Rx and operate like a monostatic radar system. When a target is defined as an active guided missile, it pursues the aim without requiring any outer sensor or RF emission. They can be launched from different platforms against various targets, such as ASM, SAM, or AAM.

The jamming methods against active-guided missiles are analyzed in two different cases. The first one, and the common one for operational purposes, is the jamming system on the target platform or self-jamming, and the other case is support

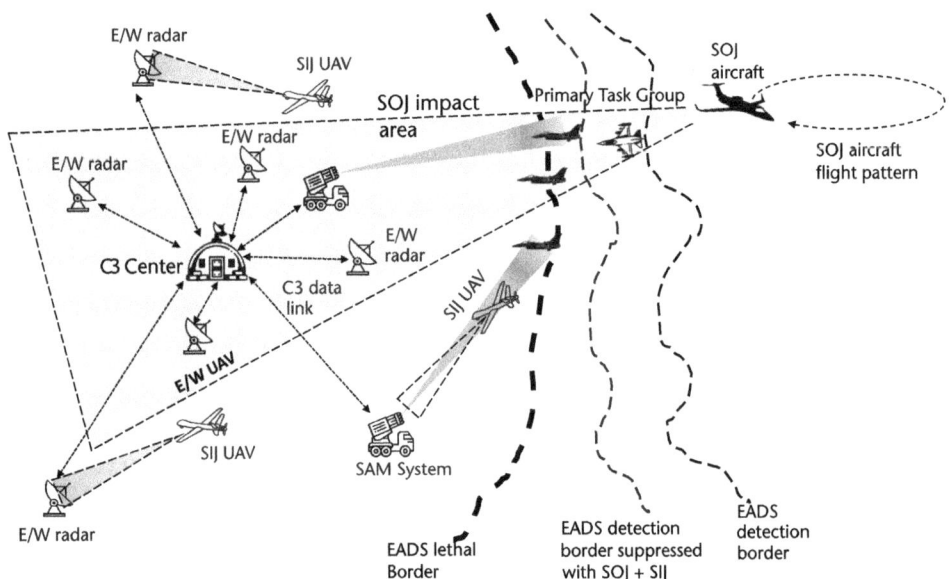

Figure 3.16 Conceptual usage of SIJ systems with a SOJ system.

jamming. Support jamming may be in the forms of escort jamming or stand-off jamming for jamming active-guided systems. The escort jamming geometry is similar to the SPJ when close to the protected aircraft. However, when the EJ systems are not close to the target platforms, the geometry will converge to SOJ. When SOJ systems protect the target platforms, the geometry differs from self-protection jamming. In this case, SOJ systems stay out of the lethal border, and active-guided missiles continuously approach the target. Since the distance between the missile and target gets smaller step by step, while SOJ-to-target distance is constant, this case generally results in ineffective jamming.

The active-guided missiles emit RF and can be detected by the victim platform. An active-guided missile can be jammed like a monostatic radar jamming from the main beam when SPJ systems are used. The model given in Case 1 of Section 3.5 is used for self-protection jamming of active-guided missiles. The J/S can be calculated as in (3.10); the only difference for active-guided missile jamming is the range between the missile and target ($R_{M\text{-}T}$) as follows:

$$\frac{J}{S} = \frac{P_J G_{Ja} 4\pi R^2_{M\text{-}T}}{P_{Tx} G \sigma} \tag{3.46}$$

Generally, J/S is defined by dB values, using the equation $10\log(J/S)$ dB, rather than the ratio of linear numbers. For close flights of EJ with the protected primary mission pack, J/S is calculated by using (3.46). The equation gives the jamming efficiency for noise jamming and provides vital information for deceptive jamming. The J/S is used to obtain the maximum benefit and to take the deceptive signal effect in the threat radar Rx. Getting high J/S rates by deceptive jamming is better than noise jamming because an experienced operator can readily determine the jamming for manual threat systems. However, detecting the occurrence of jamming is also possible for automatic systems. When deceptive jamming is revealed, the effort for it is meaningless. For this purpose, in deceptive jamming, obtaining a 0 dB or close to 0 dB for J/S should be the aim for reaching deception purposes.

When SOJ is used against an active-guided missile, this time, using the model given in Case 2 of Section 3.5 is appropriate for J/S calculations. Furthermore, the same mathematical models with SOJ suit EJ's far-distance flight from the protected aircraft. The J/S for jamming an active-guided missile by an SOJ system can be calculated with (3.21), the equation derived for monostatic radars. The difference for active-guided missile jamming is the range between the missile and target ($R_{M\text{-}T}$),

$$\frac{J}{S} = \frac{P_J G_{Ja} 4\pi R^4_{M\text{-}T} B_R}{P_{Tx} G_{Tx} \sigma R^2_{J\text{-}R} B_J} \tag{3.47}$$

3.7.2 Jamming Methods for Semiactive Guidance

As discussed in Section 3.5, semiactive missiles operate like bistatic radar systems. For the SPJ systems and close support of EJ to the primary task aircraft, the model given in Case 3 of Section 3.5 is used. However, Case 4 is suitable for SOJ systems and long-range support of EJ to mission packages. In the semiactive guidance, the target tracking radar forms a track on the target, and the tracking data is supplied to the fire control computer. Then the fire control computer directs the target illumination antenna to point at the target and radiates CW energy. So the missile passively homes on the reflected CW energy.

Semiactive guidance missiles use the target's velocity to determine the intersection point. Before launching the missile, the reference Doppler signal is defined according to the target velocity. The defined Doppler signal forms a tracking gate around the target's velocity. After the missile launch, the Tx continuously illuminates the target, and the phases can be summarized as follows:

- In the initial phase of the launch, the missile compares the predetermined reference Doppler to the target velocity Doppler signal.
- During the mid-course phase, the missile pursues the reflected CW energy by Doppler tracking, so the fire control computer does not have to send any data to the missile for course corrections. In this stage, the Tx continues illuminating the target with the CW emission.
- In the terminal phase, the missile continues to use the Doppler tracking guidance method as in the mid-course, and the launch platform retains RF illumination. At this stage, the missile steers towards the calculated intersection point with the target.

To calculate J/S for the SPJ and semiactive missile scenario, (3.33) is used without making any change. However, noise or deception techniques, such as false targets and range deception, are ineffective against semiactive missiles with Doppler trackers. For jamming Doppler tracking, phase-frequency modulation is utilized to cause a false frequency offset [9]. This technique, also known as *velocity gate pull-off* (VGPO), confuses the Doppler radar systems by altering the frequency or phase of the radar's signal to alter the target's apparent velocity.

The SPJ and SOJ systems jam the semiactive missile with the phase-frequency modulation or the VGPO technique. However, there is a second path for jamming the semiactive threat systems before launching the missile. For this purpose, the target tracker Rx of the *RF illuminator or launch platform* (RFiRx) can be jammed by applying noise, and the target platform tries to avoid being illuminated by the RF. For this purpose, a high-power noise jamming signal is required, which is not possible with SPJ. Thus, the launch platform is jammed from SOJ or EJ systems.

Figure 3.17 shows the scheme for self-protection and stand-off jamming methods against a semiactive guided missile system.

For jamming the semiactive missile with SPJ and close support EJ systems, J/S is calculated via (3.33). As stated above, the jamming technique for this case is VGPO, and for this purpose, the J/S value is adjusted according to the requirements of different jamming phases. J/S values are calculated as follows:

$$\left(\frac{J}{S}\right)_{\text{VGPO}} = \frac{P_J G_{Ja} 4\pi R_{\text{Tx}-T}^2}{P_{\text{Tx}} G_{\text{Tx}} \sigma} \tag{3.48}$$

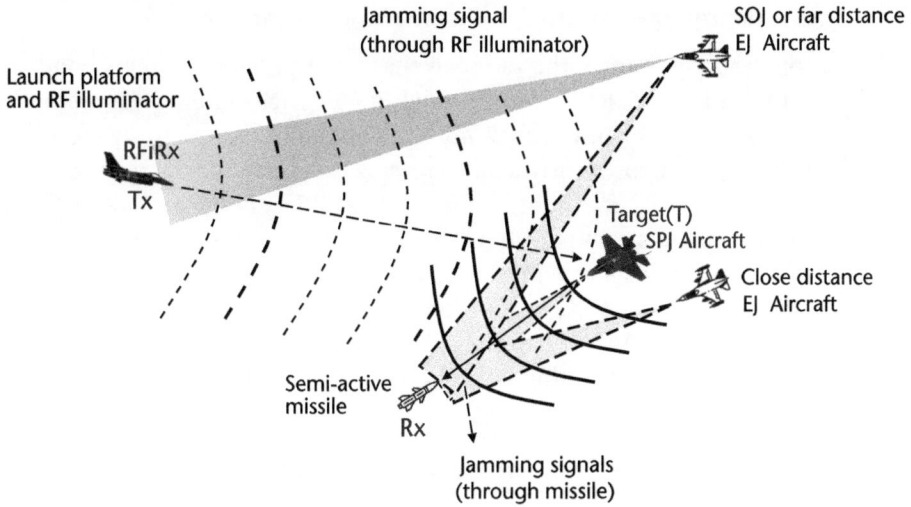

Figure 3.17   Conceptual use of SPJ, EJ, and SOJ systems against semiactive threats.

For adjusting the *J/S* value for the VGPO purposes, the jammer's power (P_J) is varied. When SOJ and far distance support EJ systems are engaged for jamming the missile from the main lobe with the VGPO technique, *J/S* is obtained as follows:

$$\left(\frac{J}{S}\right)_{\text{VGPO(ML)}} = \frac{P_J G_{Ja} 4\pi R_{Tx-T}^2 B_{Rx}}{P_{Tx} G_{Tx} \sigma B_J} \tag{3.49}$$

However, for jamming the missile with SOJ and far distance support, EJ systems from the side lobe are defined in (3.50):

$$\left(\frac{J}{S}\right)_{\text{VGPO(SL)}} = \frac{P_J G_{Ja} G_{Rx(SL)} 4\pi R_{Tx-T}^2 B_{Rx}}{P_{Tx} \ G_{Tx} G_{Rx} \sigma B_J} \tag{3.50}$$

To calculate the jamming effect of SOJ on the RF illuminator tracker, Case 2 of Section 3.5 is considered. This time, noise jamming is applied to the RF illuminator (RFi) tracker's Rx (RFiRx), and obtaining the maximum possible *J/S* is aimed. Thus, the following equations are defined for the *J/S* ratio for the main lobe and side lobe, respectively:

$$\left(\frac{J}{S}\right)_{\text{main-lobe}} = \frac{P_J G_{Ja} 4\pi R_{RFi-T}^4 B_{RFiRx}}{P_{RFiTx} G_{RFiTx} \sigma R_{J-RFi}^2 B_J} \tag{3.51}$$

$$\left(\frac{J}{S}\right)_{\text{side-lobe}} = \frac{P_J G_{Ja} G_{RFi(SL)} 4\pi R_{RFi-T}^4 B_{RFiRx}}{P_{RFiTx} \ G_{RFiTx} G_{RFiRx} \sigma R_{J-RFi}^2 B_J} \tag{3.52}$$

3.7.3 Jamming Methods for Command Guidance

In command guidance, the radar of the AAM or SAM systems conducts tracking and impact point calculation. Command guidance systems use monostatic radars, and missiles do not have any guidance capability. A fire control computer on the launch platform constantly sends steering or course correction commands to the missile throughout its flight. Thus, for the SPJ systems and close EJ support, the jamming model is compatible with Case 1 of Section 3.5. The equation (3.10) is suitable for obtaining the *J/S* for this case.

However, Case 2 is suitable for SOJ systems and long-range support of EJ to mission packages. When jamming is applied from the main beam, *J/S* is calculated using (3.21). If the jamming is applied from the side lobes, then (3.22) is proper.

3.7.4 Jamming Methods for Passive Guidance

An SPJ and an EJ should only transmit a jamming signal when they detect a threat to tactical rules and emission security. This situation is not valid for regional jammers, such as SOJ. When SPJ and EJ detect a threat emission, they start jamming. However, passive-guided missiles, homing on a specific emission, do not contain Tx and have only Rx. The target systems must cease every kind of emission to eliminate the passive-guided missiles. However, the missile must be detected for taking active or passive countermeasures, and the missile has no RF emission.

MWR systems can accomplish the detection, but this only gives information about the existence of an approaching missile. The approaching missile may be IR-guided or RF-guided. The most vital challenge for this operation is that information about the threat cannot be detected. If the missile is IR-guided, flares with evasive maneuvers may be sufficient for salvation. However, if the missile is RF-guided or HOJ, turning off all RF emissions and chaff usage with evasive maneuvers may be the only salvation. RF emissions may be radar transmission and jamming signals. For this reason, for passive guidance or RF-homing missiles, using jamming has a reverse effect and is not preferable.

3.8 Jamming Methods for Different Tracking Radars

This section explains special techniques specific radar systems use for maximum jamming efficiency.

3.8.1 Jamming Conical Scan Radar

Inverse gain and scan rate modulation techniques are used for jamming conical scan radars. The inverse gain jamming technique aims to alter the phase of the conical scan radars by injecting false signals into the conical scan radar's Rx. This technique extracts the antenna scan pattern from the received conical scan radar signal. The Tx and Rx antennas of the conical scan radars operate in coordination with each other. After determining the antenna's scan pattern, a suitable amplitude-modulated signal is generated and injected into the radar Rx. For this purpose, the jammer must determine the victim radar's frequency, PRF, and antenna scan pattern. This technique is known as inverse gain jamming.

Applying the deception signal provides to alter the amplitude of the radar's Rx signal. Since conical scan radars use the target returns' phase to calculate error signals, inverse gain jamming causes errors that can sometimes be very large in the tracking loop.

The effect of the inverse gain jamming applied to the conical scan radar Rx is shown in Figure 3.18.

The general purpose and operation of inverse gain jamming can be summarized as follows:

- One method of deceiving a radar operator or fire control radar system is to make the target appear in the wrong direction.
- For this purpose, the antenna scanning pattern, frequency, and PRF of the threat radar are detected.
- Then the jammer transmits signals that change the phase and amplitude of the target signal, resulting in a signal 180° out of phase with the actual target. This 180° error rapidly drives the antenna off the target and causes a break-lock.
- This ensures that the radar-guided weapon system is fired in a direction other than the actual target direction.

Inverse gain jamming causes the conical scan tracking radar antenna to deviate from the target. The success rate increases when the inverse gain jamming and *range gate pull-off* (RGPO) techniques are used against conical scan tracking radars.

Another jamming method that can be used against conical scan radars is scan rate modulation jamming. This angle deception technique modulates the jamming

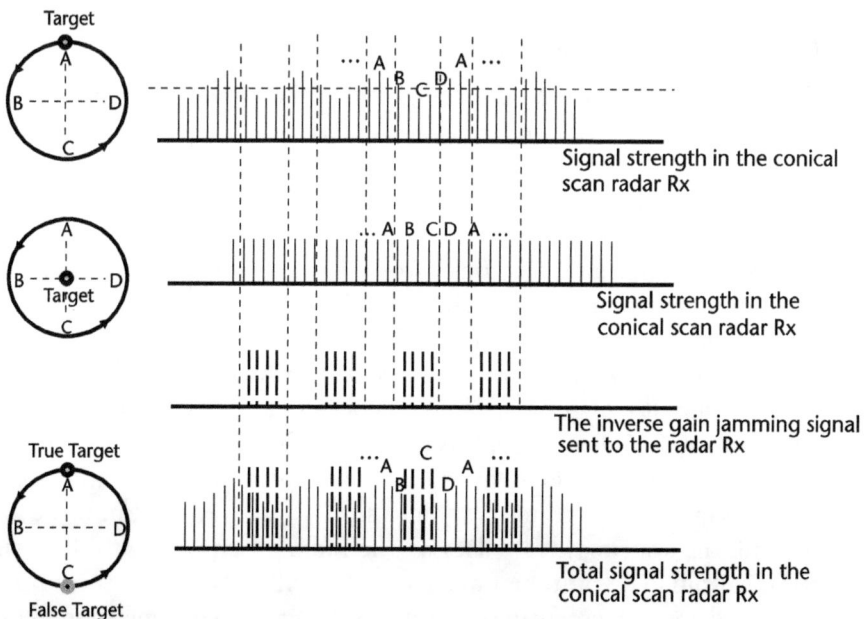

Figure 3.18 The effect of the inverse gain jamming on the conical scan radar.

pulse at or near the threat radar antenna beam's rotation frequency. When the modulation frequency approaches the beam's rotation frequency, significant error signals appear in the radar tracking process. This technique is generally applied by slowly increasing or decreasing the jamming signal's modulation frequency until it equals the rotating rate of the radar antenna beam.

3.8.2 Jamming Lobe-Switching Radar

Angle deception can be used against the specific angle measurement technique employed in the victim radar, including lobe switching. Many older type tracking radars utilize a lobe switching-type angle-tracking mechanism that induces amplitude modulation onto the target's return. As described in Chapter 2, when the sequential pulse amplitudes are the same, the radar antenna's boresight is directed precisely at the target. This type of system is easily deceived by amplitude-modulated type jamming.

Jamming a lobe-switching radar may be easy if one knows the radar's operating frequency, PRF, and lobe-switching frequency. An amplitude-modulated jamming signal at the radar's lobe switching frequency should be applied to fulfill jamming. The jammer only sends out the signal when the radar's lobe is pointed away from the aircraft to obtain success. This technique is a more elementary version of inverse gain jamming.

Another effective type of jamming induces amplitude modulation onto a repeated signal whose rate corresponds to that of the sequential lobe but whose phase is 180° concerning the modulation induced onto the target [4]. This type of jamming is similar to inverse gain jamming applied to the rudimentary structured lobe switching radars. The essential information required for effective jamming can be easily derived from the intercepted radar signal.

3.8.3 Jamming Conical Scan on Receive Only Radar

The CW jamming techniques are tuned to the operating frequency of the victim radar and can be amplitude-modulated. *Amplitude modulation* (AM) is generally applied to jam a tracking radar rather than a search radar. Amplitude-modulated CW jamming can be effective against radar sensors of the conical scan, lobe switching, COSRO, and LORO types [10].

As the inverse gain technique was very effective against early tracking radars, which were based on lobe switching, COSRO radars have mostly begun to be used. These radars typically use a fixed Tx antenna and a conically scanning Rx. The lock-on-target technique is the same as in an ordinary conical scan radar. However, the transmitted signal is constant and conceals the information about the scanning rate of the radar's Rx. Thus, this concept is also known as silent lobing.

Silent lobing techniques introduce the *swept rectangular wave* (SRW) technique [7]. This method is like inverse gain jamming but does not know the scanning rate. Instead, the system sends pulses on the radar's frequency at a PRF similar to the radar's estimated scanning rate. The radar will only receive these pulses if the Rx is pointed approximately in the aircraft's direction. The repetition frequency slowly increases and decreases so that at some point in this pattern, it briefly synchronizes with the scan rate of the antenna to ensure this will occur at some point.

When the repeater jammer signal synchronizes with the scan rate of the antenna, the victim radar receives its signal, and a second one is slightly offset in time. When fed into the phase detector, the output signal will be a double pulse rather than one, creating an error signal. Since the jammer signal sweeps in repetition frequency, it moves about the radar's signal. When the two signals are closely synchronized, it generates significant error signals that can quickly deviate the antenna away from the target. Nevertheless, the rate is constantly changing, and after a period, the tracking error may converge to zero unless the radar has entirely moved off the target. The SRW technique does not protect the platform at 100% probability, but it provides some prevention against conical scans, LOROs, or COSRO radars.

3.8.4   Jamming Lobe on Receive Only Radar

The Tx antenna illuminates the target, and the Rx antenna scans to produce the AM of the reflected signal for effective angle tracking in the LORO technique. Since the transmitting antenna does not rotate or scan, angle deception jammers cannot detect the modulation required to generate effective inverse gain modulation.

Inverse gain jamming relies on measuring the parameters of the rotating beam. An ECCM against inverse gain jamming may be developed if the radar's Rx performs the scanning motion. Using two antenna dishes, one with a fixed beam for transmission and the other with a rotating beam for reception, is a way for this purpose. This type of scanning is called COSRO or LORO. Even though radars using the LORO technique deny a jammer the information required to deceive them entirely, jamming LORO radars is still possible.

As the jammer detects that a LORO radar is around, it blindly sweeps its inverse gain signal through a range of probable scanning frequencies in a repeating cycle. This technique is called SRW, which was defined before. SRW jamming is an angle deception technique developed to counter LORO angle tracking. Based on the knowledge of the precise scan frequency of the radar (f_r), the jammer generates a false target that moves in a circle about the scan axis by transmitting a constant frequency value (f_j) very close to the radar one. The jammer signal produces a false target angle rate slow enough to be followed by the tracker, creating an equivalent error θ_j to break the lock. The maximum angle rate that can be tolerated involves f_j being very close to f_r to obtain appreciable effects.

Additional frequency modulations can be added to the *on-off modulation* (OOM) to enhance oscillation effects in the radar Rx. This technique is called the SRW deception technique. The SRW produces an on-off AM, whose repetition interval is frequency modulated around the value of the scan rate of the radar, as seen in Figure 3.19.

Figure 3.19 shows different FM types for OOM SRW signals. The fixed frequency modulated case, given in Figure 3.19(a), produces a constant rate of change between the two amplitude levels. The sawtooth modulation, shown in Figure 3.19(b), has a change rate between the amplitude levels that increases during the sweep time and then sharply switches back to the slower rate. However, the triangular

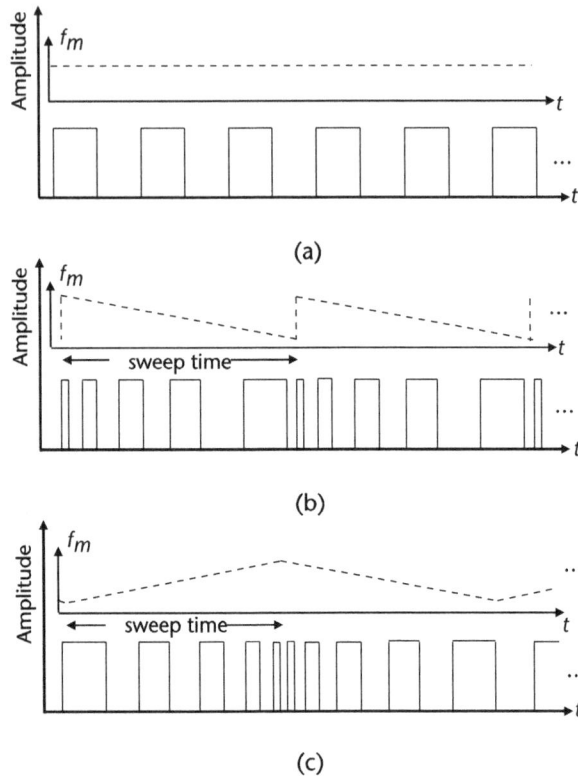

Figure 3.19 The swept rectangular wave (SRW) deception jamming signal types: (a) fixed frequency modulation, (b) sawtooth modulation, and (c) triangular modulation.

modulation in Figure 3.19(c) forms a rate of change between the amplitude levels that increases during a sweep time and decreases during the following sweep time.

Another method for jamming LORO radars is utilizing the amplitude-modulated CW jamming technique. As stated before, this technique can effectively jam the radar sensors of LORO systems.

3.8.5 Jamming Monopulse Radar

The ease with which sequential lobe radars are deceived with amplitude-modulated jamming has led to the preference for single pulse or simultaneous lobing angle tracking techniques in most modern radars. The monopulse tracking radars have an impressive capability to obtain range, azimuth, and elevation data on a pulse-by-pulse basis, making them exceptionally difficult to jam.

In a single-pulse tracking radar system, an angular error is estimated at each return pulse, making the system insensitive to amplitude fluctuations in the data. Thus, this type of radar is highly resistant to angle deception modulation from a single jamming source. Angle deception techniques such as inverse gain, used against conical scan and lobe switching radars, and SRW jamming, used against LORO or COSRO radars, highlight a target, making monopulse tracking easier. Range and velocity deception techniques, such as RGPO and VGPO, are ineffective; moreover, they serve as a beacon that aids the monopulse radar's target tracking ability. The monopulse radar can track the jammer more accurately than tracking actual radar returns because target glint effects are absent from the jamming pulse.

Several techniques apply to the jamming of monopulse systems. However, it should be remembered that monopulse radars are hard to jam, and most of these techniques are technically and operationally challenging to implement. These techniques are as follows:

- *Blinking jamming:* This jamming technique utilizes a spot or barrage noise jamming whose spectrum covers the bandpass of the victim radar. The jamming signal alternately turns on and off at approximately a 50% duty cycle. The jammer's off time should be less than when it takes the radar to reacquire the target. Effective blinking jamming maintains the radar by searching for the target or going into track-on-jam mode. Ordinary blink rates are in the low audio frequency range.

  For adequate blinking, two or more synchronized blinking jammers, which are angularly separated, are required. If airborne platforms are considered, jammers can be installed on two aircraft. These jammers are located within the radar antenna's beam but at slightly different angles. The jammers are alternately turned on and off, so the victim radar receives the strong noise signal from alternate angles around a midpoint. The antenna of a single target-tracking radar will attempt to shift its tracking direction as the jammers are turned on and off, provided that the noise jamming is of sufficient strength.

  When blinking works correctly, the victim radar will track from one jamming source to another. This confusion may cause the lock of the radar tracker to be broken. Otherwise, the radar tracker will have incorrect target information. So a missile's guidance is more complicated, and the missile may miss the target due to inaccurate target angle position information.

- *Cooperative jamming:* In this technique, jamming emissions alternate between the target aircraft. Thus, the radar is forced to wander back and forth from one plane to the other, causing a vibration in the boresight direction that can abolish weapon systems' effectiveness.

- *Cross-eye jamming:* Cross-eye is one of the techniques that can effectively deceive the angle estimation of these monopulse radars. This jamming technique induces an angular error in the radar being jammed by forming the worst-case glint angular error again.

  Figure 3.20 shows a conceptual cross-eye jammer operation. Tx antennas reradiate signals received by Rx antennas. Both signals travel the same distance from the radar, through the jammer, and back to the radar. Since the signals travel the same distance, they are generally in the phase when the radar receives them. However, in the path of Rx_1 and Tx_1, inserting a 180° phase shifter changes the signal's sign.

  These two out-of-phase signals must be adjusted in amplitude and exceed the target return's amplitude. High jamming power is required since the skin return effect is not wanted. When the signals are out of phase with 180°, the radar will track with a $0.6\theta_{3dB}$ error if there is no skin return, where θ_{3dB} is the 3-dB beamwidth of the radar [11].

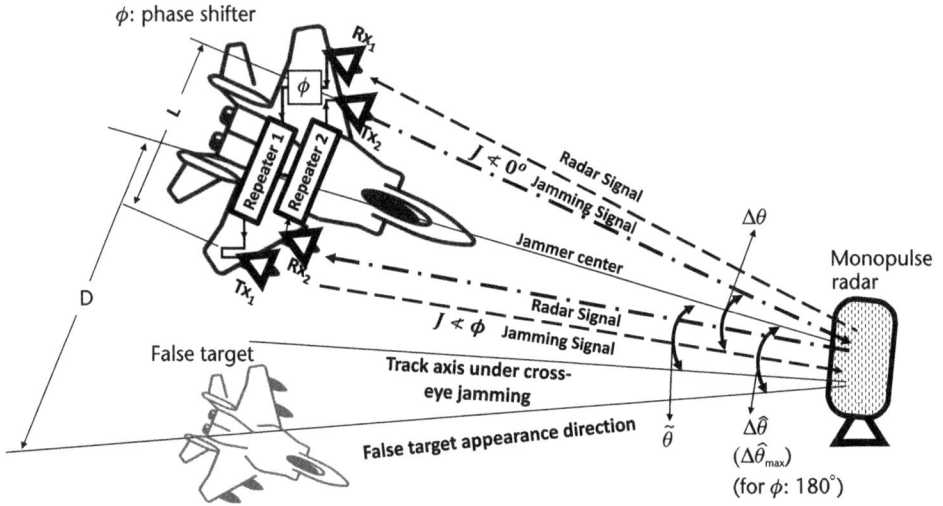

Figure 3.20 Cross-eye deception jamming concept.

The jamming signals may have different power levels, say, J and a^2J, where $0 < a \leq 1$ is an attenuation factor. Then the cross-eye sources will appear to come from a source of power $J(1 - a)^2 = P_j$ at an angle from the actual jammer center. Suppose the cross-eye sources are out of phase with ϕ. In that case, the position of this apparent jammer source or false target appearance direction is at an angle $\Delta\hat{\theta}$ defined as follows [11, 12].

$$\Delta\hat{\theta} = \frac{\Delta\theta}{2} \frac{1 - a^2}{1 + a^2 + 2a\cos\phi} \tag{3.53}$$

In (3.53), ϕ represents the phase difference between the jamming sources, which may take any phase angle value. The maximum error, $\Delta\hat{\theta}_{max}$, is achieved when the two repeater jamming signals are out of phase with $180°$, and the peak error is expressed in this case as follows.

$$\Delta\hat{\theta}_{max} = \frac{\Delta\theta}{2} \frac{1 + a}{1 - a} \quad \text{where} \quad \frac{\Delta\hat{\theta}_{max}}{\Delta\theta} = \frac{D}{L} \tag{3.54}$$

where D is the apparent position from the center of the base length and is aligned with the base length of the ghost jammer or the false target, and L is the distance between jammer antennas.

To obtain the maximum tracking error, first, the ratio of $\tilde{\theta}/\Delta\hat{\theta}_{max}$ is calculated as follows:

$$\frac{\tilde{\theta}}{\Delta\hat{\theta}_{max}} = \frac{P_J}{P_J + S} = \frac{J(1 - a)^2}{S + J(1 - a)^2} = \frac{(J/S)(1 - a)^2}{1 + (J/S)(1 - a)^2} \tag{3.55}$$

where S is the target return signal power. Thus, the tracking error, the angle between the actual jammer center and the apparent jammer source, is defined as [11]:

$$\tilde{\theta} = \frac{\Delta\theta}{2} \frac{(J/S)(1-a)^2}{1+(J/S)(1-a)^2} \tag{3.56}$$

The distance between jammer antennas (L) is an important parameter determining the effectiveness of cross-eye jamming. The wider the spacing between antenna pairs, the more distortion in the victim's wavefront near the radar return. Naturally, most fighter aircraft need more spacing between the antennas to maximize effectiveness. Cross-eye jamming effectiveness is also lost when the aircraft is out of the beam or distant from the radar.

- *Terrain bounce jamming:* This method defeats monopulse tracking radars. The technique is used primarily in low-altitude flying aircraft and is based on creating a false radar target by illuminating the ground with a jamming signal. It is effective against semiactive missiles, AAM, and monopulse tracking radars. The technique involves a repeater jammer that receives the radar or missile guidance signal. The jammer amplifies and directs this signal to illuminate the terrain directly in front of the aircraft. The conceptual approach of the terrain bounce geometry is shown in Figure 3.21. The missile or radar tracks the reflected energy from the spot on the ground instead of the target aircraft.

The antennas used for terrain bounce jamming generally have a narrow elevation beamwidth and a broad azimuth beamwidth to be effective. This transmission pattern increases the energy directed toward the ground and minimizes the direct energy transmission toward the missile or radar. The jamming system must generate high jamming power to overcome signal losses associated with uncertain terrain propagation. Thus, the power reflected from the terrain is higher than the echo return of the aircraft.

Terrain bounce jamming adds some problems to the process, including the uncertainty of the signal scattering parameters of the various terrain features

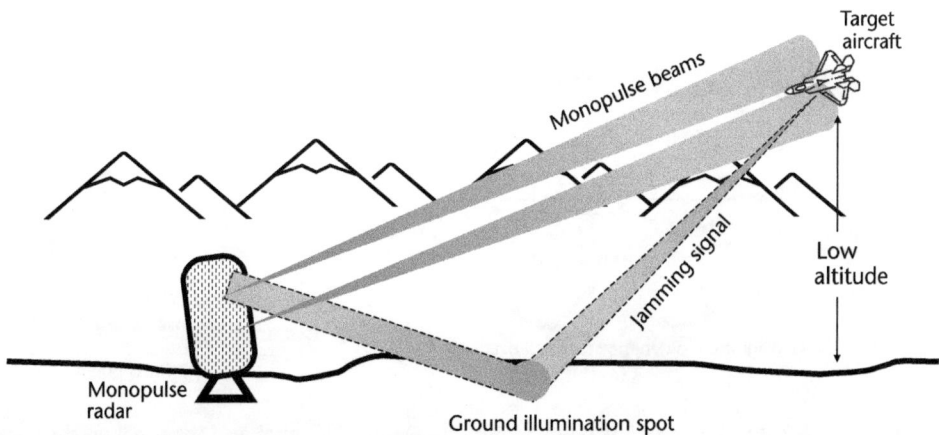

Figure 3.21   Terrain bounce jamming concept.

and the possible changes in signal polarization caused by terrain propagation. Moreover, this technique may cause maneuvering restrictions and maximum altitude limitations on the aircraft.

• *Cross-polarization jamming:* This self-protection method produces an angular error in tracking radars, including monopulse. Some monopulse radars provide erroneous angular information when the received signal is polarized at right angles to the polarization of the radar transmitter.

Before describing the cross-polarization jamming, it would be beneficial to put some definitions for monopulse radar tracking. As stated in Section 3.4, on reception, the monopulse radars can generate three signals:

- The first one, expressed as Σ, is obtained by adding up the four elementary beams, and the radar can track in range.
- The second signal, represented as $\Delta_{azimuth}$, is obtained by the difference between the suitable beams.
- The last one, shown as $\Delta_{elevation}$, is achieved by the difference between the proper beams. The two Δ signals can generate the pointing errors in azimuth and elevation necessary for angle tracking.

Figure 3.22(a) shows a conceptual repeater system on an airborne platform utilizing two separated cross-polarized receiving and transmitting antennas. The horizontally polarized signal is amplified and reradiated as a vertically polarized signal. However, the vertically received signal is amplified, phase shifted by 180°, and radiated as a horizontally polarized signal.

When applying a signal with an orthogonal polarization to its standard operational polarization, every antenna presents a pattern with a null instead of a maximum in the main lobe. The cross-polarization jamming technique exploits this fact. This technique strives to produce in the Δ signal a cross-polarized component that is large relative to the element with radar

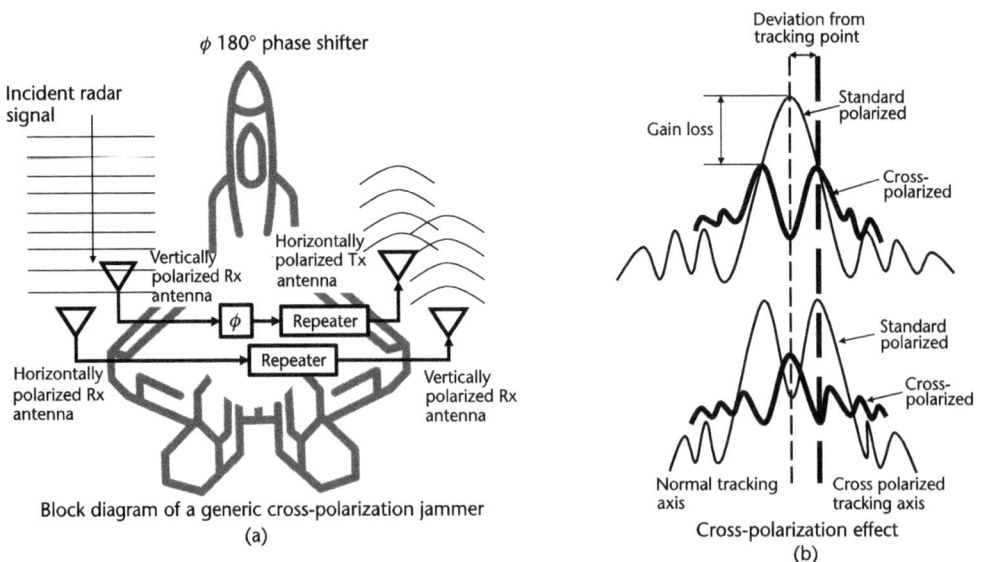

Figure 3.22 Cross-polanization jamming: (a) block diagram and (b) effect on the victim radar's tracker.

polarization. Thus, it drives the error toward or beyond the peak of the Δ lobe at approximately half of the radar antenna beam width, as in Figure 3.22(b).

The principal weakness of a monopulse radar antenna is that its response to cross-polarized signals is significantly different from that of standard polarized signals. Cross-polarization jamming makes use of this weakness. As seen in Figure 3.22(b), the cross-polarized Σ beam shows a null in the direction of the boresight in addition to a structure of side lobes higher than the standard polarized beam. Conversely, the Δ cross-polarized beam has a maximum in the boresight direction.

When the cross-polarized jammer's radiated power is sufficiently increased, and the received cross-polarized component is relatively higher than the standard polarized skin echo, the victim radar has to use the part of the cross-polarized pattern for tracking and deviates further away from the target.

- *Image frequency jamming:* This technique aims to produce a wrong angular measurement by radiating a signal at the image frequency of the victim radar's IF stage. The jammer exploits the frequency conversion process of the victim radar. The image frequency in the victim receiver is at the same distance as the target frequency from the local oscillator but on the opposite side. When the false signal beats with the local oscillator, it generates an IF with a relative phase angle between the sum and difference channels opposite to the target. This type of jamming impairs the angular error correction process by changing the sign of the correction signals. This technique assumes the knowledge of the IF of the victim's radar. Image frequency jamming is ineffective if the monopulse radar has an image rejection filter or mixer.
- *Skirt frequency jamming:* This technique aims to produce errors in the angular measurement of a tracking radar by forcing it to follow a signal on the side of the passband of its IF filter or Doppler filter. The skirt frequency jamming refers to jamming on the skirts of the frequency response curve of the threat radar Rx. Its effectiveness depends on the imbalance between the sum and difference channels at these frequencies, where rapid phase shifts are present in each channel. Radars may prevent the effectiveness of skirt frequency jamming using careful design and construction considerations.
- *Countdown jamming:* This technique attacks the threat radar's AGC. The countdown jamming forces the radar's AGC to change the value continuously and oscillate. For this purpose, the jammer swiftly changes the duty cycle of the deception pulses.

3.8.6   Jamming Track-While-Scan Radar

As stated before, track-while-scan (TWS) radars search and track simultaneously. In this method, they operate as a search-and-track radar and perform these issues in a substantial sector. A TWS radar creates angle gate pairs or pairs during its sectoral tracking. These pairs consist of early and late gates. The gate pair's centerline represents the radar antenna's boresight.

The graphical representation of the TWS jamming technique, shown in Figure 3.23, is called sliding or walking pulses [7]. Sliding pulses drift the target out of the radar's tracking area, thus avoiding the tracking of the aerial platform.

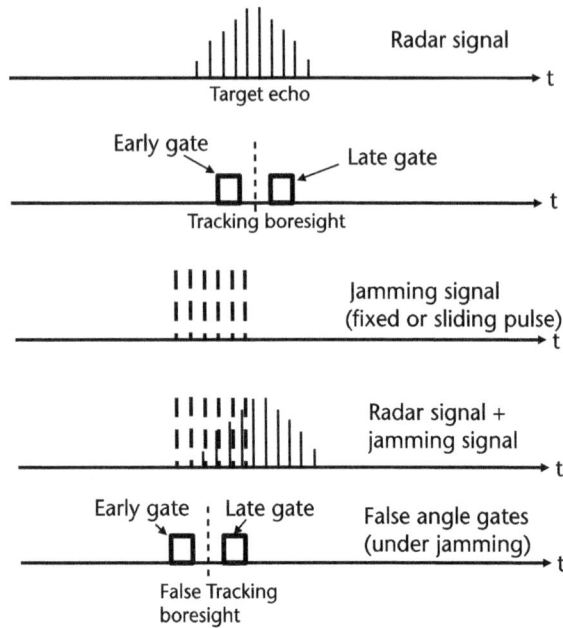

Figure 3.23   Signal representation of TWS jamming.

The gates are established using the signal return echo from the target in the radar's Rx. In the TWS jamming method, in addition to this return, the jamming pulse signal transmitted from the target platform exists. Thus, the location of the angle gates is changed involuntarily. As a result, the boresight of the radar antenna also shifts in a direction where the target is not present. The platform that performs the jamming breaks the lock and moves away from the threat by maneuvering in the time it takes until a new track forms.

3.8.7   Jamming Automatic Detection and Tracking Radar

Automatic detection and tracking (ADT) radars use adaptive tracking methods for track initiation and continuation. These methods determine separately or a combination of the target's angle, range, and velocity information by using the measurement information and estimations. These radars, like some other types of tracking radars, employ range or velocity gate forming for guidance purposes. RGPO and VGPO techniques come to mind when tracking gates are considered to avoid the threat radar systems. The RGPO technique is used against range gates, and VGPO is utilized to prevent velocity gates. These techniques may also be used along with other deceptive jamming methods. This part describes the operational details of RGPO and VGPO techniques.

- *RGPO:* When a tracking radar detects a target, range gates are placed on both sides of the range. Range gates protect the radar against asynchronous jamming pulses by eliminating all signals from distances outside a narrow window, which significantly increases the SNR.

  A radar focuses on a short-range surrounding the target's position and is not interested in the other targets. This situation is known as radar's lock-on to a target. However, formed range gates can be unlocked or stolen, known as the range gate pull-off (RGPO) technique. The RGPO aims to break the radar's lock on the target platform and escape outside the range gate. The representation of the RGPO technique on the tracking radar's Rx is given in Figure 3.24. The range gate may represent a time interval corresponding to the range or the interval between early and late gates.

Figure 3.24 explains the conceptual operational method of RGPO in stages. At the beginning of the process, a tracking radar locks on a target, the jammer part of the SPJ system is activated, and the RGPO is applied in several stages.

- *Stage 1:* Samples are taken from the radar's pulse signal, and the radar's PRI is determined. After taking the first samples, a replica of the signal is produced and transmitted back to the radar by increasing its power as soon as the subsequent pulses are received. Thus, the repeater constitutes a false echo in the radar's Rx. It can be thought that the retransmission would intensify the echo of the victim aircraft on the radar scope. After this point, the jamming power gradually increases, and the process continues until the RGPO signal is much stronger than the returning echo. At this time, the Rx sensitivity of the tracking radar is reduced by the automatic gain control (AGC) capability often found in radar systems to avoid overloading. This process causes the true target return echo to disappear below the noise floor.
- *Stage 2:* At this stage, after each false return echo, another repeat signal is sent. The false signal on the target return is then made weaker while the strength of the repeated signal is increased.
- *Stage 3:* The radar's automatic tracking system locks only on the repeat signal above the detection threshold while the target's return signal is below the noise level. Meanwhile, the repeat signal is delayed gradually, increasing with the elapsed time. The range gate then follows the fake target that is getting further away. This process continues until the range gate moves away from the target's actual location. The result is that the radar tracks a ghost target rather than the target echo.
- *Stage 4:* In the last step of the process, the jammer stops sending the repeat signal, and only noise remains inside the range gate. Thus, the lock breaking is accomplished. As a result, the tracking radar starts a search and detection

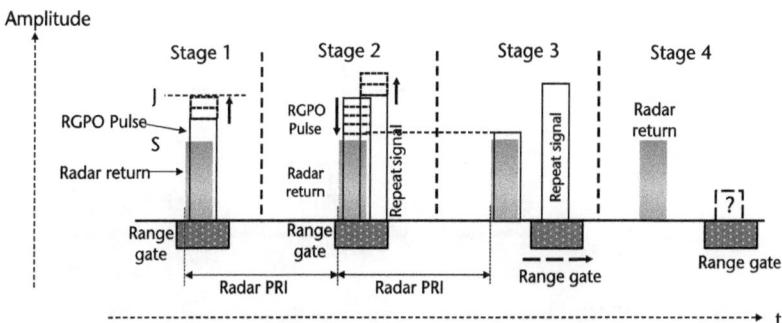

Figure 3.24 RGPO signal representation.

process again and loses time. The whole detection and tracking cycle begins again if the target is still in the detection range.

- *VGPO:* Velocity gate pull-off (VGPO) is a method used to capture the velocity gate of a Doppler radar and remove it from the target echo. It is similar to RGPO but is used against CW or Doppler velocity tracking radar systems. The conceptual representation of the VGPO technique in the Rx of the tracking radar is given in Figure 3.25.

The amplified and retransmitted CW or pulsed Doppler frequency shifts to produce an apparent Doppler shift, creating a variation impact in target velocity that is not actual. This false speed can be adjusted by either decreasing or increasing it.

3.9   Problems

Problem 3.1: A command-guided SAM system utilizes a pulse and monostatic radar with a single Tx and Rx antenna. Its operating frequency is 3.2 GHz for acquisition and 6.2 GHz for tracking. The radar's Tx peak power is 100 kW, the Rx sensitivity is −65 dBm, and the antenna gain is 30 dB for the acquisition mode. However, in track mode, the Tx peak power is 75 kW, and the antenna gain is 35 dB. An aircraft with 12-m^2 RCS uses spot noise jamming for self-screening to avoid radar locking. The jammer on the aircraft used for this purpose has 150-W peak power, its Tx antenna gain is 6 dB, and it adjusts B_R/B_J to unity.

(a) What is the maximum operational range for acquisition?

(b) If a sufficient *J/S* ratio to prevent locking is 6 dB, what is the effectiveness of the jamming to the threat radar from 6 km? (*Note:* Neglect radar, jammer, and polarization losses.)

Problem 3.2: Consider the SAM system defined in Problem 3.1. This time, the target platform is a bomber with 50-m^2 RCS, protected with an EJ on a fighter. Both aircraft are in the main beam of the radar, and the radar's angular resolution cannot discriminate their angular difference. The bomber and fighter aircraft's distances

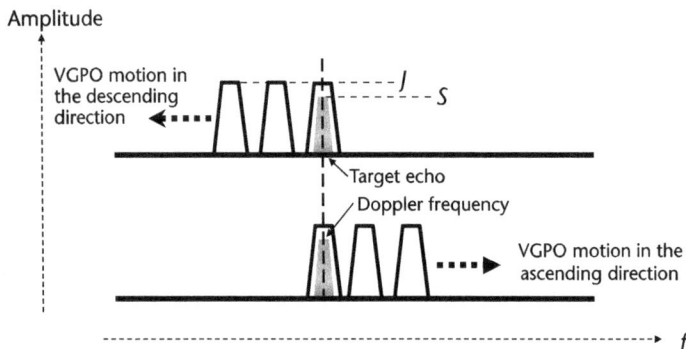

Figure 3.25 VGPO signal representation.

to the radar are 9 km and 9.3 km, respectively. The jammer's Tx peak power is 200 kW, and the antenna gain is 10 dB. It conducts escort jamming with barrage noise, and the B_R/B_J ratio is −10 dB. Inspect the jamming efficiency and the possibility of radar's passing to the tracking mode for this case. (*Note:* Neglect radar, jammer, and polarization losses.)

Problem 3.3: Again, we consider the SAM system given in Problem 3.1. This time, the SPJ system is on a 12-m^2 RCS fighter in the scenario, and the jammer applies a repeater jamming against the SAM system. The aim is to create multiple false echoes with different delays to reduce the lock-on probability. Assume that the acquisition radar has a lower threshold at 60% and an upper threshold at 140% of nominal predicted values and prunes the return echoes outside these values. If the SPJ system has a Tx antenna gain of 8 dB, an Rx antenna gain of 6 dB, an adjustable jammer processing gain between 20 and 30 dB, and the saturation power level is 30 dBm. When saturation power is reached, the jammer emits a noise signal with 50W through the Tx antenna. Let us assume that the jammer uses spot noise and adjusts B_R/B_J to unity for this case.

(a) If the fighter's distance from the radar is 10 km, what may be the jammer processing gain interval, and how many false echoes are required to reduce the lock-on probability below 15%?

(b) At what distance from the fighter to the radar does SPJ enter the saturation region, and what is the *J/S* for this case? (*Note:* Neglect radar, jammer, and polarization losses.)

Problem 3.4: A command-guided SAM system has an acquisition radar operating at 3.8 GHz and a tracking radar at 8.4 GHz. The radar specifications of the SAM system are defined as follows:

- Acquisition mode: monostatic:
  - Tx peak power: 150 kW;
  - Antenna gain (Tx and Rx): 30 dB.
- Tracking radar: monostatic:
  - Tx peak power: 75 kW;
  - Antenna gain (Tx and Rx): 35 dB.

Assume the tracking radar locks on a fighter and is ready to launch a missile. Meanwhile, the fighter's SPEW system detects the locking, and its SPJ tries to break the lock using the RGPO technique. To enable the radar's AGC system, the repeater jammer must produce a signal power level at the radar Rx 3 times higher than the target return. Assume that the RCS of the fighter is 9 m^2, and the repeater specifications of the SPJ are given as:

- Jammer processing gain G_J = 20 to 35 dB;
- Jammer Rx antenna gain $G_{Ja(Rx)}$ = 8 dB;
- Jammer Tx antenna gain $G_{Ja(Tx)}$ = 12 dB.

If the fighter-to-radar range is 8 km, find the required jamming processing gain to enable the radar's AGC system. (*Note:* Neglect radar, jammer, and polarization losses.)

Problem 3.5: An air defense radar's central beam aspect is directed to the mission pack aircraft, and the SOJ aircraft is out of the main beam. The operating specifications of the systems are given in Figure 3.26. At first, the SOJ aircraft jammed the radar from the side lobes to protect the mission pack, which consisted of several 12-m^2 RCS aircraft. The SOJ aircraft's Tx antenna gain in the radar direction is 10 dB, and the radar Rx antenna gain in this direction is −3 dB. Find the amount of change in *J/S* if the SOJ aircraft applies jamming from the same distance but from the main beam. (*Note:* Neglect radar, jammer, and polarization losses.)

Problem 3.6: A semiactive guided missile is launched from a fighter aircraft to a chopper. The loss of Tx on the fighter is 4 dB and is 10 km from the target. Tx specifications are as follows:

- Operating frequency: 5.6 GHz;
- Output power: 10 kW;
- Antenna gain: 30 dB.

The missile's Rx loss is 3 dB, and its antenna's gain is 15 dB.

(a) Calculate the return power at the Rx if the chopper's RCS is 12 m^2 and the distance between the missile and the helicopter is 3 km.

(b) The chopper's SPEW system detects the missile and applies noise jamming at this stage. The jammer's Tx power is 100W, the antenna gain is 8 dB, and the jammer loss is 3 dB. The jamming's bandwidth reduction factor (BF) is −6 dB, and there is an additional 3-dB loss due to antenna polarization mismatch. Calculate the *J/S* at the Rx for the condition given in (a).

Problem 3.7: Model four monostatic threat radars with an 8 to 12-GHz operating frequency band. For this purpose, assume these radars are at different angles than a fighter aircraft with 10-m^2 RCS and designate their Tx powers and antenna gains. Also, define the jammer's antenna gains of the SPEW system against each radar.

In light of these assumptions, find the burn-through distances of the fighter against each radar by performing a MATLAB simulation for the jamming at two different powers.

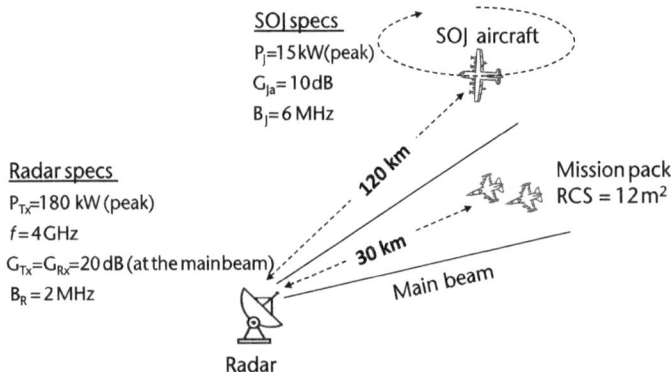

SOJ specs
$P_J = 15\,kW\,(peak)$
$G_{Ja} = 10\,dB$
$B_J = 6\,MHz$

SOJ aircraft

Radar specs
$P_{TX} = 180\,kW\,(peak)$
$f = 4\,GHz$
$G_{TX} = G_{RX} = 20\,dB\,(at\,the\,mainbeam)$
$B_R = 2\,MHz$

120 km

30 km

Mission pack
RCS = 12 m^2

Main beam

Radar

Figure 3.26 Operational specifications for Problem 3.5.

(*Notes:* 1. Neglect the variation of the aircraft's RCS in different directions in the simulation and constantly include it in the calculations. 2. Calculate the burn-through ranges according to $A = 6$ dB.)

Solutions

Solution 3.1:

(a) The first part of the solution is about the radar detection issue. In this context, all the required values given in the problem are defined in linear form,

$$\left(S_{min}\right)_{dB} = -65 \text{ dBm} = -95 \text{ dBW} \rightarrow S_{min} = 316 \text{ pW}$$

$$\left(G_{Tx}\right)_{dB} = \left(G_{Rx}\right)_{dB} = G_{dB} = 30 \text{ dB} \rightarrow G = 1{,}000$$

$$P_{Tx} = 100 \text{ kW}, \quad \sigma = 12 \text{ m}^2, \quad f = 3.2 \text{ GHz}, \quad L_R = 1$$

Equation (2.45) is used by merging L_R as given in (3.8) to obtain the maximum detection range of the two-way radar equation for monostatic radars,

$$R_{max} = \left[\frac{P_{Tx}G^2c^2\sigma}{(4\pi)^3 f^2 L_R S_{min}}\right]^{\frac{1}{4}} = \left[\frac{\left(100 \times 10^3\right)(1{,}000)^2\left(3 \times 10^8\right)^2(12)}{(4\pi)^3\left(3.2 \times 10^9\right)^2(1)\left(316 \times 10^{-12}\right)}\right]^{\frac{1}{4}}$$

$$R_{max} \simeq 11{,}393\text{m} = 11.393 \text{ km}$$

(b) For calculating J/S, Case 1 in Figure 3.9(a), and related to this situation, (3.10) is considered. The problem is defined before tracking or for the acquisition, so the operating frequency is 3.2 GHz. In addition to the above values,

$$P_J = 150\text{W}, \quad R = 10 \text{ km}, \quad L = \frac{L_J L_{pol}}{L_R} = 1$$

$$\left(G_{Ja}\right)_{dB} = 6 \text{ dB} \rightarrow G_{Ja} \simeq 4$$

Thus,

$$\frac{J}{S} = \frac{P_J G_{Ja} 4\pi R^2}{P_{Tx} G\sigma L} = \frac{(150)(4)4\pi\left(6 \times 10^3\right)^2}{\left(100 \times 10^3\right)(1{,}000)(12)(1)} = 226$$

$$\left(\frac{J}{S}\right)_{dB} = 23.54 \text{ dB}$$

The aircraft can prevent locking when J/S equals or exceeds 6 dB. The obtained value is higher than this, and it can be concluded that the jamming is effective.

Solution 3.2: Consider Case 2 in Figure 3.9(b) and (3.21) for solving this problem. The related values are defined as follows:

$$P_{Tx} = 100 \text{ kW}, \quad P_J = 200 \text{W}, \quad R_{J-Rx} = 9.3 \text{ km}, \quad R_{Tx-T} = 9 \text{ km}$$

$$G_{Tx} = G_{Rx} = G = 1000, \quad \left(G_{Ja}\right)_{dB} = 10 \text{ dB} \rightarrow G_{Ja} = 10$$

$$\left(\frac{B_R}{B_J}\right)_{dB} = -10 \text{ dB} \rightarrow \frac{B_R}{B_J} = 0.1, \quad \sigma = 50 \text{ m}^2, \quad L = \frac{L_J L_{pol}}{L_R} = 1$$

So,

$$\frac{J}{S} = \frac{P_J G_{Ja} 4\pi R_{R-T}^4 B_R}{P_{Tx} G_{Tx} \sigma R_{J-R}^2 L B_J} = \frac{(200)(10)4\pi(9,000)^4(0.1)}{\left(100 \times 10^3\right)(1,000)(50)(9,300)^2(1)} \simeq 381$$

$$\left(\frac{J}{S}\right)_{dB} = 10\log(381) = 25.81 \text{ dB}$$

The EJ prevents the radar from locking on the bomber aircraft since the *J/S* is higher than 6 dB, and the tracking mode cannot be activated.

Solution 3.3: To solve this problem, the jammer power (*J*) and the radar return signal power (*S*) are assumed, as given in Figure 3.12.

(a) First, the given values are written in linear form as follows:

$$P_{Tx} = 100 \text{ kW}, \quad P_J = 50\text{W}, \quad R = 10 \text{ km}, \quad G_{Tx} = G_{Rx} = G = 1,000$$

$$\left(G_{Ja(Tx)}\right)_{dB} = 8 \text{ dB} \rightarrow G_{Ja(Tx)} = 6.31, \quad \left(G_{Ja(Rx)}\right)_{dB} = 6 \text{ dB} \rightarrow G_{Ja(Rx)} = 4$$

$$\left(G_J\right)_{dB} = 20 \text{ to } 30 \text{ dB} \rightarrow G_J = 100 \text{ to } 1,000, \quad \sigma = 12 \text{ m}^2$$

$$\left(P_{J(Tx)sat}\right)_{dB} = 30 \text{ dBm} \rightarrow P_{J(Tx)sat} = 1\text{W}, \quad L = \frac{L_J L_{pol}}{L_R} = 1$$

The radar return signal is calculated by using (3.8),

$$P_{Rx} = \frac{P_{Tx} G^2 \sigma c^2}{(4\pi)^3 R^4 f^2 L_R} = \frac{\left(100 \times 10^3\right)(1,000)^2(12)\left(3 \times 10^8\right)^2}{(4\pi)^3\left(10 \times 10^3\right)^4\left(3.2 \times 10^9\right)^2(1)}$$

$$= 5.32 \times 10^{-10} \text{ W} = 0.532 \text{ nW}$$

and the incident radar power at the radar is calculated by using the one-way radar equation given in (2.29),

$$P_{J(Rx)} = \frac{P_{Tx} G_{Tx} G_{Ja(Rx)} c^2}{(4\pi)^2 R^2 f^2 L_J L_{pol}} = \frac{\left(100 \times 10^3\right)(1,000)(4)\left(3 \times 10^8\right)^2}{(4\pi)^2\left(10 \times 10^3\right)^2\left(3.2 \times 10^9\right)^2(1)}$$

$$= 2.23 \times 10^{-4} = 0.223 \text{ mW}$$

Then the jammer produces a replica of the signal, amplifies it, and reradiates it through the radar again. For calculating the jammer power in the radar Rx, it would be appropriate to use (2.29) by inserting the radar's antenna gain into the process by multiplying it,

$$J = \frac{P_{J(Rx)}G_J G_{Ja(Tx)}Gc^2}{(4\pi)^2 R^2 f^2} = \frac{(2.23 \times 10^{-4})G_J (6.31)(1,000)(3 \times 10^8)^2}{(4\pi)^2 (10 \times 10^3)^2 (3.2 \times 10^9)^2}$$

$$= 7.84 \times 10^{-13} G_J$$

$$\{G_J: 100 \text{ to } 1,000\}$$

Also, this result can be obtained directly using (3.17) as given here:

$$J = \frac{P_{Tx}GG_{Ja(Rx)}c^2}{(4\pi Rf)^2} \frac{G_J G_{Ja(Tx)}Gc^2}{(4\pi Rf)^2 L_J L_{pol}} = 7.84 \times 10^{-13} G_J$$

The nominal value of a 12-m² RCS target return at 10 km is 0.532 nW above; 60% of it is 0.32 nW, and 140% of it is 0.745 nW. If the replicas' return power (J) level is lower than 0.32 nW and higher than 0.745 nW, they are eliminated, and the actual target return remains under the threat of being lock-on. So the allowable jammer processing gain is calculated as follows:

$$J_{min} = 0.32 \times 10^{-9} \rightarrow G_{Jmin} = \frac{0.32 \times 10^{-9}}{7.84 \times 10^{-13}} = 408.16$$

$$J_{max} = 0.745 \times 10^{-9} \rightarrow G_{Jmax} = \frac{0.745 \times 10^{-9}}{7.84 \times 10^{-13}} = 950.25$$

To reduce the lock-on probability of the actual target below 15%,

$$\{n: \text{ no of return echoes }\} \Rightarrow 0.15 = \frac{1}{n} \rightarrow n = 6.67$$

So, the first following integer is 7, meaning six replicas with different delays must be generated. It must be noted that one return is due to the actual target.

(b) Let us find the fighter's distance to the radar so the SPJ can enter the saturation region. For this purpose, assume the minimum jammer processing gain (100) and use (3.17) by adapting to obtain jammer Tx power as follows:

$$P_{J(Tx)_{sat}} = \frac{P_{Tx}GG_{Ja(Rx)}c^2}{(4\pi Rf)^2} \frac{G_J G_{Ja(Tx)}}{L_J L_{pol}} \rightarrow \{P_{J(Tx)_{sat}} = 1W\}$$

$$R = \sqrt{\frac{(100 \times 10^3)(1,000)(4)(3 \times 10^8)^2}{(4\pi)^2 (1)(3.2 \times 10^9)^2}(100)(6.31)} \Rightarrow R = 3,750m = 3.75 \text{ km}$$

Solution 3.4: The radar return-signal power (S) and the jammer power (J) should be calculated to find the five times higher J than S. Let us write all the given values in linear form,

$$P_{\text{Tx}} = 7.5 \text{ kW}, \quad R = 8 \text{ km}, \quad G_{\text{Tx}} = G_{\text{Rx}} = G \Rightarrow G_{\text{dB}} = 35 \text{ dB} \to G = 3{,}162.3$$

$$\left(G_{\text{Ja(Tx)}}\right)_{\text{dB}} = 12 \text{ dB} \to G_{\text{Ja(Tx)}} = 15.85, \quad \left(G_{\text{Ja(Rx)}}\right)_{\text{dB}} = 8 \text{ dB} \to G_{\text{Ja(Rx)}} = 6.31$$

$$\left(G_{\text{J}}\right)_{\text{dB}} = 20 \text{ to } 35 \text{ dB} \to G_{\text{J}} = 100 \text{ to } 3{,}162.3, \quad \sigma = 9 \text{ m}^2$$

$$f = 8.4 \text{ GHz } \{\text{for tracking radar}\}, \quad L = \frac{L_{\text{J}} L_{\text{pol}}}{L_{\text{R}}} = 1$$

We can use (3.8) to obtain the radar return signal,

$$P_{\text{Rx}} = \frac{P_{\text{Tx}} G^2 \sigma c^2}{(4\pi)^3 R^4 f^2 L_{\text{R}}} = \frac{(7.5 \times 10^3)(3162.3)^2 (9)(3 \times 10^8)^2}{(4\pi)^3 (8 \times 10^3)^4 (8.4 \times 10^9)^2 (1)}$$

$$= 1.06 \times 10^{-10} \text{ W} = 0.106 \text{ nW}$$

Three times this power level is $3 \times 0.106 \text{ nW} = 0.318 \text{ nW}$; then the required processing gain is obtained using (3.17) as follows:

$$J = \frac{P_{\text{Tx}} G G_{\text{Ja(Rx)}} c^2}{(4\pi R f)^2} \frac{G_{\text{J}} G_{\text{Ja(Tx)}} G c^2}{(4\pi R f)^2 L_{\text{J}} L_{\text{pol}}}$$

$$0.318 \times 10^{-9} = \frac{(7.5 \times 10^3)(3162.3)^2 (6.31)(3 \times 10^8)^4 G_{\text{J}} (15.85)}{\left(4\pi (8 \times 10^3)(8.4 \times 10^9)\right)^4 (1)} \to G_{\text{J}}$$

$$= 2656.14$$

Thus, $\quad G_{\text{J}} = 2{,}656.14 \to \left(G_{\text{J}}\right)_{\text{dB}} = 34.24 \text{ dB}$

Solution 3.5: The situation defined in the problem is an SOJ application against a single air defense radar given in Case 2 of Figure 3.9(b). In practical applications, early warning (E/W) and acquisition mode of SAM radars have a scan pattern, and their main beams do not constantly hold on to the targets. However, SAM radars' tracking mode holds the target's main beam. In the given situation, the problem is that the radar's main beam is constantly on the targets. First, the given values are defined in linear form as follows:

$$P_{\text{Tx}} = 180 \text{ kW}, \quad P_{\text{J}} = 15 \text{ kW}, \quad R_{\text{J−Rx}} = R_{\text{J-R}} = 120 \text{ km}, \quad \frac{B_{\text{R}}}{B_{\text{J}}} = \frac{2}{6} = 0.33$$

$$R_{\text{Tx−T}} = R_{\text{Rx−T}} = R_{\text{R-T}} = 30 \text{ km}, \quad L = \frac{L_{\text{J}} L_{\text{pol}}}{L_{\text{R}}} = 1$$

$$G_{\text{Tx(dB)}} = G_{\text{Rx(dB)}} = G_{\text{dB}} = 30 \text{ dB} \to G = 1{,}000, \quad \left(G_{\text{Ja}}\right)_{\text{dB}} = 10 \text{ dB} \to G_{\text{Ja}} = 10$$

$$\left(G_{\text{Rx(SL)}}\right)_{\text{dB}} = -3 \text{ dB} \to G_{\text{Rx(SL)}} = 0.5, \quad \sigma = 12 \text{ m}^2 \quad f = 4 \text{ GHz}$$

Equation (3.27) is used to calculate the J/S for SOJ jamming from side lobe,

$$\frac{J}{S} = \frac{P_J G_{Ja} G_{Rx(SL)} 4\pi R_{R\text{-}T}^4 B_R}{P_{Tx}\ G_{Tx} G_{Rx} \sigma R_{J\text{-}R}^2 L B_J}$$

$$\frac{J}{S} = \frac{(15 \times 10^3)(10)(0.5)4\pi(30 \times 10^3)^4 (0.33)}{(180 \times 10^3)(1{,}000)^2 (12)(120 \times 10^3)^2 (1)} \approx 8.1 \rightarrow \left(\frac{J}{S}\right)_{dB}$$

$$= 9.08 \text{ dB}$$

To obtain the J/S for SOJ jamming from the main beam,

$$\frac{J}{S} = \frac{P_J G_{Ja} 4\pi R_{R\text{-}T}^4 B_R}{P_{Tx} G_{Tx} \sigma R_{J\text{-}R}^2 L B_J} = \frac{(15 \times 10^3)(10)4\pi(30 \times 10^3)^4 (0.33)}{(180 \times 10^3)(1{,}000)(12)(120 \times 10^3)^2 (1)} \approx 16{,}190$$

$$\left(\frac{J}{S}\right)_{dB} = 42.1 \text{ dB}$$

So the change rate in J/S for the main beam and sidelobe jamming is calculated as follows.

$$\left(\frac{J}{S}\right)_{dB\_ML} - \left(\frac{J}{S}\right)_{dB\_SL} = 42.1 \text{ dB} - 9.08 \text{ dB} = 33.02 \text{ dB}$$

Solution 3.6: This problem concerns a bistatic radar threat application defined in Case 3 in Figure 3.9(c). Primarily, let us define the given values in linear form,

$$P_{Tx} = 10 \text{ kW}, \quad P_J = 100\text{W}, \quad R_{J\text{-}Rx} = 3 \text{ km}, \quad R_{Tx\text{-}T} = 10 \text{ km}$$

$$L_{J(dB)} = 3 \text{ dB} \rightarrow L_J = 2, \quad L_{pol(dB)} = 3 \text{ dB} \rightarrow L_{pol} = 2$$

$$L_{R(dB)} = L_{Tx(dB)} + L_{Rx(dB)} = 4 + 3 = 7 \text{ dB} \rightarrow L_R = 5$$

$$G_{Tx(dB)} = 30 \text{ dB} \rightarrow G_{Tx} = 1{,}000, \quad G_{Rx(dB)} = 15 \text{ dB} \rightarrow G_{Rx} = 31.62$$

$$G_{Ja(dB)} = 8 \text{ dB} \rightarrow G_{Ja} = 6.31$$

$$\left(\frac{B_R}{B_J}\right) = -6 \text{ dB} \rightarrow \frac{B_R}{B_J} = 0.25, \quad \sigma = 12 \text{ m}^2, \quad f = 5.6 \text{ GHz}$$

(a) We can obtain the return power using (3.31),

$$S = \frac{P_{Tx} G_{Tx} G_{Rx} \sigma c^2}{(4\pi)^3 R_{Tx\text{-}T}^2 R_{J\text{-}Rx}^2 f^2 L_R} = \frac{(10 \times 10^3)(1{,}000)(31.62)(12)(3 \times 10^8)^2}{(1981)(10 \times 10^3)^2 (3 \times 10^3)^2 (5.6 \times 10^9)^2 (5)}$$

$$= 1.22 \text{ pW}$$

$$S_{dB} = -119 \text{ dBW} = -89 \text{ dBm}$$

(b) Now we can continue the calculation for J/S by using (3.33)

$$L = \frac{L_J L_{pol}}{L_R} = \frac{2 \times 2}{5} = 0.8$$

$$\frac{J}{S} = \frac{P_J G_{Ja} 4\pi R_{Tx-T}^2 B_R}{P_{Tx} G_{Tx} \sigma L B_J} = \frac{(100)(6.31)4\pi (10 \times 10^3)^2 (0.25)}{(10 \times 10^3)(1,000)(12)(0.8)} \approx 2,064$$

$$\left(\frac{J}{S}\right)_{dB} = 33.15 \text{ dB}$$

Solution 3.7: For this purpose, four radars are defined in power and gain values as in Table 3.1

No frequency information is required since both the jammer and radars operate at the same frequency. The assumption here is that the bandwidth ratio of B_R/B_J is adjusted to unity. The jammer power levels are 50 and 53 dBm. Also, a jammer horn antenna (MATLAB Antenna Designer) is simulated for 8 to 12 GHz. The radar azimuth locations according to the jammer and corresponding jammer antenna gains are given in Table 3.2.

The burn-through-range (BTR) calculations are made according to (3.15), and the MATLAB code given in Table 3.3 is used to obtain the BTR for two different jammer power levels.

The BTR results for two jammer power levels are in Table 3.4.

Table 3.1   Powers and Gain Values of Four Radars

Radar No.	Power (dBm)	Antenna Gain (dB)
1	80	28
2	83	30
3	70	30
4	76	37

Table 3.2   The Radar Azimuth Locations According to the Jammer

Radar No.	Azimuth Angle with Respect to Jammer Platform (Fighter) Heading	Corresponding Jammer Antenna Gain (dB) (at 10 GHz)
1	45°	−3.47
2	25°	7.48
3	350°	14.18
4	330°	4.21

Table 3.3   MATLAB Code for Burn-Through-Range Calculation

```
%% Power Definitions (dBm Tx power)
Prad_1=80; % First radar
Prad_2=83; % Second radar
Prad_3=70; % Third radar
Prad_4=76; % Fourth radar
Pj_1=50; % Jammer power first level
Pj_2=53; % Jammer power second level
%% Gain Definitions (dB)
Gj_R_1= 14.18; %jammer's Tx gain in the first radar's direction
Gj_R_2= 7.48; % in the second radar's direction
Gj_R_3= -3.46; % in the third radar's direction
Gj_R_4= 4.21; % in the fourth radar's direction
Grad_1=28; % Gain of the first radar's antenna
Grad_2=30; % the second radar's antenna
Grad_3=30; % the third radar's antenna
Grad_4=37; % the fourth radar's antenna
A_dB=6; % dB of J_mineff/S
%% BTR Calculation
RCS=10; %m2
% Tx powers of the radars in W
Prad_1_lin=10^((Prad_1-30)/10);
Prad_2_lin=10^((Prad_2-30)/10);
Prad_3_lin=10^((Prad_3-30)/10);
Prad_4_lin=10^((Prad_4-30)/10);
Prad=[Prad_1_lin, Prad_2_lin, Prad_3_lin, Prad_4_lin];
% Jammer Tx powers in W
Pj_1_lin=10^((Pj_1-30)/10);
Pj_2_lin=10^((Pj_2-30)/10);
Pj=[Pj_1_lin, Pj_2_lin];
% Radar Tx gains in ratio
Grad_1_lin=10^((Grad_1)/10);
Grad_2_lin=10^((Grad_2)/10);
Grad_3_lin=10^((Grad_3)/10);
Grad_4_lin=10^((Grad_4)/10);
Grad=[Grad_1_lin,Grad_2_lin,Grad_3_lin,Grad_4_lin];
% Jammer Tx gains in the ratio
Gj_R_1_lin= 10^((Gj_R_1)/10);
Gj_R_2_lin= 10^((Gj_R_2)/10);
Gj_R_3_lin= 10^((Gj_R_3)/10);
Gj_R_4_lin= 10^((Gj_R_4)/10);
Gj=[Gj_R_1_lin,Gj_R_2_lin,Gj_R_3_lin,Gj_R_4_lin];
% dB of J_mineff/S
A=10^(6/10);

for i=1:2 % for the loop of two different power levels of the jammer
for j=1:4 % BTR calculation loop for the four radars
BTR(i,j)=sqrt((Prad(j)*Grad(j)*RCS)*A/(Pj(i)*Gj(j)*4*pi));
end
end
BTR
```

Table 3.4 BTR Results for Two Different Jammer Power Levels

	Radar 1	*Radar 2*	*Radar 3*	*Radar 4*
Jammer Power	*BTR in m*			
50 dBm	276.3	1,062.7	838.3	1,548.4
53 dBm	195.6	752.3	593.5	1,096.2

References

[1]   James, I., "Modelling Ultraviolet Threats," *SPIE Proceedings Volume 9989, Technologies for Optical Countermeasures XIII*, Edinburgh, UK, SPIE Security & Defense, 2016.

[2]   Adam, T. E., (ed.), *Electronic Warfare*, New York: Nova Science Publishers, 2010.

[3]   AAP-6 (2021): NATO Glossary of Terms and Definitions (English and French), NATO Standardization Office (NSO), December 15, 2021.

[4]   Schleher, D. C., *Electronic Warfare in the Information Age*, Norwood, MA: Artech House, 1999.

[5]   Mahafza, B. R., *Radar Systems Analysis and Design Using MATLAB*, 3rd ed., Boca Raton, FL: CRC Press Taylor & Francis Group, 2013.

[6]   Li, H., et al., "DRFM System Based on the Principle of Radar Deception," *International Journal of Simulation Systems, Science & Technology*, Vol. 17, No. 37, 2016.

[7]   Avionics Department, *Electronic Warfare and Radar Systems Engineering Handbook*, Naval Air Warfare Center, Point Mugu, CA, 2013.

[8]   Joint Publication 3-01 (2012): Countering Air and Missile Threats, USA Joint Chiefs of Staff (CJCS), May 2, 2018.

[9]   Wiegand, J. W., *Radar Electronic Countermeasures System Design*, Norwood, MA: Artech House, 1991.

[10] Neri, F., *Introduction to Electronic Defense Systems*, 2nd ed., Norwood, MA: Artech House, 2001.

[11] Golden, A., *Radar Electronic Warfare*, New York: American Institute of Aeronautics and Astronautics, 1987.

[12] De Martino, A., *Introduction to Modern EW Systems*, 2nd ed., Norwood, MA: Artech House, 2018.

CHAPTER 4

Radar Warning Receiver Systems

The Rx are essential to any EW systems, even if they may be in EA, ES, or SPEW roles. In ES operations, the central part is Rx, and the hardware and signal processing capabilities utilized are the ultimate. However, in EA and SPEW systems, the main task is to use any form of ECM, including jamming. For this purpose, situational awareness must be provided to determine the victim or threat systems. The basic rule in EW is not to use any ECM until the presence of the victim or threat is detected. Thus, Rx systems are needed to detect the hostile weapon system.

For the SPEW systems, detecting radar-guided threats can be provided by radar warning receiver (RWR) systems. The RWR history is as old as that of the SPEW systems, and the RWR systems' development has been parallel with that of the SPEW systems. Modern RWR systems with high technology are utilized in modern SPEW systems. The RWR systems can be inspected in two main parts: the Rx section and the signal processing section. Thus, we will follow this scheme, start the RWR analysis with the Rx part, and then continue with the signal processing function of the RWR.

This chapter is dedicated to the architecture of RWR systems and their task in SPEW systems. It also explains the relationship between the RWR and SPJ. Furthermore, it contains various Rx types employed in RWR systems and its sensitivity and dynamic range concepts. The components included in common RWR architectures are another issue considered in this chapter. In the next chapter, we will discuss the signal processing part of the RWR.

4.1 RWR in Self-Protection EW System Architecture

Radar systems provide robust detection and guidance systems for mid-range and long-range, vital threats to airborne and shipborne platforms. Feeding the crew timely information about environmental signals is necessary for aircraft and vessel survival in an operational area. The RWR systems provide this vital information to the crew. They may be in various architectures, from simple to complicated and low to high cost, according to the subsystems, such as Rx, antennas, and signal processing. When designing the Rx, its components are selected considering different technologies, thus optimizing the cost and performance of the system. Depending on the technology used, the Rx components may be massive, lossy, or costly. RWR systems are installed on board platforms requiring protection against radar-guided threat systems.

The general architecture of SPEW systems is presented in Chapter 1. They are integrated systems with different subsystems designed to inform the pilots and crew about operational and situational changes and prevent their overloading. The RF band subsystem of a SPEW system mainly consists of RWR and SPJ units. Also, the chaff subunit of the CMDS can be considered a part of the RF band subsystem. The SPEW systems can detect radar-guided threats and produce countermeasures against them by the RF band subsystems. Even though using chaff and evasive maneuvers is vital for avoiding radar-guided weapon systems, the primary countermeasure is to utilize self-protection jamming.

First, to be able to use any countermeasures against a radar-guided threat, its existence must be determined. For this purpose, an RWR system, which is the essential subsystem of any SPEW system, is required. The primary purpose of an RWR system is to provide situational awareness for radar-guided threats and a regional determination of the *electronic order of battle* (EOB) for urgent action on airborne or shipborne platform survival. An RWR provides a local EOB for a single aircraft or vessel. However, threat geolocation systems can provide accurate threat location data for numerous aircraft over an entire region. Threat geolocation systems, a step above RWR systems, can deliver precise threat position information for many aircraft and vessels over a whole area to enable situational awareness [1].

Chapter 1 explained the place of RWR in the SPEW system for different platforms. The block diagram in Figure 4.1 shows the RF section in the SPEW system architecture and the functional relationship between RWR and the other RF subsystems.

In Figure 4.1, the generic SPEW systems consist of five separate functions, which may be added or removed according to the platform and operation type. These functions are classified into RF, EO, *expendable active decoy* (EAD), MWS, and EW *central management unit* (CMU) sections. Some subsystems, such as CMDS and MWS, are commonly used by different sections. When focusing on the RF section, we can summarize the functions as follows:

- After receiving signals in the dense threat environment, they are correctly separated and labeled. Then they are compared with the lists of data in the

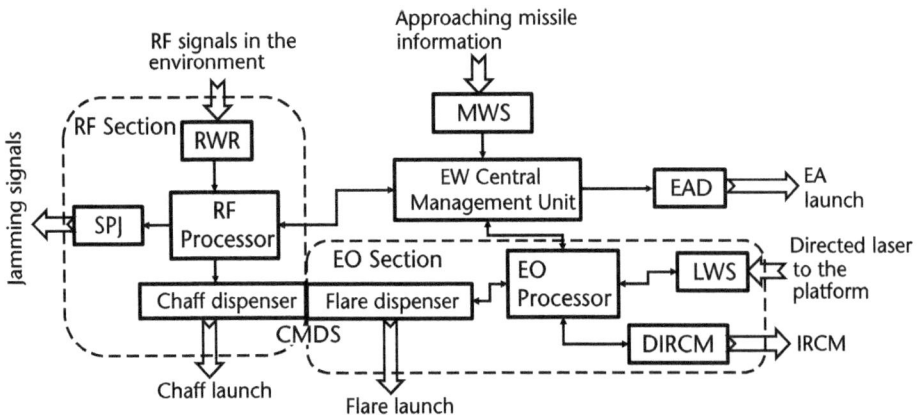

Figure 4.1   The relationship between a generic RF section and the SPEW system.

memory, called libraries, for identification. All these detection and identification processes are conducted by RWR and EW CMU subsystems.

- The MWS subsystem can detect impending missile threats. However, the threats' guidance methods cannot be confirmed when they are passive-guided systems since they do not have any indication. So the CW CMU decides on suitable countermeasures and evasive maneuvers for the situation, and both RF and EO sections are employed for avoidance.
- When the RWR detects an emission similar to one stored in its memory, it identifies the threat and gives the appropriate warning [2]. The memory and the processor may be in the EW CMU as a design consideration, but the function belongs to the RWR subsystem. In addition to the warning, the RF section provides some automatic countermeasures, such as jamming and chaff dispensing, and some manual precautions, such as directions for evasive maneuvers.

The RWR subsystem must detect and identify the threat system to activate the RF section against a threat and decide countermeasures. RWR systems also have an essential role in situational awareness. For this reason, they are indispensable main components of the SPEW systems.

4.2   Relation Between the RWR and the Self-Protection Jammer

The operational functions of the RF section's subsystems can be classified as active and passive parts. The active components are SPJ and chaff; however, passive jamming is the countermeasure obtained with chaff. The passive parts of the RF section are the RWR system, the essential passive part, and the MWS, which is used with the EO section. The passive elements detect the threat and inform the RF processor or EW CMU to use proper countermeasures. While MWS provides detection without identification, RWR fulfills deinterleaving, labeling, and identification in dense signal environments. Thus, RWR systems are the primary sensors for the RF section and the part that activates SPJ.

When using SPJ, some rules are considered, including emission control. An important rule in self-protection jamming is that jamming is to be applied when a threat is detected. This rule may not be so valid for stand-off jamming since there are many radars in an air defense system, and occasionally detecting their signal may not be possible. A predetermined jamming program is applied to suppress the air defense. Thus, detection is the first step of self-protection jamming. This detection is achieved by using RWR systems in self-protection systems.

Generally, enemy search radars detect airborne or shipborne platforms before tracking or locking on them. Radar-guided threat characteristics vary from search/acquisition to tracking radar. Different radars may be used in a missile site for search, acquisition, and tracking purposes. However, a platform may use the same radar for acquisition and tracking functions for fire control. The onboard RWR systems can distinguish between these radar modes based on the characteristics of incoming radar signals. For this purpose, the radar characteristics must be known by SIGINT studies and programmed into the SPEW system. Then, identifying the radar by RWR, SPJ takes the information and applies jamming according to the preprogrammed conditions.

The ELINT operations to obtain the required intelligence about the threat systems of the operation region are vital for preflight programming activities. The product of these activities is the *preflight message* (PFM), which consists of data to be registered in formatted tables in the main execution program of the EW CMU. The main execution program of the EW CMU is the *operational flight program* (OFP). Generally, PFPs are loaded to the EW CMU separately from the OFP according to the threats of the operation region, and for operational decisions, the OFP utilizes the loaded PFM. When considering the SPEW system's RF section, OFPs manage the RWR and SPJ systems' operations according to the PFM's data. The chaff dispenser's usage and the proper maneuvering direction are also determined similarly.

The general relationship between an RWR and an SPJ is depicted in Figure 4.2. As the figure shows, there are data and RF connections between the RWR and SPJ. The configuration and components of the systems may vary; however, the general logic and requirements are similar for the RF part of a SPEW system. Technological developments in hardware and artificial intelligence, machine learning, and deep learning methods have recently acquired some additive properties and capabilities. According to the developments in hardware, the achievements are mainly focused on simultaneously identifying the number of threats and the number of threats that applied to jam by time-sharing or frequency-sharing. The developments in artificial intelligence, machine learning, and deep learning, expected to continue at an increasing level, are reducing processing time, data pruning, decision-making, and resource allocation.

The main central unit of the SPJ system is the ECM generator, which conducts the defined jamming methods used against threats in the PFM. For this purpose, the ECM generator manages DRFM, repeater, and switch units. However, the main central unit of RWR is the analysis processor. It controls the Rx units and conducts signal detection, deinterleaving, and labeling processes. Furthermore, the analysis processor utilizes the threat library defined in the PFM to identify deinterleaved and labeled signals. The RWR system analyses processor exchanges data with control and display units via SPEW CMU or directly according to the system design.

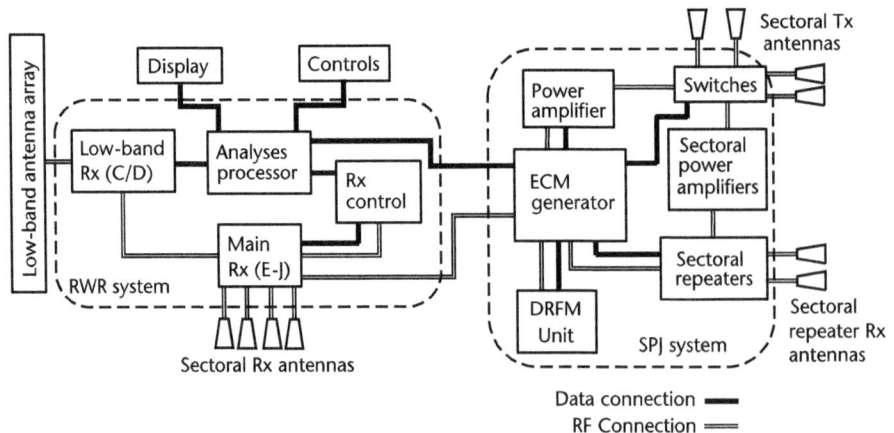

Figure 4.2 A generic block diagram of the relationship between the RWR and the SPJ.

Figure 4.2 shows the functional block diagram, and the implementation of functions may change from application to application. For example, the RWR and ECM generator functions may be in a distributed architecture, as shown in Figure 4.2, or these functions may be gathered in SPEW CMU, and all the parts are conducted from a central processor. In any case, the ECM generator function takes the active threat information in the theater from the analyses processor function of the RWR system. Also, it continuously senses the activation status of the threats via the RWR's Rx unit and the repeaters' Rx systems throughout the jamming.

The coordinated operation of RWR and SPJ systems is an important topic on which to focus. SPEW systems use different coordination methods to use RWR and SPJ simultaneously. These methods are briefly summarized as follows:

- *Look-through:* This is a time-sharing method for the joint operation of RWR and SPJ systems. When using the noise jamming method, the noise signal bandwidth is aimed to center on the frequency of the victim signal. When the Rx and Tx antennas are not sufficiently isolated from each other, a procedure should be established to interrupt the jamming and detect the environment. The SPEW systems employ a look-through mode to monitor the threat environment while simultaneously jamming multiple emitters periodically. The objective of the look-through method is to allow the jammer to update threat radar parameters and change the jamming signal to respond to changes in the signal environment while simultaneously jamming the threat emitters.
- *Look-around:* This is a frequency-sharing method for the joint operation of RWR and SPJ systems in noise-jamming mode. In this method, the jamming does not have to be interrupted since the RWR and repeaters' Rx units scan the frequency bands other than the jamming bands. The detection process continues during jamming; however, the jammed frequency bands cannot be monitored.
- *Look-over:* A different method for the joint operation of RWR and SPJ may be defined according to the RWR's Rx sensitivity adjustment. For this purpose, when the jammer is in operation, the sensitivity of the RWR's Rx gets low, which prevents detecting even high-power RF emissions. However, when the jammer is off, RWR sensitivity gets high, and the threat emissions are monitored in the operational area. As it is understood, the jamming must be interrupted for some periods, as in the time-sharing look-through method for look-over operation.

As seen from the definitions and aims of the look-through, look-around, and look-over, they have some advantages and disadvantages on each other. However, these disadvantages can be eliminated by using them together in a logical procedure.

4.3   Rx Types Used in RWRs

Rx systems for detecting radar threats participate in many ES and ELINT operations applications. Even though ELINT Rx systems do not have to make immediate, life-or-death decisions like their ES counterparts, the Rx systems used in these operations have similar properties and specifications. Furthermore, ES and ELINT systems may be tasked to gather the information that meets both requirements simultaneously. The distinction between whether a given asset is performing an ES mission or an ELINT mission is determined by who tasks or controls the collection assets, what they are tasked to provide, and for what purpose they are charged [3].

The intelligence missions aim to provide the EOB to support the reprogramming of EW systems, maintain appropriate radar threat EW data, and ensure situational awareness of EW assets by accounting for organic and requested capabilities on the intelligence, surveillance, and reconnaissance synchronization matrix. However, ES missions consist of searching for, intercepting, identifying, and locating or localizing sources of intentional or unintentional radiated EM energy. The purpose of ES tasking is immediate threat recognition, targeting, planning and conduct of future operations, and other tactical actions such as threat avoidance, targeting, and homing. ES operations intend to respond to an immediate operational requirement.

The ES applications used against radar threats may be written for different EW branches as follows:

- *For the support of EP:* Self-protection systems use ES systems against radar threats as a subunit for different airborne and shipborne platforms. These ES subunits are RWR systems.
- *For the support of EA activities:* Airborne and land-based SOJ systems utilize ES systems for detecting the RF activations in a particular region or the whole theater.
- *The ES supports different purpose platforms:* ES systems may be utilized in different platforms, such as *airborne E/W and control* (AEW&C) and maritime airborne ES systems. Furthermore, satellite-based ES systems will be engaged soon. These systems use ES to detect, identify, and track electronic transmissions from the ground, airborne, and maritime sources. Using the ES systems, AWACS can determine radar and weapons system types and immediately provide this information to the mission units for taking measures and situational awareness.

Our main aim is to discuss the RWR architectures and the Rx types used in these systems, so let us focus our inspection on the RWR systems. Several different kinds of Rx are used in the architectures of RWRs. Since the operational requirements of the RWR systems cover many other situations, using a single Rx in the RWR architecture will only partially meet the total needs. Generally, a broadband Rx is used for initial signal detection and the steering of a narrowband Rx to the correct frequency for complete signal analysis. This part will further explain the Rx architectures and operational principles.

4.3.1 Crystal Video Receiver

The most straightforward Rx architecture, most used in the old-fashioned, conventional RWR, is the crystal video receiver (CVR). This type of Rx can handle pulse signals but not CW signals. CVRs offer respectable performance despite their essential conceptual simplicity. A CVR can cover an extensive frequency bandwidth but cannot discriminate the frequency of the input signals. Since the RF bandwidth is wide, the sensitivity of the Rx is relatively low. The Rx cannot separate input signals by frequency, so detecting simultaneous signals is impossible.

The main component of the CVR is the crystal detector. Figure 4.3 shows a simplified block diagram of a high-sensitivity CVR using a bandpass filter and a preamplifier as a front component. The antenna output feeds the bandpass filter and preamplifier; however, in more superficial CVR architectures, they may not exist. The outcome of the preamplifier filter is applied to a crystal detector. Following the detector is a video amplifier that amplifies the detector output and passes the video signal to the input of a comparator. The CVR allows for measurements of *pulse width* (PW), *pulse amplitude* (PA), *time of arrival* (TOA), and *direction of arrival* (DOA).

In a CVR, the transmission signal from a radar falls upon a wideband receiving antenna, which feeds a bank of sequential devices, as shown in Figure 4.3. The critical component in a CVR is the crystal detector. The output of an antenna feeds directly to a detector. Following the detector is a video amplifier that amplifies the detector output and passes the video signal to the input of a comparator. An input signal with sufficient energy will cross the threshold of the comparator and be detected. An input signal with sufficient energy crosses the threshold of the comparator and can be detected. The video signal at the detector's output is too weak for detection. Therefore, a video amplifier is used to increase the detection probability. The receivers are each tuned to consecutive slices of the covered band, which allows simultaneous reception and discrimination of radars operating in various parts of the band.

In general, a CVR's sensitivity depends on the crystal detector's property and the noise figure of the video amplifier. From a different point of view, the sensitivity is limited by the RF gain instead of by the noise in the Rx. The typical approach to improve the sensitivity of a CVR is to enhance the property of the diode detector and reduce the video amplifier noise. The most effective way to enhance the sensitivity is to add RF amplifiers before the detector. A CVR generally covers an extensive frequency bandwidth, and finding RF amplifiers to cover the desired bandwidth is usually a challenging problem.

Some of the SPJ and RWR systems that use CVR systems in their schemes are given next [4]:

- *AN/ALQ-167(V):* The AN/ALQ-167 is a podded SPJ system for fighters and choppers with noise, deception, and combination jamming modes. It is fitted

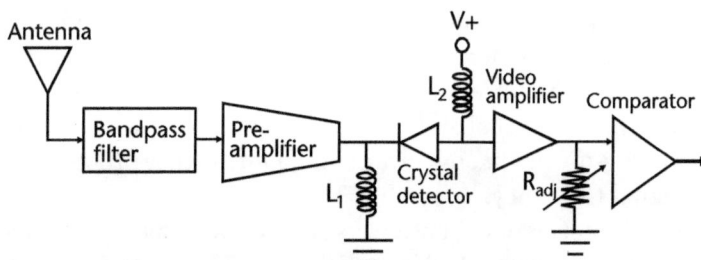

Figure 4.3   A simplified block diagram of a high-sensitivity CVR.

with a single crystal video or a dual system with crystal video and SHR for the RWR subsection. Continuous fore and aft coverage in the 2–18-GHz range.

- *AN/ALQ-184 (V):* This system consists of radar jamming and receiving pod for fighters with CVR subsystems and a signal processor. It has DF capability, rapid RF signal processing wide open in angle and frequency, and selective directional high ERP against multiple emitters.
- *AN/ALR-45/45F:* This is a family of RWR and countermeasures systems for fighters. The systems have CVR and reprogrammable software.
- *AN/ALR-46 (V):* This digital RWR system utilizes 2 to 18-GHz CVR and a reprogrammable digital processor. Its modular design allows for the analysis of CW emitters and performs C/D-band direction-finding.

The operating frequency band of the self-protection systems can be divided into different bands based on the size of the waveguide used for the transmitted and received signals. The bandwidths of each band can be very wide. The traditional band designation covers the spectrum from 2 to 18 GHz in four bands as follows [5]:

- 2–4 GHz: S or E-F-bands;
- 4–8 GHz: C or G-H-bands;
- 8–12 GHz: X or H-I-J-bands;
- 12–18 GHz: Ku or J-bands.

A design can be realized by using a CVR for each band, and the RWR covers the entire radar signal spectrum by four Rx. If there are four radar signals, each of which falls into one of these four bands, the receiver can detect four simultaneous signals.

Also, each frequency band may have more than one Rx; say, *n* Rx are used for each band. This design needs "4xn" Rx for the whole RWR system. If two radar signals of different frequencies in the same band arrive at the Rx simultaneously, the related Rx may produce false data. CVR systems can only measure the frequency in very coarse resolution, that is, the bandwidth of each band.

4.3.2 Instantaneous Frequency Measurement Receiver

A conventional instantaneous frequency measurement (IFM) Rx achieves a near unity pulse *probability of intercept* (POI) for pulse and CW signals in the frequency band without tuning. 2–18-GHz versions of IFM Rx are readily available for ES applications. Their sensitivity and dynamic range are satisfied when using a limiting amplifier at the input to the phase discriminator via a power divider. The phase discriminator's function is to accept an RF input and produce two video signals with the help of detectors, *lowpass filters* (LPFs), and amplifiers. The angular component can be scaled to read the frequency if these two signals are applied to a polar display's vertical and horizontal plates. However, the signals can be passed to an *analog-to-digital converter* (ADC) for further processing to calculate the RF input signal frequency.

The advantages of IFM receivers are their wide bandwidth and ability to measure radar frequency accurately. Nevertheless, their inability to measure multiple simultaneous signals is a severe limitation. If two signals coincide, either the IFM measures only the strongest one, or if the signals are similar in power level, then the frequency measured is an average of the frequencies of the two signals. If the IFM receiver sees a continuous signal, any subsequent weaker signals may be lost altogether.

A conventional IFM Rx has a *bandpass filter* (BPF), a *limiting amplifier* (LA), a power divider, a delay line, a discriminator circuit, four diode detectors, four LPFs, and two *differential amplifiers* (DAs) in its RF part. Also, an IFM Rx contains an ADC and processor for processing the video outputs to find the RF input signal's frequency information. Figure 4.4 shows a block diagram of a basic IFM Rx.

The bandpass filter defines the operating frequency band. Utilizing a limiter amplifier aims to solve the dynamic range and sensitivity problems. Low-power input signals can be considerably amplified, whereas high-power signals are limitedly amplified to prevent the limiter amplifier from overdriving the discriminator circuit.

The discriminator circuit is the most complex component for turning the RF signal in the output of the limiting amplifier into two video signals, one sin, and one cos component. The discriminator circuit is a two-input and four-output passive component with three 90° hybrid couplers and one 180° *hybrid coupler* (HC) [6]. A hybrid coupler, or a 3-dB coupler, has the characteristics of dividing the input power into two paths that have equal powers when appropriately terminated. The signals' amplitude decreases by a factor of $\sqrt{2}$ whenever it passes through either a power divider or a phase shifter. In Figure 4.4, the input-output relationship of a 90° HC is given. A signal going through the direct path will have a 90° phase while going through the diagonal path. Thus, the phase difference between the two outputs is 90°. However, for the 180° HC, the phase difference between the signals passing through the direct path and the diagonal path is 180°.

Let us take into account Figure 4.4 and consider a fixed frequency, constant amplitude, sinusoidal signal $2\sqrt{2}A\cos(\omega t)$ with zero phase shift at the output of the limiting amplifier, and this signal is divided into two equal portions. One-half of the signal is delayed with τ, and the outcome of the delay line is $2A\cos(\omega t - \omega \tau)$, and the other half is undelayed as $2A\cos(\omega t)$. After they pass through the 90° and 180° HCs, four terms are shown in Figure 4.4. At the output of the discriminator circuit, the signals (S_n) are defined as follows:

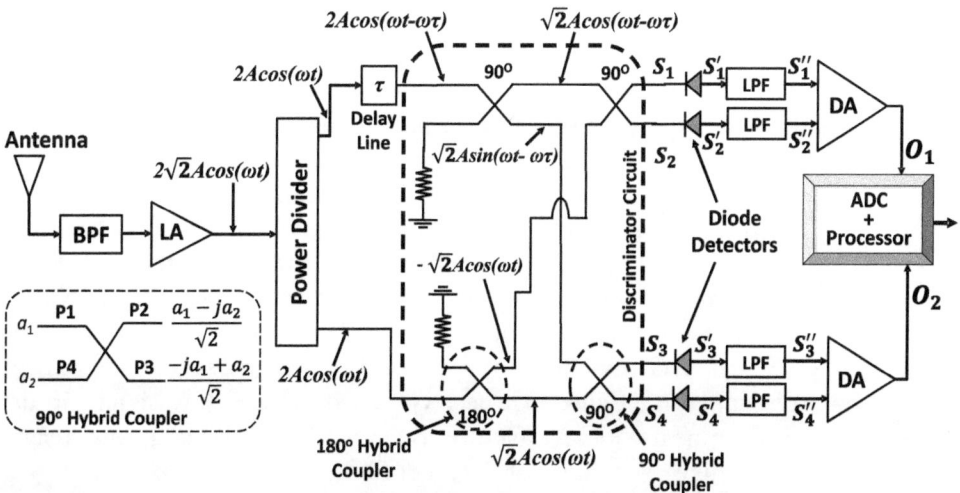

Figure 4.4   Block diagram of a conventional IFM Rx.

$$S_1 = A\cos(\omega t - \omega\tau) - A\sin(\omega t) \quad S_2 = A\sin(\omega t - \omega\tau) - A\cos(\omega t)$$
$$S_3 = A\sin(\omega t - \omega\tau) + A\sin(\omega t) \quad S_4 = -A\cos(\omega t - \omega\tau) + A\cos(\omega t)$$

$$(4.1)$$

The diode detectors are square-law devices; thus, the signals at the detectors' output (S'_n) are the squares of their counterparts given in (4.1). The signals at the detectors' output are shown below:

$$S'_1 = S_1^2 = A^2\cos^2(\omega t - \omega\tau) - 2A^2\cos(\omega t - \omega\tau)\sin(\omega t) + A^2\sin^2(\omega t)$$
$$S'_2 = S_2^2 = A^2\sin^2(\omega t - \omega\tau) - 2A^2\sin(\omega t - \omega\tau)\cos(\omega t) + A^2\cos^2(\omega t)$$
$$S'_3 = S_3^2 = A^2\sin^2(\omega t - \omega\tau) - 2A^2\sin(\omega t - \omega\tau)\sin(\omega t) + A^2\sin^2(\omega t)$$
$$S'_4 = S_4^2 = A^2\cos^2(\omega t - \omega\tau) - 2A^2\cos(\omega t - \omega\tau)\cos(\omega t) + A^2\cos^2(\omega t)$$

$$(4.2)$$

The signals are the inputs of the LPFs, and the high-frequency terms in the S'_n signals are stopped. Thus, the following signals are obtained at the LPFs' output:

$$S''_1 = A^2 - A^2\sin(\omega\tau) \quad S''_2 = A^2 + A^2\sin(\omega\tau)$$
$$S''_3 = A^2 + A^2\cos(\omega\tau) \quad S''_4 = A^2 - A^2\cos(\omega\tau)$$

$$(4.3)$$

The S''_1 and S''_2 are applied to one DA, and the signals S''_3 and S''_4 are applied to the other DA. The DAs perform mathematical subtraction, so their outputs are obtained as follows,

$$O_1 = S_{DA\_upper} = |S''_1 - S''_2| = 2A^2\sin(\omega\tau)$$
$$O_2 = S_{DA\_lower} = |S''_3 - S''_4| = 2A^2\cos(\omega\tau)$$

$$(4.4)$$

The results from (4.4) convey the frequency information for the input RF signal. An ADC digitizes these signals and then gives them to a processor to determine their frequency.

Generally, IFM Rx systems are used in collaboration with other Rx types of Rx in RWR architectures. Some examples of the RWR and ESM systems that use IFM Rx systems in their schemes are presented here [4]:

- *AN/ALR-93 (V):* The AN/ALR-93 (V) family is an RWR system developed for fixed-wing aircraft, helicopters, and UAVs. The operating frequency band is 0.5–20 GHz. Independent CVR, IFM, and superheterodyne Rx types are cooperative for detecting pulsed, pulsed Doppler, CW, LPI, jitter/stagger, pulse compression, and frequency agile radar types with 360° azimuth coverage.
- *EWS-A AIGLE:* This is an RWR and countermeasures management system developed for fighters. It detects pulsed PD and CW signals in dense EM environments. Its architecture uses IFM Rx, offering a short reaction time and a high probability of intercept (POI).

- *BLQ-355:* The BLQ-355 is a submarine ESM system with IFM Rx covering 2 to 20 GHz. It offers high POI and identifies and tracks radar emitters.
- *CS-3600:* The CS-3600 is an ESM system designed for naval and ground-based applications, covering the 2 to 18-GHz frequency range. Its dual Rx architecture meets the ESM requirements for high POI and the ability to work in dense RF environments. The CS-3600 ESM System contains IFM and channelized Rx systems.

4.3.3   Tuned RF Receiver

The general architecture of the tuned RF (TRF) Rx is simple, but they offer gradually adjustable RF and provide frequency selectivity. An Rx's selectivity measures its ability to discriminate between a wanted signal to which it is tuned and unwanted signals. In this context, avoiding interference from adjacent channels is the essential indication of selectivity. The most straightforward approach is to filter out the spectral contents outside this channel and amplify the desired signal in one or more RF amplification stages.

TRF Rx systems were used extensively in the early stages of wireless transmission technology. However, with the advancement of technology, they are rarely used today and have been replaced by other Rx types that offer much better performance. TRF Rx utilize adjustable or tunable BPFs around the desired frequency. The filters and amplifiers used in TRF Rx generally consist of many layers. The block diagram of a TRF with a two-layer RF filter is shown in Figure 4.5. These filters apply the gain directly at the received signal frequency.

The TRF Rx uses a scheme consisting of a tuned YIG bandpass filter and a CVR with a narrow frequency band, as seen in Figure 4.5. The TRF is a simple and logical Rx. The merits of this Rx are its simplicity and high sensitivity. Two, as in Figure 4.5, or sometimes three *RF amplifiers* (RFAs), all tuning together, cascade employed to select and amplify the incoming frequency and simultaneously to reject all others. After the signal is amplified to a suitable level, it is demodulated or detected.

Using multiple stages from the RF amplifier architecture with the RF filter before the CVR structure enables the CVR to detect numerous signals simultaneously. Moreover, this method improves the sensitivity of the limited RF bandwidth. An

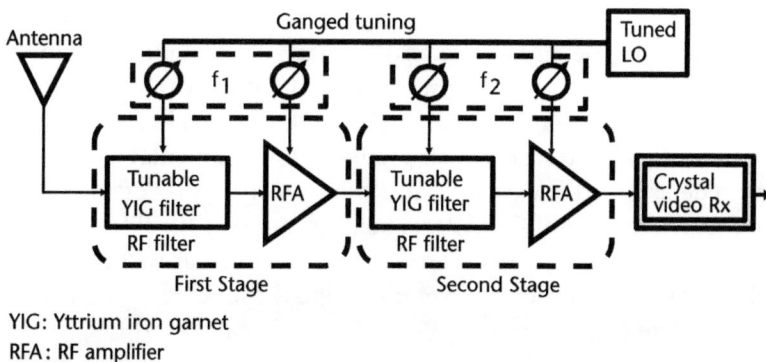

Figure 4.5   Block diagram of a conventional TRF Rx.

additional preamplifier can be placed before the TRF Rx to expand the dynamic range in applications.

The SNR and the frequency range of the Rx dominantly provide the performance specifications for TRF Rx. The selectivity of the Rx is also governed by the tuned circuit in the RFA stage. LC resonant circuits provide tuning, mostly made with tunable capacitance. The bandwidth also changes as the resonant frequency of the tuned circuit changes. This change in the bandwidth means that the Rx selectivity changes with its tuned frequency. Variable selectivity is an undesirable characteristic since the selectivity is a measurement of the Rx's ability to reject unwanted signals. Thus, the emerging problems in the TRF operation supersede the operational benefits. The issues of the TRF, such as instability, insufficient adjacent-frequency rejection, and bandwidth variation, can be solved using a superheterodyne receiver, which introduces relatively few problems.

4.3.4 Superheterodyne Receiver

The development of *superheterodyne (superhet) Rx* (SHR) is quite old, returning to the 1930s. The simplicity in the scheme of the superhet approach has replaced the TRF Rx designs, which were the most influential architecture until the development of the superhet. In the TRF Rx architecture, there are weaknesses in the selectivity feature that provide discrimination between neighboring stations and inadequacies in the system gain. These deficiencies can be overcome by using SHR.

The word heterodyne means combining two frequencies in a nonlinear device (mixer) or converting one frequency to another using a nonlinear mixer. The prefix super indicates that the difference frequency obtained from the two frequencies applied to the mixer is in the supersonic frequency level, in the supersonic region, where analog signals can be quickly processed. The quality of the SHR has been greatly improved since its initial design, and today, its basic architecture is still widely used in RF applications. Many systems use the SHRs since their gain, selectivity, and sensitivity are superior to other Rx types.

When operated in surveillance mode, narrowband SHR systems offer high sensitivity and an extended dynamic range, but there is a trade-off between resolution and POI. Precise improvements in the POI can be obtained by processing a wider bandwidth at the IF stage and using parallelism or high-speed digital processing to improve resolution. Subject to the type of postprocessing being utilized, the detection bandwidth can be far more extensive than the resolution obtained. If the detection bandwidth is larger than the resolution, the Rx is called a wideband SHR.

The block diagram of a narrowband SHR, given in Figure 4.6, is highly flexible. It uses a linear detector, which provides the best available sensitivity as a function of predetermined bandwidth and postdetection processing gain. A narrowband SHR converts part of the RF frequency range to a fixed IF level using a tuned *local oscillator* (LO). A fixed IF level effectively increases the gain and ensures filter selectivity. A frequency control unit adjusts the *RF filter bandwidth* (B_{RF}) and the *IF filter bandwidth* (B_{IF}).

The conversion from RF to the final IF may be done in two or three steps to reduce image products and spurious signals. The first IF value may depend on the RF band selected, but the final IF is constant regardless of the frequency at which

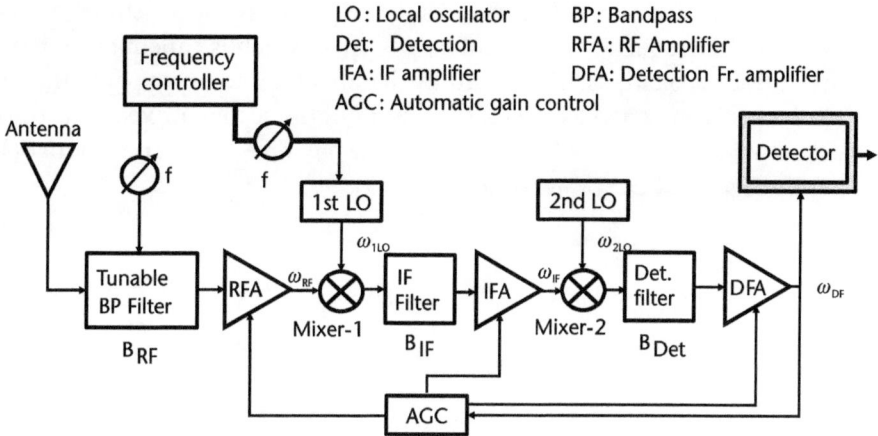

Figure 4.6   Block diagram of a narrowband SHR.

the Rx is tuned. The second IF filter is called the detection filter and has a frequency bandwidth of B_{Det}. It defines the detection bandwidth of the Rx.

An interception occurs when the detection filter is tuned to a particular frequency coincident with signal activity at that frequency. Detection occurs when the signal energy measured within the detection filter exceeds a certain threshold. To understand the detection signal obtained from the RF signal, consider a mixer's operating principle. A standard power series can approximate the nonlinear relationship between the input and output of a mixer,

$$i(t) = a_0 + a_1 v(t) + a_2 v^2(t) + a_3 v^3(t) + \ldots + a_n v^n(t) \tag{4.5}$$

where $i(t)$ is the current flowing through a diode, a_0, a_1, a_2, a_3, ..., and a_n are constants, and $v(t)$ is the applied voltage across the diode [6]. Let us assume that the input signal of the first mixer, $v_1(t)$, is the sum of the output of the RF amplifier ($V_{RF} \sin \omega_{RF} t$) and the first LO ($V_{1LO} \sin \omega_{1LO} t$).

$$v_1(t) = V_{RF} \sin \omega_{RF} t + V_{1LO} \sin \omega_{1LO} t \tag{4.6}$$

For simplicity, phase angles for both components in (4.6) are neglected. Thus, we can write the output of the mixer up to the second-order power series as follows:

$$i_1(t) = a_0 + a_1 \left(V_{RF} \sin \omega_{RF} t + V_{1LO} \sin \omega_{1LO} t \right) + a_2 \left(V_{RF} \sin \omega_{RF} t + V_{1LO} \sin \omega_{1LO} t \right)^2 \tag{4.7}$$

The first two terms in the above equation represent a DC component and the original signal, respectively. The third term is defined as follows:

$$a_2 \left(V_{RF} \sin \omega_{RF} t + V_{1LO} \sin \omega_{1LO} t \right)^2 = a_2 V_{RF}^2 \sin^2 \omega_{RF} t + a_2 V_{1LO}^2 \sin^2 \omega_{1LO} t + \ldots$$
$$\ldots \frac{a_2}{2} V_{RF} V_{1LO} \left[\cos \left(\omega_{RF} - \omega_{1LO} \right) t - \cos \left(\omega_{RF} + \omega_{1LO} \right) t \right] \tag{4.8}$$

The last term in (4.8) generates the required frequency shifts $\omega_{RF}\text{-}\omega_{1LO}$, which is called IF frequency (ω_{IF}), and $\omega_{RF} + \omega_{1LO}$, which is called image frequency. The IF filter is a bandpass filter with B_{IF} bandwidth and passes only the IF. A similar process is conducted to obtain the output of the second mixer. This time, the input signal of the mixer, $v_2(t)$, is the sum of the outcome of the IF amplifier ($V_{IF}\cos\omega_{IF}t$) and the second LO ($V_{2LO}\sin\omega_{2LO}t$),

$$v_2(t) = V_{IF}\cos\omega_{IF}t + V_{2LO}\sin\omega_{2LO}t \rightarrow \begin{cases} V_{IF} = \dfrac{a_2}{2}V_{RF}V_{1LO} \\ \omega_{IF} = \omega_{RF} - \omega_{1LO} \end{cases} \tag{4.9}$$

The output of the second mixer up to the second-order power series is as follows:

$$\begin{aligned} i_2(t) = b_0 &+ b_1\left(V_{IF}\cos\omega_{IF}t + V_{2LO}\sin\omega_{2LO}t\right) \\ &+ b_2\left(V_{IF}\cos\omega_{IF}t + V_{2LO}\sin\omega_{2LO}t\right)^2 \end{aligned} \tag{4.10}$$

As in (4.7), we continue the calculation with the third term, which is defined as

$$\begin{aligned} b_2\left(V_{IF}\cos\omega_{IF}t + V_{2LO}\sin\omega_{2LO}t\right)^2 &= b_2V_{IF}^2\cos^2\omega_{IF}t + b_2V_{2LO}^2\sin^2\omega_{2LO}t + \ldots \\ \ldots \frac{b_2}{2}V_{IF}V_{2LO}&\left[\sin\left(\omega_{IF} + \omega_{2LO}\right)t - \sin\left(\omega_{IF} - \omega_{2LO}\right)t\right] \end{aligned} \tag{4.11}$$

The last term in (4.11) generates the required frequency shift $\omega_{IF}\text{-}\omega_{2LO}$, called detection frequency (ω_{DF}), and the bandpass detection filter with B_{Det} bandwidth passes only the detection frequency.

Regarding the narrowband SHR scanning, the processing and the data collection durations are negligible. The main limiting factor to scanning a particular frequency range of interest is the settling time of the detection filter. The maximum scanning speed is proportional to the square of the detection bandwidth. The resolution is equal to the detection bandwidth during the scanning of a narrowband SHR, so the maximum scanning speed is proportional to the square of the resolution bandwidth.

Since the POI is proportional to the scanning speed, it decreases with increased resolution. However, reducing the resolution or expanding the detection bandwidth decreases the SNR within the detection bandwidth. This situation can distort the probability of detection for a given interception.

As in any Rx, for a narrowband SHR, the minimum signal detection capability depends on sensitivity, while the maximum signal detection ability is generally considered the input signal strength at which the output distortion becomes unacceptable. The input-output signal power transfer function of the RF and IF amplifier stages, assuming no *automatic gain control* (AGC), is shown in Figure 4.7(a). There are two regions of operation: the linear region and the compression region. In the linear region, a 1-dB increase in the input signal strength results in a 1-dB increase in the output signal strength. However, in the compression region, the linearity collapses.

The point at which the output deviates from the estimated linear response by 1 dB is the 1-dB compression point. Considering the input, this point is assumed to be the upper limit to the range of signals the Rx can process.

The difference between the 1-dB input compression point (in dBm) and the minimum discernible signal (in dBm) gives the single-tone dynamic range. However, this definition is gain-dependent since the 1-dB input compression point is inversely proportional to the gain. As a result, it takes work to compare amplifiers with different gains. When AGC is turned off, the sum of the dynamic range and the RF to IF gain remains constant. Employing an AGC is the prevalent method to improve the single-tone (single RF signal) dynamic range. This device monitors signal strength at some point in the Rx; in Figure 4.6, this point is the output of the detection IF amplifier and adjusts gain levels accordingly to ensure all amplifiers operate in their linear regions. The input-output transfer function of the RF to IF stage varies with the input signal level in Figure 4.7(b).

Although the two transfer function curves are very similar, the difference between them can be explained according to the compression. In the scheme of Figure 4.7(a), the amplifiers do not compress within the AGC control range; however, in Figure 4.7(b), the amplifiers apply compression within the range.

The disadvantage with narrowband SHR configuration is that, while tuned to one band, it can fail to detect a threat operating in another and transmitting for a short time. A wideband SHR is similar to its narrowband counterpart. However, in the wideband case, the detection bandwidth is larger than the ultimate frequency resolution. In most cases, it is significantly larger. In practice, this type of Rx employs a wideband filter in the RF stage and conducts further processing of the IF output. Today's ADC sample rate is high enough to directly sample the intermediate high IF following that first mixer mentioned in the narrowband SHR. The block diagram of a wideband SHR is shown in Figure 4.8.

Postprocessing in a wideband SHR conducts detecting signals and enhances the resolution. Thus, the processing is generally more complex than in the narrowband case. The processor consists of a channelized arrangement comprised of a bank of narrowband filters, each followed by an ADC followed by fast Fourier transform

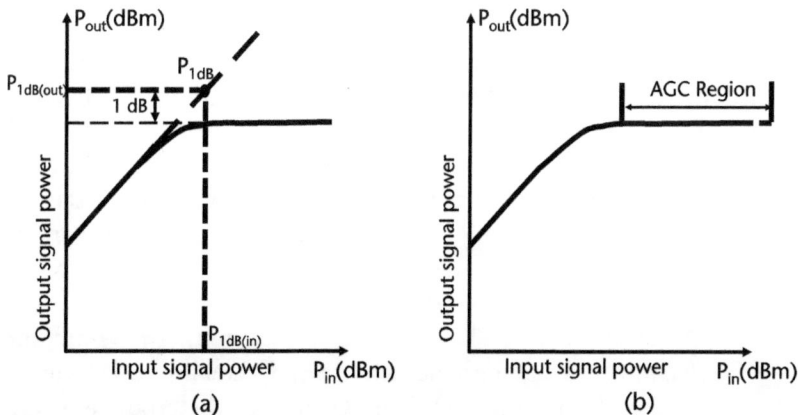

Figure 4.7   The input-output signal power transfer function of the RF and IF amplifier stages: (a) without AGC and (b) with AGC.

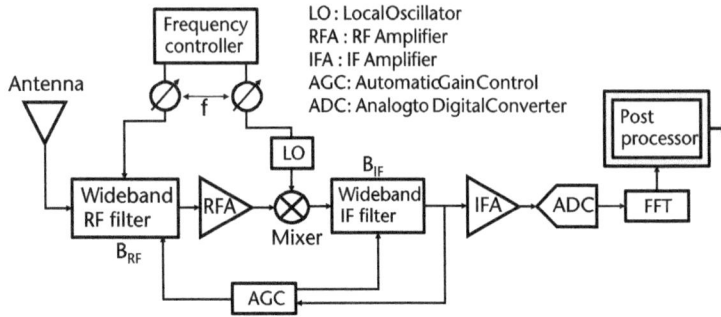

Figure 4.8 Block diagram of a wideband SHR.

(FFT) hardware. The detection in the FFT processor is performed either in hardware or software based on the FFT. The processing gain is an enhancement of SNR and plays a significant role in the dynamic range performance of the Rx.

Here are some examples of the SPEW, RWR, and ESM systems that engage SHR systems in their architectures [4]:

- *AN/ALR-56A/C:* The AN/ALR-56 family represents RWR systems for fighter aircraft that interface with an AN/ALQ-135 jammer to form a tactical EW system. The system contains a processor/low-band Rx, a high-band Rx, a display, a power supply, a control unit, a towed decoy controller, and an antenna array. The AN/ALR-56C version is digitally controlled. It contains dual-channel SHR covering E-J bands.
- *SPS-1000V-5:* This RWR system, designed for fighter and transport aircraft, incorporates an SHR-IFM Rx with a millimeter-wave option. The system receives pulse, CW, high PRF, and pulse Doppler signals.
- *EWS-16:* The EWS-16 is an internally mounted self-protection system for the F-16 fighter aircraft, including RWR, jamming, and locating subsystems. It is designed to detect, identify, and localize all modern threats without ambiguity in dense electromagnetic environments. The system includes a 32-bit processor-based management unit, CVR and high-speed wideband SHR, an instantaneous direction-finding interferometer, and an IFM Rx.
- *AN/ALQ-178(V):* The AN/ALQ-178 is a family of advanced, internally mounted self-protection systems designed and developed for fighter aircraft. It is a proven self-protection suite of integrated RWR and ECM systems, providing the pilot with situational awareness and deceptive RF jamming. Versions have wideband SHR and digital Rx capabilities to deliver precision parameter measurement and required sensitivity for timely and accurate threat detection and identification. It has separate forward and aft jammers and automatic countermeasures dispensing system. Also, its display shows the threat radar type and direction.
- *AN/ALQ-202(V):* The AN/ALQ-202 is an internally-mounted autonomous RF jammer system designed for use on fighter aircraft. It can operate in stand-alone mode or coupled with an RWR. Superheterodyne-based Rx technology provides frequency selectivity, sensitivity, and intrapulse capabilities.

- *AN/ALR-87:* The AN/ALR-87 is an advanced radar threat warning system for fighters. Its frequency coverage consists of the C-J-band. The design incorporates quadrant channelized CVR, distributed narrowband SHR, and IFM Rx systems.
- *NS-210E:* This system is a lightweight shipborne ESM for the real-time EOB generation and immediate threat warning. The NS-210E comprises an omnidirectional antenna, high-gain directional antenna, front-end RF unit, Rx/processor, and operator's console. Its features are counted as automatic detection and identification of radars, high POI, long detection range, accurate DF, fast reaction for threat warning, audio, and visual alert signals; combined digital IFM/SHR; presentation of EOB, handling of the dense environment and exotic signals; and automatic or manual operation, recording data for analysis. Its frequency coverage is 0.5 to 18 GHz.

4.3.5   Channelized Receiver

A channelized Rx is a multichannel parallel processing narrowband Rx systems architecture providing comprehensive instantaneous frequency coverage. The channelized Rx consists of parallel narrowband Rx with adjacent frequencies, such as IFM, TRF, or SHR. The RF band the receiver is required to operate (for example, 2 to 18 GHz) is typically divided into contiguous bands, each 1 or 2 GHz wide. These bands are divided into subbands (500, 200, or 100 MHz wide). Also, these subbands are further divided into channels with suitable bandwidths (such as one-tenth of the subbands) to define the frequency resolution of the Rx [7].

The frequency of the upper edge of the 3-dB bandwidth of one of the fixed frequency Rx making up the channeled Rx is tuned to be at the same frequency as the lower edge of the 3-dB bandwidth of the next Rx. The block diagram of channelized Rx is given in Figure 4.9. This Rx is one of the ideal Rx types. It provides a demodulated output for the signal on each channel. The channels are designed with narrow bands to deliver high sensitivity and selectivity. It provides 100% POI for signals in the frequency range and can receive multiple signals simultaneously thanks to different frequency channels. The circuit structure of channeled Rx and, as a result, their design and manufacture are of high complexity. For example, if a channelized Rx requires 10 MHz isolation in the 2–4-GHz frequency range, 200 channels are needed. This situation interprets that 200 narrowband (NB) Rx should be added to the system.

In Figure 4.9, the input is adjusted to 2–4 GHz with the *bandpass* (BP) filter, and the output of the BP filter is amplified by an *RF amplifier* (RFA). The Rx *multiplexer* (mux) divides the 2–4-GHz frequency band into five contiguous subbands. Each subband output has applied a mixer with an LO signal, and the mixers' outcomes are given to IF filters and *IF amplifiers* (IFA). Then each amplified IF signal is divided into 10-MHz channels, and each channel consists of a narrowband Rx.

The polyphase N-path filter bank is an alternate implementation that performs the channelization as a single merged process, as shown in Figure 4.11. The polyphase filter bank partition offers significant advantages relative to individual downconversion Rx. The primary benefit is reduced cost due to significantly reduced system

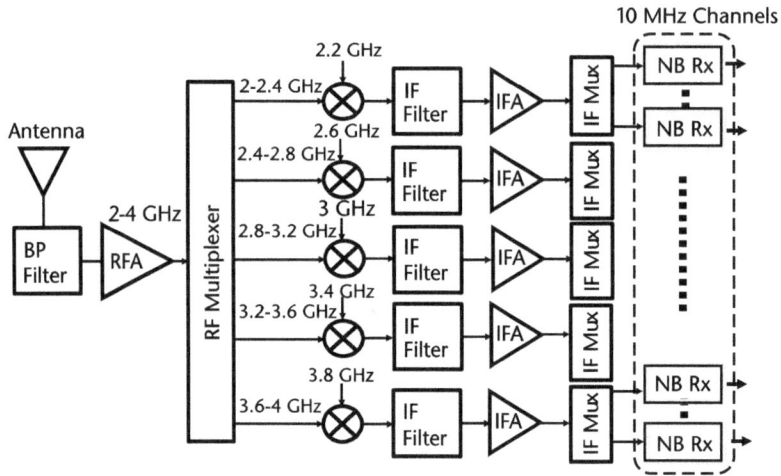

Figure 4.9   Block diagram of a channelized Rx with 10-MHz resolution.

resources required to perform the equivalent multichannel processing. This equates to decreasing computations needed for each frame [8, 9].

In Figure 4.10, there are M filters, the input data ($x(n)$) is shown, and the cycle contains M data points. The outputs are shown as $y(n)$ and used as the input of the FFT. The final results in the frequency domain are represented by $Y(k)$. In this case, the inputs are decimated by M, and the final frequency domain also has M outputs. The input data is shifted M points, the output frequency bin number. This case is referred to as the critically sampled case. A critically sampled case is one where the number of output frequency bins equals the input data shift. Thus, the output sampling rate is $1/M$ times the input sampling rate, where M is the number of input data points shifted. If one wants to increase the output sampling rate, the hardware must be modified, which is less flexible than the software approach.

In the mixer downconverter in Figure 4.9, a separate mixer pair and filter pair are assigned to each channel of the channelized Rx, and these mixers must all

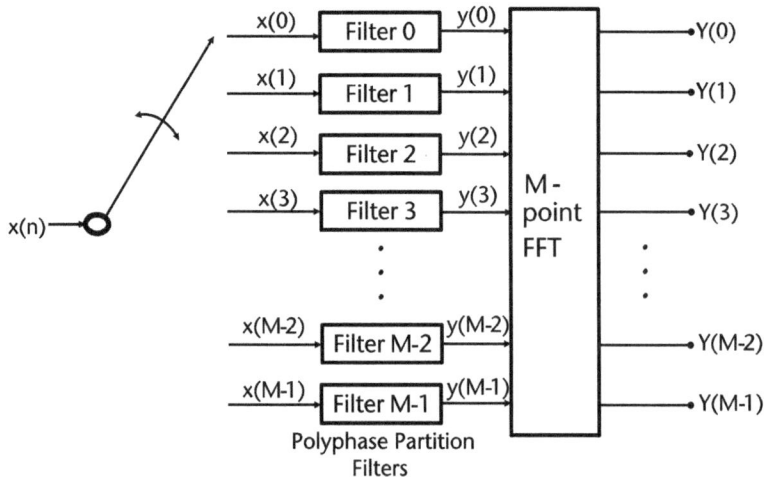

Figure 4.10   Block diagram of a polyphase channelizer.

operate at the high-input data rate before downsampling. In the polyphase filter, only one lowpass filter is required to service all the channels of the channelizer, and this single filter accommodates all the channels as co-occupying alias contributors of the baseband bandwidth. Thus, all the processing performed in the polyphase channelizer occurs at the low-output sample rate. The two processes have another significant difference when the input signal is actual. In the mixer downconverter, the signal is made complex by the input mixers entering the process, which means that the lowpass filtering task requires two filters, one for each of the quadrature components. In contrast, in the polyphase channelizer, the phase rotators make the signal complex as leaving the process. Consequently, only one partitioned lowpass filter is required to process the input signal.

When a channelizer separates adjacent communication channels, characterized by known center frequencies and known controlled nonoverlapping bandwidths, the channelizer must preserve the separation of the channel outputs. This case is typical for EW Rx, where most target environment characteristics are generally known beforehand. The polyphase filter channelizer uses the input M-to-1 resampling to alias the spectral terms residing at multiples of the output sample rate to the baseband. Thus, the output sample rate is the same as the channel spacing for the standard polyphase channelizer. Operating at this rate satisfies the Nyquist criterion, permitting the separation of the channels with an output rate that avoids band-edge aliasing.

The polyphase filter channelizer uses the input M-to-1 resampling to alias the spectral terms residing at multiples of the output sample rate to the baseband. So the output sample rate is the same as the channel spacing for the standard polyphase channelizer. Operating at this rate satisfies the Nyquist criterion, permitting the separation of the channels with an output rate that avoids band-edge aliasing.

Some examples of the RWR and ESM systems that utilize channelized Rx systems in their architectures are mentioned here [4]:

- *AN/ALR-67(V):* The system is an RWR system for tactical aircraft. It detects, identifies, and displays radars and radar-guided weapon systems in the C to J frequency range (about 0.5 to 20 GHz). The first two versions, (V) and (V)2, have been equipped with sophisticated, lightweight channelized Rx systems. The system may also take part in SPEW systems. The (V)3 version has been upgraded to a fully channelized digital architecture.
- *CS-3600 ESM systems:* The CS-3600 provides a precision ESM system featuring a unique receiver architecture covering the 2 to 18-GHz frequency range. It is designed for use in naval and ground-based applications. The dual receiver architecture was selected to meet demanding ESM requirements for a 100% probability of emitter intercept and the ability to work in dense RF environments. The CS-3600 ESM system comprises a monopulse DF, full azimuth coverage, a pulse analyzer unit, IFM, and channelized receivers.

4.3.6   Compressive Receiver

The operation of a compressive (or microscan) Rx is similar to that of SHR. A compression technique is used to increase the operating frequency band of the SHR properties. Figure 4.11 shows a compressive Rx block diagram. In a compressive Rx, the input RF signal is converted into a chirp signal through a mixer fed by an FM local oscillator. The chirp signal is compressed into short pulses through a compressive filter or dispersive delay line. These short pulses are sent to a detector, making it possible to analyze the spectrum of the signals at the device's input. The time position of each short output pulse relative to the initiation of the LO sweep represents the frequency of the corresponding input signal.

An SHR or any other narrowband Rx must dwell in that band for a period equal to or greater than its bandwidth to detect a signal. For example, an Rx with a bandwidth of 1 MHz must remain in that frequency band for at least 1 μs. The frequency tuning rate of the compressive Rx is much faster than this speed, but when performing the quick scan, its output is passed through a compressive filter with a delay proportional to the frequency [9].

The delay versus frequency slope compensates for the Rx's sweep rate. Thus, as the Rx sweeps its bandwidth across a signal, its output is coherently time-compressed to make a substantial signal spike. The resultant output is the spectral display of the Rx's full operating band. A compressive Rx has good sensitivity because the noise of the Rx is not the noise relative to the wideband input but the noise relative to the output of the compressive filter matched to the minimum duration of the pulse to be measured. The probability of intercept is approximately 100%. If the PW of a pulsed signal is longer than the sweep duration, measuring the signal will not be possible. Sensitivity is reduced for short pulses, much faster than the sweep duration. The compressive Rx's operation process can be summarized as follows:

- An RF signal existing at the input to the compressive Rx is multiplied by an LO sweeping linearly (see Figure 4.11).
- This signal is implemented to a compressive filter with an equal but opposite time-frequency slope.
- All spectral energy within the bandwidth of the compressive filter will appear in a narrow time slot at the delay line's output.
- Due to input signals with different RFs, time separation between output signals will cause the IF sweeps to be offset in frequency.

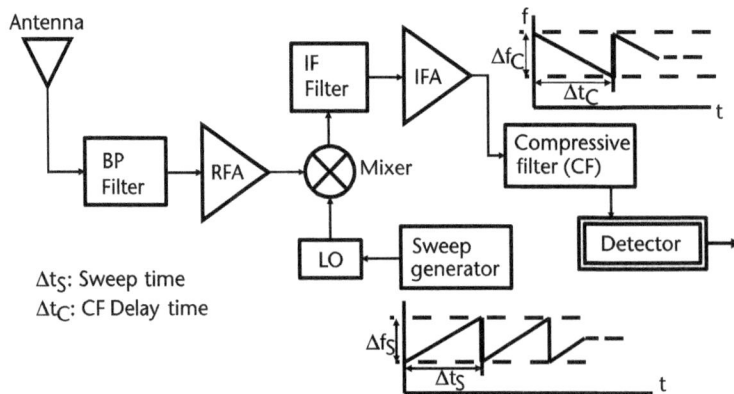

Figure 4.11 Block diagram of a compressive (microscan) Rx.

- The time separation at the output between different input signals permits parameter measurements of simultaneously received signals.

A sample ESM system that utilizes a compressive Rx system is presented here [4]:

- *Hyper-Lite:* This new-generation ES system is designed to adapt to airborne, shipboard, and ground tactical environments. It offers threat warning and situational awareness, emitter location, complex signal recognition, data collection, and signal demodulation capabilities. It covers 20 to 1,200 MHz with 25-kHz channelization and utilizes a compressive receiver with 60 MHz of instantaneous bandwidth.

4.3.7   Digital Receiver

The advancements in ADCs and increases in digital signal processing speed have changed the direction of ES and RWR Rx design and implementation. Thus, the present research has focused on digital EW Rx. A digital Rx downconverts the RF input into an IF. Then it digitizes with high-speed ADCs with many quantization levels. Digital signal processing is then used to extract the required information from the digitized signal. Since the software can functionally simulate any filter or demodulator, including those that cannot be performed in hardware, the digitized signal can be optimally filtered, demodulated, and postdetection-processed.

The ADC replaces the crystal detector and provides access to information otherwise lost after detection. The high-speed signal processing of data provided by the ADC implements sophisticated parameter measurements of signals in near-real-time possible.

Some of the significant advantages are related to digital signal processing. Once a signal is digitized, the execution of the subsequent processes will be digital. Digital signal processing is more robust because there is no temperature drifting, gain variation, or DC level shifting as in analog circuits; therefore, less calibration is required [9].

The digital Rx is mainly an instantaneous wideband (e.g., 500 MHz–4 GHz) receiving channel capable of processing all the received signals in digital form, even if received at the same time, and capable of providing for each signal all the required measurements, such as time of arrival, pulse amplitude, pulse width, and frequency. Generally, an ES or RWR system has to provide a wide frequency coverage, typically from 2 to 18 GHz or extended from 0.5 to 40 GHz.

A block diagram for a wideband digital filter for ES and RWR purposes is given in Figure 4.12. The total RF coverage typically required for ES and RWR Rx indicates that an IF sampling architecture is wanted. An RF sampling topology may be possible if a wide bandwidth is unnecessary. The superhet stage of the Rx selects instantaneous bandwidth around an RF value. It translates the RF around an IF value such that the ADC can transform the analog signal into a digital signal. The sampling rate of the ADC must be such that it respects the Shannon theory. The theory states that the sampling frequency must be at least equal to 2 times the IF bandwidth.

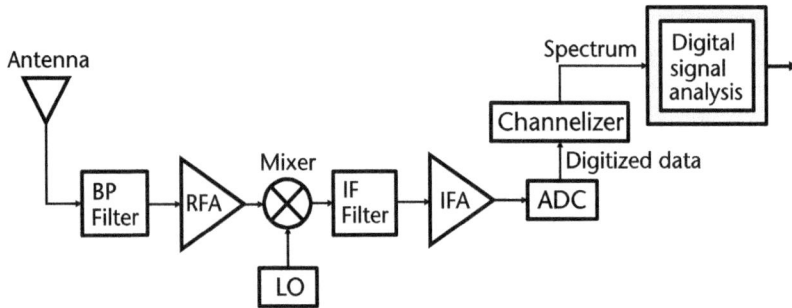

Figure 4.12 Block diagram of a wideband digital Rx.

The digital signal at the output of the ADC will be immediately processed at a very high speed with an algorithm based essentially on the repetitive use of a short FFT or, in other words, *short-time Fourier transform* (STFT). The effect of this initial digital processing is equivalent to that of an array of N adjacent filters, where N is the number of the STFT outputs. The hardware implementing this STFT processing can be called the channelizer. The channelizer outputs can be processed parallel or time-sharing in a postprocessor *digital signal processor* (DSP) unit [2].

Despite the increase in hardware and processing capacities of computers daily, they have a relatively limited processing capability. The level of processing capability limits the processing efficiency of signal data. Complex software requires a lot of storage and processor memory. Although computer technology has developed rapidly and the process size ratio has quickly increased, the processing capacity of computers is still directly related to size, weight, power, and cost. Although it varies according to needs and applications, one or more of these elements is one of the main problems to be considered in system design, sometimes directly affecting the system architecture.

Some examples of the SPEW, RWR, and ESM systems that engage digital Rx systems in their architectures are as follows [4]:

- *Advanced Digital Receiver Processor (ADRP) EW system:* This is a passive, high-performance digital Rx-based ES system that provides superior situational awareness by rapidly detecting, identifying, and locating combat threats autonomously. The system offers ELINT, ESM, and RWR functions. It can be installed on airborne, sea-based, and land-based platforms. A modular open systems architecture also enables unmanned platforms. It interfaces to any antenna configuration and has a small footprint, minimizing installation effects.
- *EAGLE:* The Electronic Acquisition Gathering Locating Equipment (EAGLE) is a family of integrated Rx systems for ELINT and ESM applications. Net-Centric uses EAGLE technology to support rapid emitter detection, DF, and signal measurement in its architecture. Wide instantaneous bandwidth operation allows the EAGLE to search thousands of frequency bins per second in any predetermined set of frequency bands over its entire frequency range. Wideband digital Rx operation combined with an advanced pulse processor

and modern software-based DSP technology ensures that the EAGLE effectively intercepts current large-time bandwidth product radar emitters and conventional radar types. Frequency coverage is typically from 0.5 to 18.0 GHz. Furthermore, special EAGLE systems can be supplied to cover 30 MHz to 40 GHz.

- *AN/ALR-69A(V):* This digital Rx-based RWR has been developed for fighter and transport aircraft. The system provides RF threat situational awareness, threat signal processing, reprogrammability, associated defensive support equipment, and training equipment. The RWR system detects, identifies, processes, and displays AI, SAM, and AAA weapon systems.
- *ALR-733 Family:* This family of ESM systems is designed to provide surveillance, ELINT–type signal analysis, and data collection for post-mission analysis while securing RWR functionality. ALR-733 family fulfills the operational requirements for maritime patrol, helicopter applications, and AEW&C missions. The upgraded versions of the system have been upgraded with digital Rx enhancements.
- *Advanced Self-Protection System (ASPS):* This SPEW system is designed for fixed-wing and rotary-wing aircraft. It integrates RWR (SPS-3000), ECM Jammer (SPJ-40), MWS (PAWS), and CFDS. The system has digital Rx in its RWR subsystem and an operating band of 0.5–18 GHz.
- *Maigret 5800:* The system has been developed for naval vessels that combine radar ESM (Maigret R 5800) and communications ESM (Maigret C 5800) inputs in the frequency range of 0.25 MHz to 40 GHz. The system can also have extended roles, such as ELINT and COMINT. Communications ESM has a frequency range from 0.25 to 3,000 MHz. For the tactical radar ESM systems, based on wideband digital Rx, the frequency ranges from 2 to 18 GHz, with the ability to extend it down to 0.5 GHz and up to 40 GHz is available.

The Rx types commonly used in RWR are described up to this point. It would be a valuable approach to see the properties of the Rx together. For this purpose, Table 4.1 compares the Rx systems' specific characteristics qualitatively. The table gives the degrees of the features for different Rx types. Degrees have been chosen as excellent (E), good (G), suitable (S), and poor (P), considering the general specifications of the Rx to be a rough guide. Thus, one can quickly compare the Rx and define the cons and pros of selecting an Rx or Rx combination.

Rx systems are the primary units of RWR systems. This section provides detailed information about the Rx systems used in the RWR system. Table 4.1 summarizes the major features of various Rx systems and the differences and similarities between their general properties.

4.4 Receiver Sensitivity Calculations

Rx sensitivity refers to the minimum signal power that must occur at the input of an Rx to perform the desired signal detection, measurements, and analysis processes. Sensitivity is a power level usually specified in dBm. Sensitivity is in negative dBm and usually has large absolute values. These values can be –90 dBm, –120 dBm,

Table 4.1   Comparison of RWR Rx

E: Excellent G: Good S: Suitable P: Poor	Instantaneous Bandwidth	Frequency Resolution	Simultaneous Signals	Sensitivity	Dynamic Range	Probability of Intercept	Selectivity	Throughput Time	Size/Weight/ Cost
CVR	E	P	P	P	S	E	P	E	E/E/E
IFM Rx	E	G	P	P	S	E	P	E	E/E/E
TRF Rx	P	S	S	S	S	S	S	P	E/E/S
Narrowband SHR	P	E	G	E	E	P	E	P	E/E/S
Wideband SHR	S	P	S	S	S	S	P	S	E/E/S
Channelized Rx	G	S	G	G	G	E	E	S	P/P/P
Compressive Rx	G	G	G	E	S	E	E	E	S/S/S
Digital Rx	G	E	E	G	G	E	E	P	E/E/S

or even lower for some systems. The sensitivity field strength can also be expressed in microvolts per meter (µV/m).

Two different situations can be considered for calculating the required Rx's sensitivity, represented S_{min} or, for simplicity, S. The first is the sensitivity level found with the following equation derived from the one-way radar equation given in (2.29).

$$S = \frac{P_{Tx} G_{Tx} G_{Rx} c^2}{\left(4\pi R_{max} f\right)^2} \tag{4.12}$$

in which P_{Tx} is the Tx peak power, c is the speed of light, and G_{Tx} and G_{Rx} are Tx and Rx antenna gains, respectively. Also, R_{max} is the maximum detection range, and f is the operating frequency. When writing the above equation in decibels, we get the following expression:

$$S = \left(P_{Tx}\right)_{dB} + \left(G_{Tx}\right)_{dB} - 32.4 - 20\log(f) - 20\log\left(R_{max}\right) + \left(G_{Rx}\right)_{dB} \tag{4.13}$$

In this equation, R_{max} is the maximum distance between Tx-Rx in kilometers; f is the Tx frequency in megahertz. In addition, 32.4 is the offset or conversion factor derived from constants and includes all necessary unit conversions to enable the response to be calculated. This value is valid only for distance in kilometers and frequency in megahertz. If the received power is equal to or greater than the sensitivity of Rx, an RF link or detection is possible between Tx and Rx. This situation should be considered when calculating the sensitivity of the RWR Rx. Another case is given in (2.38) for the reflected echo signal and returned to the radar Rx with two-way radar equality. This case is used to find the required radar's Rx sensitivity.

In both cases, from the point of view of system design, the sensitivity is calculated using the equation given earlier (2.41). There are three critical components in determining the sensitivity of an Rx system:

- The noise power, noise floor, or thermal noise floor (N_P);
- Rx system noise figure (F);
- The required SNR to separate the desired information from the received signal.

Summing these components in dB scale, the sensitivity is calculated as follows:

$$S = N_p + F + SNR \qquad (4.14)$$

In this equation, the noise power is dBm, and the noise figure and the SNR are in dB. Thus, this equation defines the sensitivity in dBm present at the demodulator for a desired SNR. It would be beneficial to inspect the sensitivity components separately.

- *The noise power (N_P):* Instead of noise power, the kTB term is often used as a word and consists of the product of the following three expressions:
  - k is the Boltzmann constant (1.38×10^{-23} J°/K).
  - T is the operating temperature in Kelvin (K°).
  - B is the Rx's effective bandwidth (Hz).

Noise power (N_P) describes an ideal Rx's thermal noise power level. Physically, the meaning of N_P shows the noise power on a resistor at a particular temperature value. The mathematical expression of the N_P is given as follows:

$$N_p = 10\log\left\{k\ (J/°K) \times T\ (°K) \times B\ (Hz)\right\} \qquad (4.15)$$

In the above equation, the operating temperature is usually considered to be 290°K (or 17°C) and used for room temperature. Regarding the noise floor for a normalized (i.e., $B = 1$ Hz) bandwidth and $T = 290$°K temperature:

$$
\begin{aligned}
N_p &= 10\log_{10}\left(k \times T \times B\right) \\
N_p &= 10\log_{10}\left(1.38 \times 10^{-23} \times 290°K \times 1\ Hz\right) \\
N_p &= -203.9\ \text{dBW/Hz} + 30\ \text{dB} \approx -174\ \text{dBm/Hz}
\end{aligned}
\qquad (4.16)
$$

If the Rx bandwidth is set to 1 MHz, which is a more suitable frequency for EW operations, the approximate value of the noise power in dBm is obtained as follows:

$$
\begin{aligned}
N_p &= 10\log\left(1.38 \times 10^{-23} \times 290°K \times 1 \times 10^6\ Hz\right) \approx -143.98\ \text{dBW} \\
N_p &= -143.98\ \text{dBW} + 30\ \text{dB} \approx -114\ \text{dBm}
\end{aligned}
\qquad (4.17)
$$

This approximate calculation value can quickly calculate any Rx bandwidth's noise power or thermal noise level. For this purpose, the dB value of the ratio of the actual Rx bandwidth to 1 MHz is added to −114 dBm as follows [10]:

$$N_{\mathrm{p}} = -114 \text{ dBm} + 10\log\left(\frac{f_{\mathrm{Rx}}}{1 \text{ MHz}}\right) \qquad (4.18)$$

- *The noise figure (NF):* Each of the systems, subsystems, or discrete devices adds noise that affects the overall performance of an Rx. This effect is known as the NF. As a general definition, the NF is a measure of how much a device degrades the SNR. When excluding the gain, the difference between the input and output SNRs gives the NF. However, when the input and output SNR is considered in the linear scale, the result is defined as the noise factor. The mathematical definition of noise factor (F_{linear}) is defined as

$$\left.\begin{array}{l} SNR_{\mathrm{in}}\left(\mathrm{linear}\right) = \dfrac{S_{\mathrm{in}}\left(\mathrm{W}\right)}{N_{\mathrm{in}}\left(\mathrm{W}\right)} \\[3mm] SNR_{\mathrm{out}}\left(\mathrm{linear}\right) = \dfrac{S_{\mathrm{out}}\left(\mathrm{W}\right)}{N_{\mathrm{out}}\left(\mathrm{W}\right)} \end{array}\right\} F_{\mathrm{linear}} = \frac{SNR_{\mathrm{in}}\left(\mathrm{linear}\right)}{SNR_{\mathrm{out}}\left(\mathrm{linear}\right)} \qquad (4.19)$$

The mathematical expression of NF F_{dB} or, in short, F is as follows:

$$\left.\begin{array}{l} SNR_{\mathrm{in}}\left(\mathrm{dB}\right) = 10\log\left(\dfrac{S_{\mathrm{in}}\left(\mathrm{W}\right)}{N_{\mathrm{in}}\left(\mathrm{W}\right)}\right) \\[3mm] SNR_{\mathrm{out}}\left(\mathrm{dB}\right) = 10\log\left(\dfrac{S_{\mathrm{out}}\left(\mathrm{W}\right)}{N_{\mathrm{out}}\left(\mathrm{W}\right)}\right) \end{array}\right\} F = SNR_{\mathrm{in}}\left(\mathrm{dB}\right) - SNR_{\mathrm{out}}\left(\mathrm{dB}\right) \qquad (4.20)$$

To understand the effect and calculation of NF, let us consider the RF amplifier component in an Rx system with 10-dB gain and 4-dB NF. The Rx frequency bandwidth is 10 MHz, and the input signal power is −15 dBm. Figure 4.13 shows the RF amplifier with input and output signal and noise powers.

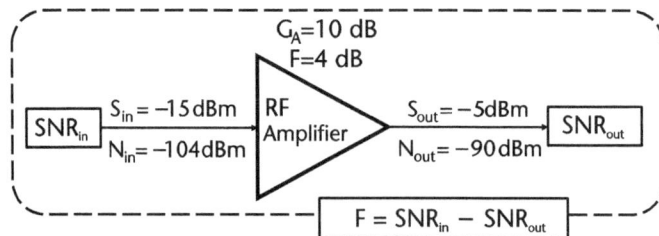

Figure 4.13 NF demonstration for an RF amplifier.

To analyze the above equation, first, determine the input noise power. Since the Rx operating bandwidth is 10 MHz (f_{Rx}), the noise power can be calculated by using (4.18) as follows:

$$N_{in} = -114 \text{ dBm} + 10\log\left(\frac{10 \text{ MHz}}{1 \text{ MHz}}\right) = -104 \text{ dBm} \tag{4.21}$$

Let us calculate the output signal and noise power by considering the amplifier's power gain (G_A) and the NF (F).

$$S_{out} = S_{in} + G_A = -15 \text{ dBm} + 10 \text{ dB} = -5 \text{ dBm}$$
$$N_{out} = N_{in} + G_A + F = -104 \text{ dBm} + 10 \text{ dB} + 4 \text{ dB} = -90 \text{ dBm} \tag{4.22}$$

The NF of an Rx's subcomponent is typically unknown if not given in the specification sheets, and one must measure it for design and implementation purposes. The NF is measured with an NF analyzer, a spectrum analyzer, and sometimes a dedicated Rx.

Also, we can express the NF in terms of the component's effective temperature, T_e. Consider the amplifier shown in Figure 4.13 and let its effective temperature be T_e. Assume the input noise temperature is T_i. Thus, the input and output noise powers are written as follows [11]:

$$N_{in} = kT_iB \tag{4.23a}$$

$$N_{out} = kT_iBG_A + kT_eBG_A \tag{4.23b}$$

The first term on the right side of the output noise power (N_{out}) in (4.23b) corresponds to the input noise, and the latter term is due to thermal noise generated inside the system. It follows that the noise factor can be expressed as

$$F_{linear} = \frac{SNR_{in} \text{ (linear)}}{SNR_{out} \text{ (linear)}} = \frac{S_i}{kT_iB}kBG_A\frac{T_i + T_e}{S_o} = 1 + \frac{T_e}{T_i}$$
$$\text{or} \quad T_e = \left(F_{linear} - 1\right)T_i \tag{4.24}$$

An Rx system includes many subsystems, such as RF filters and amplifiers, mixers, IF filters, and amplifiers. For this reason, considering cascaded systems to obtain the Rx system's noise factor will be appropriate. Let us consider a simple Rx system consisting of two cascaded subcomponents, as shown in Figure 4.14. Let us assume that the Rx bandwidth is B, the subcomponents' noise factors are $F_{linear1,2}$, the power gains are $G_{A1,2}$, and the effective temperatures are $T_{e1,2}$. The output signal power is defined as follows:

$$S_{out} = S_iG_{A1}G_{A2} \tag{4.25}$$

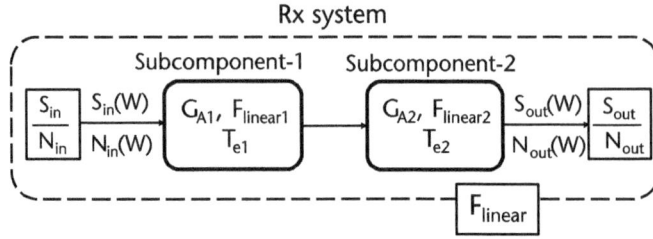

Figure 4.14 An Rx system consists of two cascaded subcomponents.

The input noise power can be calculated by using (2.23a), and the output noise power is expressed as

$$N_{out} = kT_i B G_{A1} G_{A2} + kT_{e1} B G_{A1} G_{A2} + kT_{e2} B G_{A2} \qquad (4.26)$$

The first term of the above equation's right side represents the input noise power, the second is the thermal noise generated inside subcomponent 1, and the third is the thermal noise generated inside subcomponent 2. When we consider the thermal noise found in (4.24) and the input noise power from (2.23a) and substitute these expressions in (4.26),

$$N_{out} = F_{linear1} N_{in} G_{A1} G_{A1} + \left(F_{linear2} - 1 \right) N_{in} G_{A2} \qquad (4.27)$$

Thus, the total noise factor for the cascaded system can be written as follows:

$$F_{linear} = \frac{\left(S_{in}/N_{in} \right)}{\left(S_{out}/N_{out} \right)} = F_{linear1} + \frac{F_{linear2} - 1}{G_{A1}} \qquad (4.28)$$

The generalized case of an Rx system with n subcomponents is expressed as follows:

$$F_{linear} = F_{linear1} + \frac{F_{linear2} - 1}{G_{A1}} + \frac{F_{linear3} - 1}{G_{A1} G_{A2}} + \dots + \frac{F_{linear(n)} - 1}{G_{A1} G_{A2} \dots G_{A(n-2)}} \qquad (4.29)$$

As stated before, we can write the NF by using the noise factor as

$$F = 10 \log \left(F_{linear} \right) \qquad (4.30)$$

- *SNR:* The SNR required for Rx to fulfill its duty depends on the type of information that the signal carries, the modulation that holds it, and the operation performed at the Rx output. The predetection SNR, the RF SNR, or carrier-to-noise ratio (CNR) must be defined to determine Rx sensitivity [10].

The requirements for the Rx output's SNR and the RF SNR may be different. The RF SNR and the output SNR are close for the pulse signals and the AM modulation. However, this difference may increase to 30-dB levels for FM-modulated signals. Furthermore, signal-to-quantization noise ratio (SQR) is considered in digital signals. The digitization sets the output quality of recovered digital signals and is only secondarily affected by the RF SNR. In this case, the SNR is used in place of the SQR. However, the RF SNR determines the bit error rate that will be present in the recovered digital signal.

As a result, in some modulation types, the SNR of the signals at the Rx output may need to be considerably more significant than the RF SNR. In this case, CNR is usually defined instead of RF SNR and used instead of RF SNR in calculations. Understanding the use of the required SNR in the sensitivity calculation below is essential.

Example 4.1

Let the effective bandwidth of an Rx be 100 MHz, and the noise factor of the system be 8 dB. If this Rx is designed to receive pulsed signals automatically, its required SNR level is 15 dB. Find the sensitivity of Rx for this case.

Solution 4.1

We take (4.14) and (4.18) into account and substitute the given values in the equations,

$$S = N_p + F + SNR$$

$$N_p = -114 \text{ dBm} + 10\log\left(\frac{f_{Rx}}{1 \text{ MHz}}\right)$$

$$S = (-114 \text{ dBm} + 20 \text{ dB}) + 8 \text{ dB} + 15 \text{ dB} = -71 \text{ dBm}$$

As can be seen from the result obtained, the sensitivity will improve (its value will decrease) as the SNR level drops, and it will worsen (increase in value) as the sensitivity increases. As is known, SNR indicates how high the noise level is for a signal to be detected. While this ratio is kept high to increase the detection probability of the systems required to perform automatic detection, it can be kept lower in systems that perform detection with the operator.

We add the receiver's overall noise figure to the floor to calculate the Rx sensitivity. This result gives the noise floor at the demodulator's input. The required RF SNR or CNR must be higher than the SNR at the Rx output to provide the required signal quality. Adding these together gives the power level needed for precision or the minimum power level of the Rx. Figure 4.15 shows how the noise power, NF, and CNR levels affect sensitivity in the demodulator with generic values.

The figure illustrates the effects of noise power, NF, and CNR on Rx sensitivity. The values have been selected as follows:

- The operating bandwidth is 5 MHz, and the temperature is 17°C. So the noise power from kTB is −107 dBm.

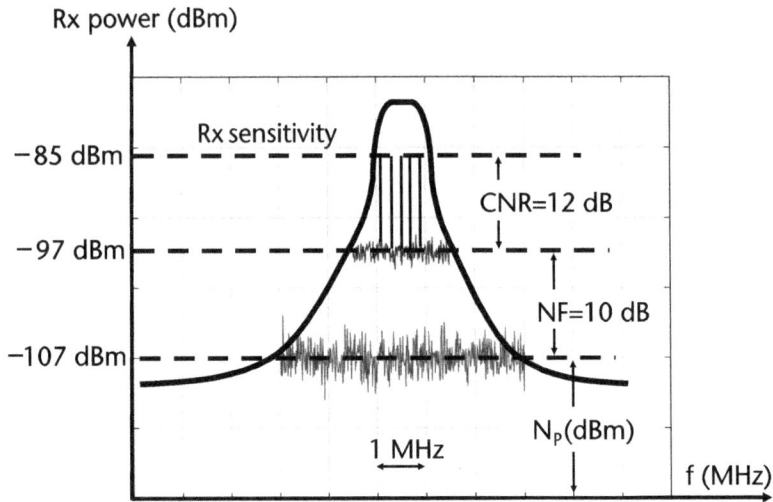

Figure 4.15 Graphical representation of Rx sensitivity.

- The NF is 10 dB, and the CNR (or RF SNR) is 12 dB.
- The resulting Rx sensitivity is −85 dBm.

4.5   Dynamic Range in Receivers

The main aim of an Rx is to detect and process the signal. However, with the proliferation of high-powered transmitters and the increasing growth of RF noise pollution, weak signal reception is often tricky, especially in dense RF operational areas. Rx dynamic range measures an Rx's ability to handle a range of signal strengths, from the weakest to the strongest. The reception of the lowest signal depends on the sensitivity, which has been inspected in detail. However, there is a maximum input signal power that an Rx can accurately detect and process. The minimum detectable signal (MDS) defines the sensitivity of the Rx. The Rx 1-dB compression point represents the saturation point of the Rx. The 1-dB compression point was described in Section 4.3.4 and depicted in Figure 4.7(a).

There are two different dynamic range concepts, which are total and instantaneous dynamic ranges. The total dynamic range is the totality of the signal power levels the receiving system can handle, including attenuators in the signal path. At the same time, the instantaneous dynamic range refers to the total range of signals that can be collected without inserting some form of attenuation in the signal path to avoid overload of subsequent processing steps. The ratio of the input saturation point and the minimum detectable signal represents the total dynamic range of the Rx, and it is defined as:

$$DR = \frac{P_{sat}}{S} \text{ (unitless)} \rightarrow DR \text{ (dB)} = 10\log\left(\frac{P_{sat}}{S}\right) = P_{sat} \text{ (dBm)} - S \text{ (dBm)} \quad (4.31)$$

4.6   RWR Architecture

RWR is a basic EW passive surveillance system that does not require high sensitivity but quick response and accuracy. Its task is to provide an instant alarm and situational awareness to the aircraft pilot or crew relevant to the RF emissions of the threats in the surrounding environment. A generic RWR block diagram in the RF part of a SPEW system architecture is given in Figure 4.2. The figure shows that an RWR system consists of four major components: antenna, Rx, processor, and control and display units. The RWR signal processing issue is discussed in Chapter 5. In this part, the remaining three components are mentioned. The following sections account for antennas, Rx architectures, controls, and displays used in the RWR systems.

4.6.1   Antennas Used in RWRs

RWR antennas are designed to receive RF pulses from threat radar systems. For this purpose, they must provide sufficient frequency and angular coverage. The antennas utilized in the RWR and ESM systems are wideband and omnidirectional. At the same time, they should have proper polarization orientation and not affect the Rx sensitivity. Let us inspect the antenna properties one by one:

- *Frequency coverage:* Generally, the wideband coverage should be 2–18 GHz (or extended to 0.5–40 GHz) to cover the RWR requirements by multi-octave bandwidth. Inherently wideband antennas that meet the frequency coverage requirements are conical dipole, monopole, biconical, spiral, and spiral conical antenna structures.
- *Angular coverage:* An omnidirectional antenna can provide 360° coverage in the azimuth plane and 30°–40° in elevation over a wide frequency bandwidth such as many octaves. The required antenna type for this case can be spiral conical, cavity-backed spiral, or biconical antennas. The physical location of the RWR antennas on the platforms can affect their ability to detect a radar signal. Let us consider a fighter aircraft as the platform since the localization, mounting, and operation are more complicated than the other platforms. The antennas are arranged to cover a predetermined horizontal and vertical area around the aircraft, and the generally used antenna placement is given in Figure 4.16.

   In high-speed aircraft RWR applications, the biconical, spiral, and spiral conical antenna structures are integrated with radome for additive protection. They meet high G shock (up to 90 G) and vibration requirements.
- *Polarization orientation:* EM waves consist of an electric field (E-field) and a magnetic field (H-field) moving in one direction. The E-field and H-field are perpendicular to each other and the direction in which the plane wave propagates. Polarization expresses the path taken by the electric field vector (E) as a function of time. More simply, it shows the orientation of E (and therefore H) waves in Tx and Rx. There are three types of polarization. These are linear (vertical, horizontal, or diagonal), circular (turning right or left), and elliptical polarizations.

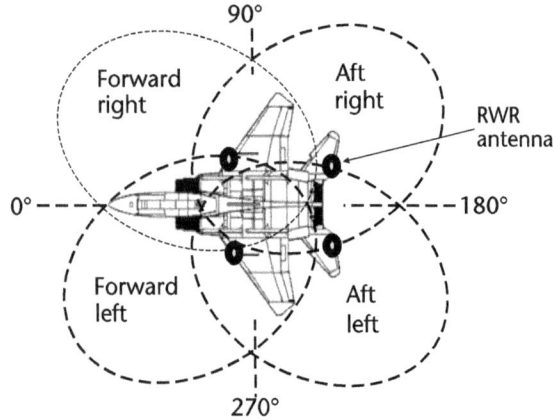

Figure 4.16 A generic antenna placement in a fighter aircraft.

Regarding EW, antenna polarization significantly affects both RWR and jammer operations. When considering the RWR operation, the received signal strength is reduced if the Rx antenna's polarization and the received signal's polarization do not match. A similar effect can be defined between the jammer and victim radar for the jammer operations. Generally, linearly polarized antennas have a linear geometry in the polarization direction. An example is that vertically shaped antennas also tend to have vertical polarizers. Similarly, round or cross-geometry antennas tend to be circularly polarized. Their geometry and phase difference formation also affect whether they are *right-hand circular* (RHC) or *left-hand circular* (LHC).

In case of polarization mismatch between Tx and Rx antennas, the received signal power level decreases. This reduction is added to the link and radar equations as a gain reduction. In this case, the question sought to be answered is which polarization types match and which do not. Furthermore, what is the degree of incompatibility of antenna pairs with polarization mismatch? The *polarization loss factor* (PLF) characterizes the EM power loss due to polarization mismatch.

If there is no polarization mismatch, the PLF is defined to reach 1 (100% or 0 dB), that is, the Rx antenna receives the maximum possible power density of the incident RF wave. If the PLF is equal to 0 (i.e., ∞ dB), it indicates complete polarization mismatch and was unable to capture power from the incident RF wave. PLF takes a value between 0 and 1, as $0 \leq PLF \leq 1$.

Polarization loss does not have a relationship with propagation. Polarization loss can be regarded as a missed opportunity to capture as much power as possible from the incident RF wave. In other words, it expresses how much Rx can receive from the emitted RF wave. Polarization efficiency has the same meaning as PLF. Polarization loss is the PLF converted to dB.

Within the scope of these explanations, the polarization losses between different polarization types are summarized in Table 4.2, which shows the theoretical ratio of transmitted power between antennas of different polarization. These ratios are rarely exact due to effects such as reflection, refraction, and interference with other waves. Therefore, for the hypothetical case where the PLF is 0, −25 dB is included in the calculations instead of −∞ dB in practice [12].

Table 4.2 Cross-Polarization Loss for Various Antenna Combinations

Polarization Type	Linear (Vertical)	Linear (Horizontal)	Slant (45° or 135°)	Circular (RHC)	Circular (LHC)
Linear (Vertical)	0 dB	−25 dB	−3 dB	−3 dB	−3 dB
Linear (Horizontal)	−25 dB	0 dB	−3 dB	−3 dB	−3 dB
Slant (45° or 135°)	−3 dB	−3 dB	0 dB	−3 dB	−3 dB
Circular (RHC)	−3 dB	−3 dB	−3 dB	0 dB	−25 dB
Circular (LHC)	−3 dB	−3 dB	−3 dB	−25 dB	0 dB

A polarization method often used in RWR systems is circular, spiral-based antennas that receive a linearly polarized signal of unknown orientation. In this case, −3-dB polarization loss can suffer, but −25-dB loss that may occur in the case of cross-polarization is avoided. When the polarization (linear or circular) of the received signal is unknown, it is common practice to make a quick measurement scan with circular polarization LHC and RHC antennas and find the stronger signal.

Also, some other designs use 45° or 135° slant polarized biconical antennas. In general, the slant polarization is preferred in ESM and RWR systems because it can detect all the possible polarizations used, whether vertical, used by tracking radars, horizontal, which is primarily used by search and acquisition radars, or right-hand or left-hand circular, often used by radars to reduce rain clutter [2].

4.6.2 Receiver Architectures in RWR

The Rx systems of the RWRs process the incident radar signals. Most RWR systems use specially defined frequency bands to distinguish environmental signals. The frequency bands used in RWR systems and their EW frequency band equivalents are listed next [1]:

- Band 0: C/D;
- Band 1: E/F;
- Band 2: G/H;
- Band 3: I/J.

The classification of the frequency spectrum in terms of EW is given in Table 2.3. It would be helpful to look at the letter codes that define the EW frequency band in the above list. In general, an RWR will analyze the following characteristics of the pulse that it detects:

- RF;
- PA;
- Direction of arrival (DOA) or angle of arrival (AOA);
- Time of arrival (TOA);
- Pulse repetition interval (PRI) or pulse repetition frequency (PRF);
- PRI type;

- Pulse width (PW);
- Antenna scan type and rate;
- Antenna main lobe (3-dB beamwidth).

Specific characteristics are measured directly, while others are obtained through additional processing. The Rx subsystems measure DOA, TOA, PW, PA, and frequency, and we will only focus on the measured parameters for now. The extraction of other features will be explored using various signal-processing techniques in the following chapter.

In conventional RWR systems, CVR-based Rx systems are utilized. The sensitivity levels of the CVR systems are in the order of −40 dBm for pulsed signals and −50 dBm for a CW signal. The sensitivity improvement for the CW source is due to the convenient narrowing of the video bandwidth [2]. For a modern CVR in a typical configuration, with an automatically processed output, the ultimate sensitivity is improved to the −65 to −70-dBm range for CW by preamplification. Figure 4.17 demonstrates a CVR-based narrowband superhet RWR architecture.

The figure shows that the measured parameters are DOA, TOA, PW, and PA. Furthermore, $\vec{\Sigma}$ shows the vector sum, and RFA and IFA represent the RF and IF amplifiers, respectively. Using this configuration, up to −80-dBm sensitivity can be obtained; however, POI is low. Figure 4.17 shows a pattern with some deterioration in sensitivity for POI increase.

In the configuration depicted in Figure 4.18, frequency (f) measurement is added to the capabilities in the architecture given in Figure 4.17. For this purpose, *adjustable bandpass* (A-BP) filters, a *frequency converter* (F/conv), and an IFM Rx exist in Figure 4.18. The main advantage of this architecture, when compared with Figure 4.17, is its flexibility for adjusting the input bandpass filters. Thus, its operational principle converges the virtues of the SHR scheme. Another advantage is to measure the frequency without so much increasing the hardware. In a straightforward architecture, each IF channel engages a separate IFM Rx, but this scheme uses only

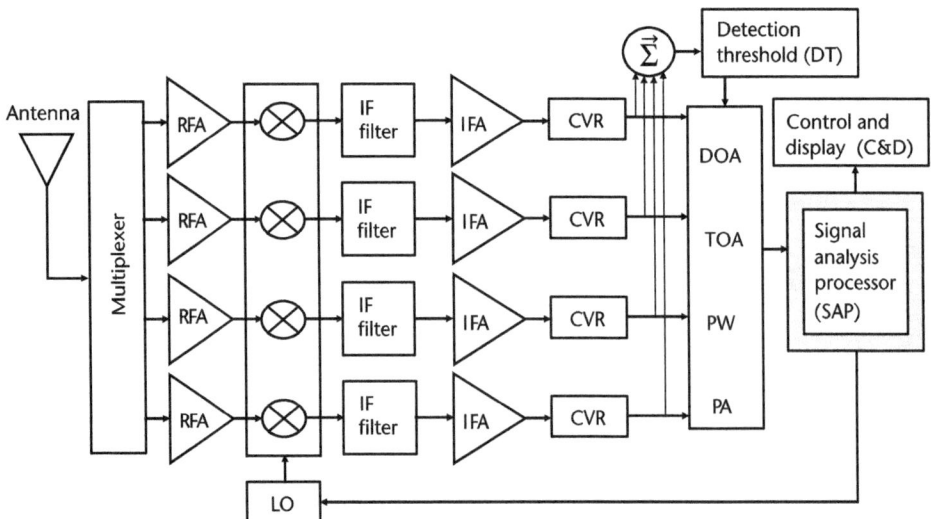

Figure 4.17 A CVR-based narrowband superhet RWR Rx architecture.

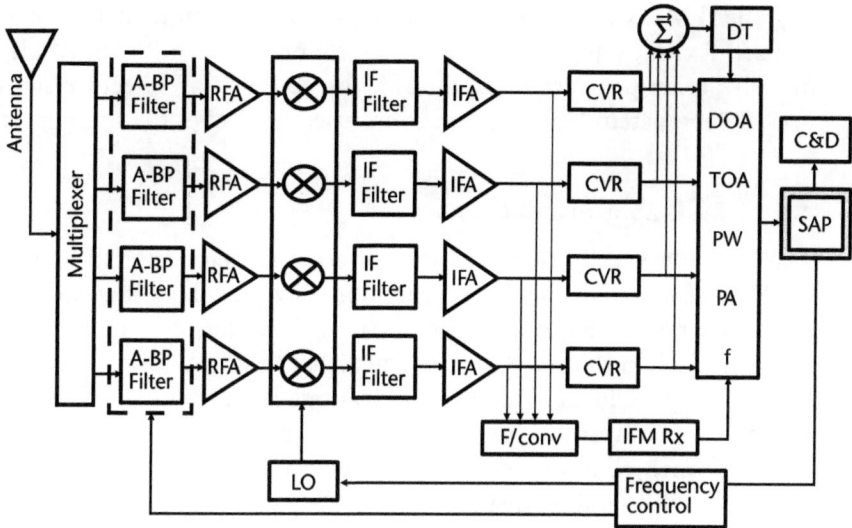

Figure 4.18 A CVR-based wideband superhet RWR Rx architecture.

one IFM Rx. F/conv provides this operational principle; each IF channel is converted to a standard frequency band. Thus, using only one IFM Rx, the entire operating frequency band measurement capability is added to the architecture.

The *signal analysis processor* (SAP) deinterleaves and sorts the signals based on the measurements. This section includes the *emitter identification database* (EIDD) or emitter library. The SAP compares the sorted signals and EID library data and conducts the threat identification process, which will be discussed in Chapter 5.

Even though the performance of the CVR-based systems is adequate against both high-powered, low-repetition pulsed weapon systems and CW systems for achieving increased POI and instantaneous frequency coverage, the IFM-based Rx schemes are an alternative for RWR systems. Figure 4.19 gives such kind of architecture. The number of the IFM Rx, or K in the figure, may be increased to obtain the required sensitivity.

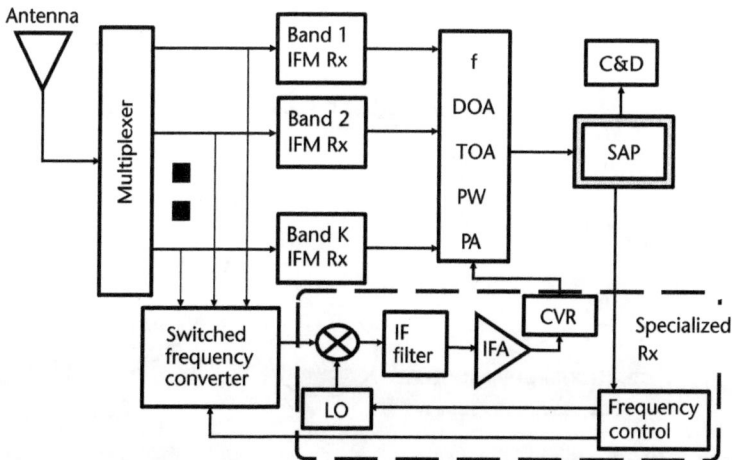

Figure 4.19 An IFM-based high POI, wideband RWR Rx architecture.

The IFM Rx measure the frequencies of signals within a specific range. A specialized Rx, such as a superheterodyne, channelized, or digital Rx, is also used to measure other parameters. In the IFM-based RWR architecture shown in Figure 4.19, a superheterodyne construction is used instead of the specialized Rx.

4.6.3 RWR Control and Displays

The SAP continually compares the sorted signal characteristics with the data in the EID tables. Once the signal processor has determined that enough signal characteristics match the information in the EID tables, it generates and positions a video symbol on the RWR scope. The airborne RWR usually has a visual display somewhere prominent or in multiple locations in the cockpit and generally generates audible tones to warn the pilots. Visual and audible warnings for multiseat aircraft are also conveyed to the other crew. The visual display often forms a polar representation, with symbols displaying the detected radars according to their direction relative to the current aircraft heading, as shown in Figure 4.20.

The display's center has an aircraft symbol representing the mother platform, and the heading displays the reference azimuth direction. Each video symbol on the display represents a specific threat, and each threat system has its unique symbol. Additionally, an audio tone is generated to alert the pilot. In Figure 4.20, the threats are arranged based on their azimuth location relative to the mother platform's heading. The display shows an AAA system (AAx), a SAM system (SAx), and two AI aircraft (Fx1) on the RWR display. The estimated distance from the threat radar or the level of danger to the aircraft determines the distance from the circle's center. The signal processor generates symbols and audio linked to specific threat systems actions such as search, track, and missile launch. These symbols may vary from system to system, and the system software may adjust them. Figure 4.20 shows some generic symbols, such as a threat in the launch mode represented by adding a circle. Furthermore, a bar appears in track mode, and nothing is added to the search mode.

Please note that not all symbols on the azimuth indicator indicate a threat. Several radars may be present in certain situations, with only a few or none representing a

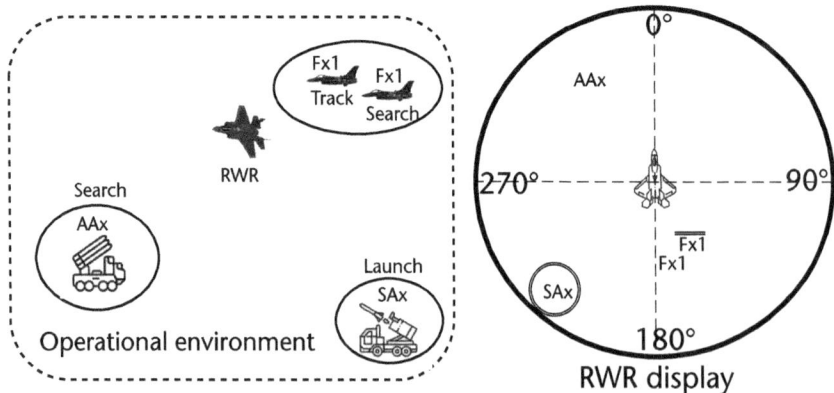

Figure 4.20 A generic RWR display.

threat. The azimuth indicator can display symbols for search, acquisition, and tracking radars simultaneously. Advanced RWR systems allow for prioritizing threats, displaying symbols for the highest priority threats only. However, older RWR systems only show symbols for tracking radars and launched missiles.

4.7 Problems

Problem 4.1: Write a MATLAB code to model a CVR for detecting a 2-GHz constant PRI radar signal.

Problem 4.2: Which of the following signal parameters can CVR not measure?
(a) Pulse amplitude;
(b) Pulse width;
(c) Frequency;
(d) Direction of arrival;
(e) Time of arrival.

Problem 4.3: What method should be applied to increase the selectivity of superhet Rx?
(a) Narrowing the frequency band of the tuned bandpass filter;
(b) Filtering the IF layer;
(c) Eliminating one of the sum and difference frequencies at the mixer output;
(d) Changing the frequency of the local oscillator;
(e) Fine-tune control.

Problem 4.4: An Rx operates with a 200-MHz bandwidth, has an overall noise factor (F) of 8 dB, and requires an SNR of 6 dB in the operating environment. Find its sensitivity.

Problem 4.5: Consider an Rx's RF amplifier at the antenna's output. The Rx's bandwidth is 300 MHz, and the RF amplifier's gain is 15 dB. If the required SNR at the output of the RF amplifier is 35 dB for −40-dBm input, find the maximum allowable noise figure for this device.

Problem 4.6: An RWR Rx consists of an antenna with cable loss of 2 dB and 1-dB NF. The cable is connected to a bandpass filter with 2-dB insertion loss; its NF is 3 dB. Following the bandpass filter, an RF amplifier takes part with 18-dB gain and 3-dB NF. After that, there is a mixer with a conversion loss of 5 dB and NF of 6 dB. Furthermore, an IF amplifier and bandpass filter integrated circuit, which has a total gain of 50 dB and NF 7 dB, is connected to the mixer. Calculate the total NF for this Rx system.

Problem 4.7: Calculate its input noise power if a channelized Rx with a 4–8-GHz frequency range has 400 channels.

Problem 4.8: An Rx's 1-dB compression point is 10 dBm, its operational bandwidth is 300 MHz, the total noise figure is 12 dB, and the required detection SNR is 10 dB. The antenna gain of the Rx is 5 dB. If a 4-GHz threat radar in the operational area of the Rx is transmitting the RF signal with a 110-dBm ERP:

(a) Find the dynamic range of the Rx.

(b) Calculate the minimum and maximum ranges for the Rx to detect the threat radar signal.

Solutions

Solution 4.1: The required MATLAB code is given in Table 4.3.

Figure 4.21(a, b) shows the signal at Rx's antenna input and the output of the Rx detector, respectively.

Table 4.3   MATLAB Code for Problem 4.1

```
%%%% CVR Modeling
clc,clear all
%%% 2 GHz Constant PRF Pulsed Radar Signal at the Rx antenna
pulse_amplitude = 2e-3; %Pa=2mW
f_carrier = 2e9;
f_sampling = 40e9;%(sampling frequency)
% Pulse generation
time_scaling_1 = 0:1/f_sampling:1e-6; % (1st Pulse Signal)
time_scaling_2 = 1.000025e-6:1/f_sampling:99.999975e-6; % (Death Time)
time_scaling_3 = 100e-6:1/f_sampling:101e-6; % (2nd Pulse Signal)
time_scaling_4 = 101.000025e-6:1/f_sampling:199.999975e-6; % (Death Time)
time_scaling_5 = 200e-6:1/f_sampling:201e-6; % (3rd Pulse Signal)
time_scaling_6 = 201.000025e-6:1/f_sampling:299.999975e-6; % (Death Time)
time_scaling_7 = 300e-6:1/f_sampling:301e-6; % (4th Pulse Signal)
total_time_scaling =0:1/f_sampling:301e-6; % Total time interval
% Sin generation
sin_1 = pulse_amplitude*sin(2*pi*f_carrier*time_scaling_1); % Pulse-1
sin_2 = pulse_amplitude*sin(2*pi*f_carrier*time_scaling_3); % Pulse-2
sin_3 = pulse_amplitude*sin(2*pi*f_carrier*time_scaling_5); % Pulse-3
sin_4 = pulse_amplitude*sin(2*pi*f_carrier*time_scaling_7); % pulse-4
% Death times
zero_1 =zeros(size(time_scaling_2));% between pulse 1-2
zero_2 =zeros(size(time_scaling_4));% between pulse 2-3
zero_3 =zeros(size(time_scaling_6));% between pulse 3-4
% 2 GHZ Pulse Radar signal
constant_PRF_radar_signal_2GHz = cat(2,sin_1,zero_1,sin_2,zero_2,sin_3,zero
_3,sin_4);
% Adding noise, determining noise floor for 2 GHz
n0=1.38e-23*290*2e9;
size_n0=max(size(constant_PRF_radar_signal_2GHz));
```

Table 4.3 *(continued)*

```
constant_PRF_radar_signal_2GHz =
constant_PRF_radar_signal_2GHz+(n0*ones(1,size_n0)+1e-4.*randn(1,size_n0));
%%% Band Pass Filter Block
filter_input_signal = constant_PRF_radar_signal_2GHz;
f_sampling_1 = 80e9; % Sampling fr.
pass_band_1 = [1e9 3e9];
% 1 - 3 GHz Band Pass Filter
filter_output=bandpass(filter_input_signal,pass_band_1,f_sampling_1);
%%% CVR Amplifier Block
amplifier_Gain = 15 ; %15 times = 11.76 dB
noise_Figure = 1e-4;
amplifier_input_signal = filter_output;
% Linear amplification
amplifier_stage= amplifier_Gain * amplifier_input_signal;
% Preamplifier NF effect
NF =(noise_Figure*rand(size(amplifier_input_signal)));
amplifier_output= amplifier_stage+NF;
%%% CVR Detector Block
detector_input_signal = amplifier_output;
%Envelope detection
%Step-1 Square law
detector_input_signal=detector_input_signal.*detector_input_signal;
%Step-2 Lowpass filter
envelope = lowpass(detector_input_signal,1e6,1e9);
%Step-3 Absolute value
detected_envelope= abs(sqrt(envelope));
% Signal in the antenna input
figure;
plot(total_time_scaling,constant_PRF_radar_signal_2GHz);
title('Antenna Input:2 GHz Radar Signal (Constant PRI)');
xlabel('t');
ylabel('Amplitude');
axis([0 350e-6 -3e-3 3e-3]);
grid;
% Detector output
figure(2);
plot(total_time_scaling,detected_envelope);
title('Detector Output:2 GHz Radar signal (Constant PRF)');
xlabel('t');
ylabel('Amplitude');
grid;
```

Solution 4.2: (c)

Solution 4.3: (b)

Solution 4.4: First, it would be appropriate to calculate the noise power (or kTB) for the given 200-MHz Rx bandwidth,

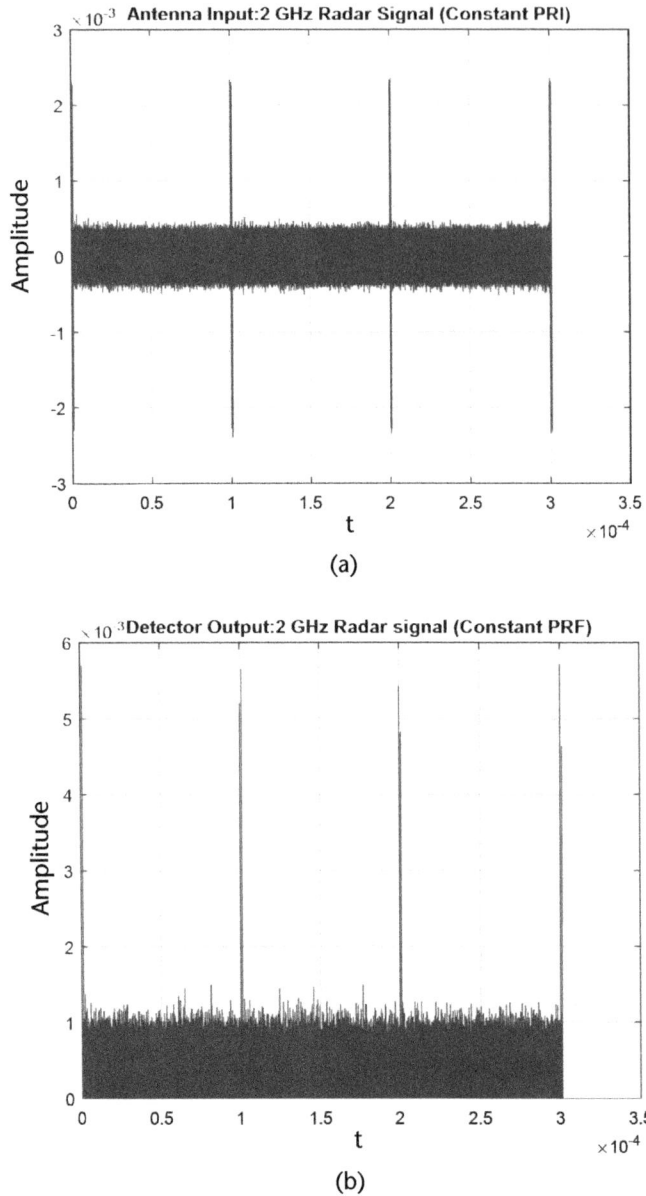

Figure 4.21 Antenna input and detector output signals for Problem 4.1: (a) at the antenna input, and (b) at the detector output.

$$N_P = k \times T \times B = \left(1.38 \times 10^{-23}\right) \times \left(290\right) \times \left(200 \times 10^6\right) = 8 \times 10^{-13}\,\text{W}$$

$$N_P\,\left(\text{dBW}\right) = \log\left(N_P\right) \approx -121\,\text{dBW} \rightarrow N_P\,\left(\text{dBm}\right)$$

$$= -121\,\text{dB} + 30\,\text{dB} = -91\,\text{dBm}$$

Also, the noise power can be calculated using the shortcut method using (4.18),

$$N_P = -114 \text{ dBm} + 10\log\left(\frac{f_{Rx}}{1 \text{ MHz}}\right)$$

$$N_P = -114 \text{ dBm} + 10\log\left(\frac{200 \text{ MHz}}{1 \text{ MHz}}\right) \approx -91 \text{ dBm}$$

Now, let us calculate the sensitivity using (4.14),

$$S = N_P + F + SNR$$

$$S = -91 \text{ dBm} + 8 \text{ dB} + 6 \text{ dB} = -77 \text{ dBm}$$

Solution 4.5: To solve this problem, we have to consider (4.20):

$$F = SNR_{in} \text{ (dB)} - SNR_{out} \text{ (dB)}$$

and Figure 4.13. First, let us find the input noise by using (4.18):

$$N_P = -114 \text{ dBm} + 10\log\left(\frac{f_{Rx}}{1 \text{ MHz}}\right)$$

$$N_P = -114 \text{ dBm} + 10\log\left(\frac{300 \text{ MHz}}{1 \text{ MHz}}\right) \approx -89.23 \text{ dBm}$$

Now we can calculate output power by using the first equation in (4.22):

$$S_{out} = S_{in} + G_A = -40 \text{ dBm} + 15 \text{ dB} = -25 \text{ dBm}$$

and using the second equation in (4.22), we can calculate the output noise,

$$N_{out} = N_{in} + G_A + F = -89.23 \text{ dBm} + 15 \text{ dB} + F$$
$$= -74.23 \text{ dBm} + F$$

However, we cannot obtain the exact output noise power value since NF still needs to be discovered. Now let us write the equation for output SNR and equalize it for the required 35 dB:

$$SNR_{out} = S_{out} - N_{out} = -25 \text{ dBm} - \left(-74.23 \text{ dBm} + F\right)$$
$$= 35 \text{ dB}$$

Thus, the maximum value of the RF amplifier's NF is obtained as

$$F = -35 \text{ dB} - 25 \text{ dBm} + 74.23 \text{ dBm} = 14.23 \text{ dB}$$

Solution 4.6: To solve this problem, the generalized case of an Rx system with n subcomponents expressed with (4.29) is used:

$$F_{\text{linear}} = F_{\text{linear1}} + \frac{F_{\text{linear2}} - 1}{G_{A1}} + \frac{F_{\text{linear3}} - 1}{G_{A1}G_{A2}} + \dots$$
$$+ \frac{F_{\text{linear}(n)} - 1}{G_{A1}G_{A2}\dots G_{A(n-2)}}$$

The question gives five components; thus, we can adapt the above equation to find the noise factor for each element as follows:

$$F_{\text{linear}} = F_{\text{linear1}} + \frac{F_{\text{linear2}} - 1}{G_{A1}} + \frac{F_{\text{linear3}} - 1}{G_{A1}G_{A2}}$$
$$+ \frac{F_{\text{linear4}} - 1}{G_{A1}G_{A2}G_{A3}} + \frac{F_{\text{linear5}} - 1}{G_{A1}G_{A2}G_{A3}G_{A4}}$$

Now, let us define each component in the equation separately. Note that F_n is the noise figure, and $F_{\text{linear}(n)}$ is the noise factor. Furthermore, if any G_{An} is positive, it represents a gain; conversely, if it is negative, it defines a loss. For applying the values to the above equations, all the gains must be transformed into a linear case, a unitless ratio.

Antenna with cable: G_{A1} (dB) = −2 dB → G_{A1} = 0.631

F_1 = 1 dB → F_{linear1} = 1.259

Bandpass filter: G_{A2} (dB) = −2 dB → G_{A2} = 0.631

F_2 = 3 dB → F_{linear2} = 2

RF amplifier: G_{A3} (dB) = 18 dB → G_{A3} = 63.1

F_3 = 3 dB → F_{linear3} = 2

Mixer: G_{A4} (dB) = −5 dB → G_{A4} = 0.316

F_4 = 6 dB → F_{linear4} = 4

IF integrated circuit: G_{A5} (dB) = 50 dB → G_{A5} = 10^5

F_5 = 7 dB → F_{linear5} = 5

Let us substitute the numerical values in the equation to find the total noise factor,

$$F_{\text{linear}} = 1.259 + \frac{2 - 1}{0.631} + \frac{2 - 1}{(0.631)(0.631)}$$
$$+ \frac{4 - 1}{(0.631)(0.631)(63.1)} + \frac{5 - 1}{(0.631)(0.631)(63.1)(0.316)}$$
$$F_{\text{linear}} = 1.259 + 1.584 + 2.51 + 0.12 + 0.504 \approx 6$$

Thus, the total NF of the Rx system is obtained as follows:

$$F = 10\log\left(F_{\text{linear}}\right) = 10\log(6) = 7.78 \text{ dB}$$

Solution 4.7: First, let us find the bandwidth of one channel of the Rx:

$$B_{\text{one\_chan}} = \frac{(8-4) \times 10^9}{400} = 10 \text{ MHz}$$

Now we can calculate the input noise power for the Rx by using (4.18):

$$N_{\text{P}} = -114 \text{ dBm} + 10\log\left(\frac{f_{\text{Rx}}}{1 \text{ MHz}}\right)$$

$$N_{\text{P}} = -114 \text{ dBm} + 10\log\left(\frac{10 \text{ MHz}}{1 \text{ MHz}}\right) \approx -104 \text{ dBm}$$

Solution 4.8:

(a) The Rx's saturation point is reached at the 1-dB compression point, while the minimum detectable signal is called sensitivity. Divide the saturation point by the sensitivity to calculate the total dynamic range. This calculation can be done using the formula (4.31).

$$DR = \frac{P_{\text{sat}}}{S}(\text{unitless}) \rightarrow DR \text{ (dB)} = 10\log\left(\frac{P_{\text{sat}}}{S}\right)$$

$$= P_{\text{sat}} \text{ (dBm)} - S \text{ (dBm)}$$

The saturation point is given in the question as 10 dBm, and the required component is the sensitivity of the Rx, which can be calculated as follows:

$$N_{\text{P}} = -114 \text{ dBm} + 10\log\left(\frac{300 \text{ MHz}}{1 \text{ MHz}}\right) \approx -89.23 \text{ dBm}$$

$$S = N_{\text{P}} + F + SNR$$

$$S = -89.23 \text{ dBm} + 12 \text{ dB} + 10 \text{ dB} = -67.23 \text{ dBm}$$

Let us calculate the dynamic range as follows:

$$DR \text{ (dB)} = P_{\text{sat}} \text{ (dBm)} - S \text{ (dBm)} = 10 \text{ dBm} - \left(-67.23 \text{ dBm}\right)$$

$$= 77.23 \text{ dB}$$

(b) To calculate the minimum and maximum ranges that the Rx can detect the threat radar signal, the one-way radar equation given in (4.13) is used:

$$S = \left(P_{\text{Tx}}\right)_{\text{dB}} + \left(G_{\text{Tx}}\right)_{\text{dB}} - 32.4 - 20\log(f) - 20\log\left(R_{\text{max}}\right) + \left(G_{\text{Rx}}\right)_{\text{dB}}$$

The saturation power is used to calculate the minimum range as follows,

$$P_{sat} = \left(P_{Tx}\right)_{dB} + \left(G_{Tx}\right)_{dB} - 32.4 - 20\log(f) - 20\log\left(R_{min}\right) + \left(G_{Rx}\right)_{dB}$$

$$10 \text{ dBm} = 110 \text{ dBm} - 32.4 - 20\log(4000) - 20\log\left(R_{min}\right) + 5 \text{ dB}$$

$$20\log\left(R_{min}\right) = 0.56 \rightarrow R_{min} = 1.06 \text{ km}$$

For the maximum range, the sensitivity is considered,

$$S = \left(P_{Tx}\right)_{dB} + \left(G_{Tx}\right)_{dB} - 32.4 - 20\log(f) - 20\log\left(R_{max}\right) + \left(G_{Rx}\right)_{dB}$$

$$-67.23 \text{ dBm} = 110 \text{ dBm} - 32.4 - 20\log(4000) - 20\log\left(R_{min}\right) + 5 \text{ dB}$$

$$20\log\left(R_{max}\right) = 77.79 \rightarrow R_{min} = 7753.5 \text{ km}$$

The maximum distance detected in this situation using RF technology is 7,753.5 km. However, the range is determined by the line of sight between the threat radar and the Rx platform.

References

[1]   Air Combat Command Training Support Squadron (ACC TRSS), *Electronic Warfare Fundamentals*, Nellis AFB, NV, 2000.

[2]   Neri, F., *Introduction to Electronic Defense Systems*, 2nd ed., Norwood, MA: Artech House, 2001.

[3]   Joint Publication 3-13.1, *Electronic Warfare*, Joint Chiefs of Staff, USA, 2012.

[4]   Donaldson, P., (ed.), *Electronic Warfare Handbook 2008*, Berkshire, UK: The Shephard Press, 2008.

[5]   Cheng, C. H., and J. Tsui, *An Introduction to Electronic Warfare: From the First Jamming to Machine Learning Techniques*, Gistrup, Denmark: River Publishers, 2021.

[6]   Tsui, J. B., *Microwave Receivers with Electronic Warfare Applications*, New York: Wiley-Interscience, 1986.

[7]   Robertson, S., *Practical ESM Analysis*, Norwood, MA: Artech House, 2019.

[8]   Poisel, R. A., *Electronic Warfare Receivers and Receiving Systems*, Norwood, MA: Artech House, 2014.

[9]   James, T., and C. Cheng, *Digital Techniques for Wideband Receivers*, 3rd ed., Raleigh, NC: Scitech Publishing, 2015.

[10] Adamy, D. L., *EW 101: A First Course in Electronic Warfare*, Norwood, MA: Artech House, 2001.

[11] Mahafza, B. R., *Radar Systems Analysis and Design Using MATLAB*, 3rd ed., Boca Raton, FL: CRC Press/Taylor & Francis Group, 2013.

[12] Avionics Department, *Electronic Warfare and Radar Systems Engineering Handbook*, Naval Air Warfare Center, Point Mugu, CA, 2013.

CHAPTER 5

RWR Signal Processing

Chapter 4 provided the subcomponent properties of the RWR systems and discusses the signal-processing techniques. During the initial phase of implementing RWR, the systems were run manually, with trained operators responsible for detecting threat signals. Because of this, the signal processing techniques used in the first RWR systems have been refined to ensure that the operator can readily identify received signals.

The signal environment became more complex as the effective use of radar-controlled weapons in the operational environment increased. Real-time and automatic detection and identification of these weapon systems have become important. Threat detection has been one of the most crucial signal-processing tasks of all EW systems since its inception and will continue to maintain its importance in the future. Threat Tx location detection is another task that is fundamental to EW operations. Tx location detection is difficult and requires complex and high-level signal processing.

In SPEW systems, RWRs are responsible for detecting and identifying potential threats, and the information they supply forms the basis for using the related countermeasures. Modern RWR systems must handle many signals, often in crowded environments with millions of pulses per second. Identifying threat signals from received RF energy is a crucial signal-processing task involving various functions. Figure 5.1 gives a block diagram of a generic RWR system signal-processing function.

At the Rx output, radar pulse measurements are combined into a single data packet known as a *pulse descriptor word* (PDW). PDWs usually include basic measured RF signal parameters such as *pulse width* (PW), *radar operating frequency* (ROF), *time of arrival* (TOA), *time difference of arrival* (TDOA), horizontal and vertical *angle of arrivals* (AOAs), *amplitude* (A), and in some systems *polarization* (P). An RWR's Rx system generates a PDW following the end of a received pulse; say that the maximum pulse width is 600 μs. Under CW conditions, a *signal descriptor word* (SDW) is generated after a pre-settable time with the CW present tag set, considering the maximum pulse width (higher than 600 μs for the example). SDW consists of AOA, ROF, TOA, and A. SDWs continue to be generated with the same interval for the duration of the CW signal.

The TOA and AOA information is used for deinterleaving and emitter localization purposes. For deinterleaving, the information is essential for the dense RF environments. However, different techniques are used for localization purposes. These techniques are triangulation, interferometry, and TOA. They are summarized as follows:

- *Triangulation* is the most basic form of DF available. This method comprises taking direction measurements from multiple sensors simultaneously or from one sensor sequentially. For this reason, triangulation can be performed by various platforms simultaneously or by one platform transiting some distance to get multiple azimuth measurements. The intersection of the azimuth measurements, called bearing lines, is the likely location of the emitter. The participating platform must have accurate data on their current positions to get the bearing lines effectively. In SPEW systems, the RWR units use one platform triangulation method.

- The *interferometry method* uses phase difference of arrival. In this method, platforms utilize multiple Rx antennas for predetermined sectors, and 360° azimuth coverage is essential for the RWRs' of SPEW systems. Figure 5.1 shows the sectoral RWR antennas for the front, rear, left (or port), and right (or starboard). These systems utilize RWR antennas as DF antennas. The interferometry method operates by comparing the phase of a radar wave as it impacts two or more DF antennas. Then the phase difference is used to obtain AOA.

- The *TOA technique* is another type of location technique. The method depends on the radar signals traveling at approximately the speed of light. The method solves for the distance that the emitter is away from the Rx using the equation distance equals the speed of light multiplied by time. Because there is no directional information, the equation represents the radius of a circle around the receiving antenna. When multiple platforms are used, simultaneous multiple-distance measurements are taken. Otherwise, multiple-distance measurements are achieved sequentially in time using one moving platform with an Rx. The intersection of the circles is the position of the emitter.

The PDW and SDW obtained at the Rx output are conveyed to the deinterleaving block. As a result of the clustering and analysis made during signal deinterleaving, critical parameters are extracted. RWR systems utilize various deinterleaving techniques, either conventional or modern, to sort pulses into clusters based on their parameters. This process aims to generate tracks, ideally one for each radar. New pulses are grouped based on their similarity in parameters. PRI determination is

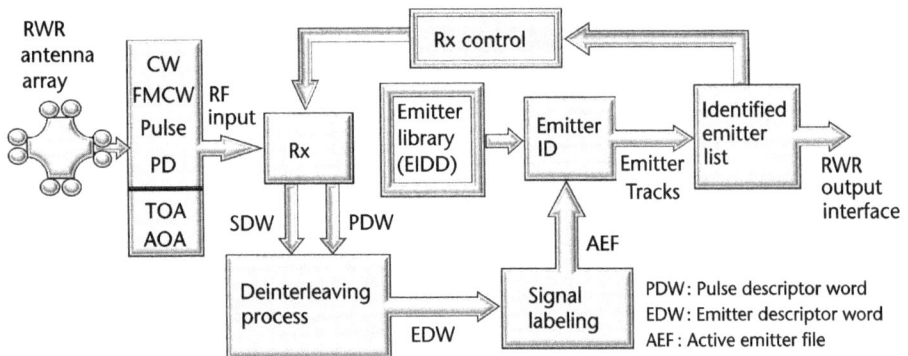

Figure 5.1 Functional block diagram of a generic RWR system.

performed, and the deinterleaving process results in a sequence of pulses [1]. The deinterleaving process naturally clusters the pulses so that different emitter tracks are formed for various modes of a single emitter. Separated necessary and sufficient information for threat identification is called an *emitter descriptor word* (EDW). EDW is composed of clustered pulses that are considered to have emerged from an emitter. In addition to the clustering, PRI information is added to the PDW to obtain the EDW of pulsed signals. In comparison, the EDW of CW signals consists of only clustering. PRI agility properties, such as stagger and jitter, are also defined here. After deinterleaving, each EDW is labeled to determine the *active emitter file* (AEF) in another process function. The AEF also contains information on active tracks. The AEF is passed to the *emitter identification* (EID) block for a specific EID. At this stage, tracking EWDs are compared with the emitter library or the *emitter identification database* (EIDD), and the emitter tracks are defined. Emitter tracks form an identified emitter list passed to the RWR output interface for display and ECM control purposes. Since the identified emitter list contains the emitter information in the environment or situational awareness, it controls the Rx operation according to the situation.

5.1   Radar Signals

Radar systems utilize different modulation types and designs according to the operational requirements. Detailed explanations of varying radar types, operating methods, and radar-guided threat systems were given in Chapter 2. This section provides information about the impact of pulsed and CW radars on the signal-processing processes of RWR systems. To summarize, the Doppler velocity information of the target is obtained from the frequency difference between the transmitted signal and the received signal in CW radars used only in velocity measurement. In CW radars, speed and distance information can be determined by modulating the signal.

Pulsed radars obtain information about target distance, direction, and physical properties. As stated before, these radars may use modulated or unmodulated pulses. Using unmodulated pulsed radar signals with a large pulse width can help to detect and measure targets from long distances. However, it may be challenging to distinguish between close targets due to poor distance resolution. Radars with narrow pulsed signals carry out target detection and tracking in short ranges with better distance resolutions. Furthermore, by modulating the pulses, pulse compression techniques yield long-range detection without deteriorating the distance resolution. The RWR systems used on the airborne platforms and ESM systems on the shipborne platforms can simultaneously receive different radar signals, such as pulsed, CW, and their combination.

5.1.1   Pulsed Radar Signals

The pulsed radars periodically broadcast at high frequency and power throughout the pulse width and listen for the echoes from the targets. Thus, the PRI time consists of a pulse width (Tx on) and the receiving time (Tx off) until the next pulse is sent. The PRI definition was given in Chapter 2 and depicted in Figure 2.14.

Pulsed radars use a train of pulse waveforms with different modulations. The modulation process can be applied to the envelopes or the RF signal inside the envelopes. Envelope modulation is amplitude modulation and is called *on-off keyed* (OOK). Envelope modulation may be considered as PRI type, such as fixed, stagger, or jitter. The modulation of the RF signal is frequency modulation, which is in chirp form. It may be applied to each pulse (intrapulse modulation) or from pulse-to-pulse (interpulse modulation) for pulse compression purposes. In addition to chirp, the operating frequency may be fixed (unmodulated), agile, or hopper. Several consecutive pulses will appear at the same RF for the frequency hopper or batch hopper case. Meanwhile, for agile frequencies, the radar is pulse-to-pulse agile in RF.

Based on the PRF, pulsed radar systems can be classified as low PRF, medium PRF, and high PRF radars. Low PRF radars are primarily used to range where target velocity or Doppler shift is not the main interest. High PRF radars aim to measure target velocity. CW and pulsed radars can measure target range and radial velocity using different modulation schemes [2].

In the dated generation RWR systems, PRI measurements of pulsed signals were conducted with digital filters. These devices aim to detect the presence of a particular pulse interval. The digital filter creates a predictive gate for a fixed period after a received pulse. If a pulse is received while the gate is open, the digital filter searches for another pulse in the same range as the next gate. When sufficient pulses are received, the PRI value is calculated, and the presence of a signal with the specified PRI in the environment is determined [3].

The main advantage of this approach is that pulses from a single signal can be detected by deinterleaving the signals from interleaved pulse trains of many signals in a wideband Rx. Computer-based systems can hold the arrival times of the leading edges of many pulses in their memory. Thus, they can mathematically identify fixed, staggered, jittered PRIs and *pulse group repetition intervals* (PGRI).

- *Fixed PRI:* This structure is PRI's primary or unprocessed form. If pulse signals are produced at the same periodic time, pulsed radar signals have a fixed PRI structure. Furthermore, the fixed PRI structure does not increase the system's complexity. However, fixed PRI is less preferred in operational radar systems, as this will facilitate the radar signal analysis in RWR and ESM systems. Figure 5.2 shows a radar signal set with a fixed PRI pattern.

  Many radars operate with a simple fixed PRI. The range of PRI values for different radar types varies from tens of microseconds to milliseconds. Types of radar working with fixed PRI are ship navigation and weather radars. In many RWR and ESM systems, the classification of fixed PRI is encompassed in the staggered PRI definition. Thus, a fixed PRI signal is considered a one-position, one-level stagger [4].

- *Staggered PRI:* An important point that adds complexity to the PRI measurement and deinterleaving is using more than one PRI or corresponding PRF, a technique developed for MTI radars. This technique offers additional flexibility in the design of MTI Doppler filters. It reduces the effect of the blind speeds and allows a sharper low-frequency cutoff in the frequency response than might be obtained with a cascade of single-delay-line cancelers.

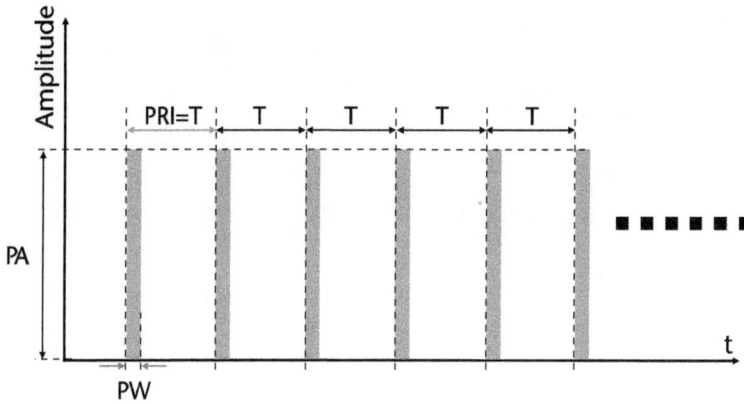

Figure 5.2   Fixed PRI radar signal structure.

The blind speeds of two independent radars operating at the same frequency will differ if their PRFs differ. Therefore, if one radar were blind to moving targets, it would be unlikely that the other radar would be blind also. Instead of using two separate radars, the same result can be obtained with one radar, which time-shares its PRF between two or more different values. For this purpose, the PRIs can be changed from batch to batch of pulses or from pulse to pulse, called staggered PRI. ATC radar and surveillance radars generally utilize this technique. Using multiple PRIs may be required for MTI radars' operational purposes; however, it may also be engaged in different radars as an ECCM technique.

Thus, staggered PRI can be defined as using two or more PRIs in a particular order. Generally, it can be defined as changing PRI from pulse to pulse. Pulse signals have a gradual feature when multiple PRI values are periodically generated in a level-staggered fashion. Staggered pulse signals increase the value of maximum unambiguous range; thus, the radar system can detect at long range. According to the fixed PRI signal, resolving staggered pulse signals is difficult for RWR and ESM systems. Staggered PRI is the most prevalent used structure pulse radars as it does not significantly increase the system complexity. Figure 5.3 shows the two-level and three-level PRI structure.

However, staggered PRI appears as an additional ECCM method in radars; it is a beneficial waveform signature for the radar identification process operated by the RWR and ESM systems [5].

- *Jittered PRI:* The jittered PRI is a single-level PRI that varies by a percent from the mean value of the emission of each pulse. The variation is up to 30% of the mean of the PRI at each repetition interval between pulses. This change is expressed as the vibration limits ($2 \times J$ in the range of $-J$ to J), and the constant PRI value at each step plus the sum of this value gives the PRI vibration.

$$T_{r\_Jn} = T_r \pm J_n \rightarrow \left\{ J_n = \left(J_n \% \right) \times T_{r\_mean} \right\} \text{ and } \begin{cases} J_{n\_max} \% = 0.3 \\ 0 < J_n \% \le J_{n\_max} \% \end{cases} \quad (5.1)$$

Two-level staggered PRI

Three-level staggered PRI

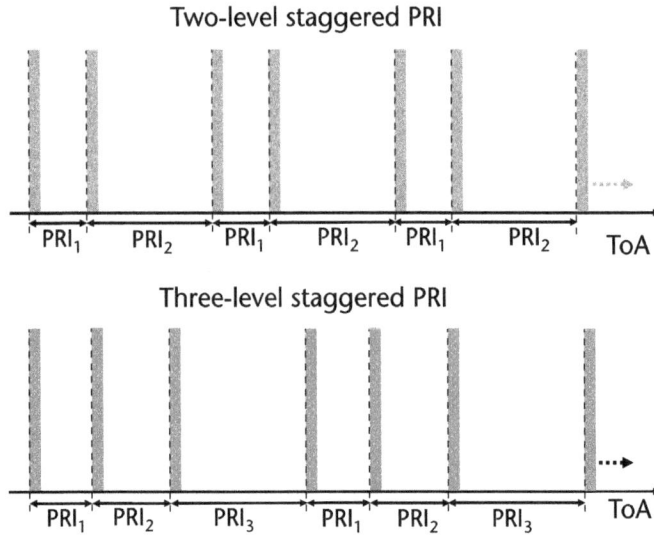

Figure 5.3 Two-level and three-level staggered PRI radar signal structure.

In the equation, T_r is the PRI, T_{r_Jn} is the PRI vibration at the nth step, and T_{r_mean} is the PRI's mean, which is defined as

$$T_{r\_mean} = \frac{1}{N}\sum_{n=1}^{N} T_{r\_Jn} \tag{5.2}$$

T_{r_mean} changes with each new step, affecting the PRI rate changes calculated in (5.1) throughout the process.

As a result, the PRI rate takes a value between J more and J less than the mean value of the previous PRI rate at each step. $J\%$ is the percentage of vibration, and it is multiplied by the mean value of the PRI to obtain the vibration ratio (J). This type of PRI could distort the operation of predictive gate processors. In predictive gate systems, the limits set up to find the next pulse from the radar should be wide enough to handle the potential PRI jitter but narrow enough to prevent any confusion in the deinterleaving and labeling processes. Calculating PRI is challenging in a crowded RF environment where pulses are received every few tens of microseconds.

Jittered PRI is generally used for ECCM purposes, and exact detection and labeling of pulses from radar systems using this method can be difficult. The jittered PRI radar signal structure is shown in Figure 5.4.

• *Pulse group repetition interval (PGRI):* The PGRI is the sum of all PRI elements before a repeat in the PRI sequence occurs. This method is helpful as some RWR and ESM pulse deinterleaving algorithms can determine the PGRI but cannot identify the individual elements. A PGRI is a pulse train in which groups of closely spaced pulses are separated by much longer times between these pulse groups. PGRI is usually defined by the number of *pulses per group* (PPG) and *pulse groups per second* (PGPS).

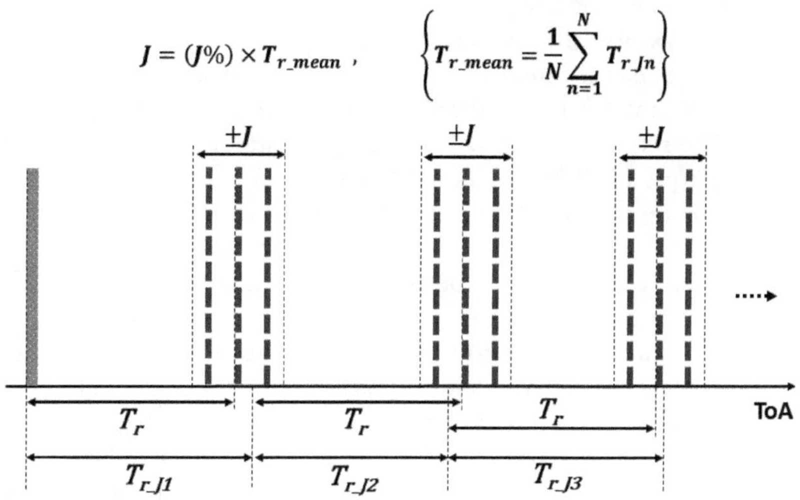
$J = (J\%) \times T_{r_mean}$, $\left\{ T_{r_mean} = \frac{1}{N}\sum_{n=1}^{N} T_{r_Jn} \right\}$
±J
±J
±J
T_r
T_r
T_r
ToA
T_{r_J1}
T_{r_J2}
T_{r_J3}

Figure 5.4 Jittered PRI radar signal structure.

The PGRI may comprise only a few elements, such as four or five for a radar with a simple stagger, to several tens of components for a radar with a more complex PRI sequence. Figure 5.5 demonstrates five PRI elements for radar with a simple stagger.

As stated before, we can mention the different RF types for pulsed radar signals and PRI agility. These methods are fixed, chirp, agile, and hopper RF types. As understood from its name, the frequencies are identical for all the pulses in fixed RF signals. Let us briefly explain the variable RF types.

- *RF chirp:* This method is utilized for pulse compression to improve the range resolution of pulse radars. Pulse compression can be conducted in a pulse (intrapulse modulation) or a group of pulses (interpulse). The intrapulse modulation can be either FM with linear or nonlinear (symmetrical or nonsymmetrical) FM for chirp radars. Furthermore, it can be intrapulse code with time-frequency-coded waveforms. Interpulse modulation is an LFM modulation that is distributed to many sequential pulses. The general application term for intrapulse modulation is *stepped frequency waveform* (SFW). The SFW class of radar waveforms is used in wide bandwidth applications requiring an

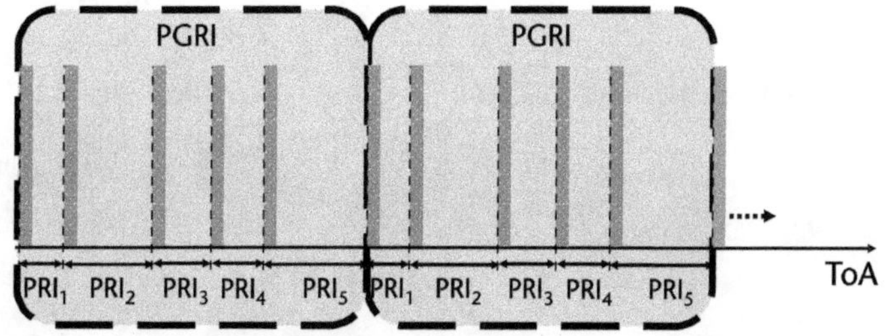
PGRI = PRI$_1$ + PRI$_2$ + PRI$_3$ + PRI$_4$ + PRI$_5$
PGRI
PGRI
PRI$_1$ PRI$_2$ PRI$_3$ PRI$_4$ PRI$_5$
PRI$_1$ PRI$_2$ PRI$_3$ PRI$_4$ PRI$_5$
ToA

Figure 5.5 PGRI radar signal structure.

extensive compression ratio. SFW is a particular case of an extremely wide bandwidth LFM waveform [6].

The constant and chirp RF structures are depicted in Figure 5.6. The constant RF has the same frequency in each pulse, given in the first line. The second line shows the intrapulse modulation for up-chirps and down-chirps. The FM occurs inside the pulses, and LFM is applied to each pulse. The modulation frequency of the LFM is FM. The third line represents the interpulse modulation for up-chirps and down-chirps. Here, the LFM modulation process is distributed to n pulses, and then the modulation process is repeated continuously.

- *RF agile:* This method presents an agile RF emission from pulse to pulse. Some agile signals follow a specific pattern of changing RF from pulse to pulse, whereas others change RF on a pseudorandom basis. The extent of RF agility varies from just a few megahertz differences between the RF levels to hundreds of megahertz differences between maximum and minimum values.
- *RF hopper:* In this method, RF emission changes from batch to batch of pulses. Like the RF agile signals, changing RF from pulse to pulse occurs in a specific pattern or on a pseudorandom basis. Figure 5.7 shows the structures of agile and hopper RF signals.

5.1.2 CW Radar Signals

CW radars continuously emit electromagnetic energy and use separate transmit and receive antennas. CW radars continuously transmit high-frequency signals. The reflected energy is also received and processed consistently. An unmodulated CW radar transmits at constant amplitude and frequency. A CW radar transmitting an

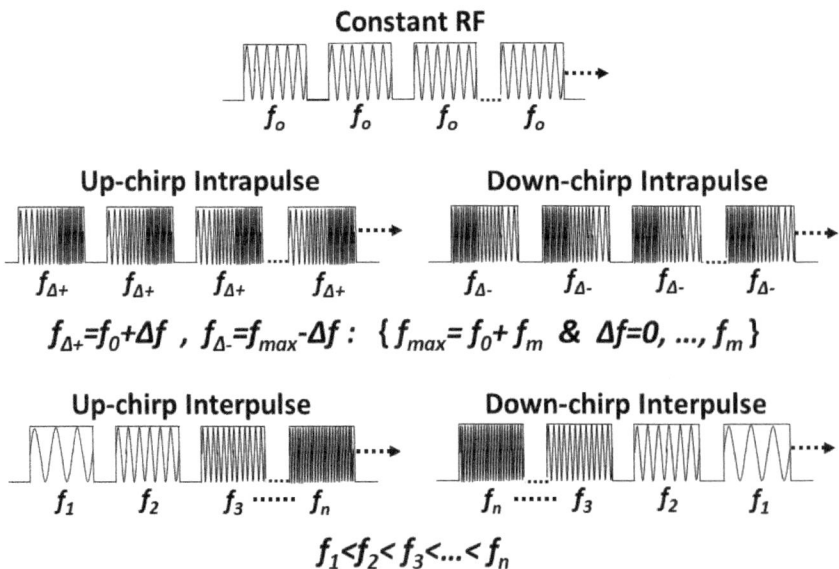

Figure 5.6 Constant and chirp RF radar signal structures.

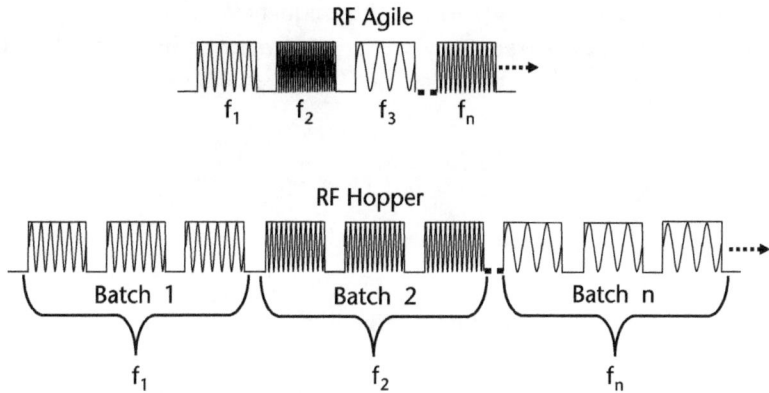

Figure 5.7 Agile and hopper RF radar signal structures.

unmodulated signal can measure speed using the Doppler effect and angular position. However, range measurement is only possible with modifications to the radar operations and waveforms. In other words, some form of modulation is necessary to extract target range information. The primary usage of CW radars is in target velocity search, track, and missile guidance operations.

Modulating with FM, CW radars turn to FMCW radars, and they can measure distance by measuring the time delay between the received and transmitted RF signals. FMCW radars may use LFM waveforms so that both range and Doppler information can be measured. Practical CW radars use triangular LFM waveforms, which can be changed in modulation in one-direction or two-direction periodicity. Figure 5.8 shows a one-directional periodic or sawtooth linear FMCW radar signal. However, two-direction periodic or triangular linear FMCW has been analyzed in Section 2.5.2.2 and demonstrated in Figure 2.19.

In the sawtooth linear FMCW signal structure, frequency values vary from an initial or the operating frequency value in the bandwidth range. In Figure 5.8, f_0 shows the initial frequency value, Δf shows the modulation bandwidth, and T_r shows

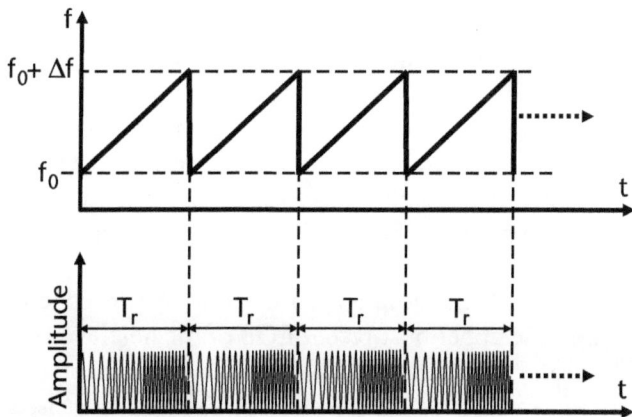

Figure 5.8 Sawtooth linear FMCW radar signal structure.

the chirp period. In this method, the frequency change direction is only upwards. The time-dependent frequency variation is defined in (5.3). The t value represents the instantaneous time value, while f' denotes the frequency change slope.

$$f(t) = f_0 + (f' \times t), \quad \left\{ f' = \left(\frac{\Delta f}{T_r} \right) \right\} \tag{5.3}$$

A mixed form of linear FMCW may also be possible besides sawtooth and triangular linear FMCW signals. Segmented linear FMCW can be given as an example [7]. In this case, after a triangular period (up-chirp and down-chirp), a nonmodulated signal interval at f_0 frequency occurs. Then again, a triangular period arises before a nonmodulated period, and the process repeats itself in this sequence.

5.2   Emitter Identification Database

SPEW systems are produced for general purposes rather than specific cases and engage missions in different operational regions. Other theater regions mean different threat system combinations, and the expectation from a SPEW system is to detect them all and activate the appropriate ECM. Chapter 1 stated that a SPEW system comprises different RF, IR, and laser subsystems. However, this book restricts the inspection to RF.

In the RF part of a SPEW system, detection is conducted with RWR or ESM systems, and ECM is fulfilled by utilizing a jammer and chaff dispenser. In most cases, proper maneuvering supports the used ECM. One expectation from the RWR or ESM part of the SPEW system is to detect every threat without exceptions. Let us assume that the capability of an RWR system is technically sufficient for detecting and identifying any RF-guided threats. This time, the emerging problem is to prepare a suitable emitter identification databse (EIDD) or emitter library for the situation. Preparing an emitter library for general usage is improper, but prepping it for the mission area is preferable. For this purpose, different EIDDs are packed for other territories, and the existing EIDD is generally called an *emitter program library* (EPL) or *preflight message* (PFM). EIDDs have limited capability regarding the number of radar parametric mode lines or entries available.

In the environment where the RWR operates, the threat radar types and numbers vary in a wide range, which may be tens to several hundreds. Several thousand parametric mode data entries are required for their collective specification when considering several hundred radars. However, the capabilities of current RWR libraries are insufficient and need to be improved. Thus, a meticulous study should be conducted to prepare PFMs. This study must contain the following:

- Realistic and reliable intelligence operations are conducted to collect information on the uppermost recent EOB of the hostile or neutral regions.
- The EOB data obtained from intelligence operations are analyzed to determine specific and distinguishing characteristics of the threats to prevent excessive PFM data.

- A study is diverted for the operational requirements, and the threat priorities are determined.
- Based on analyzed intelligence data and threat priorities, an optimization study is fulfilled to fit the required data into the EIDD's radar parametric mode lines.

5.2.1   EIDD Parameters

The emitter identification is a comparison process. The signal characteristics in each track file are filled with processed data and are constantly updated based on the TOA and location of the received signals. These track files are consistently compared to the parameters of the EIDD table. Thus, the EIDD parameters are essential since they are the base for the EIDD table. Furthermore, properly defining the EIDD parameters is a crucial RWR signal processing part design criterion.

At this point, mentioning the similarities and differences between a SIGINT system and an RWR or ESM system would be beneficial. The primary operational purposes of a SIGINT system are detecting the existence and location of a known emitter and obtaining the parameters of the unknown emitters. The latter task of the SIGINT systems is about technical intelligence duty that requires further studies after the mission. This property sometimes and limitedly exists in the ESM systems. The SIGINT systems' signal processing parts for the first task contain EIDD tables similar to ESM systems. However, their parameters are more detailed since they do not have to identify the emitter rapidly, such as the ESM systems. Because the EIDD parameters are detailed, they have larger EIDD tables and far greater data than ESM systems. These properties meet the SIGINT systems' operational requirements. They conduct the emitter identification process longer and aim to achieve high accuracy.

The EIDD parameters are not comprehensive in conventional RWR systems. Other emitter parameters are also available, which may be necessary to characterize the threat system. The typical emitter characteristics that an RWR system can measure and calculate by further processing for a pulsed radar include ROF, signal power amplitude, AOA or DOA, TOA, PRI, PRI type, PW, scan rate and type, and beamwidth. However, these parameters do not include all elements. Other emitter parameters are also required to characterize the threat system. Modern RWR systems can measure additional parameters, such as PRI modulation characteristics, interpulse and intrapulse modulation, and missile guidance characteristics. Furthermore, contemporary RWR systems can classify the beam scan types, such as circular, sectoral, raster, spiral, conical, and lobe switching, through the magnitude values associated with each pulse of an emitter. Other parameters that can describe an EM wave but are currently not commonly used for EID include polarization and phase.

These parameters are the primary reference designators for defining a specific threat. There are two criteria for determining a threat emitter: the specifications' accuracy and the second not to intersect with the other systems' technical properties.

Some emitter parameters can be measured using a single pulse, called monopulse parameters. The monopulse parameters include ROF, PW, AOA, amplitude, and TOA. RF can be determined pulse by pulse by Rx that can measure frequency. Frequency is beneficial for EID since most radars operate at a single frequency. Most real-time systems measure PW instead of pulse shape because the latter is more difficult to characterize mathematically. PW information is a suitable discriminator parameter for resolving identity ambiguities between radars in this context. However, multipath effects may alter the received PW or cause modulation in the pulse profile, leading to pulse width measurement problems.

For CW radars, the RWR can measure and calculate by processing the RF, signal power amplitude, AOA, TOA, modulation type, scan rate and type, missile guidance characteristics, and polarization.

AOA cannot be used for EID but is excellent for sorting signals. Some RWR systems use frequency and AOA information to distinguish new signals from old ones. Amplitude can also not be used for emitter identification. However, it can be used for sorting and coarse distance estimation using a defined emitter's ERP. Moreover, amplitude in conjunction with TOA can be used to determine the emitters' scan characteristics.

Other emitter parameters such as PRI, guidance, and scan characteristics can be determined only by analyzing a group of pulses. All these parameters are helpful for EID; unfortunately, they require time for data collection and analysis and call for sophisticated signal processing algorithms.

Using scan period as an identifier generally causes problems for RWR systems that are not wide open to all RFs within their range, as some radar scan peaks may be missed altogether.

Some radars can operate with various PRI stagger values, and specifying them all precisely in an EIDD is impossible. However, some radars have relatively few PRI options and can be more tightly defined. Other data, such as the platform type of the radar, such as airborne, shipborne, and land-based, can be used to increase the accuracy level for the EID. The clock period and PRI stability are beneficial parameters for distinguishing between radars, but information on these parameter values is only readily available for some radar types.

5.2.2 EIDD Table

RF-guided threat radars generally operate in more than one mode, and even simple fixed RF radars have different combinations of PRI and PW according to the range at which they are performing. Each variety of PRI and PW is a separate mode and should be specified separately in the RWR emitter library (or EIDD). The essential data fields in EIDD for a mode specification are ROF max, ROF min, PRI max, PRI min, PW max, and PW min. Other parameters may include TOA, AOA, power amplitude (A), PRI type, scan rate and modulation, 3-dB beamwidth, interpulse, and intrapulse modulation. Also, EIDD is a standard table for pulsed and CW emitters. For this reason, the emitter type is first defined as pulsed or CW, then the related fields (pulsed-P or CW) are filled. An example of a generic unfilled EIDD is given in Table 5.1.

The signal characteristics in each track file are registered with processed data and are constantly updated based on the arrival and location of the received signals. The track files are compared to the EIDD table installed in the memory of the signal processor computer. The EIDD is a predefined technical specifications table for the known threat radar systems.

As can be seen from Table 5.1, EIDD specifies each radar and the mode of each radar, if any. This specification is defined for different parameters as follows:

- The maximum and minimum value expected for some set of parameters (i.e., ROF, PRI, PW).
- A definite value for another parameter set (i.e., TOA, AOA, A, scan rate, 3-dB beamwidth).
- The remaining set has mathematical definitions of the required properties.
- Additive notes define the additive properties, such as harmonics or PGRI modes.

The generic EIDD parameters given in Table 5.1 may be extended in some systems. The used parameters depend on the system design and operational requirements, but the standard parameter entries should exist in the systems. The EID table is created from information gathered from ES systems and intelligence sources (SIGINT). This table can be modified and updated to reflect the most up-to-date radar features available for threats expected to be in the planned operational area. Each RWR system has specific procedures for reprogramming the signal processor and updating the EIDD (or PFM). The immediate reprogramming action that must be taken when a new threat emerges that is not defined in the current PFM is called pacer ware [8].

Table 5.1   An Example for a Generic EIDD

Parameters	*Threat (T) Numbers*						
	T1 (Mode 1)	$\vdots$	*T1 (Mode n)*	*T2 (Mode 1)*	$\vdots$	*T2 (Mode n)*	$\vdots$	*TN (Mode1 to Mode n)*
Signal type (Pulsed (*P*) or *CW*)							
ROF max-ROF min (*P* + *CW*) (MHz)							
PRI max-PRI min (*P*) (ms)							
PW max-PW min (*P*) (µs)							
PRI type (*P*) (fixed, staggered, jittered)							
Antenna scan rate (s) (*P* + *CW*)							
Antenna scan modulation (*P* + *CW*) (mathematical definition)							
Modulation type (*CW*) (mathematical definition)							
3-dB beamwidth (*P* + *CW*) (degree)							
Interpulse modulation (*P*) (mathematical definition)							
Intrapulse modulation (*P*) (mathematical definition)							
Additive notes (*P* + *CW*)							

5.3   Signal Deinterleaving

In RWR systems, RF/digital signal processing blocks in the Rx subarchitecture constituted the PDW and SDW, the input signals of the deinterleaving processes. The Rx digital signal processing blocks form the PDW by performing the following steps:

- Frequency spectrum scanning and signal detection;
- AOA estimation of the detection signals;
- Time labeling (or TOA) of the detection signals;
- Amplitude (A) measurement of the detection signals;
- Pulse width (PW) measurement for the detected pulse signals.

The primary purpose of the deinterleaving is to separate ambiguous signals in frequency and AOA and calculate the PRI. The PDW comprises PW, ROF, TOA, AOA, and A. Note that we exclude the P to simplify the analysis. However, for CW signals, SDW consists of ROF, TOA, AOA, and A by following the first four steps of the PDW above. The deinterleaving process aims to obtain the EDWs using the sequential PDWs and SDWs and feed the labeling process with them. The EDW includes PRI, PRI modes, and RF modes in addition to the PDW for pulsed radar signals. However, for CW radar signals, EDW consists of modulation information in addition to the SDW. EDW involves antenna scan and pattern properties for both and contains further specifications if they exist.

Since the deinterleaving process forms the source data for the emitter tracks and the identified emitter list, it must handle the interleaved signals even with the *pulse-on-pulse* (PoP) phenomenon.

5.3.1   Received Signal Complexity

The dense RF environments are the main received signal complexity problem for airborne and shipborne ESM systems. Although the quality of the RF threat systems increases the jeopardy of the theaters in modern battlefields, the quantity is still significant. For this reason, the IADSs have comprised many RF threats since the beginning of radar usage for air defense, and it will probably be the future approach. When the number of RF systems for the enemy and allied forces increases, the signal complexity will increase simultaneously.

Generally, RWR systems used in airborne platforms can simultaneously coincide with pulsed and CW/FMCW (or clustered in CW for simplicity) radar signals. The deinterleaving process can be conducted separately for pulsed and CW signals in two parallel processes. A classification of pulsed and CW signals may be realized for a predetermined duty cycle threshold or bandwidth threshold. This classification is conducted in the Rx's RF/digital signal processing blocks, and the PDWs for pulsed and SDWs for CW signals are composed separately. Simultaneously, existing pulsed and CW signals add additional workload to the deinterleaving process and increase the computational burden.

Another reason for signal complexity is the similar technical specifications of the same mission radar systems due to the physical limitations. The predominant discriminative properties of the PDW are frequency and AOA for deinterleaving.

The AOA depends on the locations and operational array of the radars and victim platforms. More than one radar may be in the same direction or sector in operational and environmental conditions. If these radars are the same type, such as land-based air defense, airborne surveillance, maritime patrol, airborne fire control, battlefield, or naval fire control, the operational frequencies will also be in the same ranges. Furthermore, the same types of ally radars may exist in addition to hostile radar systems in the same direction. A similar PDW for this case will be obtained for different radar systems, and the deinterleaving process will be more confusing. As a demonstration of how close frequencies are in the same mission radars, a rough frequency information list of the radars engaged for the same type of mission is presented in Table 5.2 [9].

The table generally considers frequency bands used for the related radar types. Some radars that operate out of the given limits are possible. For instance, land-based air defense radars like Giraffe AMB and TRM-L 3D's operating range is 4 to 8 GHz. However, some land-based air defense radars like Over-the-Horizon-Backscatter P-10 (Knife Best), P-12 (Spoon Best), and P-14 (Tall King) use RF in the megahertz levels. These examples can be extended to include different types of radar.

The main point here is to focus on the values at which the frequencies concentrate. Also, some radars use different frequencies in different operational modes. The radar systems with the same missions from the same directions have similar ROF and AOA information in their PDW. This situation causes uncertainties and increases the complexity of the deinterleaving process.

In addition to operating in a dense RF environment, simultaneously receiving pulsed and CW signals and similar technical specifications of the same mission radar systems, another essential issue for the complexity of the receiving signal is the probability of intercept (POI). POI represents the amount of time a signal must be present; thus, there is a probability that the signal will be intercepted and adequately captured for analysis. When a high POI is required, discussing the related issues affecting the detection process is vital.

A wide beamwidth receiving antenna means low antenna gain, resulting in low POI. Furthermore, a wide beamwidth Rx antenna means poor AOA estimation accuracy. However, using a high-gain Rx antenna requires a long search time for angular search volume, which reduces POI. Another impact on POI comes from the operating frequency bands. RWR systems are wide-frequency range systems requiring a long search time, which reduces the POI. A wide bandwidth Rx, which causes low sensitivity, is engaged to shorten the search time. Although the POI increases up to a level, this time, the low sensitivity reduces the POI. Another reason for lowering POI is to use narrow beamwidth (or high gain) Rx antennas to increase sensitivity. A further reason to reduce POI is low-duty cycle signals, which can be detected only when they are present in the Rx. The POI reduction stems from the required search time being extended due to narrow beamwidth antennas or using narrow bandwidth Rx.

5.3.2   Deinterleaving Pulsed Signals

The increased Rx bandwidth in ESM applications allows an increased POI with a wide-open Rx, providing the ultimate frequency-versus-time performance. Likewise,

Table 5.2 Operating Frequencies of Different Mission-Type Radars

Radar's Mission Type	Radar Systems (with model name)	Operational RF Range
Land-based air defense	5N69 (Big Back), 67N6E GAMMA-DE, AN/FPS-124(V), AN/FPS-8, AN/FPS-88, ARSR-4, AN/TPS-44, DR-162 ADV-ER, ESR-220, ESR 380, ESR 360L, GRL 600/610, INDRA II, DR 151, DR 162, TRS 205X, TYPE 571/581.	1 to 2 GHz
	AN/FPS-108 COBRA DANE, AN/FPS-117, AN/TPS-59, AN/TPS-63, DR172S	1.215 to 1.4 GHz
	AN/FPS-6, AN/TPS-70, AN/TPS-32, AN/TPS-78	2.7 to 3.1 GHz
	Big-Bird, GERFAUT, HR-3000, JY-9F, MASTER-A/M/ M3R/S/T, RAT 31DL, TRS 2140, TRS 22XX	2 to 4 GHz
Airborne surveillance and maritime patrol	AN/APS-125/138/139/145	0.3 to 1 GHz
	AN/APY-1/2, EL/M-2075, ERIEYE, L-88(V)3, MESA	1 to 4 GHz
	AN Series, AN/APR-241/242, AN/APQ-164, AN/APY-3, APS-504(V), GUKOL-1/2/3, SEA DRAGON	8 to 10 GHz
	AN/APQ-174/186, AGRION, AN/APQ-156/170/181, AN/ ZPQ-1, ANEMONE, HORIZON, IGUANE	8 to 20 GHz
Airborne fire control	AN/APG-66, AN/APQ-153/157/159, ZASLON M	6 to 10 GHz
	AESA, AN/APG-63 /65/67/68/71/73, KOMAR, KOPYO, OSA, N019 SAPFIR-29, RDI, PS-46A	8 to 12 GHz
	AN/APG-63(V)1-2/70, AN/APQ-126, CYRANO IV, ECR 90 CAPTOR, PS-05/A, UAP 13	8 to 20 GHz
Battlefield	AN/TPQ-47, CROTALE, FAN SONG	2 to 4 GHz
	ARK-1M, AN/MPQ-53, 3D RAC, DOG EAR	4 to 8 GHz
	AN/PPS-5C, BFSR-MR, ARSS-1, BOR-A 550, ARABEL, CYMBELINE, DR Series, EL/M-2129, FLYCATCHER MK2, MODEL ST-312, LOW BOW, JY-17/A	8 to 12 GHz
	ARINE	12 to 18 GHz
	ARS2000, CREDO, ASKARAD, FLAPWHEEL, GS-11/13, J/MPQ-7	8 to 20 GHz
	EAGLE, EDT-FILA	Higher than 20 GHz

the systems' increased instantaneous angular coverage, with 360° coverage, enhances the POI. One of the problems with growing bandwidth and instantaneous angular coverage, particularly in a dense environment, is that the systems must deal with multiple simultaneous signals.

In ESM applications, an SHR is often used with other wideband Rx, such as digital Rx. These Rx receive the signals from the SHR as an IF and are digitized. In general, all the signals are from the SHR in continuous form, so all the signal structures are clustered or interleaved. The digital Rx represents the signals in discrete form, making it easy to identify the signal characteristics. ESM wideband digital Rx spend much of their effort deinterleaving, identifying, and tracking simultaneous incident radar pulse trains. The problem of determining the presence of a specific emitter in the environment is a problem of detecting a consistent pulse sequence in the incoming stream of interleaved pulses.

The pulse sequence deinterleaving algorithm is based on five parameters of the intercepted pulse stream: the AOA, ROF, PW, PA, and TOA. These parameters are determined for a single pulse sequence, and PRI information, which is very important to identify the emitter, is obtained with these parameters. The detection parameters that make up the PDW table are transmitted to the signal deinterleaving block with a particular analysis time dimension. Table 5.3 shows a sample set of PDWs recorded by an RWR system.

Table 5.3 contains the pulse definitions of two radar signals in an example PDW set. Since the number of emitters will be much higher in the operational environment, the diversity in the PDW table will increase, making the deinterleaving process difficult. The PoP problem occurs during deinterleaving due to the structure of the interleaved signal. The interleaved signal is obtained by aligning the PDW sets on the arrival time axis. Therefore, clustering and analysis studies should be performed with high accuracy and performance.

While the pulse signals are separated in the RWR, the preclustering problem is defined by plotting the interleaved signal status in the time axis. Figure 5.9 shows four pulsed emitter signals and the interleaved signal with the PoP problem. The figure shows that the interleaved signal (Figure 5.9(b)) is exposed to a remarkable PoP phenomenon even for four simultaneous pulsed radar signals. Overlapping pulses can only be separated by a clustering operation to extract PRI information.

RWR systems generally begin deinterleaving by clustering pulses regarding ROF and AOA. Clustering is followed by PRI extraction to arrive at separate intercepts for each radar the RWR has deemed in the cluster. Attempts are made to correlate each intercept to an existing track or create a new track if no suitable tracks exist. Thus, the deinterleaving process has two main subprocesses: clustering and PRI extracting subprocesses, and the utilized methods are as follows.

5.3.2.1 Clustering Methods

Signals are grouped after coming to the deinterleaving block, and PRI and parameter information are extracted based on the periodicity analysis. The grouping, or the clustering operation, must be performed before parameter extraction. In clustering, signals with the same pattern are brought together. After the signals with the same characteristics in the PRI list are collected in a cluster, PRI clustering is performed in each group.

Table 5.3 An Example of a PDW Set

PDW	AOA (°)	ROF (GHz)	PW (μs)	TOA	PA (dBm)
1	55.6	10.6	0.5	12:25:18.1284	−52.3
2	5.8	9.5	1.2	12:25:18.12852	−45.6
3	55.7	10.6	0.5	12:25:18.12863	−62.2
4	5.9	9.5	1.2	12:25:18.12871	−55.4
...	...	...	...	...	...
N-1	55.8	10.6	0.5	12:25:18.129	−45.5
N	56	10.6	0.5	12:25:18.1296	−48.6

Pulsed signals from different emitters

(a)

Interleaved signal

(b)

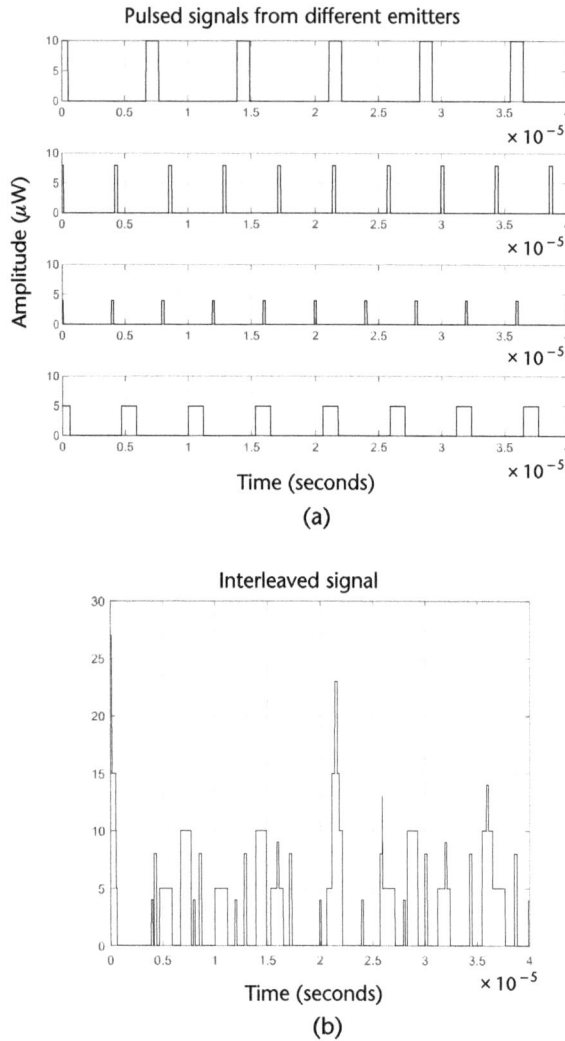

Figure 5.9 PoP phenomenon: (a) Four different pulsed signals and (b) interleaved signal with PoP.

In clustering, close signals are collected around a cluster by determining the proximity/distance over the feature parameters. Calculating an average value and standard deviation of the collected signals in terms of proximity, the signals that still need to be clustered are gathered to the nearest cluster.

Various clustering methods aim to cluster nested data successfully. Apart from the two basic approaches, distance-based clustering, and K-means clustering, there are nearest neighbor-based clustering [10], hidden Markov model (HMM)-based clustering [11], machine learning, and deep learning-based clustering methods.

1. *Distance-based clustering:* Depending on the parameter distance, the primary approach in clustering methods is to collect the unclustered data in the interleaved signal to the nearest cluster and update the cluster's mean value and standard deviation. Two prevalent distance functions used in distance-based clustering are as follows:

- *Euclidean (Euclidean) distance:* The distance between two vectors or values is calculated according to the parameter weight.
- *Minkowski distance:* This is the general form of the Euclidean and Manhattan distances. It is a distance measured between two points in N-dimensional space. It is widely used in machine learning, especially in finding the optimal correlation or data classification.

2. *K-means clustering:* One of the most frequently used clustering algorithms, the K-means clustering method, divides N data into K clusters. The primary purpose of this method is to minimize the similarity between sets by providing maximum similarity within the cluster. The technique detects K clusters, making the mean square error the smallest. Although the problem is complex, the algorithm is implemented with an iterative approach. The K-means algorithm includes four basic steps:
   - Creation of cluster (group) centers;
   - Grouping objects outside the center according to their distance values;
   - Updating the centers after each grouping;
   - Grouping and updates are continued until the clustering becomes stable.

When grouping with the K-means clustering algorithm, the computational load increases as the number of clusters increases. K-means-based clustering is used for machine learning-based classification and anomaly detection. The PRI extracting process, which conducts detailed analyses within each group, follows the clustering process. For this reason, the performance of the clustering process directly affects the signal deinterleaving results.

5.3.2.2   PRI Extracting Methods:

While performing analyses for signal deinterleaving in RWRs, parameters such as PRI value, PRI type, and intrapulse modulation are extracted. Signals and data with similarities within the cluster give results according to the periodicity analysis. Histogram methods, frequently used in data analysis, are also used in PRI decomposition. This section mentions the histogram-based and PRI conversion methods commonly used in deinterleaving pulsed radar signals.

1. *Sequential difference histogram (SDIF):* This method extracts differential histograms by taking consecutive derivatives on the arrival time vectors. In the case of fixed PRI, a single histogram size is expected. However, histogram thresholding takes the second and third derivatives to control the staggered PRI states. Deinterleaving takes the relevant PRI for the value or values whose derivative result is above the threshold. The SDIF method offers high performance in fixed and cascading PRI signals. Nevertheless, its performance is low for a jittered PRI signal.
   The SDIF process flow is performed in four stages for the automatic classification and determination of fixed and staggered PRI values:

- The derivative is taken on the TOA vector.
- The histogram of the derivative vector is taken.

- The peaks of TOA difference values are determined by adaptive thresholding on the histogram.
- A staggered PRI decision is made if there is more than one peak and the histogram values exceed the threshold. If there is only one peak, a fixed PRI decision is made.

2. *Cumulant difference histogram (CDIF):* In this method, the harmonics of the PRI value are inspected by taking the first-order and second-order derivatives. PRI values are determined for fixed and staggered PRI states by performing cumulative operations for signals within the cluster. The CDIF method, which offers high performance for fixed PRIs, decreases performance for staggered PRIs.

3. *PRI transform method:* The peak points in the histogram spectrum are determined after obtaining the derivative histogram. A dynamic sequence search is performed for the PRI value at the maximum peak. This search aims to produce a reliable result by determining the PRI values. However, the computational burden increases due to the dynamic search for the staggered PRI signals. Since it gives more reliable results in jittered PRI structures, it provides the RWR with a critical capability.

5.3.3   Deinterleaving CW and FMCW Signals

In the Rx block, SDWs are generated with the AOA, ROF, A, and TOA parameters of CW and FMCW (or CW for short) radar signals. The SDW table is an essential input in signal decomposition, clustering of CW signals, and parameter estimation. Table 5.4 shows an example SDW list for CW signals.

Table 5.4 contains the signal definitions of two CW radar signals in an example SDW set. The number of emitters will be much higher in the operational environment. Thus, the complexity of the SDW table will increase, making the deinterleaving process difficult.

The fundamental parameter values of CW radar signals are center frequency, modulation bandwidth, chirp period, and chirp direction. The values obtained in the SDW are used to determine these parameters. The first step of the decomposition is to use the clustering methods described in the clustering of pulsed signals.

Table 5.4   An Example of an SDW Set

SDW	AOA (°)	ROF (GHz)	A (dBm)	TOA
1	16.3	9.5	−52.3	14:33:18.12852
2	35.7	10.3	−45.1	14:33:18.12866
3	16.15	9.7	−52.5	14:33:18.12872
4	35.65	10.4	−45.2	14:33:18.12881
5	16.14	9.8	−52.4	14:33:18.12892
6	35.7	10.5	−45.4	14:33:18.1291
...	...	...		...
N–1	35.8	10.4	−45.15	14:33:18.3458
N	35.95	10.3	−45.3	14:33:18.3466

A connected component labeling-based clustering approach is also possible for clustering CW signals [12]. Then, parameter extraction of the signals is performed using time and frequency vectors for each cluster.

The operations for each parameter extraction are explained in the items:

- *Center frequency:* This is calculated by averaging the frequency vector formed from the frequency values in the cluster.
- *Modulation bandwidth:* The modulation bandwidth is obtained by taking the difference between the maximum and minimum values of the frequency vector. Without modulation, this bandwidth will narrow and represent only the frequency oscillation.
- *Chirp period:* Frequency change analysis is performed by taking the derivative of the frequency vector. The frequency change peaks in both positive and negative directions are to be found. The peak analysis calculates the chirp period using the histogram-based threshold operation.
- *Chirp direction:* The direction at each chirp transition is determined by determining the minimum and maximum points with the frequency change vector.

The third and fourth items are conducted by utilizing time-frequency analysis techniques. An example parameter extraction process can be summarized as follows:

- Each clustering result is determined as a new radar signal if parameter extraction is performed in the first analysis window.
- In sequential analysis processes, update steps are performed using pulsed radar signals.
- The update step calculates the closest cluster with ROF, AOA, and bandwidth parameters.
- Cluster parameter values are updated by making associations with the closest cluster.
- Before the update, control is performed with the center frequency variance.
- If the new data is not associated with any cluster, it is defined as a new radar signal.

In addition to parameter determination, the signal detection process is also a problem in LPI radars, which mainly use CW transmissions. For this purpose, time-frequency analysis techniques are the main signal processing techniques. Wigner-Ville distribution, Choi-Williams distribution, quadrature mirror filtering, cyclostationary spectral analysis, wavelet transform, and short-time Fourier transform are the time-frequency analysis techniques used in LPI radar signal detection. In recent studies, machine learning methods have also automatically detected and classified the LPI radar signals.

5.4 Signal Labeling and Emitter Identification

The output of the deinterleaving process is EDWs. The EDWs feed the signal labeling or active track file block. In this block, signal labeling is fulfilled for the potential tracks, and active tracks (or AEF) are defined. The emitter identification process is conducted after labeling the signals and obtaining the AEF. Thus, emitter tracks are formed, and the identified emitter list is obtained. The identified emitter list usage is twofold. The first one is for the interface of the RWR subsystem with the other subsystems, such as display and ECM. Moreover, the second one is for the Rx scan control to adjust the future scan regime.

5.4.1 Signal Labeling

When the EWD is composed and parameters are determined, the RWR processor first attempts to correlate the EWD to an existing track containing EWDs from the same radar. Parameter tolerances are to be defined in the RWR systems to allow for correlation. These tolerances must be within acceptable limits for correlation purposes. The tolerance of AOA may be less than 1°; similarly, ROF may be within the tolerance limits of a few megahertz. The DOA tolerance may be variable depending on the geometry and the operational conditions of the RWR or ESM platform relative to the radar or the classification of the mother platform of the radar, such as air, ship, and land.

A new track is created if the EWD cannot correlate to an existing one. A predetermined pulse number (i.e., 5 to 8 pulses) is needed to make a good intercept in a typical ROF environment. RF is usually well measured by RWR systems, but there may be reasons for PRI miscalculations, such as missing pulses and complex PRI structures. PW is often measured in error for pulses that have intentional modulation. Contrary to traditional thinking, DOA is the most likely to be mismeasured parameter. All these reasons mean that there are likely to be many more tracks created than radars in the RF environment [1].

If the EWD can be correlated to an existing track, the track's parameters, such as ROF, PW, and PRI, are updated using the new information available in the EWD. The maximum and minimum values allowed for future track correlation are also updated, taking account of the parameter tolerances.

If the EWD can be correlated to an existing pulsed radar track, the track's parameters, such as ROF, PW, and PRI, are updated using the new information available in the EWD. However, updating conducts in the center frequency and modulation properties with the new EWD information for a CW radar. The maximum and minimum values allowed for future track correlation are also updated, taking account of the parameter tolerances.

The signal labeling process for pulsed radars starts with EDW being formed, and then the RWR attempts to correlate it into an existing track. It fulfills this by considering each parameter and comparing the EDW parameters with those of the tracks. The first parameter considered is usually the DOA. If the DOA of the intercept is within a few degrees of any of the existing tracks (typically up to 5°), then the track is considered a candidate for further consideration.

Generally, the following parameter to be compared is ROF, typically with a tolerance of ±2 MHz. Then the process continues by comparing the PRI. PW may also be used as a correlation parameter. Parameter types, such as fixed, hopper, agile for RF, and fixed, stagger, and jitter for PRI, are another step in the correlation process.

Once the list of candidate tracks matching the EDW's identity has been generated, the next step is to perform the parameter tests for each candidate. The advantage of using the identity in the correlation process is that different parameter tolerances can be set for each radar type.

In cases where more than one possible identity has been determined for the EDW, the list of available correlation tracks increases as those with all the possible EDW identities can be considered candidates for correlation. Suppose an EDW cannot be identified and is labeled as unknown. In that case, two methods can be pursued: testing all the tracks or only the unknown tracks in the correlation process.

The most challenging part of the signal labeling process is that in some RF bands, there are likely to be many types of radar with overlapping parameters. Therefore, using identity as a correlation parameter restricts the list of tracks available for correlation and may lead to the creation of extra tracks.

When the EDW matches more than one track, it correlates to the track with negligible parameter differences. Once the best match is found, the EDW parameters update the track parameters. Suppose the identification has yet to be used in the correlation process. In that case, the track identity is checked against the RWR radar library, and the track identity is updated. For RWR systems that produce location fixes for radars, the geolocation algorithm, the last part of the correlation process, will check whether the threshold for an initial location estimate has been met for the track and, if so, perform it. A successful correlation would yield an update if the track selected for correlation already has a location.

At the end of the signal labeling process, the active tracks or AEF are obtained and conveyed to the emitter identification block. A sorted emitter is reported in the AEF with the parallel parameters with EIDD. Figure 5.10 presents the block diagram of the signal labeling process.

5.4.2   Emitter Identification

After an emitter has been sorted by signal labeling and reported in the RWR system's AEF, it must be compared with the emitter data recorded in the library (relevant to

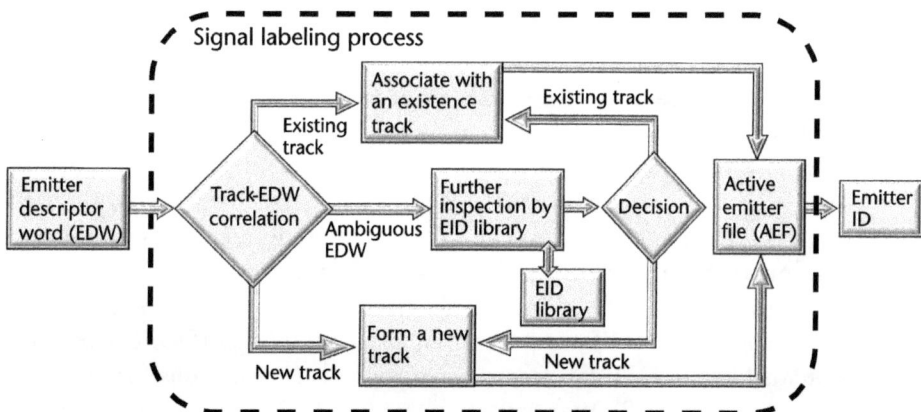

Figure 5.10 Block diagram of the signal labeling process.

known emitters) to determine its most likely identity. This phase relates to EIDD or emitter library entries matching the active tracks in the AEF.

Correctly identifying radars is the primary function of RWR and ESM systems but is one of the major causes of poor performance. The identification process relies on the correct measurement of the radar parameters. While RF is usually measured with sufficient accuracy, other parameters, such as PW and PRI, suffer from measurement errors or calculation errors for various reasons. In addition to this difficulty, lots of different types of radars exist, many of which have overlapping parameter ranges. Similar radars operate in a few RF ranges, as shown in Table 5.2. With so many radar types in narrow RF ranges, the likelihood of generating an EPL or EIDD to identify all radars of interest in a particular mission unambiguously is minimal.

The most challenging processing function for an RWR system is the formation and effective operation of the EIDD. Current EIDD (sometimes called EPL or PFM) has limited capability regarding the number of radar parametric mode lines available. In the environment where RWR systems operate, several hundred potential radar types may need several thousand parametric mode lines for their collective specification. However, the capabilities of current RWR libraries need to be improved. The most spacious library can currently retain about 10K mode lines, but this is an extensive RWR library, with most EIDDs having a capacity of only about 1K mode lines.

Many radars can operate in more than one mode. Even simple fixed RF radars have different combinations of PRI and PW according to the range at which they work. Each variety of PRI and PW is a separate mode and should be specified separately in the EIDD. The generally used data fields for the specification of a mode are given in Table 5.1. Some of them are indispensable parts, such as RF (max, min), PRI (max, min), and PW (max, min). Other parameters may be complementary and discriminative parts for similar radars and modes.

PW is generally utilized as a parameter to resolve identity ambiguities between radars. However, multipath effects and fading are essential since they may alter the received PW, cause additive modulation in the pulse profile, and yield PW measurement problems.

The EIDD specifies each radar as a series of modes, with each mode containing the maximum and minimum values expected for each parameter. Sometimes, the parameter dispersions are included in the mode. The mode lines in the EIDD may differ slightly from the specified modes for the radar; the emitter library will almost always require more mode lines to describe the radar than the number of radar specification modes.

The block diagram of an emitter ID process is demonstrated in Figure 5.11. It is seen from the figure that the AEF data is compared with EIDD, and two different outputs are obtained. One output is the matched emitter, which we will discuss in the continuation part of this part. The matched emitter means the AEF data is compatible with the data defined in the EIDD. The other output is an unmatched emitter, which shows that AEF data is incompatible with any EIDD. Contemporary RWR systems like SIGINT can record the technical specifications for further analysis.

Conventional EID (or emitter library matching) methods can be classified as parameter weighting and scoring [1]. The usage of machine learning and deep

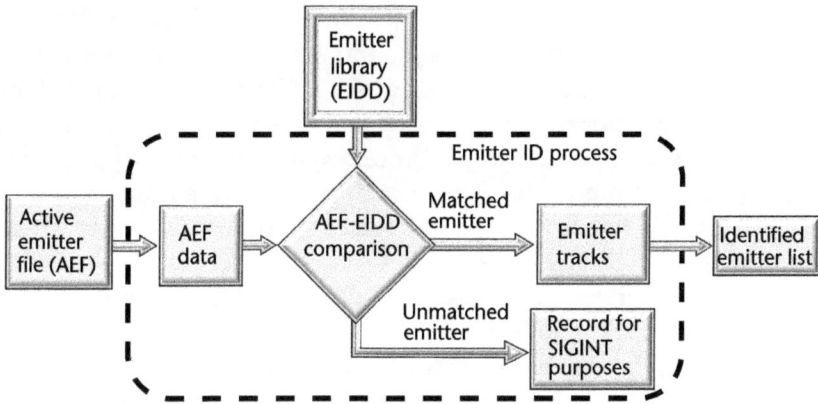

Figure 5.11   Block diagram of the emitter ID process.

learning for EID techniques are discussed in Section 5.6. At this point, explaining the conventional methods would be beneficial.

- *Parameter weighting:* The simplest mode for EID is to define the weights of each radar parameter to match the radar modes with the EIDD. The minimum interval for the required parameter ranges is determined in the first phase of the process. It must be noted that this definition is not unique to a particular radar. It follows that determining a weight between minimum and maximum for narrowing the range. Furthermore, the maximum parameter weight is achieved as accurately as possible by identifying a radar in the last step.

  Table 5.5 shows a generic parameter weighting process for EID. In the table, a sample radar with four modes is given. Also, it shows parameter weighting included in the mode lines. Data for an active track from AEF is also shown in the table to demonstrate the library matching process. The

Table 5.5   An Example of Parameter Weighting for Library Matching

| Parameters | *The Radar's Code Name and Modes* | | | | |
	Radar X Mode 1	*Radar X Mode 2*	*Radar X Mode 3*	*Radar X Mode 4*	*Active Track 1*
ROF range (MHz)	9,150–9,450	9,150–9,450	9,650–9,950	9,650–9,950	9,850
ROF weight (0–1)	0.1	0.1	0.4	0.4	
PRI range (ms)	2.6–3.2	0.5–1.0	2.8–3.4	1.5–2.7	2.6
PRI weight (0–1)	0.3	0.05	0.15	0.5	
PW range (μs)	3.4–7.0	8.2–12.1	2.1–4.2	5.4–8.1	7.2
PW weight (0–1)	0.2	0.15	0.05	0.70	
Scan rate (s)	0.8–1.5	2.2–4.2	3.1–4.6	4.3–5.6	4.8
Scan weight (0–1)	0.03	0.07	0.3	0.6	
Total score for active track 1	0.63	0.37	0.9	2.2	

radar's code name is Radar X, and each mode is considered separately. The weights are between 0 and 1, and radar mode with the active track is more likely to be graded at a higher weight. The sum of weights for one line must be one. Another assumption is that the radar is a pulsed radar.

The library matching process assigns the identity with the highest total weight score to the active track. As seen from Table 5.5, a weighting process is conducted, and at the end of the process, the most probable matching with Active Track 1 is Mode 4 since the total score is the highest at 2.2. The weighting values are assigned according to the corresponding values of the active track and the predetermined ranges.

- *Parameter scoring:* In this EID method, each active track is compared with every mode line in the EIDD, and a score is calculated. The active track identification with the highest score is revealed by lining up the matching radar types. Consider the example in Figure 5.12 and focus on ROF and PRI. The radar mode values from the EIDD are drawn on the related axis, so drawing is conducted for RF ranges on the frequency axis and PRI ranges on the time axis. Then, score scales are defined from 0 to a maximum value (i.e., 0 to 10). Then the active trace measurements on the relevant axis take their places on the axis, and the intersection areas between the EIDD and active tracks are calculated. A score exists for each active track based on the maximum intersection areas. Thus, the parameter scoring gives the most probable matching according to the total maximum.

5.5   Preprogramming of RWR for Operational Requirements

Equip RWR with the required EIDD for the operational requirements, which is called preprogramming or reprogramming. The preprogramming task is to prepare the necessary threat information given in Table 5.1. The data in this table may be extended or scant, and the format may change, but the aim will not change. The

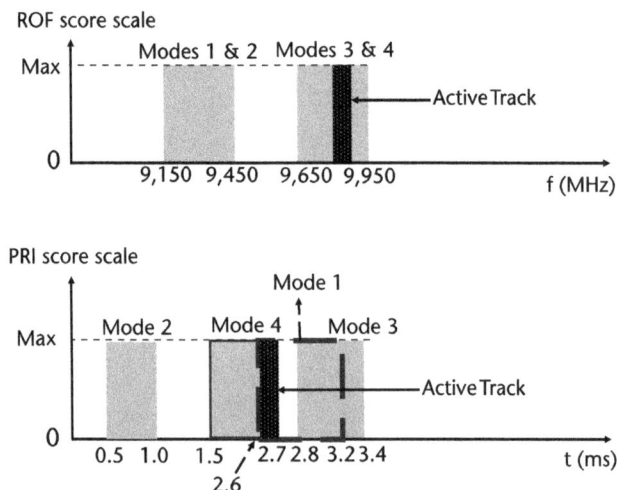

Figure 5.12   Parameter scoring example for EID with two parameters.

RWR system is not a self-separate system but a subsystem of the SPEW system. Furthermore, it operates in coordination with the other subsystems of the SPEW system. The coordination, resource allocation, and data flow optimization are provided by the *operational flight program* (OFP), which is the principal program that operates the SPEW system. Therefore, RWR uses the OFP like the other subsystems of the SPEW system. A compatible program with OFP is used to preprogram RWR suitably for the operational requirements. This program is the PFM. Even though there may be differences from system to system, some common properties of the PFM may be clustered as follows:

- The OFP makes operational decisions using PFM, which consists of preflight tables.
- The PFMs may be prepared for operational purposes, such as PFM-A or PFM-B, or training purposes, such as PFM-trial or PFM-train. Operational PFMs are formed for regional or task-specific purposes. In other words, regional threats and task requirements shape the PFMs.
- PFM is installed on the system separately from the OFP.

The importance and usage purposes of PFMs are twofold, one for RWR and the other for ECM systems. Thus, the resource allocation algorithms in OFP are developed for RWR and ECM applications by defining threat priority in PFMs. In the PFMs, the signal parameters required for threat detection and identification, emitter/mode-specific jamming, and dispenser techniques exist. However, we only discuss the RWR part here.

For preprogramming the RWR, an optimal PFM is to be prepared, and then it is loaded to the OFP via user interfaces. Thus, preprogramming must be inspected in two parts: preparing and loading an optimal PFM.

- *Optimal PFM preparation:* Optimal PFM means operating the RWR system with a fast response time for different operational conditions and accuracy. The response time of an RWR system is the time between when a threat radar signal intercepts the mother platform and after the RWR signal processing phases discussed in the previous subsections appear on its display. A more specific definition can be made from the operational perspective as the time interval between the threat locks onto the mother platform and appears on its display. As stated, each threat is looked at based on the emitter table (or EIDD) entries defined in the PFMs. The maximum response time for each threat is the time to look at the threat successively. The response time depends on the following parameters:
  - The emitter density in the operational environment;
  - If jamming is done in that band, the allowed look-through time;
  - The period when the threat scans back to the mother platform, such as sporadic illumination (SI) threats, TWS, and track-and-scan (TAS);
  - The number and types of threats in the PFM;
  - Distribution of scan types in the emitter tables.
    Response time can be optimized for each PFM by considering all threats. Response times for some PFM threats can be explicitly reduced by priority

scanning. This process may result in increased response times for other threats. Special PFMs can be prepared with reduced threat types according to the operational regions. Response time can be reduced by dividing large PFMs into smaller segments (e.g., air, land, surface) using the multi-PFM feature. It is worth noting that the accuracy of PFMs depends on the accuracy of information about the threats. Moreover, this accuracy affects not only the performance but also the optimal structure of the PFM.

• *PFM loading:* Loading PFM to the OFP is another preprogramming phase. Since the EID process is such a difficult task, updating the emitter library or PFM preparation as often as possible is necessary. Not only do the storage constraints on EIDD mean that there is a limit to the number of mode lines they can contain, but different missions mean that various versions of the EIDD containing different radar types must be produced. As stated before, updating the emitter library is called RWR preprogramming. The need for PFM loading means a suitable user interface should be available to update the EIDD or emitter library. Even though different systems use different interfaces and EIDD formats, the main aim is to fill in the information in Table 5.1.

When looking at the preprogramming requirements from an operational point of view, the maximum number of threats will be put in the PFMs. For this purpose, the number of mode lines of the EIDD must also be unlimited. The mode line capacity is the limitation of computer technology in terms of *hardware* (HW) and *software* (SW). However, the increasing number of threats results in a high processing time, which causes a longer response time. Thus, preprogramming is a complicated process that must be studied by an extensive team consisting of intelligence and operational officers and engineers.

5.6 Machine Learning and Deep Learning in RWR Processes

The density and interpenetration of radar signals in the operative field make it difficult for RWR systems to distinguish and identify signals accurately. Therefore, it is necessary to deinterleave and label the signals using their properties. After these processes, a comparison of the labeled signals with the EIDD occurs. This process is another essential part of RWR signal processing and is conducted with mathematical models in conventional systems. However, recent studies have shown that *deep learning* (DL) algorithms are good alternatives to traditional signal processing procedures for deinterleaving and labeling. However, *machine learning* (ML) algorithms seem more suitable for the *emitter ID* (EID) process. Thus, ML and DL techniques will soon be used in the developed RWR systems. Their most essential advantage over the current methods is that they reduce transaction accuracy and volume. Although DL is a subfield of ML that structures algorithms in layers to create an artificial neural network that can learn and make intelligent decisions independently, their contributions to RWR signal processing are discussed separately here.

The usage of neural networks in pulse stream processing dates to the early 1990s. In early systems, relatively simple networks were reported to extract features of each pulse very well and succeed in categorizing pulses according to separated or standard features. However, the problem of pulse categorization is relatively simple compared to the problem of deinterleaving and can be well solved by conventional numerical methods. Recent developments in ML demonstrate that deep neural networks have enhanced capabilities than shallow ones. A neural network is a series of algorithms that endeavors to recognize underlying relationships in a data set through a process miming how the human brain operates. In a regular neural network, there are three types of layers:

- *Input layer:* The input is fed to this layer, which has several neurons equal to the total number of features in our data.
- *Hidden layer:* The input from the input layer is passed into the hidden layer. There can be many hidden layers depending on the model and data size. Each hidden layer can have different numbers of neurons, generally more remarkable than the number of features. The output from each layer is computed by matrix multiplication of the previous layer's outcome with learnable weights of that layer and then by adding learnable biases followed by the activation function, which makes the network nonlinear.
- *Output layer:* The output from the hidden layer is conveyed into a logistic function such as SoftMax or Sigmoid, which converts each class's result into its probability score.

The primary DL method applied to deinterleaving and labeling is the *convolutional neural network* (CNN). This is because CNN has been applied to classify LPI radar signals, a formidable threat to EW systems, and has achieved good results. In addition, CNN is frequently used in introductory ML workshops as the first practice example, thus making it a natural starting point for our discussion on ML. As CNN is often used to classify images, to apply the CNN to classify LPI signals, the first task is to convert a one-dimensional signal into an image. Time-frequency signal processing techniques can be used to analyze how the signal's frequency component changes over time. Its results can be displayed as an image known as a spectrogram. Different signals have different spectrograms used to deinterleave many pulses from various radars [13].

Shallow neural networks with a single or limited number of layers can represent unpretentious functions to solve simple classification problems. However, when facing more complex classification tasks, such networks suffer from overfitting, making the need for deeper networks more urgent. Adding more layers to the network in a CNN model makes it more in-depth, but such a problem as gradient vanishing has more chances to appear. Gradient vanishing emerges in the backpropagation stage of training, leading to rapid degradation in model performance or saturation in better cases.

CNN is predominantly used to extract the feature from the grid-like matrix dataset. CNN can extract features from an input sample at different levels. They take advantage of local spatial coherence in the input sample, allowing them to share weight, resulting in fewer parameters. One limitation of CNNs is that there is no memory associated with the model to remember the case of the previous samples.

The *recurrent neural network* (RNN) technique deals with this issue by including a feedback loop as a memory. So all the previous inputs to the model leave its footprint [14].

An RNN is an artificial neural network that uses sequential or time series data. These DL algorithms are commonly used for ordinal or temporal problems, such as language modeling, text generating, speech recognition, machine translation, image recognition, face detection, and time series forecasting. Like feedforward and CNNs, RNNs utilize training data to learn. Their memory distinguishes them as they take information from prior inputs to influence the current input and output.

Although traditional deep neural networks assume that inputs and outputs are independent, the production of RNNs depends on the primary elements within the sequence. While future events would also help to determine the output of a given sequence, unidirectional RNNs cannot account for these events in their predictions. Another distinguishing characteristic of RNNs is that they share parameters across each layer. While feedforward networks have different weights across each node, RNNs share the same weight parameter within each network layer. These weights are still adjusted through backpropagation and gradient descent to facilitate reinforcement learning.

For LPI radars, utilizing FMCW signals, parameter extraction adds computational complexity to the deinterleaving and labeling processes. Parameter extraction for FMCW signals is a parallel process conducted separately from the pulse deinterleaving, but it conveys information to the labeling phase. In addition to the conventional techniques, some studies have been conducted using ML methods. In the studies, feature extraction is performed using Wigner-Ville distribution, Pseudo-Wigner-Ville distribution, and Choi-Williams distribution time-frequency analysis, and a time-frequency image is obtained. CNN or its variants' pretraining model utilizes the acquired images for feature extraction. After the feature extraction, the feature vector is input to the *support vector machine* (SVM) classifier for offline training [15].

RWR analyzes ROF, pulse width, PRI, interpulse and intrapulse modulations, CW modulations, TOA, AOA, and antenna scan characteristics to identify threats to mother platforms. In modern RWR systems, the computer determines the closest fit of these parameters to those found in an emitter library to force an identification. Even modern RWR systems do not use identification algorithms that go well beyond schemes based on matching these parameters. Thus, it makes sense to explore more sophisticated techniques.

Applications of ML and DL to image and speech recognition in everyday life situations are now increasingly common, and their usage in the RF domain has been pioneering applications. However, little work has been done to explore the connection between RF signal processing and ML/DL. Moreover, only a few of these studies cover the EID process supported by ML/DL. Furthermore, most of them are at the theoretical level and unsuitable for application to real-life systems.

Let us make a future projection for correlating the RWR EID process and ML/DL algorithms' properties. The nature of the EID process must have the highest accuracy possible, and the processing time must be as short as possible. However, the radar signal specifications are very close to each other in terms of ROF, PRI, modulations, PW, and antenna scan properties. These specification convergences are much closer for the same type of radars.

First, let us consider the nature of the EID process with priory information since the EIDD (or PFM) is loaded to the RWR system according to the mission area and related threats. Thus, considering the supervised methods will increase the accuracy and decrease the identification process time. Therefore, unsupervised methods, such as K-means, K-medoids, Fuzzy C-means, hierarchical, *Gaussian mixture model* (GMM), and hidden Markov model (HMM), are less suitable than the supervised methods for the case. Their validation results and accuracies could be significantly higher. There is a particular situation in the GMM method. It assumes that the data are generated from Gaussian distributions. However, the radar signal parameters' distribution characteristics used in different missions are generally incompatible with the Gaussian distribution. Therefore, GMM may be suitable for the EID process only with the predetermined limitations of the radar types.

Second, classification techniques predict categorical responses, and classification models classify input data. However, in the emitter identification process, RF, PRI, modulations, PW, and antenna scan properties are close, and one or two can be classified together for different emitters. This wrong classification may yield the total process and result in incorrect identification. Thus, ML methods used only for classification, such as nearest neighbor (NN) and naive Bayes (NB), will not be appropriate for dense emitter environments. Especially when the number of parameters in the database is getting larger, inconsistencies are expected in the NN and NB algorithms' results.

The regression techniques may also be considered. They predict continuous responses and are suitable for continuously comparing the labeled emitter data with EIDD. Some regression methods, such as linear regression, generalized linear model (GLM), and Gaussian process, have only regression properties. However, some approaches, such as SVM, decision trees, ensemble methods, and neural networks, have characteristics of classification and regression methods. The ML and DL methods with classification and regression methods are good candidates for the EID process shortly RWR systems.

However, the *logistic regression algorithm* (LGA) is particularly unsuitable for the EID process. It is mainly used for classification tasks where the goal is to predict the probability of an instance belonging to a given class. The LGA assumes that the dependent variable must be binary or it can take only two values. Nonetheless, the data set of the different types of radar parameters consists of signals in a wide range, and variables have more than two values. Therefore, using the LGA may result in low validation rates for the EID process.

Discussing the computational requirements related to ML and DL as a last issue would benefit this part. Some recent studies mention the problem of partitioning training data into homogeneous subsets. Although these studies are on radar systems, as a projection, they may be applied to the RWR processes. Some of these studies assume that only partial information about the environment is available at the radar Rx, namely, a given number of clutter boundaries is present. These studies propose several algorithms to classify clutter radar echoes, intending to partition the possibly heterogeneous training dataset into homogeneous subsets, which can be used for estimation and detection purposes. For this purpose, the *expectation-maximization* (E/M) algorithm alone or in conjunction with other algorithms, such as the latent variable model. These schemes can be used as the preliminary stage of a detection architecture, where the detection stage exploits the information the classifier provides to process homogeneous data [16].

Another study presents radar scenario classifiers that can detect possible inhomogeneities within the reference window and are used to form a suitable estimate of the interference covariance matrix [17]. Radars face challenging scenarios characterized by high levels of heterogeneity that make conventional detection and estimation algorithms no longer effective. This ineffectiveness is because most of them rely on a well-defined homogeneous environment. The study considers operating situations encompassing a clutter edge, multiple discrete clutter clements, outliers, or both. In this context, sparsity-promoting priors are used to model the behavior of discrete clutter and outliers. Thus, the unknown parameters are estimated using cyclic procedures. The sparsity-promotion prior techniques are used to classify radar clutter and meet the computational requirements of the ML and DL. However, these techniques are likely to be adaptable for the deinterleaving process. They provide a further stage that refines the classification results and are good candidates for increasing the success of ML and DL techniques.

5.7   Future Trends in RWR Projection

In Chapter 4 and this chapter, a detailed analysis of the HW structure and signal processing techniques of RWR systems has been realized. The future trends of any design can be defined as the ideal case, the conditions of the present system, and where it has come from. Furthermore, the main factors for future projection are the deficiencies of the current systems, the main requirements, and the situation for technological developments in the area.

In this context, not to be reached to the ideal system, but a logical expectation for the future trends in RWR projection can be summarized in the following items:

1. Design and implement Rx systems with high sensitivity, high dynamic range, high instantaneous bandwidth, and frequency coverage without violating the SWaP requirements of the platforms. When surveying the present RWR systems, Table 5.6 is obtained.

   All these specifications are related to the technological situation of the HW, semiconductors, and physical restrictions. One way to enhance the

Table 5.6   Some Specifications of Present RWR Systems

Parameter	*Specification Limits*
Sensitivity	Up to −75 dBm
Dynamic range	50 to 60 dB
Instantaneous bandwidth	Up to 1.5 GHz
Frequency coverage	0.5 to 18 GHz (extended 20 GHz for some systems)

sensitivity is to decrease the instantaneous bandwidth, which is not wanted for RWR systems. The other way is to reduce the noise figure, which depends on the technological development of electronic devices such as amplifiers, mixers, and filters. The more capable low-noise electronic devices are developed, the more reduction in the noise figure will exist. The other component that affects the sensitivity is the required SNR. Although special signal processing techniques can reduce SNR levels, an actual SNR threshold for different applications cannot be reduced further. Moreover, it will have an additive adverse effect on the sensitivity. As a result, with technological developments in the future, sensitivity may converge to −80 dBm for the given instantaneous bandwidth (1.5 GHz); however, it is not logical to enhance it further.

After discussing the sensitivity, it would be proper to explain the dynamic range. As mentioned, dynamic range is the receiving signal power range between the sensitivity and 1-dB compression point or saturation level. The limits of the dynamic range depend on the technological developments of semiconductor devices. However, expecting more than 5-dB enhancement is not a logical prediction.

Frequency coverage and instantaneous *bandwidth* (BW) are two vital parameters of the RWR systems. For the present systems, the frequency coverage is 2 to 18 GHz, which may be extended to 0.5 to 20 GHz. Shortly, the upper limit must increase to 40 GHz or higher. Studies are focused on this direction. Although frequency coverage is viewed as a single issue, it affects many parts of the RWR system and its SWaP performance. Also, instantaneous BW is another dominant factor influencing the SWaP.

The vital matter related to frequency coverage is the Rx architecture of RWR. This subject is a crucial topic that has been worked on at various times and aspects. Also, RWR's Rx architecture, which uses different Rx together, will continue to be studied and developed. Furthermore, another important subject is the Rx antenna structures that cover the required BW without violating the SWaP requirements. Using phased array antennas by directed ML algorithms may be a very effective technique for RWR applications.

2. PW measurement capability is another critical Rx parameter for the RWR. The present RWR systems' PW measurement limits are 75 to 100 ns. The current systems cannot measure the lower PW levels. In conventional RWR systems, these measurements were conducted using highpass filters to find the initial pulse and the ending of the pulse by determining positive and negative spikes, respectively. However, the present systems digitize the pulse at a high sampling rate and perform analysis to determine the pulse width. This time, sampling ratio limits are encountered. For the time being, measuring 75 ns is possible. It would be reduced with the technological developments of SW and HW. However, the main limitation of this process comes from the transaction throughput, and with the increasing rate of the emitter density in the future operational environment, reducing it may not be possible.

3. The response time is a core parameter for RWR systems. It depends on the overall system's HW and SW properties. A good design and utilizing high-quality electronic devices provide a shorter response time. The present

systems have a response time level of 100 ms. The part that wastes the most time is the Rx scanning process. The time after taking the signal with Rx to appear on its display takes less than a microsecond. Thus, the primary delaying source is the Rx's operating band spectrum scanning process. This process is represented in Figure 5.1 as an Rx control block controlled by an identified emitter list block. Using ML and DL techniques to optimize the scanning process may reduce the response time to a certain level.

4. Also, using ML and DL in deinterleaving and EID processes will reduce the response time and increase the accuracy of the overall RWR systems. These processes were more summarized in Section 5.6.

5. Another critical point for potential development is using ML and DL in the PFM preparation process. Optimizing PFMs is essential for RWR's accurate operation and response time reduction. PFMs with many emitters will demand optimization, merging, and pruning studies. For this purpose, using ML and DL techniques in developing and testing PFMs may be an essential issue to study.

6. Finally, the ultimate aim should be to use a general EIDD without using different regional or operational PFMs for every mission. For this purpose, a general EIDD can be constituted, and by using ML and DL methods, the RWR systems decide the operational procedures and resource allocation methods.

5.8   Problems

Problem 5.1: Write MATLAB code to create 10 pulsed signals with different PRIs. Select the signal's frequency of 4 MHz (so as not to overload the program while running). Draw the pulsed signals on the same scale. Also, sketch the positive parts of the interleaved signal.

Problem 5.2: Repeat Problem 5.1, but this time, do not use the carrier frequency for producing the pulses and take only the positive envelopes of the pulsed signals.

Problem 5.3: A part of an RWR system's EIDD (or emitter library) is given in Table 5.7. The parameters provided by EIDD are for pulsed radars, and their parameters are very close to each other. At a time, a signal is detected by the Rx of the RWR, and after a sequence of processes, an active track is defined and registered to the AEF. The active track data is as follows:

Table 5.7   Part of an EIDD for Problem 5.3

	The Radar's Code Name		
Parameters	*Radar 1*	*Radar 2*	*Radar 3*
ROF range (MHz)	4,350–4,600	4,500–4,700	4,200–4,450
PRI range (ms)	2.6–3.2	3.0–3.8	2.4–2.7
PW range (μs)	1.4–2.0	1.6–2.1	1.8–2.4
Scan rate (s)	0.8–1.5	1.4–2.2	1.8–2.6

$$ROF = 4.6 \text{ GHz}; PRI = 3.1 \text{ ms}; PW = 1.9 \text{ } \mu s; \text{Scan rate} = 1.9s$$

Please use the parameter weighting technique for EID or library matching and find the most probable matching between the active track and the radar in the library.

Problem 5.4: A part of an RWR system's EIDD (or emitter library) is given in Table 5.8. The parameters provided by EIDD are for CW radars, and they all use sawtooth linear FM modulation. At a time, a signal is detected by the Rx of the RWR, and after a sequence of processes, an active track is defined and registered to the AEF. The active track data is as follows:

$$ROF = 8.3 \text{ GHz}; BW = 21 \text{ MHz}; \text{Scan rate} = 1.2s$$

Please use the parameter weighting technique for EID or library matching and find the most probable matching between the active track and the radar in the library.

Solutions

Solution 5.1: The required MATLAB code is as shown in Table 5.9.
The pulsed signals are given in Figure 5.13(a, b).
Figure 5.14 shows the interleaved signal for the above 10 pulsed signals.

Solution 5.2: The MATLAB code is the same as the previous solution. The only difference between the codes is in producing the signals as shown in Table 5.10.
The pulsed signals are given in Figure 5.15(a, b).
Figure 5.16 shows the interleaved signal for the above 10 pulsed signals.

Solution 5.3: Form the parameter-weighting EID table to find the best matching. Table 5.11 is established for this purpose.
After applying the parameter weighting technique, all the corresponding weightings are defined for every radar-track pair. After determining the weightings, the total score is calculated, and the most probable matching is obtained between the active track and Radar 2 with 1.69.

Solution 5.4: Establish the parameter weighting EID table for the CW active track. Thus, we can obtain the probable best matching. Table 5.12 is formed for this purpose.

Table 5.8 Part of an EIDD for Problem 5.4

	The Radar's Code Name		
Parameters	*Radar 1*	*Radar 2*	*Radar 3*
ROF range (MHz)	8,230–8,450	8,000–8,320	8,300–8,550
Modulation BW (MHz)	15.4–21.0	12.6–16.1	20.1–24.4
Scan rate (s)	0.8–1.5	0.6–1.2	1.8–2.6

Table 5.9   The Required MATLAB Code for Problem 5.1

```
%% Problem 5.1
clear, clc, close all
f_signal = 4e6;
duration = (2e-3);
A = 1;
f_sampling = 32e6;
no_samples = 0:round(duration*f_sampling)-1;
time_axis = 0:1/f_sampling:duration-1/f_sampling;
sine_wave = A*sin(2*pi*f_signal*time_axis);
waveform = phased.RectangularWaveform('SampleRate',f_sampling,'PulseWidth',
100e-6,...
'PRF',1e3,'OutputFormat','Pulses','NumPulses',2);
waveform1 = phased.RectangularWaveform('SampleRate',f_sampling,'PulseWidth'
,400e-6,...
'PRF',0.5e3,'OutputFormat','Pulses','NumPulses',1);
waveform2 = phased.RectangularWaveform('SampleRate',f_sampling,'PulseWidth'
,200e-6,...
'PRF',2e3,'OutputFormat','Pulses','NumPulses',4);
waveform3 = phased.RectangularWaveform('SampleRate',f_sampling,'PulseWidth'
,500e-6,...
'PRF',0.5e3,'OutputFormat','Pulses','NumPulses',1);
waveform4 = phased.RectangularWaveform('SampleRate
',f_sampling,'PulseWidth',30e-5,...
'PRF',2e3,'OutputFormat','Pulses','NumPulses',4);
waveform5 = phased.RectangularWaveform('SampleRate
',f_sampling,'PulseWidth',1e-4,...
'PRF',5e3,'OutputFormat','Pulses','NumPulses',10);
waveform6 = phased.RectangularWaveform('SampleRate
',f_sampling,'PulseWidth',1e-4,...
'PRF',5e3,'OutputFormat','Pulses','NumPulses',10);
waveform7 = phased.RectangularWaveform('SampleRate
',f_sampling,'PulseWidth',1e-7,...
'PRF',4e3,'OutputFormat','Pulses','NumPulses',8);
waveform8 = phased.RectangularWaveform('SampleRate
',f_sampling,'PulseWidth',1e-5,...
'PRF',1e3,'OutputFormat','Pulses','NumPulses',2);
waveform9 = phased.RectangularWaveform('SampleRate
',f_sampling,'PulseWidth',2e-5,...
'PRF',2e3,'OutputFormat','Pulses','NumPulses',4);
y1= 4*waveform1();
y2 =3*waveform2();
y3 =3*waveform3();
y4= 3*waveform4();
y5=2*waveform5();
y6=waveform6();
y7=4*waveform7();
y8=3*waveform8();
y9=waveform9();
y10 =3*waveform();
```

Table 5.9 *(continued)*

```
sig1=y1.*sine_wave';
sig2=y2.*sine_wave';
sig3=y3.*sine_wave';
sig4=y4.*sine_wave';
sig5=y5.*sine_wave';
sig6=y6.*sine_wave';
sig7=y7.*sine_wave';
sig8=y8.*sine_wave';
sig9=y9.*sine_wave';
sig10=y10.*sine_wave';
totaldur = 2*1/waveform.PRF;
totnumsamp = totaldur*waveform.SampleRate;
t = unigrid(0,1/waveform.SampleRate,totaldur,'[)');
figure
subplot(5,1,1)
plot(t.*1000,real(sig6),'k');
title('P6');
set(gca,'XTick',0:0.2:totaldur*1e3);
subplot(5,1,2)
plot(t.*1000,real(sig7),'k')
title('P7')
set(gca,'XTick',0:0.2:totaldur*1e3)
subplot(5,1,3)
plot(t.*1000,real(sig8),'k')
title('P8')
set(gca,'XTick',0:0.2:totaldur*1e3)
subplot(5,1,4)
plot(t.*1000,real(sig9),'k')
title('P9')
set(gca,'XTick',0:0.2:totaldur*1e3)
subplot(5,1,5)
plot(t.*1000,real(sig10),'k')
title('P10')
set(gca,'XTick',0:0.2:totaldur*1e3)
xlabel('Milliseconds')
figure
subplot(5,1,1)
plot(t.*1000,real(sig1),'k')
title('P1')
set(gca,'XTick',0:0.2:totaldur*1e3);
subplot(5,1,2)
plot(t.*1000,real(sig2),'k')
title('P2')
set(gca,'XTick',0:0.2:totaldur*1e3);
subplot(5,1,3)
plot(t.*1000,real(sig3),'k')
title('P3')
set(gca,'XTick',0:0.2:totaldur*1e3);
subplot(5,1,4)
```

Table 5.9 *(continued)*

```
plot(t.*1000,real(sig4),'k');
title('P4');
set(gca,'XTick',0:0.2:totaldur*1e3);
subplot(5,1,5)
plot(t.*1000,real(sig5),'k')
title('P5')
set(gca,'XTick',0:0.2:totaldur*1e3)
xlabel('Milliseconds');
figure
plot(t.*1000,real(sig1+sig2+sig3+sig4+sig5+sig6+sig7+sig8+sig9+sig10),'k');
axis([0 totaldur*1e3 0 20]);
xlabel('Milliseconds');
ylabel('Amplitude');
title('Interleaved Signal');
set(gca,'XTick',0:0.2:totaldur*1e3);
%Finish
```

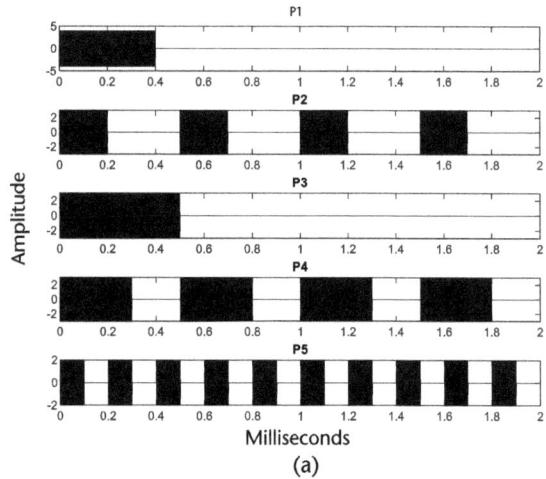

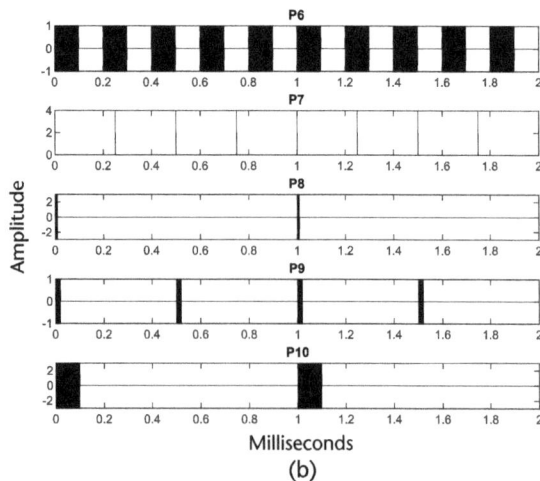

Figure 5.13 Drawing pulsed signals for Problem 5.1.

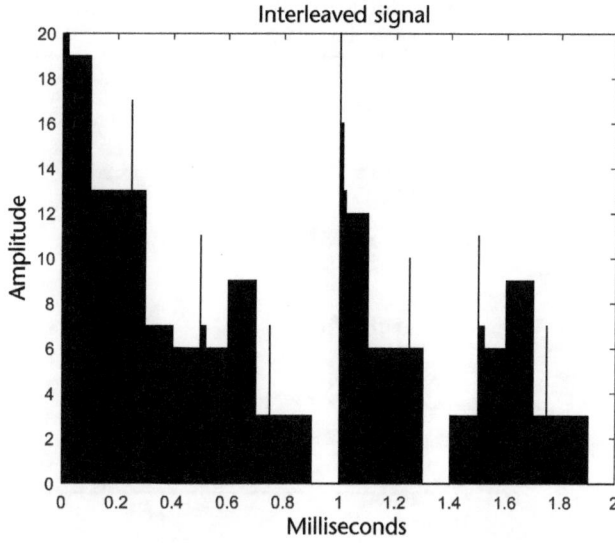

Figure 5.14 Drawing interleaved signal for Problem 5.1.

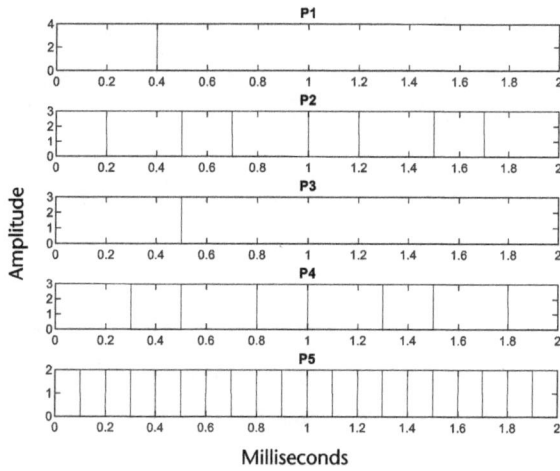

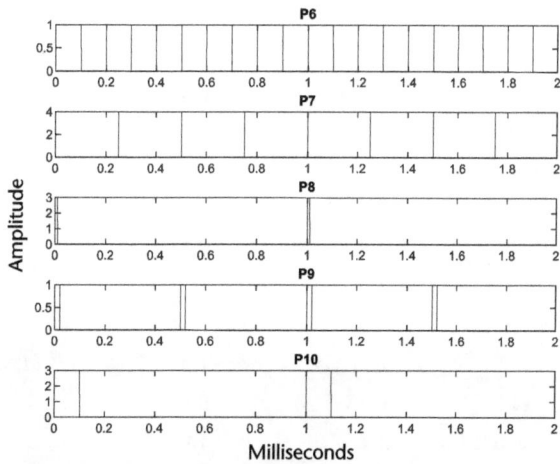

Figure 5.15 Drawing pulsed signals for Problem 5.2.

Table 5.10 The Required MATLAB Code for Problem 5.2

```
...
sig1=y1;
sig2=y2;
sig3=y3;
sig4=y4;
sig5=y5;
sig6=y6;
sig7=y7;
sig8=y8;
sig9=y9;
sig10=y10;...
```

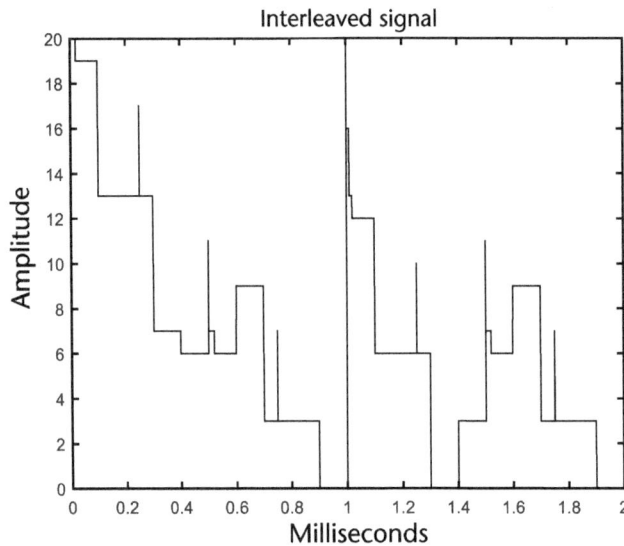

Figure 5.16 Drawing interleaved signal for Problem 5.2.

Table 5.11 Part of an EIDD for Problem 5.3

Parameters	The Radar's Code Name			
	Radar 1	Radar 2	Radar 3	Active Track
ROF range (MHz)	4,350–4,600	4,500–4,700	4,200–4,450	4,600
ROF weight (0–1)	0.35	0.5	0.15	
PRI range (ms)	2.6–3.2	3.0–3.8	2.4–2.7	3.1
PRI weight (0–1)	0.4	0.45	0.15	
PW range (µs)	1.4–2.0	1.6–2.1	1.8–2.4	1.9
PW weight (0–1)	0.33	0.34	0.33	
Scan rate (s)	0.8–1.5	1.4–2.2	1.8–2.6	1.9
Scan weight (0–1)	0.15	0.4	0.35	
Total score for active track	1.23	**1.69**	0.98	

Table 5.12 Part of an EIDD for Problem 5.4

Parameters	The Radar's Code Name			
	Radar 1	Radar 2	Radar 3	Active Track
ROF range (MHz)	8,230–8,450	8,000–8,320	8,300–8,550	8,300
ROF weight (0–1)	0.4	0.3	0.3	
Modulation BW (MHz)	15.4–21.0	12.6–16.1	20.1–24.4	21
BW weight (0–1)	0.3	0.15	0.55	
Scan rate (s)	0.8–1.5	0.6–1.2	1.8–2.6	1.2
Scan weight (0–1)	0.6	0.3	0.1	
Total score for active track	**1.3**	0.75	0.95	

After applying the parameter weighting technique, the corresponding weightings are defined for every radar-track pair. The weightings are determined, the total score is calculated, and the most probable matching is obtained between the active track and Radar 2 with 1.3.

References

[1] Robertson, S., *Practical ESM Analysis*, Norwood, MA: Artech House, 2019.

[2] Mahafza, B. R., *Radar Systems Analysis and Design Using MATLAB*, 3rd ed., Boca Raton, FL: CRC Press/Taylor & Francis Group, 2013.

[3] Adamy, D. L., *EW 101: A First Course in Electronic Warfare*, Norwood, MA: Artech House, 2000.

[4] Robertson, S., *Practical ESM Analysis*, Norwood, MA: Artech House, 2019.

[5] De Martino, A., *Introduction to Modern EW Systems*, 2nd ed., Norwood, MA: Artech House, 2018.

[6] Mahafza, B. R., *Radar Systems Analysis and Design Using MATLAB*, 3rd ed., Boca Raton, FL: CRC Press/Taylor & Francis Group, 2013.

[7] Jankiraman, M., *FMCW Radar Design*, Norwood, MA: Artech House, 2018.

[8] Air Combat Command Training Support Squadron (ACC TRSS), *Electronic Warfare Fundamentals*, Nellis AFB, NV, 2000.

[9] Donaldson, P., (ed.), *Electronic Warfare Handbook 2008*, Berkshire, U.K.: The Shephard Press, 2008.

[10] Shakhnarovich, G., T. Darrell, and P. Indyk, *Nearest-Neighbor Methods in Learning and Vision*, Cambridge, MA: MIT Press, 2006.

[11] Coelho, J. P., T. M. Pinho, and J. Boaventura-Cunha, *Hidden Markov Models Theory and Implementation using MATLAB*, Boca Raton, FL: CRC Press/Taylor & Francis Group, 2019.

[12] Fişne, N., and A. G. Pakfiliz, "Parameter Extraction for Frequency Modulated Continuous Wave Radar Signals with Connected Component Labeling-Based Clustering," *2022 30th Signal Processing and Communications Applications Conference (SIU)*, Safranbolu, Turkey, 2022, pp. 1–4.

[13] Cheng, C., and J. Tsui, *An Introduction to Electronic Warfare: from the First Jamming to Machine Learning Techniques*, Gistrup, Denmark: River Publishers, 2021.

[14] Al-Malahi, A., et al., "An Intelligent Radar Signal Classification and Deinterleaving Method with Unified Residual Recurrent Neural Network," *IET Radar, Sonar & Navigation*, 2023, pp. 1–18.

[15] Guo, Q., X. Yu, and G. Ruan, "LPI Radar Waveform Recognition Based on Deep Convolutional Neural Network Transfer Learning," *Symmetry*, Vol. 11, No. 540, 2019.

[16] Addabbo, P., et al., "Learning Strategies for Radar Clutter Classification," *IEEE Transactions on Signal Processing*, Vol. 69, 2021, pp. 1070–1082.

[17] Han, S., et al., "Sparsity-Based Classification Approaches for Radar Data in the Presence of Clutter Edges and Discretes," *IEEE Transactions on Aerospace and Electronic Systems*, Vol. 59, No. 3, 2023, pp. 2141–2162.

CHAPTER 6

Self-Protection Jammer Systems

High technology and invaluable military assets, such as fighter aircraft, bombers, helicopters, assault transport aircraft, middle and large UAVs, and warships, are not just tools of war but the very keys to victory. These assets, primarily designed for offensive roles, are also the prime targets of enemy units. Therefore, it is not just a matter of choice but an absolute necessity for these assets, or dedicated units, to be equipped with robust self-protection systems. The absence of such systems can jeopardize these assets' safety, and dedicating other units to protect them significantly complicates operational planning.

Electronic and software edge technology, derived from the requirements of the weapon systems, is the backbone of developing self-protection systems. This is valid for guidance systems in missile systems and self-protection systems. As stated, the guidance methods may be RF and EO, which include IR, visual band, and UV. With the developments of the guidance methods, countermeasures have been developed in parallel. So valuable military assets protect themselves via cutting-edge electronics and software technology. Let us narrow our scope with the RF-guided threats and countermeasures.

A fighter may be on an attack mission to destroy a target or in a defensive role, such as an AI mission. The first aim must be to protect the fighter aircraft during the mission and ensure its safe return to the base. For this reason, it must be equipped with a SPEW system, a critical component in the aircraft's defense. Otherwise, the pilot cannot be aware of being tracked and take precautions in time. This is also valid for other aircraft types, including UAVs and warships.

The main aim is always to accomplish the duty safely, and situational awareness, the ability to perceive and understand the environment, is critical. When failing to detect environmental threats, a question may arise: Who is the prey and the hunter? Without situational awareness, the pilot or the crew may assume that they are the hunter when they are the prey. Situational awareness is only sufficient with countermeasures. Against RF-guided threats, the military assets must have RWR systems for situational awareness and chaff dispenser and jammer systems for countermeasures. In other words, they must have SPEW systems to protect themselves in the dense RF threat environment.

The previous two chapters deeply inspect the structures and signal-processing techniques of the RWR systems. This chapter is dedicated to *self-protection jammer* systems. The jamming principles of the self-protection systems are given, and the equations used with them are derived. The structures of the jammer systems and the types of SPJ methods are discussed elaborately.

6.1   SPJ Principles

Jammer systems can be diversified by their platforms and roles. Platforms can be land-based, shipborne, or airborne. However, the jammers' roles may be self-protection and support jamming. Land-based systems can apply very high jamming powers in support of jammer roles. However, they may not be engaged in a self-protection role since land-based systems are not the target of radar-guided threat systems. The shipborne jammer systems can reach high jamming powers according to the available volume and power consumption. The requirements for shipborne jammer systems are sometimes self-protection and sometimes support jamming. In both cases, they require high power since their low maneuvering capability can compensate for using high powers.

The airborne jammer systems may be in self-protection and support jamming roles. This time, the systems' dimensions, weight, and power needs are getting important since the SWaP requirements are limited. The support jammer systems aim to jam the threat systems for different platforms, and their primary operational purpose is jamming. Thus, they do not carry payloads or ammunition other than jammer systems. Furthermore, like SOJ aircraft, they do not enter the enemy IADS's lethal border, and they may be larger aircraft than fighters, such as business jets. They can provide more SWaP capacity for the jammer systems. However, for airborne SPJ systems, the limited space, carrying capacity, and power supply drive the design engineers into compromises about operational capabilities and technical properties. When comparing cargo aircraft, the compromise study becomes essential for fighters and helicopters. However, the trade-off will take the most drastic state for UAVs since their SWaP opportunities are minimal. So one of the SPJ principles is the balance of the platform's SWaP capabilities and the jammer system's requirements.

Fighter aircraft and UAVs have an essential advantage when they use the SPEW system. This advantage is their extreme maneuvering capability. Their maneuvers support self-protection jamming capabilities. These properties make them the most successful vehicle for using RF countermeasures for self-protection. Also, their maneuvering capabilities yield the use of various jamming techniques. They can use noise and different deception techniques. However, medium-speed and low-maneuver air vehicles such as helicopters, cargo aircraft, and nonjet engine UAVs can use fewer techniques than fighter aircraft and UAVs. For naval vessels, the SPJ concept covers noise and deceptive jamming techniques. The speed and maneuverability of warships are insufficient compared to air platforms. However, their SWaP-providing capabilities for the jammer systems are very high. Thus, according to the air platforms, their SPJ capabilities in power and frequency bandwidth are relatively increased. The shipborne platforms close the maneuver's disadvantages gap with competent systems with large systems compared with the air platform systems. Thus, another SPJ principle can be counted as the jammer system platform adaptation. Furthermore, this principle determines the jammer usage tactics, jamming techniques, and operational planning.

The SPJ systems must conform to the operational region's threats. For this reason, technical properties of the SPJ systems and threat analysis of the operating area are essential. The technical properties of the SPJ system are defined, designed, and implemented during the procurement project phase. For this reason, the technical requirements must be well-defined in this phase. A technically qualified SPJ system can be handled with proper preprogramming during the operation phase. Thus, a detailed threat intelligence and technical study should be conducted before operations. Intelligence information must be collected through SIGINT facilities, and detailed technical analysis of threat systems is required. Then, jamming techniques must be developed for the threat systems. Each study requires complex engineering skills; if these skills are absent or the processes are conducted by untrained staff, the result may be catastrophic for the task pack. Thus, another SPJ principle is the suitability of the jammer to the environment, the threats, and proper preprogramming.

Another critical point for the SPJ systems is the degradation of the threat radar systems and jamming effectiveness. Degradation is to reduce the effectiveness or efficiency of adversary EM spectrum-dependent systems [1]. Employing EM jamming, EM deception, and EM intrusion is intended to degrade adversary systems, thus confusing or delaying the actions of adversary operators. However, a proficient operator can work around the effects to reduce or eliminate their impact. The impact of degradation may last a few seconds or remain throughout the entire operation. Degradation is accomplished with EM jamming and EM deception. However, jamming effectiveness shows the impact level of the jamming on the victim radar operation. The mathematical expression of the jamming effect is measured by the *jamming-to-signal ratio* (*J/S*). SPJ aims to degrade the threat radar functions and prevent being detected and tracked. For this reason, another SPJ principle may be assumed to be the effectiveness of the applied jamming.

The last principle of SPJ may be assumed to be a link between the operational plans and the role of the platforms with SPJ systems. The operational plans define the platforms' roles and tasks. These plans represent the engagement rules and the usage of missiles, ammunition, bombs, and the SPEW system, including the SPJ system. These usage procedures are preprogrammed to the SPEW system. Thus, the last principle of SPJ is the relationship between the operational plans and jamming usage.

6.2   Derivation of SPJ Equations

The RF part of the SPEW systems aims to protect the mother platform from radar-guided threat systems. For this purpose, they try to prevent being detected and then tracked. Different jamming techniques are required to avoid detection and tracking. In SPJ systems, jamming falls into two main categories: noise and deception, each containing several subcategories. This section will discuss the subcategories and derive their related equations. Although the derivation of the equations is separated according to the main categories, such as noise and deception, their subcategories will also be given as a base for future discussions. It must be noted that all the jamming applications are conducted according to the preprogramming codes or the PFMs. For this reason, the jamming activities are not a coincidence, but they are studied and programmed into the SPEW system before every operation. The PFMs are prepared to consider RWR and SPJ together, and in this arrangement, an SPJ starts activation when detecting a threat signal. Different jamming techniques are developed and used against various threats. In this context, for neutralizing threat radars, sometimes noise, sometimes deception, and sometimes both are used to generate the proper jamming techniques.

6.2.1 Noise Jamming Equations in SPJ

The noise jamming used in SPJ systems can be classified into two groups. The first is preset noise jamming, and the other is noncoherent repeater jamming. Each has noise-modulated signals; however, their noise-generating processes and operational usages are different and discussed separately.

6.2.1.1 Preset Noise Jamming

Preset noise jamming is a self-running jamming method. Blocking some frequency bands might be effective for some combination of threats, especially well-known and obsolete ones. Moreover, this method may be used for more contemporary threats, such as frequency-agile radars. For this purpose, PFM is adjusted to some center frequencies and bandwidths to block the required frequencies. However, power management may be a complicated process with this technique [2]. Preset noise jamming is usually used in overlapping frequency bands or barrage-noise jamming over a wide band. However, sweep-spot noise jamming may also be possible when a jammer's full power is shifted from one frequency to another and from one jamming antenna to another. While this has the advantage of being able to jam multiple frequencies quickly, it does not affect them all simultaneously and thus limits the effectiveness of this type of jamming.

The first technique used in the preset noise jammers is barrage-noise jamming. In this technique, many radars with overlapping frequencies are jammed simultaneously. This jamming method is demonstrated for the frequency domain in Figure 6.1. The barrage jamming is applied between f_{min} and f_{max} frequencies.

As can be seen from Figure 6.1, there are five victim (or threat up to the looking point) radars with different operating frequencies and bandwidths. To apply barrage jamming, all the victim radars must be in the jammer antennas' main beam. Assuming four jammer antennas cover all the 360° azimuth in an SPJ system, radar aspect angles will be critical. The jammer power is high when all the radars drop in one antenna's main beam since all the power is transmitted from one antenna. However, when the radars are in different directions, the power will be divided into more than one antenna, and the power level will drop.

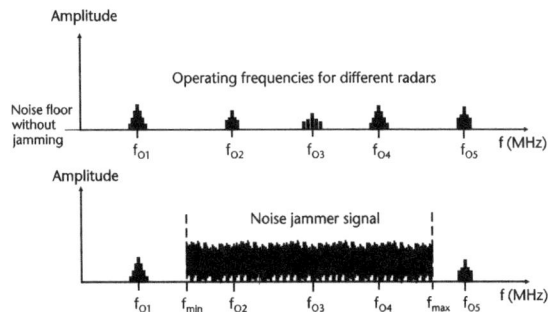

Figure 6.1 Barrage-noise jamming application for three victim radars.

Let us assume that the preset jammer is preprogrammed to apply effective jamming, which aims to obtain sufficient J/S for the existing radars. The jammer considers the power management and uses barrage noise between f_{min} and f_{max} frequencies. The top of Figure 6.1 shows the radars in the environment with no jamming, and f_{on} represents the operating frequency for the corresponding radar number ($n = 1, 2, ..., 5$). Thus, three radars ($n = 2, 3, 4$) can be jammed effectively. The remaining two ($n = 1$ and 5) are unaffected by the jamming since they are out of the jammer bandwidth.

When considering one of the radar system's Rx inputs without and with jamming in the time domain, we obtain Figure 6.2. The left side of the figure demonstrates the radar return envelope signal under no jamming case. Two return pulses are available, and when operated with the operator, the pulses can be distinguished from noise. Moreover, an automatic determination is also possible if the detection S/N level is sufficient. Under the jamming condition of the same radar, the return and noise jammer signals are demonstrated on the right side. It can be seen from the figure that the return pulse amplitudes are below the amplitude level of the noise jammer signal, and they are far from detection.

As a general concept, the J/S must be sufficient to prevent the radar from detecting the return pulse for effective jamming. The J/S calculation of barrage-noise jamming against each monostatic radar (n) is done separately and can be written as follows:

$$\left(\frac{J}{S}\right)_n = \frac{P_J G_{Ja} 4\pi R_n^2 B_{Rn}}{P_{Tx} G_n \sigma_n L_n B_J} \quad \left\{ n = 1, 2, ...; \quad L_n = \frac{L_J L_{pol(n)}}{L_{Rn}} \right\} \tag{6.1}$$

The above equation is written for equal jammer antenna gains in the case of more than one antenna covering the whole azimuth. When more than one antenna is used for barrage jamming, the jammer power will divide the number of used antennas. In (6.1), the quantities are defined as follows:

- P_J and P_{Tx}: jammer and radar Tx's peak powers;
- G_{Ja}: jammer's antenna gain;
- G_n: the nth radar Rx and Tx antenna gain;
- B_{Rn}: the nth radar Rx bandwidth;
- B_J: jammer bandwidth;

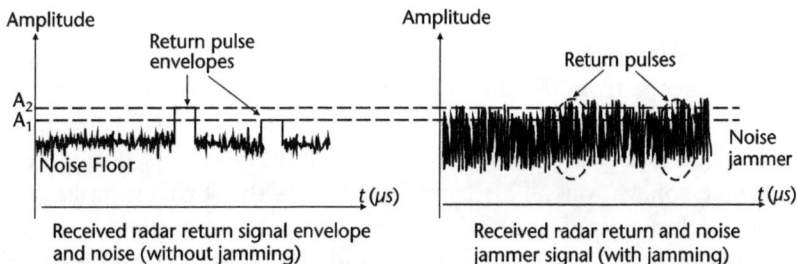

Figure 6.2 Preset jamming effect on a radar Rx.

- R_n: the distance between the nth radar and the jammer platform;
- L_n: the total losses related to the nth radar consist of jammer losses (L_J), nth radar losses (L_{Rn}), and antenna polarization losses ($L_{pol(n)}$).
- σ_n: The RCS of the jammer's mother platform according to the nth radar.

Equation (6.1) is derived from (3.10), and mathematically, the only difference between them is adding B_{Rn} and B_J to the calculations since the magnitudes of B_{Rn} and B_J differ in this condition. They are close to each other in (3.10); however, the B_J may be tens of times greater than the radar's widest bandwidth ($B_{R(max)}$) in the jamming frequency ranges.

When the sweep-spot noise jamming technique is utilized for preset noise jamming, the jammer jumps the preset frequencies defined in PFM in a predetermined sequence. Also, which antenna must be used is determined according to the directions of the threat radar. As a result, this case equation (6.1) is still valid, but this time, only some of the radars operating in the sweep frequency range can be jammed simultaneously. The B_R and B_J are close in this case, and the jamming effectiveness is high.

6.2.1.2 Noncoherent Repeater Jamming

This is a repeater technique that depends on the RWR information and the accuracy of the repeater Rx. This method is used when a narrowband noise jamming is required. Another utilization of this method is to be a backup for coherent repeater jamming. In the coherent operation, SPJ captures the radar signal, delays it by employing an analog or digital memory loop, and modulates and amplifies. In this method, repeater jammers apply constant amplification to the received radar signals and retransmit them to the radar. When the received signal power is higher than a threshold (i.e., radar is too close to the SPJ platform), input power is amplified more than the Tx transmitting capability. Thus, the constant gain produces a saturated output for high input power, as shown in Figure 3.2.

The noncoherent repeater jammers shift the noise power to the radar Rx bandwidth, which might be assumed to be a tricky method without transmitting wideband signals. From a technical point of view, this technique is similar to sweep-spot noise jamming. In the noncoherent repeater method, a repeater jammer works as a transponder. It receives several radar pulses and determines the PRI of the threat radar. It then uses this information to predict when the next radar pulse should arrive. A noise-modulated signal is amplified and transmitted using an oscillator gated for a time based on the expected pulse arrival time.

Another essential point in noncoherent repeater jamming is the look-through concept. Applying uninterrupted jamming to a threat system is not a suitable solution to escape from the threats due to the HOJ technique. Therefore, the jamming transmission instants must coincide with the instants when the radar receiver is active. For this purpose, the look-through method is used in noncoherent repeater jammer applications. This technique allows the Rx to sample the signal environment periodically. The objective of the look-through mode is to enable the jammer to update victim radar parameters and change the jamming signal to respond to changes in the signal environment.

The operating frequency of the threat radar is detected and regenerated with an *automatic frequency control* (AFC) structure to apply noncoherent repeater jamming. Then a signal at the radar operating frequency is produced by a kind of LO, such as a *voltage-controlled oscillator* (VCO) or *direct digital synthesizer* (DDS). In the last step, the obtained signal is modulated by noise. The bandwidth is adjusted according to the radar signal and retransmitted to the radar. This time, the radar signal is not sent back; a new signal is generated and transmitted to the radar, and the signal is not coherent or not phase coherent with the radar. As a result, the operation method is noncoherent and uses a repeater jammer. When the noise jamming signal is centered on the frequency and bandwidth of the threat radar, the jamming reaches the highest efficiency. The ability of a noncoherent repeater jammer to intensify the jamming signal depends on the ability of the jammer to identify the exact frequency and bandwidth of the radar.

Jamming efficiency can be calculated using (6.1) in this operation mode. As stated, the magnitudes of B_R and B_J are close in this case, and the B_J/B_R ratio aims to keep a limit close to unity. Thus, the J/S equation converges to (3.10) as follows:

$$\frac{J}{S} = \frac{P_J G_{J_a} 4\pi R^2}{P_{Tx} G \sigma L} \tag{6.2}$$

The quantities of the above equation are defined as follows:

- P_J and P_{Tx}: jammer and radar Tx's peak powers;
- J: jammer signal power at the radar Rx;
- S: target's backscattered echo at the radar Rx;
- G_{J_a}: jammer's antenna gain;
- G: radar's Rx antenna gain;
- R_n: the distance between the nth radar and the jammer platform;
- L: the total losses, calculated as (6.1), are related to the radar (L_R), jammer (L_J), and antenna polarization (L_{pol}) losses;
- σ: the RCS of the jammer's mother platform according to the radar.

6.2.2 Deceptive Jamming Equations in SPJ

Deception jamming for SPJ systems, the coherent repeater is generally used for deception. However, sometimes noise techniques, such as narrow pulse and skirt jamming, may also be used for deceptive techniques. Early repeater jammers used a *frequency memory loop* (FML) but could not provide a coherent waveform. In present-day technology, repeater jammers utilize a *digital RF memory* (DRFM) to function as a coherent repeater without delay restrictions. Additionally, it can produce jammer signals for a noncoherent repeater. The noise jamming equations have been discussed previously. Here, the coherent repeater equations are derived. For this purpose, let us consider two cases, one for monostatic and one for bistatic radars. First, take the following figure into account for the monostatic case.

In Figure 6.3, G_{Tx} and G_{Rx} are the radar's antenna gains for Tx and Rx, and they are equal for monostatic single antenna systems. S_a is the backscattered signal

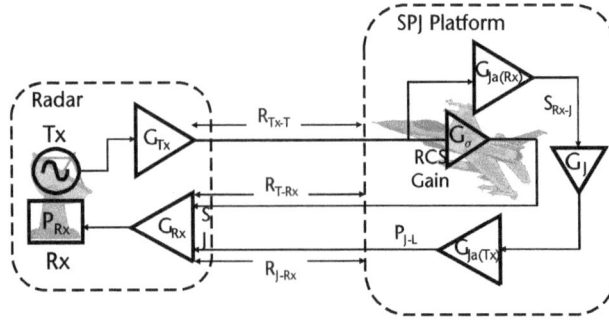

Figure 6.3 Coherent repeater jamming operation against monostatic radar.

power at the antenna's input, and S is the power at the antenna's output. J_a is the jammer power at the antenna's input, and J is the power at the antenna's output. Moreover, the Rx power (P_{Rx}) can be defined as the total of J and S. The ranges are shown with R, and the indices show T for the target or the mother platform of the SPJ, Rx for the radar receiver, and J for the SPJ. $G_{Ja(Rx)}$ and $G_{Ja(Tx)}$ represent the antenna gains for Rx and the Tx of the repeater jammer. Finally, G_J shows the jammer process gain, and G_σ is the RCS gain of the target.

There are two return signals in the radar Rx, one from the target skin return and the other from the repeater jammer. Using a DRFM, the signal from the repeater jammer may be delayed and amplified. Also, more than two signals with different delays can be sent to the radar Rx as false targets. This time, the high J/S ratio, vital for noise jamming, loses its importance; however, J/S optimization is getting an essential meaning. Let us consider no delay for the repeater jammer signal and calculate the J/S. For this purpose, the target's backscatter radar signal power (S) is calculated using (2.33), obtained in Section 2.4. This time, the total radar loss (L_R) is incorporated into the equation.

$$S = \frac{P_{Tx}G^2\sigma c^2}{(4\pi)^3 R^4 f^2 L_R} \quad \{G = G_{Tx} = G_{Rx}; \ R_{T-Rx} = R\} \tag{6.3}$$

The jamming signal power at the Rx can be obtained using (3.17) as follows:

$$J = \frac{P_{Tx}G G_{Ja(Rx)}c^2}{\left(4\pi R_{Tx-T}f\right)^2} \frac{G_J G_{Ja(Tx)}G c^2}{\left(4\pi R_{T-Rx}f\right)^2 L_J L_{pol}}$$

$$R_{Tx-T} = R_{T-Rx} = R \Rightarrow J = \frac{P_{Tx}G G_{Ja(Rx)}c^2}{\left(4\pi R f\right)^2} \frac{G_J G_{Ja(Tx)}G c^2}{\left(4\pi R f\right)^2 L_J L_{pol}} \tag{6.4}$$

To obtain J/S, we can divide (6.4) by (6.3) side by side,

$$\frac{J}{S} = \frac{G_{Ja(Rx)}G_J G_{Ja(Tx)}c^2 L_R}{4\pi\sigma f^2 L_J L_{pol}} \tag{6.5}$$

The above equation represents the jammer efficiency of SPJ against monostatic radar systems. The Rx and Tx locations differ for bistatic radar threats, and the SPJ operation scheme is given in Figure 6.4.

In the bistatic case, the Tx is on the launch platform, and the Rx is on the missile. Thus, the antenna gains for the Tx and Rx are different, and the distance between the launch platform and the target ($R_{Tx\text{-}T}$) differs from the distance between the target and Rx ($R_{T\text{-}Rx}$). For the bistatic case, the backscattered radar signal from the target at the Rx is defined as

$$S = \frac{P_{Tx}G_{Tx}G_{Rx}\sigma_{Rx}c^2}{\left(4\pi\right)^3 R_{Tx\text{-}T}^2 R_{T\text{-}Rx}^2 f^2 L_R} \tag{6.6}$$

In (6.6), σ_{Rx} is the RCS seen from the missile or Rx. The jamming signal power at the Rx is written as

$$J = \frac{P_{Tx}GG_{Ja(Rx)}c^2}{\left(4\pi R_{Tx\text{-}T}f\right)^2} \frac{G_J G_{Ja(Tx)}Gc^2}{\left(4\pi R_{T\text{-}Rx}f\right)^2 L_J L_{pol}} \tag{6.7}$$

and J/S can be defined as follows:

$$\frac{J}{S} = \frac{G_{Ja(Rx)}G_J G_{Ja(Tx)}c^2 L_R}{4\pi\sigma_{Rx}f^2 L_J L_{pol}} \tag{6.8}$$

As in the monostatic case, in the bistatic case, J/S does not show the efficiency of the jamming. The aim is not to increase the J/S but to generate the optimal value. Thus, a coherent repeater jammer requires significantly less power than a noncoherent noise jamming system. It gains this advantage using an identical waveform as the radar, and the matched filter at the radar's Rx cannot feel the difference between the skin return and the repeater jamming return. Moreover, the victim radar Rx amplifies the jamming signal, increasing its effectiveness.

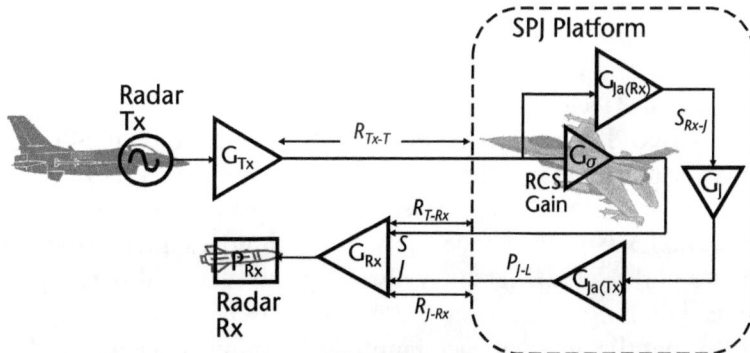

Figure 6.4 Coherent repeater jamming operation against bistatic radar.

6.3   SPJ Architecture

This section discusses the structure of the SPJ components. An SPJ system mainly consists of jammer Tx antennas, RF power amplifiers, and jamming waveform generators. Before these components are discussed separately, two significant points are considered. First, the relationship between the RWR and SPJ has been clarified up to this point. However, from the SPJ point of view, this relationship is mentioned with brief information. Second, the general structures for noise and repeater jammers are inspected.

6.3.1   Relationship Between the RWR and the Jammer

For the SPJ to operate effectively, the essential factor is PFM. The jamming techniques against different threats are defined in the PFMs. However, the performance of the RWR and the information it provides are other vital factors. RWR detects and identifies threats by monitoring the EM threat landscape. It provides the pilot with the latest information on the threat environment and sends the necessary information to the SPJ and CMDS systems—the pilots control the CMDS and SPJ. In general SPEW systems, pilots can select CMDS usage manually. Furthermore, the RWR system runs CMDS functions automatically when needed.

Pilots can also select noise mode or SPJ mode. In noise mode, the preset jamming is applied according to the preset frequencies in the PFMs. In this mode, all the actions are defined in the PFMs, and without taking any information from the RWR SPEW system, it starts to jam. Naturally, this reveals the airborne platform's position and sometimes gives valuable information to the enemy when the enemy radars are not in operation, that is, when jamming is not necessary.

In the repeater or SPJ mode, the jamming is accomplished after taking the signal of the threats, which means it depends on the RWR information. This time, the jamming operations are automatic and do not depend on the pilots' decisions. The jamming process is conducted by the definitions given in the PFMs. The prestudied threats are defined accurately and in detail for being detected and identified by the RWRs and properly jammed or deceived by the SPJs. As seen in Figure 6.5, PFM operation is a series process that starts with detection and identification in the RWR subprocess. The product of the RWR process is EID or an identified emitter list. Then, SPJ produces noise or deceptive jamming according to the threat radar type.

Underestimating the importance and complexity of developing jamming techniques may be a big mistake. The primary noise techniques used for deception in SPJ are cover pulse, fountain, narrow pulse, and skirt jamming. However, the general repeater techniques are direct repeat (pulsed, PD, or CW), RGPO, *range gate pull-in* (RGPI), narrowband noise, and VGPO. Although SPJ can produce noise or repeater techniques, sometimes variations and sometimes mixtures of these techniques may be used for different threats. While the number of threats increases, the complexity of the applied jamming techniques increases exponentially. Various modes of the same threat may also vary and require different jamming techniques. Thus, jamming technique development may be a challenging problem. It must be noted that the usage of each jamming technique depends on the EID information taken from the RWR system.

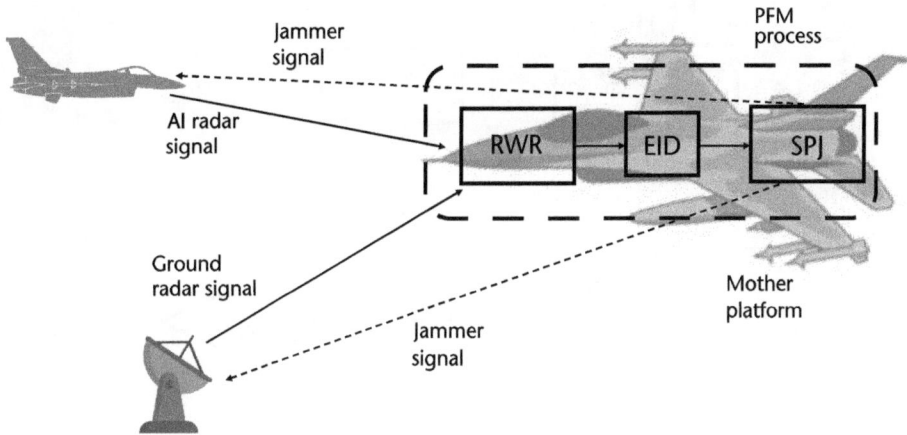

Figure 6.5 PFM process for SPJ.

6.3.2 SPJ Structure

SPJ hardware can be examined for two different architectures. The first is the noise jammer scheme, which produces noise for preset and repeater noise purposes. The preset noise mode does not utilize the received threat radar signal to produce the jamming signal. In the repeater noise or noncoherent repeater mode, the SPJ generates noise, but this time, noise jamming is applied after detecting the threat radar signal. The two noise jammer types need different generation structures; thus, they are inspected separately. The second is the repeater jammer architecture, which has a different scheme than the noise jammers. Also, considering different coherent repeater architectures is another issue.

6.3.2.1 Noise Jammer Structure

Noise jammer architectures used in SPJ systems can be classified into two groups. First, the preset noise jammer architecture, given in Figure 6.6, is considered. This scheme has a simple structure, and digital signal processing is not required. The preset frequency information is in the PFM and taken from the SPEW CMU. The jamming carrier signal or jamming signal is generated at the DDS. With the prevalent use of digital techniques in radar, communications, and jammer systems, a digitally controlled method of generating multiple frequencies from a reference frequency source, DDS, has evolved.

The DDS is a frequency synthesizer technique that is becoming more widespread. Frequency synthesizers employ this method to create arbitrary waveforms from a single, fixed-frequency reference clock. DDS takes a different approach to the more usual indirect frequency synthesis techniques using *phase-locked loops* (PLLs) by directly synthesizing the waveform from a digital map of the waveform stored in a memory. The jamming and jammer noise signals are mixed at the jammer modulator. The noise must be white Gaussian and have suitable characteristics and properties. The most critical part of these systems is to obtain the optimal combination of the

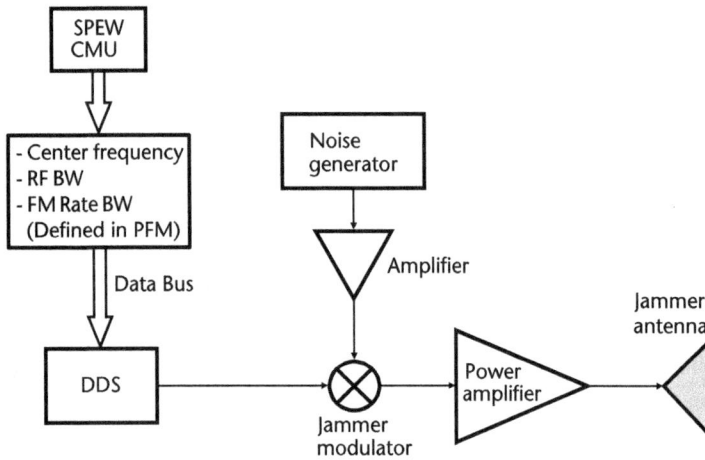

Figure 6.6 Preset noise jammer architecture with DDS.

bandwidth requirement, effective power, and antenna coverage since brutal force is required from these systems.

The output of the jammer modulator produces a jammer signal. This signal is amplified with an RF power amplifier, such as a *traveling wave tube* (TWT), solid-state amplifiers, or, for obsolete systems, klystron tubes. The jammer antenna emits an amplified signal to the environment. Figure 6.7 demonstrates another architecture for the preset noise jammer.

The scheme in Figure 6.7 differs from the one in Figure 6.6 in some aspects. The input data streams are the same for both architectures, and the amplifications and the emissions at the outputs are the same. The main difference is in generating the jammer signal. The scheme given in Figure 6.7 uses a *digital noise generator* (DNG) and an RF VCO to generate the jammer signal. For this purpose, the required frequencies and BW information are sent to DNG from CMU according to the PFM. DNG produces noise in a digital form, and this noise turns into an analog signal using a *digital-to-analog converter* (DAC). The analog noise signal is applied to the RF VCO, and the jammer signal is obtained. Then it is amplified with a power amplifier and sent to the antenna.

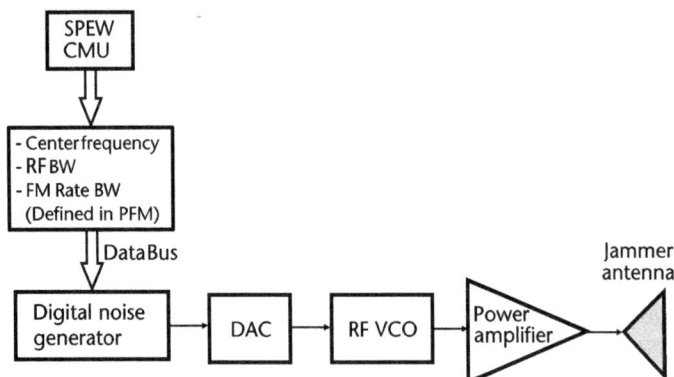

Figure 6.7 Preset noise jammer architecture with RF VCO.

Another noise jammer, the noncoherent repeater jammer scheme, is shown in Figure 6.8. In this structure, the incident threat radar signal is detected, and the same carrier frequency signal is generated, modulated with noise, and sent back to the radar.

A *low noise amplifier* (LNA) amplifies the received signal to generate a signal with the incident signal carrier frequency. Then an RWR and SPJ coordination generator controls the LNA output. The coordinated operations between RWR and SPJ, described in Chapter 4, are look-through, look-around, and look-over. Thus, the SPEW systems utilize RWR and SPJ subsystems simultaneously. After the coordination generator, an *automatic frequency control* (AFC) unit detects the carrier frequency and produces a voltage. An AFC circuit is used when one must accurately control the frequency of an oscillator by some external signal. The repeater Rx uses an AFC circuit to tune the RF VCO to the operating frequency of the threat radar. At the output of the RF VCO, a signal at the threat radar's operating frequency is produced. This signal is modulated by the amplified noise made by the noise generator. The output is the jammer noise signal, and it is controlled via RWR and SPJ coordination generator as the amplified received signal. Then the signal power is amplified and sent to the jammer antenna.

Figure 6.9 shows another repeater architecture. This scheme deviates from the one in Figure 6.8 in producing threat radar operating frequency signals. This time, the signal is generated using DDC. First, an LNA amplifies the received signal, and then the amplified signal is digitized by an ADC and sent to the DDC. The other process is the same as the RF VCO used architecture.

6.3.2.2 Coherent Repeater Jammer Structure

The coherent repeater jammers are essential to generating false target echo for deception jamming. There is a drastic difference between noncoherent and coherent repeater jamming techniques. In the coherent repeater, jamming noise is not utilized. In this technique, the SPJ captures the radar signal and then delays it by employing an analog memory loop or DRFM. After that, SPJ modulates the delayed signal if

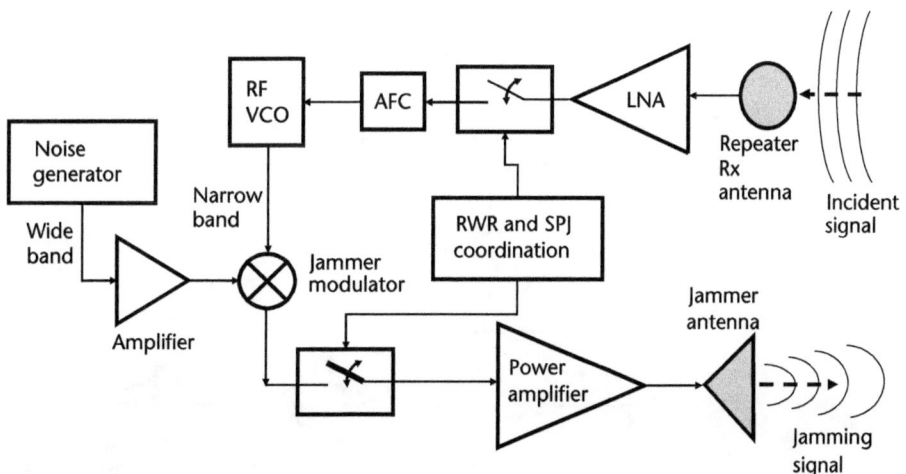

Figure 6.8 Noncoherent repeater jammer architecture with RF VCO.

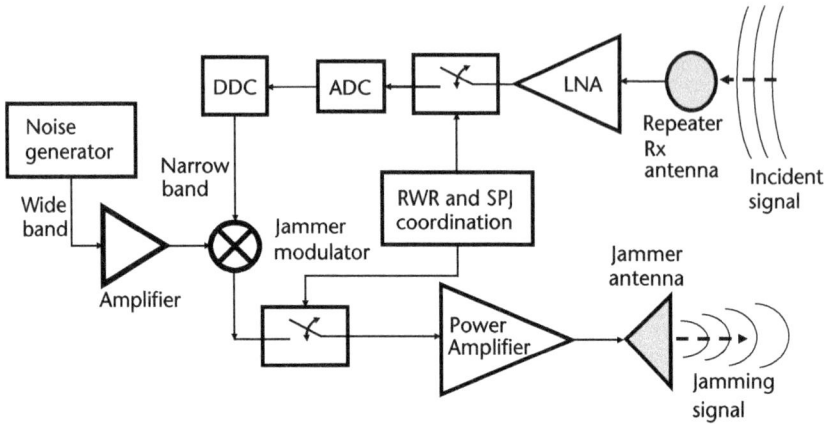

Figure 6.9 Noncoherent repeater jammer architecture with DDC.

required and amplifies it before retransmitting it to the radar. A coherent repeater with an analog memory loop or FML is given in Figure 6.10.

Even though FMLs are obsolete components, it would be beneficial to understand their operation. The FML is on standby when the RF switch is closed, and the loop switch is open. SPEW CMU controls both switches according to the PFMs. Thus, the FML operation process starts with the command that comes from the PFM. When an incident threat radar pulse signal, defined in the PFM, is amplified with an LNA and sent to the input port of the FML, the leading edge of the pulse propagates through the amplifier and the delay line to the position of the loop switch. Meanwhile, the loop is full of RF by considering a sufficient wide pulse width. Then the switch positions are reversed by changing the RF switch from close to open and the loop switch from open to close. Now the device is a closed loop with gain. The signal circulates through the loop until saturation or unity loop gain is reached. The loop is held in this position throughout the operation of the repeater jammer system. Output power is coupled from the loop, and the jammer system amplifies

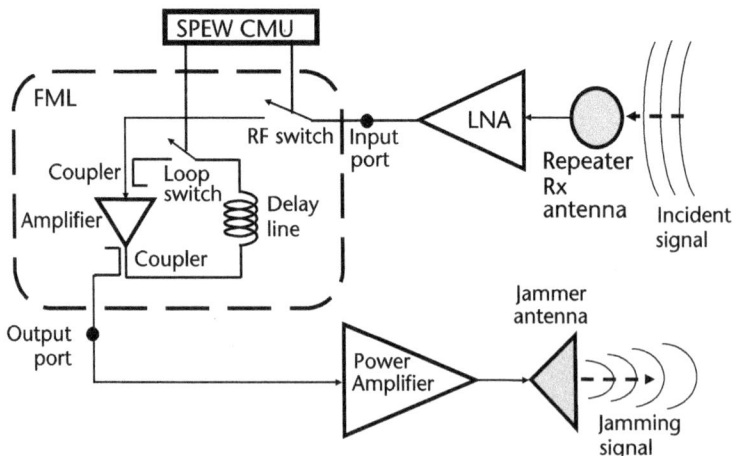

Figure 6.10 Coherent repeater jammer with an FML.

and retransmits the pulse to the radar with a delay. The coupling position before the delay line determines the minimum through-time delay, and the coupling position after the amplifier defines the power level.

A general RF memory subsystem's objective is to store an incoming RF pulse over time with minimal RF pulse degradation. The performance of an RF memory is determined by its maximum storage time, instantaneous BW, SNR, and phase distortion. The disadvantageous of the FML may be summarized as follows:

- Maximum storage or delay time is limited in FMLs.
- The quality of the stored signal degrades as the delay time extends. This degradation means lower SNR and high phase distortion.
- Even though the storage time is an essential measure of the FML effectiveness, it is another figure of merit for the system's capability to work effectively in an SPJ system. The instantaneous bandwidth changes significantly when the desired duration of storage is exceeded.
- The storage time of an FML device would be very long, but the FML would nonetheless be ineffective against the threat radar because there would be only a tiny fraction of the loop energy near the RF input frequency. This ineffectiveness is due to the low coherency of the FML. Coherency is a fraction of the total FML output power within some specified bandwidth of the input RF. Coherency and the delay time are inversely proportional, and the coherency reduces when the delay time increases.

These limitations for the delay time and the performance of FML systems necessitate a new solution, which is called DRFM systems. These systems create cohesive deception, jamming the enemy's radar system. The typical characteristics of the DRFM systems can be summarized as follows:

- DRFM systems can generate the RF signal's coherent time delay.
- They can capture and replay the radar pulses with brief and long delays and only a small signal degradation. The quality of the reproduced signal does not depend on the delay time.
- The DRFM systems can change the fake target's latency and cause it to appear to be moving.
- Modulating the amplitude, frequency, and phase of the collected pulse data can create other interference effects.
- DRFM systems can generate multiple replays of the collected pulse, which results in many false targets.
- ADCs usually quantize the in-phase (I) and quadrature (Q) signals with 4 to 6 bits. This reduces spurious signal content and avoids suppressing small signals if multiple signals are stored simultaneously [3].
- Other spurious sources for a DRFM signal are the imbalance between the quadrature channels and the intermodulation products due to the simultaneous presence of two or more intercepted signals [4].
- By using files in the PFM, the DRFM may generate arbitrary waveforms.
- DRFM jammers will also be effective when dealing with FMCW radars, not just pulse waveforms.

A coherent repeater with a DRFM is given in Figure 6.11. Both channels are digitized to preserve the phase of the RF signal. At the beginning of the DRFM process, the input RF signal is converted into equivalent lowpass I and Q components. This is accomplished using LOs that multiply the input RF signal with quadrature reference signals whose frequency is centered in the middle of the band to be sorted. The SPEW system's CPU controls the LO frequencies via PFMs. After multiplying the input signal by quadrature LO signals, the signals in I and Q channels are applied to *lowpass filters* (LPFs), and the quadrature components are obtained. The discrimination of I and Q components is mentioned in Chapter 2. If the signal is band-limited to less than one-half the sampling frequency in the baseband, the lowpass I and Q components contain all the necessary information within the signal.

The next step is to digitize the I and Q components using an ADC. This digital conversion can be accomplished by sampling the lowpass signal at the Nyquist frequency, at least twice the highest-frequency component. This phenomenon indicates the advantage of converting to baseband signals since this results in the lowest possible sampling rate consistent with the signal's bandwidth. Separate I and Q converters are also necessary to digitize the signal.

Assume that a 0.1-µs pulse is to be stored and that the frequency band coverage of the DRFM is 1 GHz. The 1-GHz band is converted into I and Q channels with 500-MHz bandwidth. The sampling frequency of the ADC is then 1 GHz to conserve all the information in the pulse. As stated, ADCs generally quantize the I and Q signals using 4 to 6 bits to reduce spurious signals. Increasing the bit number will not help but will raise the computational losses. The signal is then represented by 100 (1 GHz by 0.1 µs) I and 100 Q, 4-bit to 6-bit samples, or a total of 200 samples, that are stored in digital memory.

When the signal is needed to reproduce, the I-channel and Q-channel *random access memories* (RAMs) are sent the pulse signal at the 1-GHz rate, and the resulting signals are turned into an analog signal by DAC. Then they are passed through lowpass filters, and the resulting lowpass filtered signal is mixed with the LO signal

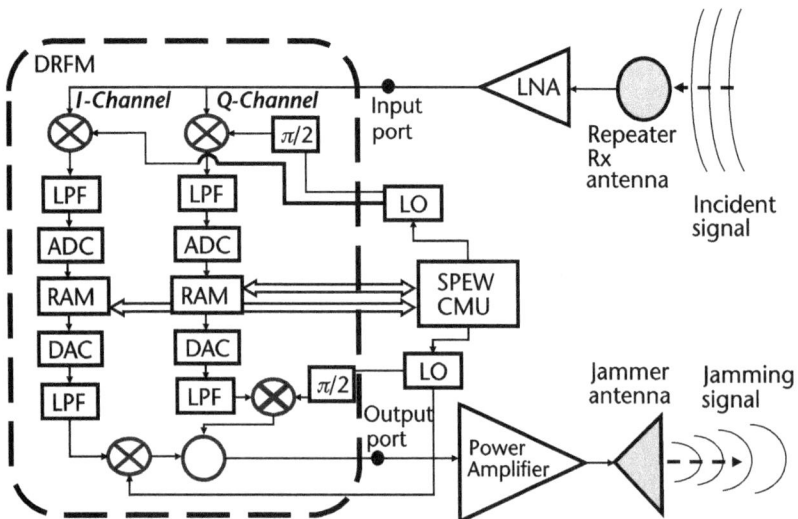

Figure 6.11 Coherent repeater jammer with a DRFM.

for upconverting to the original RF. Thus, the reproduction process is accomplished, and the RF signal is amplified and transmitted.

6.3.3 Jammer Transmitter Antennas

The antennas used in RWR and ES systems must be *ultrawideband* (UWB) and omnidirectional to cover the 360° azimuth and the entire operation frequency range with the minimum number of antennas. Another limitation in RWR and ES antenna selection is polarization. The polarization losses are aimed to keep within 3 dB in many scenarios where the polarization of the threat signal is unknown. For this reason, circular polarization is generally used, as described in Chapter 4. The antennas used in SPJ systems have the same properties for UWB and polarization. The main difference between the SPJ system antennas and the Rx counterparts is that they are directed and require higher gains than Rx systems.

Conventional SPJ systems use jamming systems for sectoral-directed antennas rather than highly directed structures. The jamming is aimed to cover 360° in azimuth, and this coverage is provided with several antennas depending on the application. Provided that the antenna coverages are sufficient, the number of antennas determines the sectors in the azimuth. The SPEW system processor controls the antenna switching circuits to ensure that the transmitted jamming pulse is radiated at the proper azimuth. Furthermore, frequency coverage is essential; different antennas may be used for various frequency bands. To cover all the sectors and frequencies, additive hardware is required. When the location of the threat is determined, applying the jamming from the related sector is an advantage for this structure. The direction is known in the SPEW systems due to the RWR system.

In modern SPJ systems, phased array antenna structures are used for jamming. In this case, the hardware being used is reduced, and due to high directivity, jamming power may be used effectively. However, the precise direction of the missile is to be detected to direct the jammer signal. The missiles used in the semiactive guided systems are passive, and sometimes, it is hard to detect their approaching angle. If the preinformation about the semiactive threat system is known, and the Rx is appropriately programmed, the missile identification is fulfilled using the technical specifications. This process yields identification but will not provide a way to determine the direction of the missile approach. For this purpose, using an MWS is suitable, and the information exchange between the RWR and the MWS is essential for utilizing countermeasures.

UWB directional antennas can be used in SPJ systems for shipborne and airborne platforms. This section will discuss the essential antenna elements most widely used in conventional SPJ systems, such as the spiral, blade, horn, and Vivaldi. Also, antenna arrays, which are used in modern SPJ systems, are mentioned. The dimensions of the antennas and radomes depend on the operating frequency range, mounting location, and platform properties.

6.3.3.1 Spiral Antennas

There are different types of spiral antennas, such as cavity-backed, conical, 4-arm, and planar spiral. These antennas are generally broadband, with a bandwidth of

5:1 to 9:1, and low-power radiators. They can be used in frequencies ranging from 500-MHz to 18-GHz applications. The cavity-backed dipole element spiral antennas are linearly polarized, low-gain, and unsuitable for RWR and SPJ applications. The 4-arm and cavity-backed circular element spiral antennas are low-gain and used in RWR and ES Rx. However, conical spiral, cavity-backed spiral, and planar spiral antennas are directional and can be engaged for jammer applications. A conical spiral antenna's gain is between 5 and 8 dBi, and a cavity-backed spiral antenna's gain is better than this. Furthermore, these antennas are circularly polarized and suitable for airborne SPJ applications. The planar spiral antennas are appropriate for aerodynamic structures of internally installed onboard systems. However, their gain is approximately 4 dB lower than the equivalent size cavity-backed spiral. The MATLAB simulation result for a planar spiral (Archimedean) antenna at 3 GHz is given in Figure 6.12. The figure shows the 3-D radiation pattern; the maximum gain is 5.17 dBi. Moreover, an actual planar spiral antenna is shown in the figure.

A cavity-backed spiral antenna with the same dimension can be compared with the planar spiral antenna. This time, the MATLAB simulation result for a cavity-backed spiral antenna at 3 GHz for a radiation pattern is considered. Figure 6.13 demonstrates the obtained 3-D radiation pattern from the simulation. The planar spiral antenna's maximum gain is 5.17 dBi, and the cavity-backed spiral antenna's maximum gain is 9.27 dBi. When comparing the two results, one can confirm the theoretical 4-dB gain between the two antennas. Figure 6.13 also shows an actual cavity-backed spiral antenna.

6.3.3.2 Blade Antennas

Blade antennas used for jamming are generally broadband with continuous coverage from 500 to 2,000 MHz. A blade antenna is comparable to a very thick dipole, which is one-half of a dipole or a monopole. Because of this thickness, it functions over

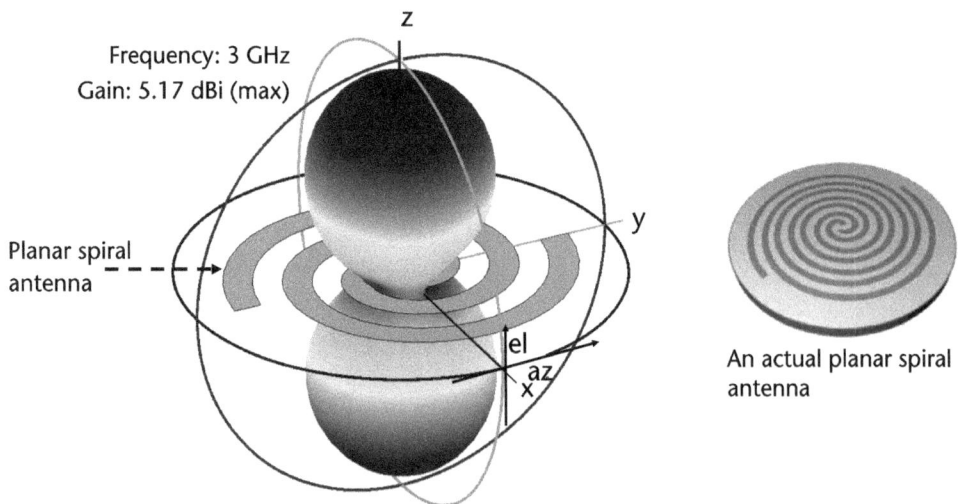

Figure 6.12 The 3-D radiation pattern for a planar spiral (Archimedean) antenna and an actual planar spiral antenna image.

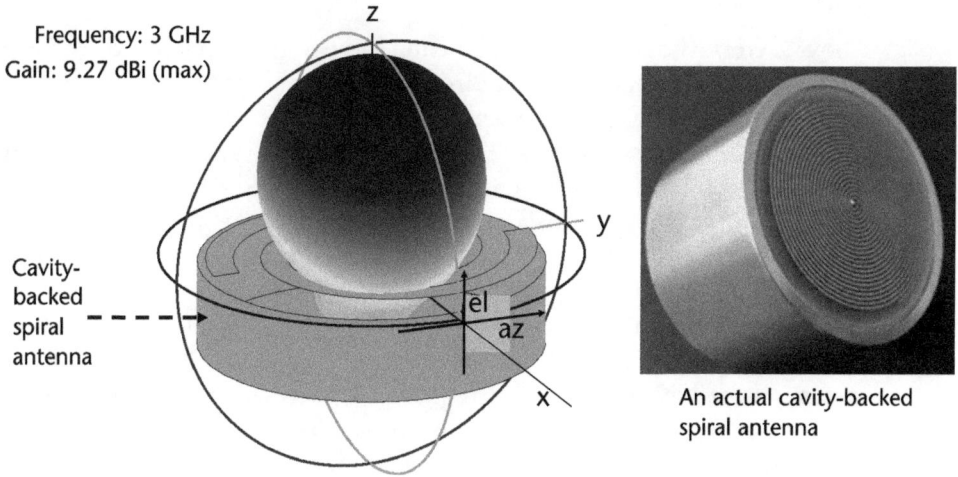

Frequency: 3 GHz
Gain: 9.27 dBi (max)

Cavity-
backed
spiral
antenna

An actual cavity-backed
spiral antenna

Figure 6.13 The 3-D radiation pattern for a cavity-backed spiral antenna and an image of the actual one.

a greater bandwidth than a thin dipole and must also be aerodynamically shaped [5]. Their gains are similar to the dipole antennas. Blade antennas are usually rugged, lightweight, fully qualified blade antennas designed for supersonic aircraft and are coated in rain erosion paint. Figure 6.14 shows the radiation pattern obtained by the MATLAB simulation for a blade antenna. The simulation is conducted at 1.5 GHz for the blade antenna, and a 360° coverage for azimuth is obtained. The maximum gain (at the most comprehensive circle of the azimuth) is 2.95 dBi. Figure 6.14 also demonstrates an actual blade antenna.

The blade antenna is suitable for high-power noise jamming against low-frequency radars, such as long-range military E/W, air traffic control, and search and surveillance radars.

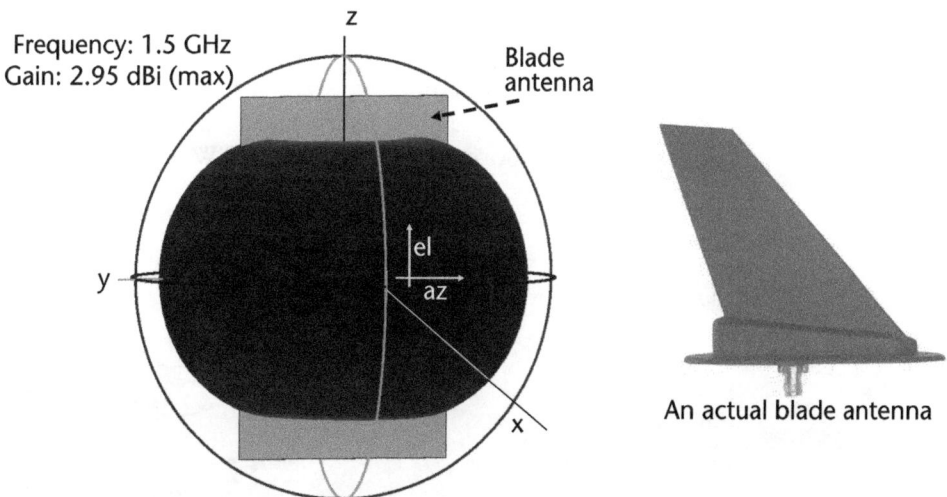

Frequency: 1.5 GHz
Gain: 2.95 dBi (max)

Blade
antenna

An actual blade antenna

Figure 6.14 The 3-D radiation pattern for a blade antenna and an actual blade antenna image.

6.3.3.3 Horn Antennas

Above 2 GHz, the typical SPJ Tx antenna is the horn, a very broadband device that operates. Horn antennas have a directional antenna pattern, high gain, and wide bandwidth. The gain of horn antennas increases as the operating frequency increases and the beamwidth decreases. There are two basic types of horn antennas. The first is the regular horn antenna, whose structure is shown in Figure 6.15(a). The polarization of regular horn antennas is linear, and the antenna gain is around 5 to 20 dB. The regular horn antennas can be used in frequency applications ranging from 50 MHz to 40 GHz.

The most serious difficulty is making the desired incident signal propagate down the feeding waveguide over the octave bandwidth of a typical SPJ system. A system whose highest operating frequency is twice its lowest operating frequency is said to operate over an octave. This problem is solved using a ridged waveguide; the operating frequency bandwidth can be up to 4:1 with a ridged. Larger bandwidths are possible, but at higher power levels, they are much reduced from regular waveguides. The shape of the radiation pattern does not remain constant as the frequency changes but varies from a broad pattern to a narrower pattern as the frequency increases.

Another horn antenna type is the polarized horn antenna. The structure and the beamwidths of the polarized horn antenna are given in Figure 6.15(b). The polarization of polarized horn antennas is circular. Antenna gain is lower than a regular horn antenna, around 5 to 10 dB. The operating frequency bandwidth is also reduced to 2:1 compared to regular horns. Polarized horn antennas can be used in frequency applications ranging from 2 to 18 GHz.

Figure 6.16 shows the radiation pattern obtained by the MATLAB simulation for a regular horn antenna. The simulation is conducted at 5 GHz for the horn antenna, and the maximum gain is 15.5 dBi. The figure also shows an actual horn antenna.

6.3.3.4 Vivaldi Antennas

The Vivaldi, or tapered slot, antenna is a linear-polarized planar antenna. The Vivaldi design is usually low-cost and possesses excellent radiation characteristics, such as high gain, broadband performance, constant beamwidth, and low side lobes. The directivity of Vivaldi antennas increases as the length L of the antenna increases, achieving gains up to 17 dB [6]. As a single element, an impedance bandwidth from below 2 GHz to above 40 GHz covers the EW band. A typical Vivaldi

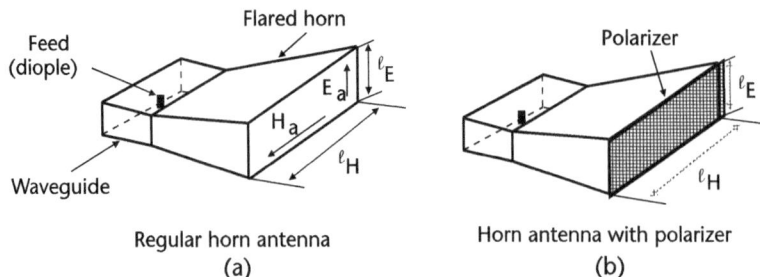

Figure 6.15 Horn antenna structures: (a) regular horn and (b) horn with polarizer.

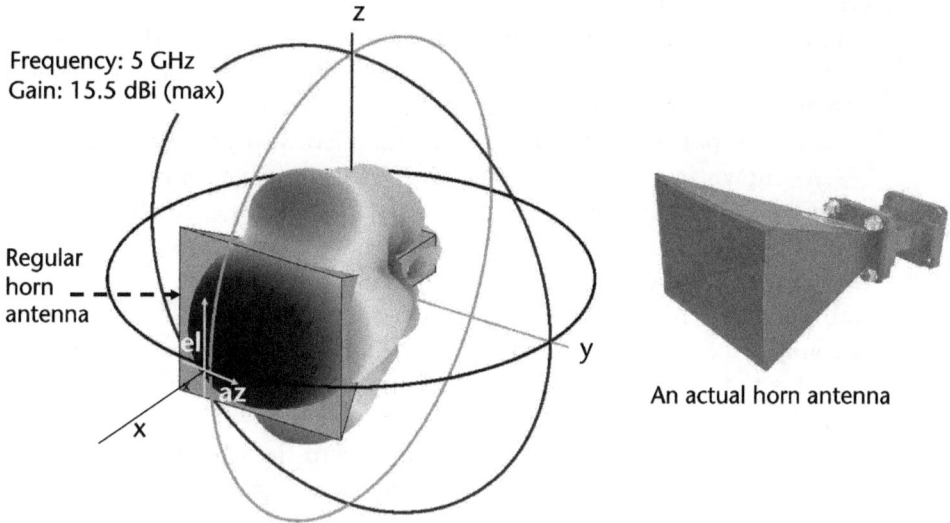

Figure 6.16 The 3-D radiation pattern for a regular horn antenna and an image of an actual one.

element gains 10 dB, and its side lobes are −20 dB. Subsequent wide bandwidth applications have used the Vivaldi antenna geometry in arrays, including active electronically scanned arrays [7].

Figure 6.17 shows the basic geometry of a typical exponentially tapered Vivaldi antenna whose center frequency is 3 GHz. This antenna is usually implemented on a substrate with the Vivaldi design etched on the upper cladding of the substrate. Vivaldi antennas are broadband or frequency-independent antennas, and they can achieve up to 10:1 or greater with a *voltage standing wave ratio* (VSWR) lower than 2. The bandwidth is limited by the antenna's opening width and the aperture width.

The gap's opening width (W_{min}) defines a Vivaldi antenna's upper-frequency limit, while the size of the aperture width (W_{max}) is the lower-frequency limit. The shape and size of the slot also determine the antenna's radiation pattern. Vivaldi antennas are called tapered slot antennae because their symmetrical radiating

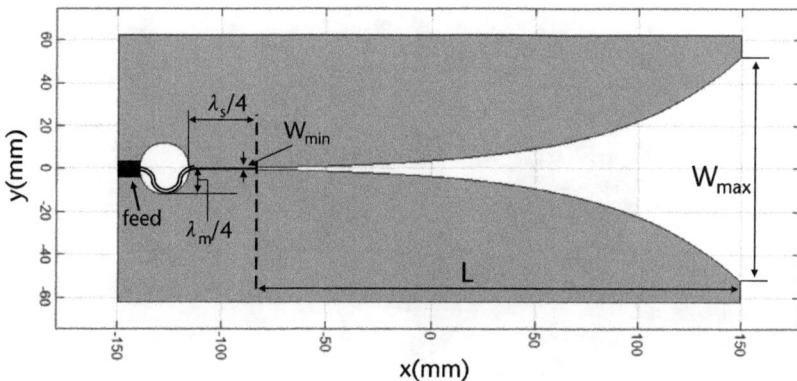

Figure 6.17 Basic geometry of a typical exponentially tapered Vivaldi antenna.

elements taper outward from their slot line. Tapered slot antenna types are exponentially tapered slot, linear tapered slot, continuous-width slot, and dual exponentially tapered slot antennas. Vivaldi antennas are helpful for any frequency, as all antennas are scalable in size for use at any wavelength. Printed circuit technology makes this type of antenna cost-effective for microwave frequencies 1 GHz or higher.

As shown in Figure 6.17, the basic structure consists of a $\lambda_s/4$ uniform slot connected to an exponentially tapered slot. The slot is excited/fed by a microstrip transmission line from the undersurface of the substrate, as shown in the figure. An alternate design uses a resonant area, typically square or circular, instead of the $\lambda_s/4$ uniform slot, which is also usually excited by a microstrip line. The resonant area is connected to an exponentially tapered slot, with or without a transmission line.

This type of antenna is typically combined with *monolithic microwave integrated circuits* (MMICs). The antenna's bandwidth is limited by the transition between the microstrip line, which connects to the MMICs, and the slot line of the antenna. The transition must be designed correctly to achieve good broadband performance matching the antenna's performance. The simplest way to accomplish this is by using a $\lambda_m/4$ open microstrip and a uniform $\lambda_s/4$ slot line, where λ_m and λ_s are the wavelengths at the center frequency; λ_m is used to identify the microstrip line. Another option is to use a coplanar waveguide feed, which provides a wider bandwidth. Baluns can also be employed with Vivaldi antennas to make them more compatible when connected to strip lines and microstrips [7].

Antennas with multi-octave bandwidth are widely used for phased arrays, EW, and radar navigation due to wide bandwidth, high gain, wide beamwidth, and flat conformal characteristics. Vivaldi antennas and their variants are classic examples of such multi-octave radiators. Figure 6.18 shows the radiation pattern obtained by the MATLAB simulation for an exponentially tapered Vivaldi antenna. The simulation is conducted at 5 GHz for the exponentially tapered Vivaldi antenna, and the maximum gain is 6.32 dBi. The figure also shows an exponentially tapered Vivaldi antenna etched upon a printed circuit board.

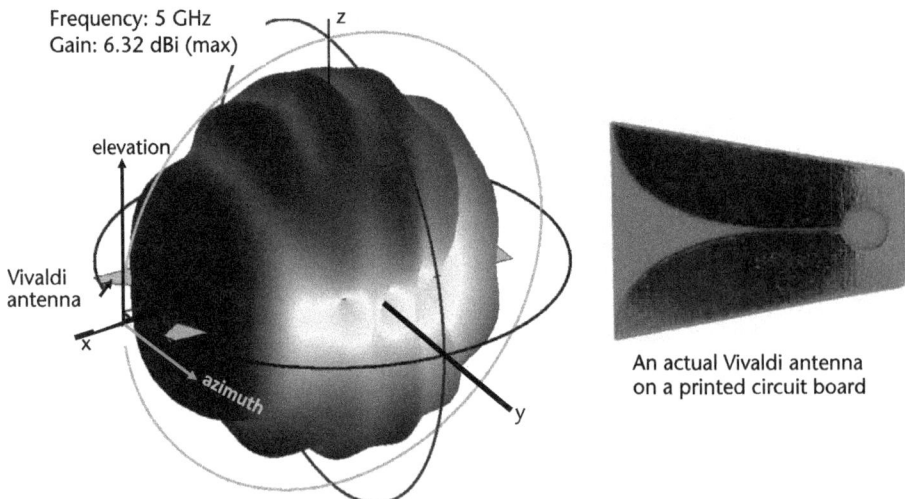

Figure 6.18 The 3-D radiation pattern for a Vivaldi antenna and an image of an actual one.

6.3.3.5 Antenna Arrays

Two possible architectures exist for the phased array antennas: linear and planar arrays. Linear phased array antennas consist of lines commonly controlled by a single-phase shifter. In other terms, only one phase shifter is needed per group of radiators in this line. Several linear arrays arranged vertically on top of each other form a flat antenna. The advantage of linear phased array antennas is their simple structure; conversely, their disadvantage is low beam steering capability, which is only in a single plane. The planar phased array antennas consist of single elements with a phase shifter per element. The components are arranged like a matrix; the flat arrangement of all elements forms the entire antenna. The advantage of the planar phased array antennas is the possibility of beam deflection in two planes; however, their disadvantage is the requirement for many phase shifters.

Antenna arrays, especially planar phased array antennas, are becoming increasingly crucial for EW applications. The planar phased arrays are focused on for the following part; for this reason, only phased array expressions will be used instead of planar phased arrays. In some examples, one-line phased array antennas will be given; however, each element has a different phase shifter. The electronic steering will be in one direction for these examples. Using the phased array antennas on radars, one can instantly switch from one target to another to track multiple targets efficiently. From the point of view of EW (especially ES), this operating regime makes it almost impossible to determine the antenna parameters of the threat radar from the analysis of the time history of the received signal strength, as it is not scanning.

The constant requirement for higher ERP for SPJ applications and the finite power sources, even at warships, always have been a trade-off study. This study is more challenging for airborne platforms, such as aircraft, helicopters, and UAVs. The best candidate factors for increasing the jammer ERP are the Tx antenna's gain and the power amplifier's gain. Let us turn our focus to the antenna's gain at this point.

Since an increase in antenna gain reduces beamwidth or coverage, determining the direction of the threat will be essential. The RWR system provides this information. The requirement for antenna coverage around the entire platform dictates using a rapidly steerable antenna to counter multiple threats. The emerging question is about the method of meeting the requirements of the SPJ system.

Phased array antennas may satisfy these requirements. These antennas consist of many elements, which are generally identical, each consisting of a spiral, horn, or dipole for SPJ applications. The block diagram of the beam routing method for a phased array antenna is given in Figure 6.19. As shown in the figure, the power divider unit at the power amplifier's output is connected to the signal attenuator, phase shifter, and antenna element (or radiator) groups operating in parallel. The direction of the resultant wave is changed by the signal attenuator and phase changer elements, which adjust the signal at the antenna element output.

Phased array antennas contain antenna elements connected to a separately controlled phase shifter. In the Tx operation, the divided signal in the power divider is adjusted to a selected angle with phase shifters. The polarization of phased array antennas depends on their element. The beamwidth is 0.5° in height and 30° in azimuth, and the antenna gain is around 10 to 40 dB. The operating frequency depends on the state of the bandwidth elements. Depending on the elements, phased

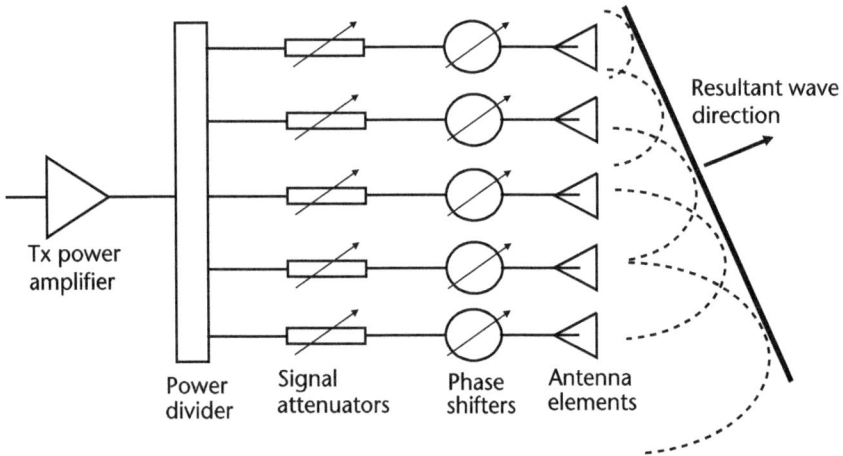

Figure 6.19 Block diagram of a phased array antenna.

array antennas can operate to frequency applications ranging from 10 MHz to the EHF (30 to 300 GHz) band.

When phased array antennas are used for RWR or jamming purposes, the EW system can flexibly utilize its signal processing and power resources. Thus, the Rx part of an RWR system can detect different radar threats from different directions without adhering to the conventional antennas' field of view or main lobe. Moreover, a jamming system can divide its power among multiple threats and allocate resources using phased array antennas. That is, it can perform jamming simultaneously or time-sharing. Another advantage of phased array antennas is that they can be adapted to the shape of the vehicle carrying them.

Each element that makes up the phased array is like an independent antenna, and they are spaced by half the wavelength of the highest frequency in the operating frequency range. This prevents grating lobes that degrade the phased array antennas' performance. Grating lobes are duplications of the main beam, as the pattern multiplication theorem predicted. The pattern multiplication theorem expresses that the radiation pattern of an array of N identical antennas ($P(\theta)$) can be obtained as follows,

$$P(\theta) = P_e(\theta) \times P_a(\theta)$$ (6.9)

In the equation, $P_e(\theta)$ represents the pattern of one of the antennas, and $P_a(\theta)$ is the pattern obtained upon replacing all the actual antennas with isotropic sources.

When the space between the array elements (d) is less than or equal to one-half of the wavelength ($\lambda/2$), only the main lobe is present in the visible area, and no other grating lobe is present. Grating lobes begin to appear when the space is greater than $\lambda/2$. The number of grating lobes (N_{GL}) is obtained as follows:

$$N_{GL} = \frac{2d}{\lambda}$$ (6.10)

Figure 6.20 demonstrates the effect of grating lobes on a phased array system. The array has five blade antenna elements. The operating frequency is 2 GHz, and the half-wavelength is 0.075m. The space between the array elements on the left side of the phased array equals the wavelength, and no grating lobes exist. However, the space between the array elements on the right side of the phased array is higher than the half-wavelength, and grating lobes start to form.

The 3-dB beamwidth (θ_{3dB}) of the dipole element phased array antenna is determined by the formula:

$$\theta_{3dB} = \frac{102}{N} \tag{6.11}$$

N is the number of elements in the array, and θ_{3dB} is in degrees. For example, a horizontal one-line array with 20 pieces will have a horizontal beamwidth of 5.1°. This angle is the 3-dB beamwidth of the beam formed in the direction perpendicular to the direction of the antenna array. The 3-dB beamwidth is obtained for higher gain antenna arrays by dividing one-element beamwidth, $(\theta_{3dB})_e$, by N.

$$\theta_{3dB} = \frac{\left(\theta_{3dB}\right)_e}{N} \tag{6.12}$$

As shown in Figure 6.21, the beamwidth increases with the cosine of the direction angle as the antenna array is oriented such that the beam direction deviates from the centerline. If the beamwidth of 5.1° on the centerline is directed to 30°, the beamwidth increases to 5.89°. Enlarging the beam angle (θ_{φ_s}) means expanding the beamwidth and decreasing the antenna gain. This angle value is found from the ratio of the beam angle to the deviation angle (φ_s) from the boresight as follows:

$$\theta_{\varphi_s} = \frac{\theta_{3dB}}{\cos\varphi_s} = \frac{5.1°}{\cos(30°)} = 5.89° \tag{6.13}$$

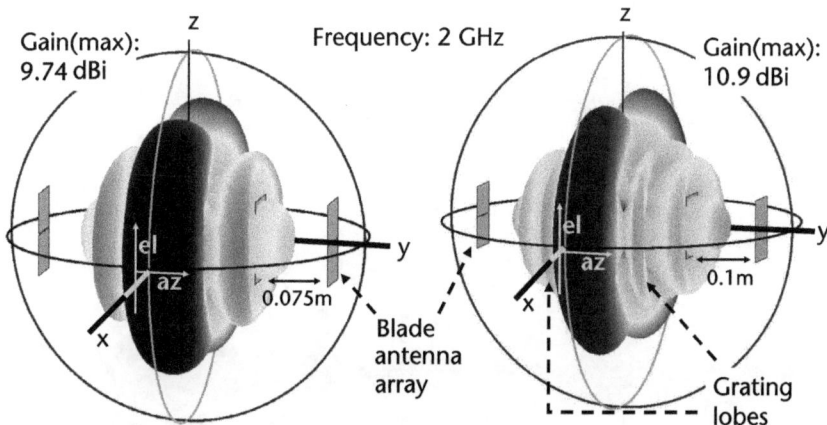

Figure 6.20 Grating lobe effect on a phased array antenna.

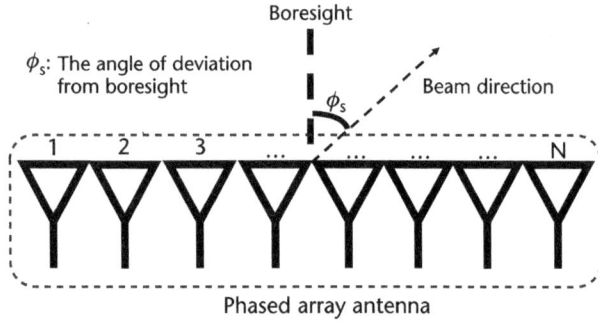

Figure 6.21 Beam direction deviation from boresight in phased array antennas.

The boresight gain of a half-wavelength element spacing in a phased array antenna in dB is found using the following equation,

$$G_{BS} = 10\log_{10}(N) + G_e \qquad (6.14)$$

In the equation, G_{BS} is the boresight gain or the antenna gain when its main lobe is directed 90° to the antenna plane, N is the number of elements in the array, and G_e is the gain of an individual component.

Let us confirm (6.14) with a simulation study. Figure 6.22 shows the 3-D antenna pattern for a regular horn antenna array. The antenna array has a 3×3-dimensional matrix structure in the planar architecture. The conditions are similar to those in the one-element horn antenna case in Figure 6.16. The MATLAB simulation is conducted for nine regular horn antenna elements at 5 GHz. The result shows that the maximum gain of the planar phased antenna array is 25.3 dBi.

Compared with the results of Figure 6.16, the gain for one regular horn antenna is 15.5 dBi. Moreover, the following is obtained if we apply the phase array situation to (6.14).

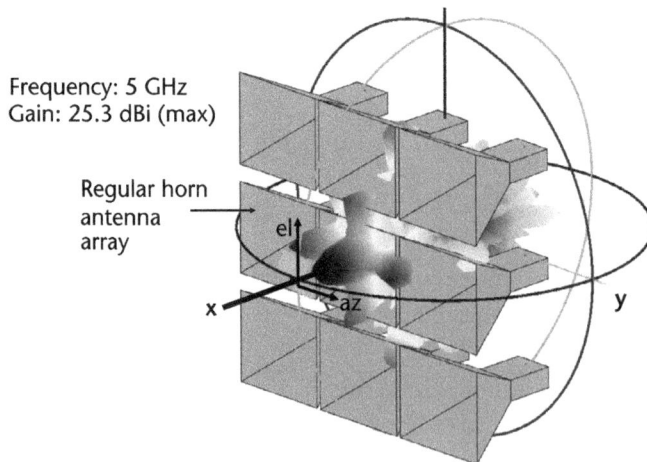

Figure 6.22 The 3-D radiation pattern for a regular horn antenna array.

$$G_{\mathrm{BS}} = 10\log_{10}(N) + G_e = 10\log_{10}(9) + 15.5 = 25.1 \text{ dBi}$$

It can be seen from the results of the MATLAB simulation that the antenna array gain is 25.3 dBi. Similarly, when we use (6.14), we obtain 25.1 dBi, which is a very close result to the one obtained from the simulation.

A planar array antenna in combination with solid-state MMIC-based Tx and Rx (Tx/Rx) modules is a suitable choice for a shared aperture antenna. Vivaldi planar array antenna suitable for the Tx/Rx module-based active aperture phased array EW systems working in the 4 to 20-GHz (G-J) band. This system can use a planar array as a shared aperture for multiple ESM and jamming functions against modern threat radars. Furthermore, installing Vivaldi elements in a planar array provides more flexible placements than the other 3-D antennas, such as horn, blade, and cavity-backed spiral antennas. For these reasons, a separate simulation study is conducted for a Vivaldi planar antenna array.

Figure 6.23 shows the 3-D antenna pattern for a planar array antenna consisting of Vivaldi elements. The antenna array has a 3×3 dimensional matrix structure in the planar architecture. The conditions are like the one-element Vivaldi antenna case in Figure 6.18. The MATLAB simulation is conducted for nine Vivaldi antenna elements at 5 GHz. The result shows that the maximum gain of the planar phased antenna array is 16.1 dBi.

Compared with Figure 6.18's results, the gain for one regular Vivaldi antenna is 6.32 dBi. Moreover, the following is obtained if we apply the phase array situation to (6.14).

$$G_{\mathrm{BS}} = 10\log_{10}(N) + G_e = 10\log_{10}(9) + 6.32 = 15.86 \text{ dBi}$$

The results of the MATLAB simulation show that the antenna array gain is 16.1 dBi. Similarly, when we use (6.14), we obtain 15.86 dBi, which is very close to the one obtained from the simulation.

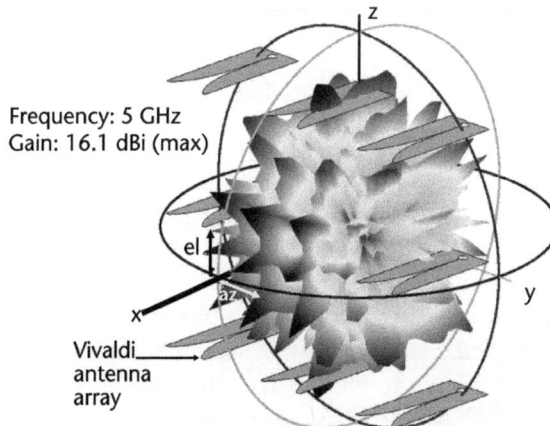

Figure 6.23 The 3-D radiation pattern for a Vivaldi antenna array.

Let us consider an example to understand the phased array antenna's steering. A 20-element antenna array with a gain of 5 dB for each element has a gain of 25 dB. The gain decreases with the cosine of the angle between the boresight and the beam direction in Figure 6.21. This gain reduction is not a dB but a gain reduction factor as a ratio. The gain reduction factor (L_{φ_s}) is defined in dB as follows:

$$L_{\varphi_s} = 10\log_{10}\left(\cos\varphi_s\right) \tag{6.15}$$

In this case, for a 30° deviation from the centerline, the reduction would be 0.866 or 0.624 dB. A phased array antenna with a half-wavelength element pitch can be oriented to deviate from the centerline by up to 45°. If the antenna array elements are closer than a half-wavelength (with a decreasing centerline gain), the directional limit can be increased to 60° [8]. Deviation from the centerline causes a reduction in gain with the cosine of the steering angle (φ_s). The effect of this decrease is given as follows:

$$G_{\varphi_s} = 10\log_{10}\left(N\right) + G_e + 10\log_{10}\left(\cos\varphi_s\right) \tag{6.16}$$

If the beamwidths of the phased array antennas in two orthogonal planes are known, it would be better to use the directivity first instead of gain. In this case, a phased array antenna's directivity is defined as follows:

$$D_0 \simeq \frac{32,400}{\Omega_A} = \frac{32,400}{\Theta_{1d}\Theta_{2d}} \tag{6.17}$$

where Ω_A is the solid angle in steradian units, and Θ_{1d} and Θ_{2d} are the angles (degrees) in the orthogonal planes. In the equation, D_0 represents the maximum directivity level.

The antenna gain (G) is lower than the directivity (D) due to the ohmic losses of the antenna and its radome (if any). During transmission, these losses include part of the supply power, which is not radiated but raises the temperature in the antenna. The mismatching between the antenna and the transmission line can also reduce the gain. The relationship between the gain and directivity of an antenna can be established by the efficiency of the antenna (η_R) as follows.

$$G_{BS} = \eta_R D_0 \tag{6.18}$$

One of the significant problems associated with forming phased array antennas that cover a wide frequency band is the physical spacing of the antennas and the resulting beam steering by applying the phase shifts to the corresponding antennas. These radiating elements that cover such a wide frequency range are the physical spacing of the antennas in phased array placements and the resulting beam steering by applying the phase shifts to the corresponding antennas.

The array itself generally consists of several elements distributed on a planar surface. Each element at least consists of a phase shifter and radiating element. The components are grouped into sets called subarrays. Subarray steering is generally used to minimize the number of ports in the monopulse networks, and this also allows large arrays to support wide bandwidth waveforms. Active and dummy elements can be distributed across the planar surface to control the aperture illumination function, which determines the antenna's radiation pattern and side lobe structure.

The conventional phase shifters have been implemented as delay line elements with mechanically or electronically tunable delays. They act more like timed arrays than phased arrays since an explicit time delay is introduced in the signal path. There are different kinds of RF phase shifters, as follows:

- Ferrite phase shifters with slow response;
- MEMS-based phase shifters with medium response;
- MMIC phase shifters with fast response.

These passive phase shifters formed the backbone of earlier radar and communication phased array solutions, integrated with other off-the-shelf Tx/Rx modules, such as LNAs and PAs. However, a single-chip integrated solution with these phase shifters is impossible due to their size and speed.

It is also essential to consider the type of the antenna element. The size of a single antenna determines the minimum antenna separation. When an antenna array is established using wideband antenna elements, such as spirals, blades, or horns, the required space between the array elements (d) may not satisfy the condition of less than or equal to $\lambda/2$. However, in the case of using Vivaldi antennas, reaching the required spacing would be more feasible.

In high-frequency and wideband applications, antenna arrays cause another problem. Phased array antennas cause pulse dispersion when the wideband signals' received pulses are longer than the transmitted pulses. We can consider two forms of pulse dispersion that occur in a phased array antenna. The first emerges from the separation distance between the Tx and Rx antennas and impacts the definition of the far field in the time domain. The second is a function of beam scanning and array size. Time delay units at the element and subarrays limit the pulse dispersion [9].

A phased array uses a phase shifter at each element to align the signal phases and coherently add all the pulses. Phase shifters have a constant phase shift across their operational bandwidth. As a result, a constant phase shift over frequency means that the scan angle changes with frequency. Due to the frequency-steering phenomenon for phased arrays operated over very wide bandwidths, a loss referred to as dispersion loss results from the antenna main lobe scanning across a target versus instantaneous waveform frequency. When wideband operation is required, some combination of time-delay and phase-shift steering is needed.

6.3.4 RF Power Amplifiers

Various technologies have been deployed during the RF *power amplifier* (PA) design history. One of the oldest technologies for RF PA realization is vacuum electronic devices such as magnetrons and klystrons. Vacuum devices have been utilized for amplification, high-speed switching, and rectifying applications. Their ability to handle high power makes vacuum electronic devices good candidates for military and space applications such as radar signaling and satellite communications. Along with their benefits, some drawbacks, such as excessive heat dissipation, bulky structure, high power supply requirement, high noise figure, and complex grid control management, are their disadvantages. TWTs, invented while seeking an alternative with better noise performance than the Klystron tubes, have found a wide range of application areas in radar and satellite communications applications in the 1960s. TWT technology has been chosen over solid-state PAs in high-power and frequency applications. However, solid-state PAs have rapidly displaced TWT counterparts in mid-power radar and jamming applications.

Factors such as operating frequency band, BW, RF power output, gain, efficiency, noise figure level, ruggedness, thermal management, and cost should be considered when choosing an RF PA. The TWT and solid-state amplifiers are the RF PA types utilized in SPJ applications.

6.3.4.1 TWT Power Amplifiers

A TWT performs as a wideband RF PA. This wideband amplification feature is obtained by using an interaction circuit, essentially a transmission line that does not usually contain any resonance. The TWT can be split into several major elements: the vacuum tube, electron gun, magnet and focusing structure, RF input, helix, RF output, and collector. The TWT is contained within a vacuum tube, generally made up of glass. This maintains the vacuum that is required for the operation of the TWT. The physical structure of a TWT is given in Figure 6.24.

The first element is the electron gun, primarily composed of a heated cathode and grids. It also has anode plates, a helix, and a collector. The RF input is sent to one end of the helix, and the output is drawn from the other.

A magnet and focusing structure are included so the electrons can travel as a tight or narrow beam along the length of the TWT. The field from the magnet keeps the beam as narrow as required and, in this way, ensures that the beam travels along the length of the TWT. The RF input consists of a direction coupler, a waveguide, or an electromagnetic coil. This is positioned near the electron gun emitter and induces current into the helix.

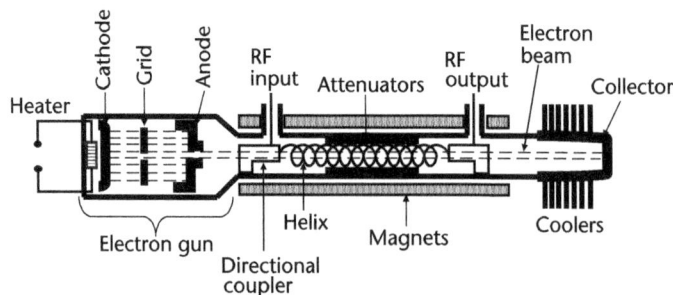

Figure 6.24 Structure of a TWT power amplifier.

A helix is a vital component of the TWT. It provides a delay line in which the RF signal travels at nearly the same speed along the tube as the electron beam. The EM field due to the current in the helix interacts with the electron beam, causing the bunching of the electrons in effect known as velocity modulation, and the electromagnetic field resulting from the beam current then induces more current back into the helix. In this way, the current builds up, and the signal is amplified.

The RF output from the TWT consists of a second directional coupler, which may be an EM coil of a waveguide positioned near the collector. It receives the amplified signal from the electron gun or emitter from the far end of the helix. An attenuator is included on the helix, usually between the input and output sections of the TWT helix. The helix is essential to prevent the reflected wave from traveling back to the cathode of the electron gun. The collector finally collects and absorbs the electron beam. In this area, one expects high levels of power dissipation; therefore, this part of the TWT can become very hot and require cooling, and coolers are engaged for this purpose.

The TWT PAs used in SPJ systems have wide BW and high-power ratings. The BW of the off-the-shelf TWT PAs is up to 12 GHz—the average power output ratings of several hundred watts in most operating BW for CW TWT PAs. Also, there are some types of TWT for pulsed operations. Pulsed TWT amplifiers have instantaneous power of up to several kilowatts. The TWT output VSWR is generally poor and is typically 2.5:1.

The main uses of TWT PAs in SPJ systems are preamplifiers and output power amplifiers. The main factor for CW TWT effectiveness for an SPJ application was the ability to modulate the tube's electrical phase length, provided by the external phase shifters. The main parameters for pulsed TWT amplifiers are the pulse-up maximum length and the pulse-up duty cycle.

6.3.4.2 Solid-State Amplifiers

The recent developments in high-power, high-frequency, solid-state transistors led to the expansion of solid-state technology in the RF amplifiers' market. RF solid-state technology is an excellent alternative to traditional vacuum tube-based RF/microwave systems. This technology offers advanced control, reliability, and ease of use. The ability to dynamically adjust power and frequency helps optimize transmitted energy. The appropriate RF power transistors must be employed when designing high-performance, high-power, solid-state PAs. RF power transistors can be categorized according to the semiconductors from which they are built. Early military-use transistors were made from *gallium arsenide* (GaAs), but nowadays, commercial solutions are based on *silicon* (Si) and, more recently, *gallium nitride* (GaN). *Gallium nitride on silicon-carbide* (GaN-on-SiC) has been utilized because GaN can maintain a high field for a small geometry, and SiC has better thermal performance. *Silicon-based laterally diffused MOS* (Si-LDMOS) is a crucial device technology used in high-power RF PA applications for frequencies ranging from 1 MHz to greater than 3.5 GHz.

General operational ranges of different PAs in terms of frequency and power rate can be given in Figure 6.25. One can conclude the functional properties of

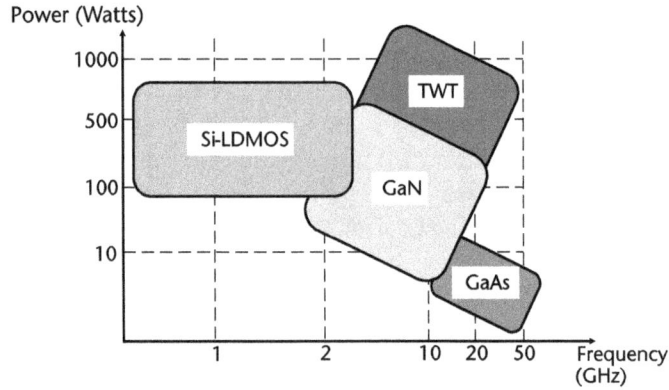

Figure 6.25 Operational ranges of different PAs.

PAs by looking at the figure. The TWT PAs are utilized in high power and high frequencies. Si-LDMOS technology is eligible for systems with low frequencies and high powers. However, GaAs PAs are for high-frequency, low-power applications. The most probable solid-state PA technology used in SPJ systems is the GaN. This technology presents a wide frequency operating band and a varying power range from low to high.

A general high-power RF PA system consists of vacuum tube technology, the dominant power source technology, but needs to be better adapted to output power control. It requires costly regular maintenance due to the limited lifetime of vacuum tubes. Moreover, vacuum tube systems are bulky and need additional high-voltage supplies, leading to massive systems. RF solid-state technology is an excellent alternative to traditional vacuum tube-based RF systems. This technology offers advanced control, reliability, and ease of use. The ability to dynamically adjust power and frequency helps to optimize transmitted energy.

Another advantage of solid-state devices is their extremely long life if properly used and cooled. When employed in adequately designed equipment, they should last their entire useful life, commonly 100,000 hours or more. Vacuum tubes wear out as their filaments (and sometimes other parts) deteriorate with time in regular operation; a typical vacuum tube's useful life may be 10,000 to 20,000 hours.

Figure 6.26 shows a generic architecture for an RF solid-state PA. The solid-state transistor alternative is based on many power modules and can be more compact and adjustable. Each module or *RF power transistor* (RFPT) is limited to a kilowatt level in a single chip if it adopts current solid-state technologies, and the output power can be adjusted by supply voltage control. Power combining can be accomplished in fully isolated integrating systems that allow hot swapping of RF-PA units during system operation.

The RF input power is amplified by the driver amplifier first, and then the amplified RF power is split and distributed to N RF-PA modules. Power splitters are critical components designed to distribute the RF power from the driver amplifiers to the individual RF modules with optimal amplitude and phase dispersion. The essential element of a solid-state RF-PA module is the transistor technology that predetermines the limits of the design parameters. The main design parameters of an RF are frequency band, BW, RF power output, gain, efficiency, ruggedness,

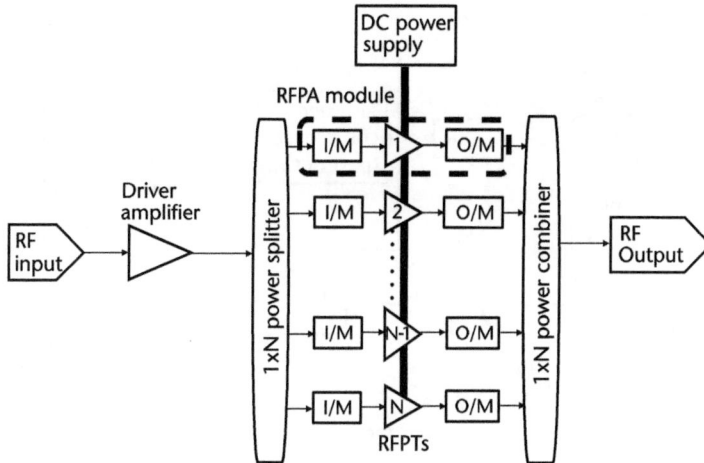

Figure 6.26   A generic block diagram of an RF solid-state PA.

thermal management, and cost. The outputs of RF-PA modules are recombined and coupled to the RF waveguide feeder. A power converter system supplies the amplifier modules with DC power.

A linear PA generates RF power most efficiently and with a minor distortion when it delivers RF current into an optimum load resistance value for a given active device. For an amplifier, the output network transforms the load impedance, such as an antenna, into an optimum load resistance value for the active device. *Input matching* (I/M) and *output matching* (O/M) networks differentiate active power amplifiers from passive transmission lines.

Another advantage of solid-state RF PAs for SPJ applications is that they do not require tuning up. Their operations lend themselves to low-impedance broadband operations. Fixed-tuned filters with readily available components can suppress harmonics and other spurious signals. Band switching of such filters is easily accomplished when necessary, often using solid-state switches. However, tube PAs must be retuned on each band, even for significant frequency movement within a band.

6.3.5   Jamming Waveform Generator

The jamming waveform generators generate the transmitting signal on the IF. Section 6.3.2 mentioned different jamming waveform generators. The primary purpose of this section is to describe the functional meaning of the ECM generator and the jamming waveform generator's technical properties. For this purpose, Figure 6.27 is given. In general understanding, the ECM generator produces the jamming signal according to the threat systems described in the PFMs. An ECM generator consists of a jammer controller and a jamming waveform generator. The jammer controller is an operational part of the CMU. It takes the information from RWR and repeater Rx and controls the waveform generator according to the PFM. However, the jamming waveform generator is a physical component that produces noise and coherent and noncoherent repeater jamming signals.

Different jamming waveform generator types are considered for producing radar jamming signals. As stated, their structural and operational details were

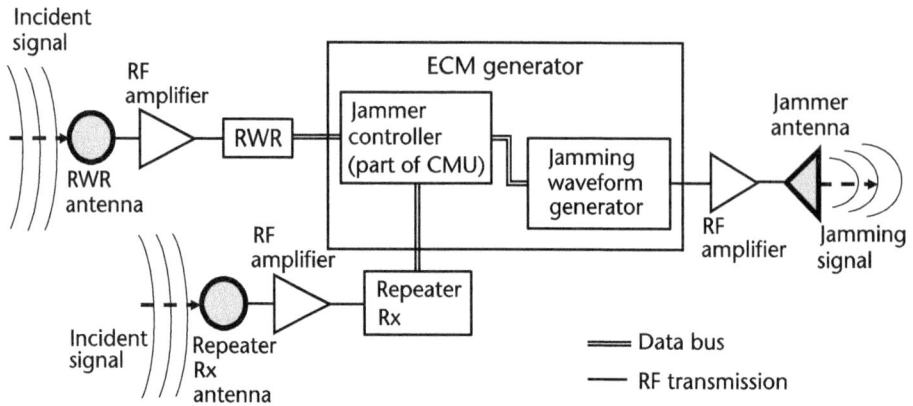

Figure 6.27 Functional block diagram of an ECM generator.

given in Section 6.3.2. Designing and implementing various jamming waveform generator schemes and the control unit are also important issues. Noise-jamming techniques, including noncoherent repeaters, are accomplished using DDSs, VCOs, and noise generators. These structures can generate noise, single-point frequency, multipoint frequency, and other jamming modes. However, recently, the jamming waveform generator parts have been realized using FPGAs instead of DDSs, VCOs, and noise generators.

The technical properties of the jamming waveform generator are mainly related to the operational function and jamming pattern requirements of the SPJ system. When looking for a radar jammer system that only needs to perform one function, has a simple structure, and requires only a few jamming methods. This type of jammer system would include using active decoys. If a multifunctional jamming source with multiple jamming generation methods is to be utilized, it brings more complex system design problems and more significant SWaP needs. These requirements are crucial for airborne SPJ systems. For large platforms and special EW equipment such as SOJ aircraft, the radar active jamming system often requires that it achieve various jamming effects on various radars of different systems. Hence, multiple jamming generation technologies must be considered to meet the requirements.

Another factor that affects the operational capabilities of the jamming waveform generator is the parameter measurement accuracy the jamming system Rx can provide. The parameters of noise jamming signals mainly include center frequency and bandwidth. When the Rx cannot accurately measure the operating frequency of the radar signal, it is necessary to increase the bandwidth of the jamming signal to ensure the effective suppression of jamming to target radar. When the Rx has high measurement accuracy for the radar signal parameters, such as operating frequency, pulse width, and intrapulse modulation, the SPJ can generate jamming signals independently according to these parameters. However, obtaining the intrapulse modulation information is almost impossible by an Rx with low measurement accuracy. In this case, the jamming signal is generated by coherent repeater jammers [10].

The jamming signal generation method is another vital factor that affects the jamming waveform generator specifications. For example, the spectral structure of the noise jamming signal generated analogously will usually be closer to the noise of the radar Rx and have a better jamming effect than the noise signal generated digitally. However, the stability and flexibility of analogously generated signals still need to be improved. When high coherency is required between the jamming and radar signals, it will provide a high jamming effect by using coherent repeater noise instead of noncoherent and saturated noise jamming, which generates synthetic noise.

6.4 Noise Jamming in Self-Protection

Up to this point, the technical aspects of noise jamming in SPJ systems have been discussed. The jamming efficiency equation is derived in this context, and the jammer architectures are discussed. This and the following sections aim to merge the technical explanations with operational requirements and give a different point of view. Although the inspections are conducted separately for noise and deceptive jamming, the primary explanations for the mixture are valid. The reason for this is that, most of the time, they are used together simultaneously or sequentially.

To prepare a successful PFM, all the environmental threats should be detected and effectively jammed or denied. Both requirements are essential, and making concessions to them may result in a catastrophic outcome. The requirements for detecting the threats were discussed in Chapter 5. For applying ECM against threats, the threats' technical properties must be known and related to this, and the methods of jamming against them should be studied before preparing the PFMs. Furthermore, the disruptive effect of using multiple simultaneous or sequential different noise and deceptive jamming techniques must be investigated. The PFM preparation is a multipartite process and may require burdensome studies regarding intelligence, operation, and techniques.

The PFM preparation is a subbranch of preprogramming. The preprogramming activities include planning, preparing, and modifying the PFMs. In the planning phase, the requirement determination and technical analysis are conducted, and this stage occurs in all the preprogramming activities. The other activities define two different processes: preparing a new PFM and modifying an existing PFM. Figure 6.28 gives the PFM preparation cycle. Even though the process may vary for different organizations, each party's main milestones are similar.

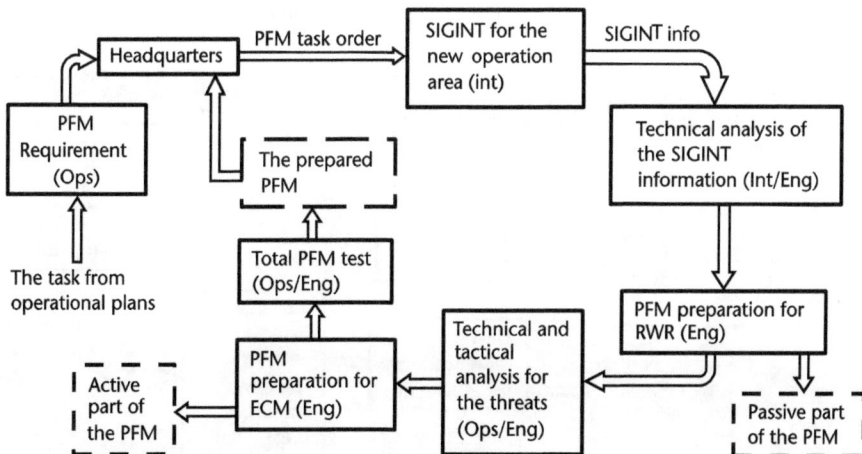

Figure 6.28 The PFM preparation process cycle.

The PFM preparation process starts with the tasks from the operational plans. The PFM requirement is sent to the headquarters from the operational (ops) branch. Headquarters sends the requirement to the intelligence (int) branch to obtain technical intelligence information by employing the SIGINT facilities. The intelligence and engineering (eng) branches analyze the information obtained from SIGINT. The information received at the end of this process is used to prepare the PFM's passive part. After this point, operational and engineering branches analyze the threats. The engineering branch studies effective jamming methods for each threat. Also, the mutual effect of the different jamming techniques is inspected, and an optimal PFM for the ECM part is developed. After development, the total PFM is tested by using a hot mock-up. After validating the PFM, it is returned to the headquarters for operational usage.

Figure 6.29 gives the PFM modification cycle. The PFM preparation process starts with the new SIGINT information for an operation area with an existing PFM. The PFM modification requirement is sent to the headquarters from the operational (ops) branch. Headquarters sends the requirement to the intelligence (int) branch to obtain technical intelligence information by employing the SIGINT facilities. The intelligence and engineering (eng) branches analyze the obtained new and existing SIGINT information together. Then stages similar to the preparation process are followed to get the modified PFM, as shown in Figure 6.29.

The PFMs accommodate noise and deceptive jamming in them. For this reason, the above explanations are valid for noise and deceptive jamming techniques. Now we can turn our attention to noise techniques. The types of noise jamming that are used with SPJ are inspected separately.

- *Cover pulse (CP):* CP jamming can be interpreted according to the time and frequency domains. The CP generates gated noise with a noise-like spectral content. It aims to cover the target's skin echo with jamming pulses synchronized with the PRF of the threat radar. In the time domain, the jamming pulses return to the threat system so that the noise covers the skin return. The main point here is that the width of the cover pulse should be greater

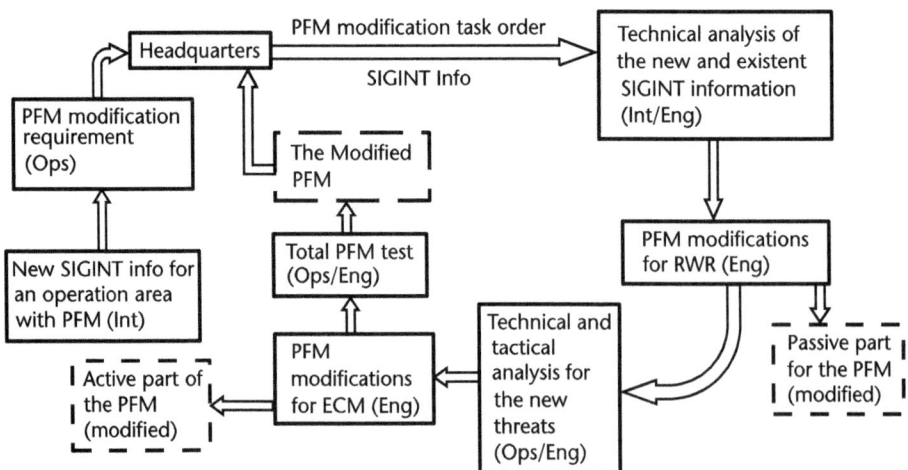

Figure 6.29 The PFM modification process cycle.

than the width of the radar pulse. Thus, the skin echo at the radar Rx can be covered. However, in the frequency domain, the jammer produces a spot or barrage noise jamming around the center frequency of the radar signal. Figure 6.30 demonstrates the CP jamming effect on the radar Rx in the time and frequency domains when target echo signals exist.

Jamming pulses are transmitted with a programmed time anticipation and duration based on the predicted arrival time of threat pulses. Say that the pulse detection of a radar's incident pulse is taken at a time. Then, the skin echo (se) is predicted, and an estimated delay ($\hat{\tau}_{se}$) is calculated. After this point, an appropriate calculation is conducted for the CP's delay (τ_{CP}) and the PW of the CP. This situation is shown in Figure 6.30(a). The threat radar's center frequency and bandwidth (BW_R) are determined in the frequency domain. After that, a jammer bandwidth (BW_J) is determined according to the threat radar's BW, jammer ERP, and the number and their frequencies spreading over the spectrum. Jammer BW determines the type of noise jamming. In some applications, spot noise jamming is sufficient; however, barrage jamming may be more suitable for other situations. Figure 6.30(b) demonstrates the frequency domain CP jamming activities.

The CP jamming has a notable superiority over the CW noise waveform. Typically, the *home-on-jam* (HOJ) technique is engaged against CW noise. However, CP noise is an effective jamming technique without incurring HOJ threats. Another advantage of this jamming method is that it is a multithreat technique since it utilizes a time-sharing method. However, it has some disadvantages, such as complexity and pulse synchronization. Pulse detection and deinterleaving processes are conducted using TOA and frequency information. A TOA expectation process is necessary to predict the sequential pulses and determine the threat radar's PRI, also called PRI synchronization.

The PRI synchronization process may be straightforward for one threat radar with fixed PRI. However, jittered and staggered PRIs add complexity to the process, and multiple threat radars with close operating frequencies increase the difficulty of the situation. The jamming signal can be generated according to the threat radar's carrier by coherent and noncoherent techniques. CP technique may be a narrow pulse or wide pulse noise jamming. The type of CP is determined by the length of the jammer bandwidth (BW_J). When the BW_J is only sufficient for covering the target echo pulse, this is a *narrow*

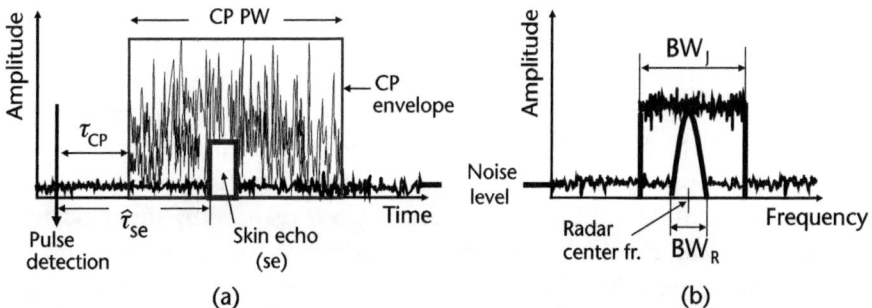

Figure 6.30   Cover pulse jamming: (a) in the time domain and (b) in the frequency domain.

pulse (NaP) noise jamming. However, when BW_J gets larger, wide pulses are obtained, and each covers more than one return pulse, which may be named fountain noise jamming. The fountain jamming may be spot or barrage noise from the frequency domain point of view. When fountain noise jamming is used for multiple pulses from the same target, say, that spot noise, then the jamming power between two or more pulses can be assumed as a loss. When fountain noise is used with barrage jamming against simultaneous threat radars, the jamming power between multiple pulses cannot be considered a loss. However, using barrage jamming results in a power decrease.

- *Continuous noise:* The continuous noise is the most ordinary noise-jamming method. In the general sense, the noise jamming increases the noise level or lowers the SNR of the radar Rx. As a result, low SNR degrades the target detection capability of a radar. The continuous noise covers all the operation time when looking at the time domain. As stated before, spot, barrage, and sweep jamming techniques are the continuous noise's classification in the frequency domain. Spot jamming focuses all its power on a single frequency with a narrow BW; however, this technique is ineffective against frequency-agile radars. Barrage jamming is the jamming of multiple frequencies simultaneously by a single jammer. The main drawback of this technique is that the jammer spreads its power across various frequencies, making it comparatively less potent at a single frequency. Sweep jamming shifts a jammer's full power from one frequency to another. This shifting motion jams multiple frequencies quickly, although not all simultaneously.

6.5   Deceptive Jamming in Self-Protection

Deceptive jamming involves manipulating radar systems to produce incorrect target information. This is achieved by emitting, retransmitting, or reflecting electromagnetic waves to generate signals that resemble the target's echo. Deceptive jamming can be classified based on the content of the false information type. In this context, the types of deceptive jamming may be classified as range, velocity, angle, and false target.

Before a detailed explanation, we briefly mention the deceptive jamming classes that would be suitable. Range deception causes the radar to produce false range information about the target, which the RGPO and RGPI techniques can accomplish. Velocity deception generates incorrect target velocity information, usually realized by the VGPO technique. However, angle deception is a more prevalent application area. It causes radar to generate incorrect angle information of the target. This technique is generally conducted by inverse gain jamming for conical scan radar and cross-eye jamming or cross-polarization jamming for monopulse radar. False target deception adds forgery targets to radar detection results to prevent detection and track initiation. False target deception is realized by sophisticated modulation in range, angle, velocity, and energy dimension for the generated jamming signal, like the target echo. The next section explains the applications of deceptive jamming techniques related to the classes.

6.5.1   Range Deception

The operational principles of the tracking radars are explained in Section 2.6. The single-target range tracking principles are described there. The radar angle tracking loop operates only for signals arriving at the radar during the radar's range tracking gate. The objective of the range deceptive jammer system designed to interfere with radar tracking operation is to move the range tracking gate away from the target position to where only the jamming signal is present. To successfully jam a radar signal, the jamming signal must simultaneously enter the Rx as the target signal. The process can be interpreted as the SPJ system must initially transmit jamming signals as coincidentally as possible with the signal reflected from the vehicle being protected. The range measurement function of search radars is to inject signals before the arrival of, or in exact synchronism with, the pulse at the radar. The SPJ system is required to rely on second time-around signals. However, these signals are ineffective when the radar uses carrier frequency or PRF agility.

The primary methods for the deception of the range tracking gate are applying noncoherent repeater noise jamming, RGPO, and RGPI. This time, the cover pulse method uses repeater noise jamming for deceptive purposes. The coherent repeater, the other method for range deception, is used to realize RGPO and RGPI. Both methods are mentioned next.

- *Noise method:* The SPJ system can transmit a noise signal produced by a noncoherent repeater to mask the target signal and the regions on either side of the target position. In this case, if the SNR is sufficiently lower than the required detection level, the radar range tracking loop tends to wander in range. It may reach an equilibrium position at one or the other of the edges of the noise burst. Assuming that this phenomenon occurs, the range tracking gate is removed from the actual target position to one at which only the jamming signal is in the angle tracking loop. Appropriate angle jamming modulation of the noise signal will affect the operation of that loop in the radar [11].

  As shown in Figure 6.30, the success of the CP noise signal is based on the fact that the victim radar does not use carrier frequency or PRF agility. Otherwise, it would be challenging to generate a noise signal whose leading edge is at a nearer range to the radar than the actual target. In that case, the radar range tracking gate would reach equilibrium at the leading edge of the noise signal at the true target position.

- *RGPO:* The most prevalent self-protection deceptive jamming technique is *range gate pull-off* (RGPO), which is based on manipulating the TOA of the target return echo at the radar Rx. The main difference between the RGPO and the noise techniques is that they use the original signal source. In the noise technique, the signal source is the noise generator, while the original radar signal is in the RGPO. RGPO uses a DRFM to process the stored signal by using processes such as delaying the signal, amplifying the signal, directly repeating the signal, and overlapping two amplified and delayed signals.

  RGPO technique is utilized against automatic range-tracking radars. When a tracking radar detects a target, range gates are placed on both sides of the range. Range gates protect the radar against asynchronous jamming pulses by eliminating all signals from distances outside of a narrow window or, in other words, significantly increasing the SNR. Radar concentrates in a short range surrounding the target's position and does not look at other targets. This process is known as locking. However, the distance gates can be unlocked or stolen, known as the RGPO technique. The RGPO aims to break the radar's lock on the target platform and provide its escape outside the distance gate.

The RGPO process can be summarized as capturing the threat radar's range gate, pulling (walking) it off in range, turning it off, and leaving it without an actual signal. In general, automatic tracking radars use early and late tracking gates that cover the return echo. A range servo removes the gates by following the target and predicting its future position using a range gate or velocity estimate developed in the range tracker. The range tracking circuit excludes all other returns except those returned within the tracking gates. This operation prevents spurious signals from entering the range-tracking circuitry and distorting the range estimate. Still, it also allows a deception jammer to operate in a range-stealing or break-lock mode.

It would be appropriate to explain the operational mechanism of the RGPO and then clarify its phases by considering its effect on the tracking radar Rx. The RGPO technique causes the radar to get false target range errors, which can result in significant aiming guidance errors for SAM and AAA systems that use command guidance.

The DRFM receives the threat radar's signal, amplifies it, and retransmits it with a minimum delay to provide a false target signal to the radar. The false target return signal with high power intensity causes the radar Rx's gain to decrease because of AGC action. As a result, the actual target RCS signal is suppressed, and the strong beacon signal captures the radar range gate circuit. Note that the primary purpose of AGC is to fully utilize the available dynamic range. The time delay of the repeated signal is progressively increased on a pulse-to-pulse basis from the actual target position to a time equivalent to many radar range gate widths. Upon reaching the outer pull-off limit, the repeater jammer is turned off.

Figure 6.31 demonstrates the baseband waveform programming of the RGPO technique for a typical tracking radar Rx. Figure 6.32 depicts the pulse structure for a generic RGPO delay. The time offset between the backscattered echo from the target and the jamming pulse as a function of time.

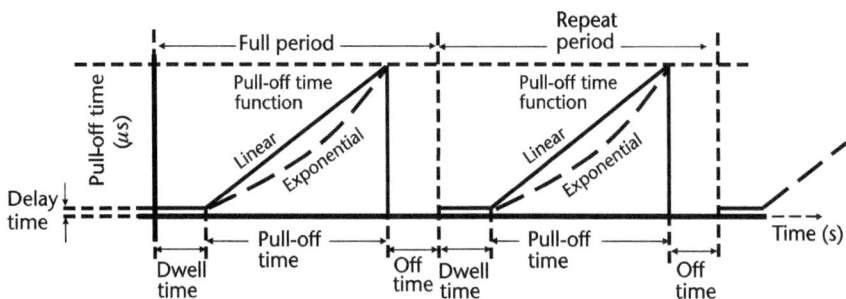

Figure 6.31 RGPO baseband waveform programming.

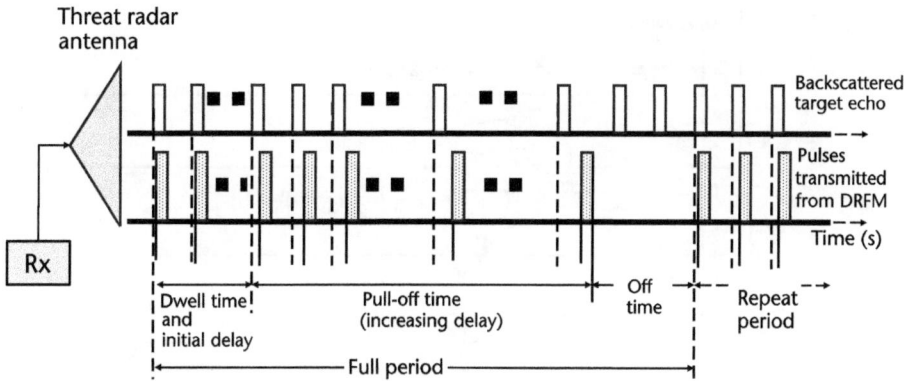

Figure 6.32   Pulse structure for a generic RGPO delay.

The parameters demonstrated in Figures 6.31 and 6.32 require further explanation. Also, these brief definitions are essential explanations to understand the RGPO technique.

- *Dwell time:* The radar PRI should synchronize the pulses generated by the DRFM. The time required for the synchronization is dwell time. Each full period or RGPO cycle can be set to a different value.
- *Pull-off time:* This phase is the primary interval of the RGPO process. In pull-off time, the TOA of the received jamming pulses pull or walk off. It shall step away from the echo pulse at least at a duration of 10 PW. It may be different for each period or RGPO cycle. During the pull-off time, false targets can be generated using RGPO by hold-out or overlapped RGPO techniques. The maximum acceleration rate must be below the threat radar's range tracking limits.
- *Off time:* At this process interval, the DRFM stops transmitting the repeater signal upon reaching the next full period starting instant or outer pull-off limit. Each RGPO cycle can be set to a different value.
- *Pull-off time function:* This function defines the time delay variation pattern. It may be parabolic or linear.

Now let us clarify the phases of a generic scenario for one target. As stated, this technique generates the deception signal that relies on an RF memory, tuned pulse-by-pulse. It is also effective against frequency-agile radars. Furthermore, it is effective even if the radar uses random PRF. Sometimes, amplitude modulation is added to the RGPO. This modulation simultaneously generates an angular error in a scanning tracking radar but is scarcely effective against a monopulse radar [12].

Figure 6.33 summarizes the RGPO process in five sequential stages. It demonstrates the baseband programming waveform RGPO technique for the sequential time instants on the tracking radar Rx. DRFM units generate the coherent pulses. The given situation is for constant PRI, but this technique may also be applied to agile PRIs. In this context, the following articles provide detailed explanations of each stage.

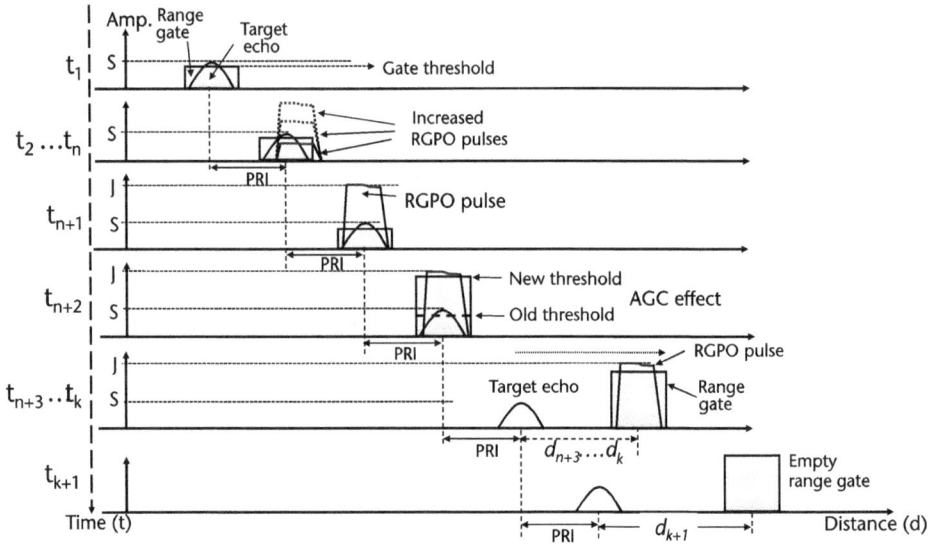

Figure 6.33 RGPO effect at a tracking radar Rx.

- *Stage 1 (t_1 instant):* At the beginning of the process, the tracking radar detects the target return echo (or skin return), and the range gate is on the echo. As seen in Figure 6.33, the radar's automatic gain control property adjusts the sensitivity threshold or gate threshold according to the echo power's amplitude (amp.). There is no deceptive jamming signal at this instant, and the tracking goes on without any problem.
- *Stage 2 (t_2 to t_n instants):* This stage is an interval rather than an instant. In this stage, samples are taken from the pulse signal from the radar. Without delay, the signals are transmitted to the radar by increasing their power as soon as the successive pulses are received. Thus, the victim radar's Rx detects a nonexistent echo. From this point of view, the mother platform strengthens its echo in the radar scope with RGPO jamming capability. The RF memory unit (i.e., DRFM) gradually increases the repeater jamming signal power in the n consecutive instants.
- *Stage 3 (t_{n+1} instant):* The Stage 2 process continues until the RGPO signal is much stronger than the skin return. In this stage, both the target echo and the RGPO signals are in the range gate.
- *Stage 4 (t_{n+2} instant):* In this stage, the radar's Rx sensitivity is reduced by the *automatic gain control* (AGC) capability, shown as the AGC effect in Figure 6.33, usually found in radar systems, to avoid overloading. This process causes the skin return echo to disappear below the noise floor. As a result, the tracker has locked onto the delayed replica.
- *Stage 5 (t_{n+3} to t_k instants):* This stage is an interval like Stage 2. In this stage, the RF memory unit delays the replicas. Concerning each of the radar's pulses, the replica is delayed by small but increasing amounts of time. The range gate follows the fake target, which appears to be receding. This process lasts until the range gate drifts from the target's position. Thus, the radar tracks a ghost target, and the range gate conceals the skin return.

- *Stage 6 (t_{k+1} instant):* In the final stage, the jammer is turned off and leaves the radar with nothing but noise inside its range gate. Thus, the break-lock is accomplished. As a result, the tracking radar starts a search and detection process again and loses time. If the target is still in range, the detection and acquisition cycles start again for track initiation.

Even though for no delay replicas, the repeater jammer systems cause microscale delays due to the detection and amplification processes of the radar-intercepted signals. The leading edge of the delayed signal from the RF memory unit arrives at the radar as much as 0.1 μs after that of the skin return. The aim of designing SPJ systems is to minimize the TOA delay at the radar of the deception signal relative to the echo signal. Radar designers have focused on this effort in their antijamming designs since realizing this vulnerability in deceptive jammer systems. Leading-edge range tracking is an EP technique used to defeat RGPO jamming.

The leading-edge tracker, as can be considered by its name, obtains all range data from the leading edge of the target return. All RGPO cover pulse jamming tends to lag the target return by some increment of time. The target return can be separated from the jamming pulse by differentiating the entire return concerning time. By utilizing a split-gate tracker that is electronically positioned at the leading edge of the returning pulse, the range tracking loop can accurately track the target return while disregarding any jamming signals. The range tracking loop then uses split-gate tracking logic to determine the magnitude and direction of range tracking errors and reposition the range gate.

The primary purpose of the RGPO technique is to break the range track circuit of the tracking radar. The ground-based computation of a weapon system's angle tracker is strongly influenced by the target range as determined by the radar. However, the tracking radar angle circuitry still functions to point the radar antenna's boresight in the direction of the target. This angle information is still sufficient for missile guidance. Hence, RGPO alone is insufficient to prevent hits by the weapon system. For this reason, RGPO is generally used with spot noise or angle deception techniques.

- *RGPI:* The range deception jammer memorizes the radar signal using DRFM and then amplifies and retransmits the signal, much stronger than the return signal, with a specific time delay. Increasing these time delays, the range gate will detect an increase in range and automatically move off to a false range. This range deception method, which pulls the false target away, is called RGPO, as was explained before. On the contrary, the range deception method that the range gate will detect a decrease in range and move into a false range is called the RGPI method. Figure 6.34 demonstrates the target, RGPO, and RGPI echoes on a generic radar scope.

The RGPI application is the same as the RGPO technique, except the false target seems to be approaching. RGPI simulating an inbound target is applied chiefly against leading-edge range trackers. The repeater should anticipate the radar pulse's reception and have sufficient storage capability to store the radar pulse for one PRI to apply the RGPI technique. For this purpose, the repeater timing should be

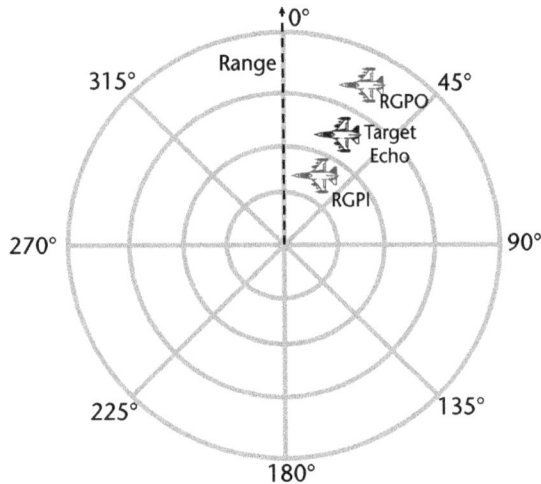

Figure 6.34 Target, RGPO, and RGPI echoes on a generic radar scope.

initiated from the previously received pulse. That is provided by using trackers in the jammer system.

6.5.2 Velocity Deception

Using the same deception method as RGPO and RGPO in the frequency domain, we can implement velocity deceptive jamming such as *velocity gate pull-off* (VGPO) and velocity gate pull-in (VGPI). The signal received in a coherent radar exhibits a Doppler shift whenever there is a time-varying path length between the Tx and Rx. Many radars employ coherent operation modes, allowing them to distinguish targets from background clutter based on the relative Doppler shifts. Furthermore, we can implement the range-velocity deceptive complex jamming by transmitting the frequency-shifted signal with a corresponding time delay.

The *CW Doppler* (CWD) and *pulsed Doppler* (PD) radars use Doppler shift and establish velocity gate rather than range gate for tracking. In the coherent radar operation, the signal is exposed to a Doppler shift if there is a change in the distance between the Tx and Rx locations over time. Many radars employ coherent operation modes, allowing them to distinguish targets from background clutter based on the relative Doppler shifts. Focusing on the radar threat types that use CWD and the PD effect is essential in this context. Semiactive homing missiles are a prevalent class of radar tracking systems that are examples of CWD radars. Moreover, PD radars detect low-flying targets from ground or AI aircraft missions.

The deception technique against these radars is VGPO and VGPI. Since the radars are coherent in this operation, generating noise in the related frequency, composed of the sum of operating frequency and Doppler frequency $(f_0 + f_D)$, will not be a proper solution. Using DRFM is an effective method for producing velocity deception like the range deceptive counterpart of it. Doppler frequency may be negative or positive according to the moving direction. Similarly, the deception signal can be generated as if the ghost target is approaching (VGPO) or receding (VGPI). In these methods, frequency shift directions are considered for nomenclature. The out (VGPO) is used to increase the frequency, and the in (VGPI) is used to reduce the frequency. The VGPO and VGPI are deception techniques that operate in CWD with relatively low peak power.

The VGPO counters CWD and PD radars by stealing the velocity gate of their automatic tracking loop. VGPO aims to capture the Doppler velocity tracking gate by transmitting an intense false Doppler signal. Then the frequency of the fake signal is changed to move the tracking gate away from the target Doppler. This technique is similar to the RGPO against the range gate tracking loop.

Similar to the RGPO analysis, we discuss the operational mechanism of the VGPO first, then clarify its phases by considering its effect on the tracking radar Rx. VGPO technique causes the radar to get false target velocity errors. These velocity errors can result in significant aiming guidance errors for SAM and AAA systems.

Figure 6.35 shows the modulating waveforms for the programming purposes of the VGPO technique for a typical tracking radar Rx. This technique operates against coherent radar velocity gates and can generate different sweep types. These sweeps can comprise simple LFM or piecewise (stepped) LFM signals.

The parameters demonstrated in Figure 6.35 require further explanation. These definitions are also fundamental to understanding the VGPO technique. The VGPO application phases are dwell time, sweep or pull-off time, and hold time.

- *Dwell time:* In this stage, the DRFM detects the signal and generates a repeated jammer signal with a programmed frequency shift to put it within the passband of the Doppler filter containing the target return. Then the jammer increases the amplitude of the generated signal to capture the filter containing the target through the radar's AGC action.
- *Sweep time:* In this interval, the radar AGC is captured, and the radar locks on to the false Doppler signal. The modulation may be LFM or programmed as various stepped LFM. The jammer slowly sweeps the false signal's frequency away from the actual Doppler frequency of the target. This fakes the radar into calculating an incorrect Doppler shift for the target.
- *Hold time:* Later in the process, when the radar is moved far enough away in frequency, the jammer signal gradually weakens and eventually turns off.

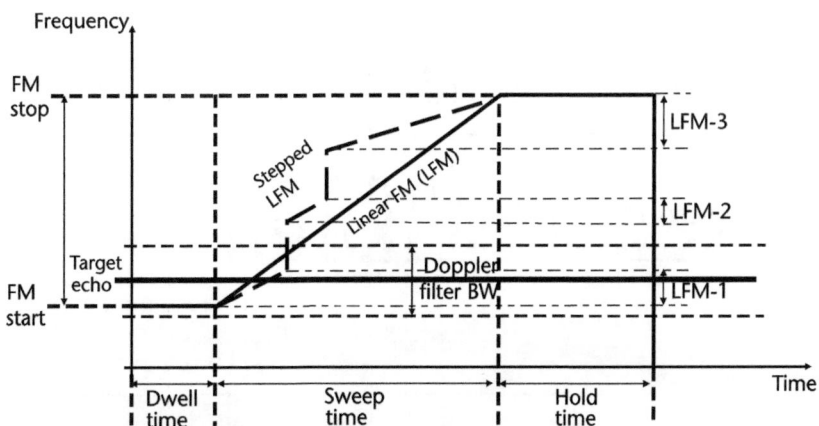

Figure 6.35   VGPO modulating waveforms.

The radar is left without a target. The threat radar must restart its acquisition process before starting the tracking process again.

Now, let us clarify the application stages of a generic scenario for one target. VGPO is employed by repeating a frequency-shifted replica of the received radar signal, and the frequency of the fake return is gradually changed to interfere with the true Doppler shift. This technique operates against coherent radar velocity gates and sets up a gate-stealing method, allowing it to fulfill the three consequent stages shown in Figure 6.36.

No jamming signal exists at the beginning of the process, and the velocity gate locks on the target return Doppler frequency. The AGC facility also adjusts the sensitivity threshold or the amplitude of the velocity gate. Let us assume that the VGPO process starts at this point.

- *Stage 1:* The jammer RF memory detects the signal and produces a replica signal with the same Doppler frequency to put it within the passband of the Doppler filter containing the target return frequency. The target return frequency consists of operating frequency (f_0) and Doppler frequency (f_D). The frequency of the initial jamming pulse must appear within the same velocity tracking filters as the target return; otherwise, the victim radar will discard it. The bandwidths of the velocity tracking gates are generally between 100 and 500 Hz. These bandwidths are relatively narrow for the radar operation frequencies. Thus, the accurate frequency band information of the Doppler tracking filters is vital intelligence for the PFM preparation process. Then, the jammer amplifies the power amplitude of the regenerated signal. Once the tracking gate captures the jamming pulse, the AGC increases the Rx sensitivity to prevent overloading, and the target skin is concealed under the jamming replica.

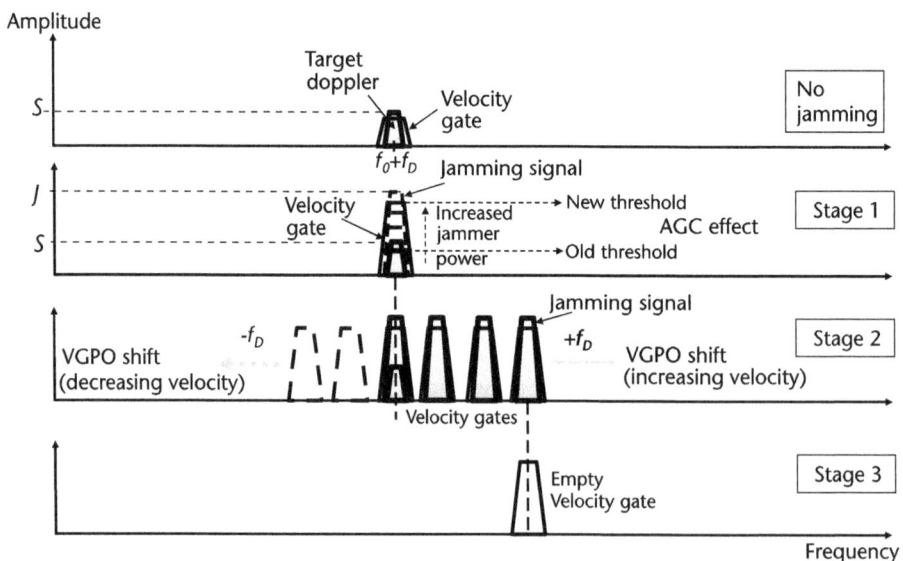

Figure 6.36 VGPO effect at a tracking radar Rx.

- *Stage 2:* After the radar locks on to the false Doppler signal, the jammer slowly shifts the false signal's frequency away from the actual Doppler frequency of the target. This shifting may be in an increasing or decreasing direction. The increasing direction shows increasing velocity. However, the reducing direction represents the decreasing velocity. In Figure 6.28, the target is assumed to have increased velocity. The most common method is adding linear FM on the target's Doppler frequency. Thus, a false movement can be added to the jamming signal. As a result, the VGPO deceives the radar by calculating an incorrect Doppler shift for the target.
- *Stage 3:* The jammer signal is progressively attenuated when the radar's velocity gate wanders far enough in frequency. Finally, it is turned off, leaving the radar without a target. The detection and acquisition process must be restarted again to track initiation.

In summary, the VGPO technique entails presetting sensors around the gate where tracking is performed. The tracking system switches to memory for a short time and reacquires the old target as soon as an additional echo is detected. Therefore, the actual and deceptive echoes are separated when a deception jammer tries to attract the tracking gate to a fake target. The true echo enters the guard gate and blocks the tracking gate. The gates return to regular operation when the sensors indicate that the deceptive echo has disappeared. The VGPO technique sweeps the frequency from low to high, generating a fake closing target at the threat radar Rx. However, the VGPI technique's LFM or stepped LFM sweeps the frequency from high to low and produces a fake receding target at the threat radar Rx.

6.5.3   Angle Deception

Angle deception is a jamming method that exploits weaknesses in the angle-tracking loop of the threat radar. The technique aims to generate false targets that appear to come from a target at some different azimuth and elevation from the actual target. For this purpose, the deception jammer generates fake targets in some directions other than the target's bearing and elevation. The angle deception techniques are classified as amplitude modulation (AM), monopulse, and frequency offset techniques.

6.5.3.1   AM Deception Techniques

AM techniques are also known as envelope modulation techniques. They aim to produce angular tracking errors in passively scanned radars by distorting the information of amplitude necessary to track the target accurately. The types of AM deception techniques are given next.

- *Inverse gain jamming:* This angle deception technique is engaged against conical scan and lobe switching radars, described in detail in Section 3.6.1. An adaptive threshold circuit determines the modulation induced by the radar and coherently generates an on-off modulation. This modulation can cause the threat radar to move in the opposite direction to what is required for accurate tracking, thereby breaking the lock.

- *Swept rectangular wave (SRW):* This technique is the primary angle deception method against COSRO and can also be used against LORO and conical scan radars. The details of the process are presented in Section 3.6.3. The COSRO radars typically used a fixed Tx antenna and a conically scanning Rx antenna. The lock-on technique works the same as in a typical conical scan radar, but the transmitted signal is constant and denies showing the radar's scanning rate information. For this reason, this concept is also known as silent lobing.

  The SRW technique, known as scan rate modulation, deceived these radars. This technique is generally like the inverse gain method, but the scanning rate is unknown. Instead, the system sends pulses on the radar's frequency at a PRF similar to the radar's estimated scanning rate. The radar will only receive these pulses if the receiver is pointed roughly in the aircraft's direction. To ensure this will occur at some point, the PRF is slowly increased and decreased so that at some point in this pattern, it synchronizes with the scan rate of the antenna for a short time. In this case, the radar receives its signal and the deception signal, which drifted slightly in time. When fed into the phase detector, the output signal will be two pulses rather than one, creating an error signal. When the two pulses are closely synchronized, it generates significant error signals that can quickly drive the antenna away from the target. Since the rate is constantly changing, after a period, the error will drop to zero again, potentially before the radar has entirely moved off the target.
- *Sliding pulses:* This method is used against the TWS radars and is described in Section 3.6.6. A TWS radar creates angle gate-pair or pairs during its sectoral tracking. These pairs consist of early and late gates. The gate pair's centerline represents the radar antenna's boresight. Sliding pulses drift the target out of the radar's tracking area, thus avoiding the tracking of the aerial platform.

6.5.3.2 Monopulse Techniques

These techniques are engaged against monopulse radars. As explained in Chapter 2, monopulse radars use a simultaneous beam lobing technique of angle measurement that uses the relative amplitudes (or phases) of the simultaneously received skin return by a couple of beams. The measurement is unaffected by the time variation of the received target return amplitude. The AM is ineffective against these radars because the jamming signal provides an equally acceptable signal for monopulse angle measurement. Techniques effective against monopulse radar are cross-polarization and cross-eye jamming techniques.

- *Cross-polarization jamming:* As described in Section 3.6.5, this technique utilizes the difference in the monopulse antenna pattern for a jamming pulse that is polarized orthogonal to the design polarization. The antenna pattern for a two-channel monopulse radar using sigma and delta beams shows the tracking point between the two beams. This situation is valid when the radar uses the design polarization. However, the radar antenna has a receiving pattern for a cross-polarized signal with its design frequency. The tracking point is shifted one beamwidth to the right for a cross-polarized signal. This shift in the tracking point results in a target tracking signal 180° out of phase with the actual signal. A jamming signal polarized orthogonally to the radar's operating frequency must be 25 to 30 dB stronger than the radar signal to be effective.

A cross-polarized jammer must receive and measure the polarization of the victim monopulse radar. Then the jammer transmits a high-power jamming signal to the victim radar at the same orthogonally polarized frequency. In addition, the jamming signal must be 25 to 30 dB stronger than the target return and be as purely orthogonal to the design polarization as possible. A cross-polarized jammer must generate a powerful jamming pulse that is polarized orthogonal to the threat radar.

- *Cross-eye jamming:* Cross-eye jamming is another technique used against monopulse radars, as explained in Section 3.6.5. In a monopulse radar, this method aims to alter the wavefronts of the beams, causing angle tracking errors. The two basic operational principles of monopulse tracking logic are as follows:

  - A target return will always be a normal radar pulse echo.
  - Any shift in amplitude or phase in a target return is due to the tracking antenna not pointing directly at a target.

  Cross-eye jamming exploits these two properties by receiving and transmitting jamming pulses from different antennas separated as far apart as possible. The phase front of a monopulse signal is received by one of the Rx antennae, amplified by the repeater, and transmitted by the other transmit antenna. The same phase front hits the other Rx antenna. Then it is shifted by 180°, amplified by the repeater, and transmitted by the other Tx antenna. These two out-of-phase signals must have the same amplitude and exceed the target return's amplitude. When jamming signals are sent to a threat radar's Rx, they cause the tracking loop to eliminate any differences in amplitude and phase. Using two jamming sources that are widely spaced and have different stages, it becomes impossible for the antenna to reach a null position or tracking solution. A leading-edge tracker can reject jamming signals arriving at the antenna after the target returns to defeat cross-eye jamming. Figure 3.20 gives the cross-eye jamming concept.

6.5.3.3   Frequency Offset Techniques

These techniques comprise skirt and image frequency jamming, as mentioned in Section 3.6.5. Frequency offset methods aim to intervene in the operating frequencies of the threat radar systems to produce an erroneous angular measurement on them. It would be proper to mention the techniques.

- *Skirt frequency jamming:* This method is used against the monopulse radars. The skirt frequency-jamming technique exploits the discrepancy of the monopulse radars' Rx IF filters. The filters should be correctly tuned to the transmitting frequency. If a monopulse system is appropriately designed and regularly maintained, it will not experience frequency imbalances. The Tx and IF filter frequencies will be the same. Jamming signals outside this specific frequency range will not disrupt the radar's monopulse tracking ability. This situation is shown in Figure 6.37(a).

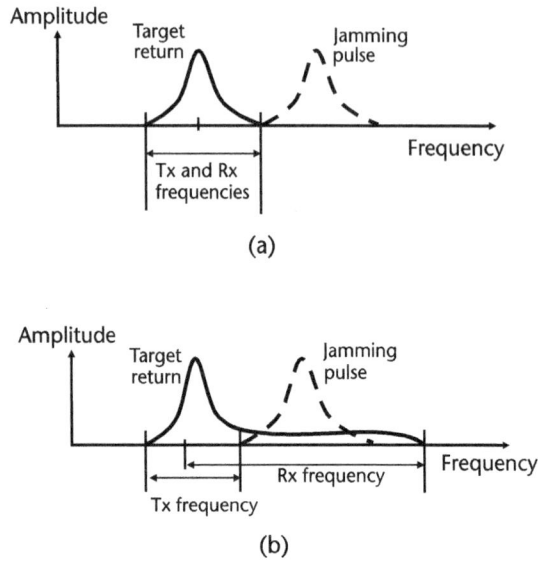

Figure 6.37   Skirt frequency jamming: (a) ineffective and (b) effective.

If the two components of the monopulse Rx are not precisely tuned, the target echo signal may be presented on the edge of the Rx's IF filter. Thus, the method offers an opportunity to inject a jamming signal into the edge, and the technique takes its name skirt due to this phenomenon. The skirt frequency jamming exploits this imbalance by sending a repeater jamming pulse tuned slightly off the radar's transmitted frequency and in the middle of the Rx IF filter, as seen in Figure 6.37(b). The jamming pulse produces a false error signal, driving the antenna away from the target return.

Skirt frequency jamming can be effective when precise knowledge of the internal operation of the IF filter is given. Typically, this information can only be obtained through exploiting the system. Differences in radar and frequency imbalance between various radar IF filters create uncertainty about the effectiveness of this technique.

• *Image frequency jamming:* Another technique that uses the weakness of monopulse radars is image frequency jamming. This method involves transmitting a signal at the image frequency of the radar's IF stage to generate an incorrect angular measurement. The jammer takes advantage of the frequency conversion process the targeted radar uses. The image frequency in the radar Rx is located at the same distance from the local oscillator as the target frequency but on the opposite side, as seen in Figure 6.38, which only shows the positive part of the frequency axis.

When the jamming false signal beats with the local oscillator (LO), the system produces an IF with a phase angle difference between the sum and difference channels of the monopulse radars that is opposite to the target return. This kind of jamming alters the polarity of the angular error correction voltages. This technique requires to know the IF information of the threat radar.

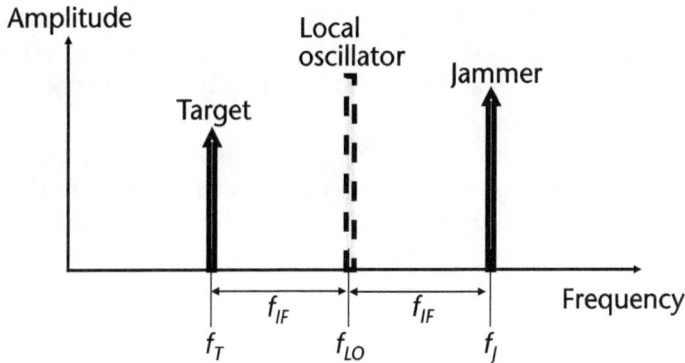

Figure 6.38   Frequency spectrum of image frequency jamming.

6.5.4   False Target

Generating false targets for jamming aims to confuse the threat radar systems. For this purpose, many false target returns are injected into the threat radar's detection and tracking systems. However, it must be noted that one of the target echoes emerges from the real target. False target deception jamming aims to cause the radar operator to not distinguish between false and actual targets for manual systems. However, the technique tries to prevent detection and track initiation and continuation in automatic systems.

At the technological level, the deceptive jammers that generate false targets are RF memory units, such as FML and DRFM. The deceptive jammer must be adjusted to the threat radar's frequency, PRF, and scan rate to produce false targets. The false target must appear on the radar scope exactly like a radar echo from the mother platform. Multiple false targets can be created farther away than the jammer by delaying the transmission of a jamming pulse until after receiving a radar pulse. False targets that are closer in the range can be generated by anticipating the arrival of a radar pulse and transmitting a jamming pulse before the victim radar pulse hits the mother platform. However, if the victim radar uses a jittered PRF, only targets that are farther away can be produced [13].

The deception jammer synchronizes its transmitted pulse with the victim radar's antenna pattern to generate different azimuth false targets. Thus, the jammer Rx must sense the main lobe and the sidelobes. Compared to the main beam, sidelobes are challenging to detect and analyze. For this reason, the Rx in the deceptive jammer must be sensitive enough to detect these sidelobes and not be saturated by the power in the main radar beam.

A false target deception jammer must inject a jamming pulse that looks like a target return into these side lobes. However, penetrating the radar side lobes requires much power. This unbalanced power utilization brings the power management process down. The false targets will be easily detected if a powerful jamming pulse is injected into the main beam. Generally, the jammers used for producing false targets vary the power in the jamming pulse inversely with the power in the received signal on a pulse-by-pulse basis. Thus, the weak signal powers from sidelobes are highly amplified, and the strong signal power that emerges from the main lobe will be moderately amplified. To generate false azimuth targets effectively, the repeater jammer must have an Rx with a wide dynamic range to detect both the main beam and the side lobes. In addition, the jamming system must generate high power that the Rx can effectively control.

The effectiveness of false target jamming is based on the reliability of the generated false radar returns. The technique will not be effective if the threat radar can easily distinguish between false and target returns. The false returns must look identical to the mother platform's echo. Thus, the false returns on the threat radar display should have the same intensity, depth, and width as a target echo. The signal properties of the fraudulent returns and target echo must be similar for automatic detection and tracking systems. The signal properties of the false returns and target echo must be identical for automatic detection and tracking systems.

6.6   Self-Protection Jammer Constraints in Dense Threat Environment

Generally, the radars engaged for the same missions use similar signal specifications regarding the time and frequency domains. Furthermore, the systems with the same roles produced in a country have similar operational specifications and signal processing characteristics. Different companies in the same country may build various systems. Also, their specifications may vary from one another. The resemblance in the radar systems' specifications gives some advantages to the logistics support of the systems. In this context, maintenance, repair, and spare parts support will be easier for the same or similar systems.

Facilitate warfighting readiness for the current armed forces by executing repair parts in the planning and supply chain management for more than a few thousand weapon systems. This situation is valid for almost all the modern armed and multinational allied forces. Thus, logistics support of the assets will be a challenging problem. Establishing an optimal, controllable, sustainable, and computer-aided integrated logistic system (ILS) is essential. The diversity of the assets is critical for obtaining such a logistics structure and supply chain. Optimization can be provided when the asset variety and computation losses are low. Furthermore, the physical warehouses' size and number will be reasonable. The less developed and developing countries generally obtain military assets from developed countries through procurement projects. For this reason, they are forced to follow the same logistic rules as the modern ones.

Radar systems, air platforms, and warships sometimes use their own ILS structure rather than the general ILS pool for high-technology assets. Air platforms and warships may be considered a system of systems since they are composed of many subsystems that are systems in themselves. However, radar systems are different than others. Air defense systems consist of many radar systems. Additionally, they have a basic C3 or more complex C4ISR structure. C3 systems have communication infrastructure and data exchange capability to form the regional radar picture. However, C4ISR systems add additive sensors and assets to gather intelligence, surveillance, and reconnaissance data and combine the data into the regional radar picture. Thus, these systems provide more information to the decision-makers.

The primary sensors of any air defense system are land-based, shipborne, airborne E/W, and coastal surveillance radars. Also, the surveillance mode of the SAM radars supports these sensors. The tracking mode of the SAM and AI radars are engaged for the active part of the air defense systems. From this point of view, it would be suitable to deal with IADS systems composed of many radar systems rather than consider them separately. There are a high number of radar systems regarding the total IADS system. When all the radars in the same roles have the same or similar technical properties, an undeniable advantage for the logistics support of the defense system will be obtained. Now let us inspect the operational aspect of the situation.

In light of the explanations up to this point, detecting and jamming the same threat systems by SPEW systems require the same processing techniques. Thus, preprogramming of the RWR and SPJ systems becomes a resolvable problem. This is because the detection, identification, and jamming processes are similar for all radar systems in the operation region. However, establishing an IADS system based on radars with various technical specifications will be a challenging opponent for the SPEW systems. From the defense point of view, an IADS system with multiple radars and different technical specifications is a disadvantage for logistics but an advantage for operational purposes. However, the situation is reversed for the SPEW systems. Thus, a standardized defense system is an advantage, and a nonstandard system is a disadvantage for the SPEW systems.

A wide variety of threat radar systems are expected in a multinational, dense operational environment. Especially in crisis and war situations, what is known about their functional techniques and operational tactics will be more limited and more challenging to overcome. In addition, situations where existing old and modern systems are frequently encountered together, and SPEW systems have to cope with this situation.

Considering these preconditions and assumptions, the SPJ systems' constraints in a dense threat environment are digested in the following items:

1. Present-day air defense systems have very different radar systems with different ECCM techniques. Some are low probability of intercept (LPI), and others are agile in frequency and PRF. Changes in the antenna and transmitter design of the radar system gain LPI features. Low side beam level, low output power, power management, high duty cycle, irregular scan patterns, coded pulses, high processing gain, and wide bandwidth are among the features of LPI radar systems. Although these ECCM techniques affect the RWR and Rx systems, it will not be possible to jam without detecting the threat.

2. Occasionally, modern radars use different ECCM techniques against jamming, and different methods are required to jam them. When multiple threat systems track the SPJ mother platform, various jamming techniques must be applied simultaneously. As a result, the technical capabilities of SPEW systems and optimal PFM design are of paramount importance. For this purpose, the effectiveness of the jamming techniques must be studied separately and simultaneously in an EW support center before operations.

3. Using the old and modern radars together in the operation environment requires using ordinary noise and high-level deceptive jamming techniques

sequentially or simultaneously. This situation forces the SPJ system to have brilliant power management methods.

4. As stated before, similar roles bring similar technical properties to the radar systems due to physical limitations. This analysis is presented in Table 5.2. The emitter identification process gets harder when threat radar systems increase in number. If the PFMs have not been adequately prepared and similar threats are not defined with their detailed technical properties, they cannot be appropriately identified. This situation will result in RWR ambiguities and the application of wrong jamming techniques or jamming the allied systems (fratricide).

5. SWaP opportunities are higher in warships than on airborne platforms but are also limited. Fighters, helicopters, and UAVs mainly have limited SWaP capabilities. When the operation is conducted in a dense threat environment, obtaining sufficient jamming power amplitude and power management will be a compelling problem.

6. Although jamming a limited number of threats with the SPJ system is successful, the increased number of threats will reduce the success of SPJ systems. This ineffectiveness is not only due to the power level but also due to specification limitations of switching components and antennas.

6.7   Future Trends in SPJ Systems

With the development of technology, primary weapon systems, such as fighter aircraft and UAVs, helicopters, bombers, and warships, are becoming more expensive. Because of this, they and their crew must be protected at all costs. They are the main attack and interception assets and are occasionally the primary targets for hostile forces. The most dangerous threat systems are radar-guided missile systems for medium-range and long-range. For this reason, as of the time of this writing, this is not expected to change. In parallel with the technological developments of aircraft, helicopters, UAVs, and warships, radar and missile systems come a long way. As a result, the technological evolvement of the SPEW and SPJ systems must follow radar, air vehicle, and warship technologies.

Some properties are SPJ-specific, while others are related to SPEW systems. As stated, SPEW systems have some limitations due to the mother platform's computerization capabilities and SWaP capabilities. Due to the unprecedented pace of technological developments, a SPEW system's hardware is expeditiously obsolete. As a result, it is almost impossible to upgrade the operating system and the PFM structure of a SPEW system without changing the version of it.

Another critical disadvantage of the SPEW system is related to the mother platform. Different platforms have various SWaP-providing capabilities, and in terms of the radar section of SPEW systems, most of the SWaP requirement is due to the SPJ systems. The design of the SPJ systems is platform-dependent, and the operational requirements and potential threats are considered. For this reason, using a SPEW in a different platform is almost impossible. This situation concerns different platform types such as fighters, bombers, cargo aircraft, helicopters, UAVs, warships, and the various models of the same types. Say that it is impossible to use an SPJ system that is designed for a warship in a fighter. However, a modification process would be required to adapt an SPJ system intended for a specific fighter to use in another one.

The modification process may be challenging for on-board systems due to the placing of the subcomponents. The SPEW systems' input-output (I/O) requirements are compatible with the avionics data buses (i.e., MIL-STD-1553), so encountering an I/O adaptability problem is not expected in standard and modern systems. However, when an absolute SPEW system says that a system produced during the 1980s is used in a relatively new fighter, such as the 4.5 or fifth generation, this time, not only the I/O structure but also the computer and electronic structures are out of date. It will not be compatible with the aircraft's electronics and avionics systems. Furthermore, these SPEW systems' RWR limitations and the SPJ systems' response time will not conform with the maneuver capabilities of modern fighters. In these conditions, upgrading the SPEW system requires development studies and may be more complicated than designing and implementing a new system.

Thus, a significant future trend for SPEW systems may be a standardization for similar platforms, even though they are podded or onboard. It may not be possible to use SPEW systems developed for different generation platforms, but for similar platforms, such as fighters, helicopters, and warships, the I/O should be platform-independent. However, the fifth-generation and some fourth-generation fighters, and the new-generation helicopters use platform-specific SPEW systems. For this reason, when a country needs to start a SPEW system procurement project for an off-the-shelf system, they have to pay both for the system and adaption modifications. Alternatively, they must create a development project for design and implementation.

Another critical issue for developing or modifying the SPEWs system is the platform. The right to mount any subcomponents on the platforms belongs to the platform's manufacturer country. Thus, without permission from the producers, SPEW systems cannot be mounted on a fighter, helicopter, UAV, or warship. For this reason, making a future projection of a SPEW system is only possible by owning a country's self-systems. The future forecast for SPJ systems can be clustered within these constraints as functional and operational issues for SPJ systems:

1. The jamming processes for each threat system are defined in the PFMs, and PFMs are prepared and loaded according to the operation region. The main reason for this is the limited computer storage devices. However, high mobility platforms, such as fighters, bombers, and helicopters, may change the operational theater frequently. Especially for multinational operations and cross-border antiterrorist actions, their roles and task regions may vary daily and sometimes within a day. Thus, loading different PFMs between the tasks for these platforms may delay the overall operation and overload the planning process.

   To overcome this problem, a general PFM that contains the jamming techniques of all the possible threats may be a suitable solution. In this case, the PFM must include threat identification, which means a more extensive emitter library and a more complex signal processing method. When considering a general PFM, more than expanding the SPEW system's computer structure properties will be required. This requirement implies that the electronic components and the RF capabilities must be compatible with the defined PFMs. The first and most critical functional future trend for SPJ (also SPEW) systems is the capability to develop a general PFM. Every newly detected threat can be added to the PFM, and the SPEW system architecture is expected to have the capacity to extend to cover the latest threats.

2. When the number of threats increases, the simultaneous and sequential jamming of the threats will be a problem due to the power management and the application of different jamming techniques. Self-protection jamming is applied against specific threat radars in the PFMs. Most of the up-to-date SPEW systems can effectively jam a limited number of threats simultaneously. Even though this number may be stated by the producers up to 20 or 30 threats, there are also some physical limitations. From the computer's point of view, the hardware and software may be sufficient for the signal processing part of the system and can produce the required jamming signals at the necessary time intervals. Computer architecture may be examined in a simulation environment, but real-life limitations cannot be modeled in a simulation study.

Nonetheless, simultaneously jamming radars in different frequencies and from various aspects is a challenging problem for the RF and electronic components, even though computer architecture suits the situation. Simultaneous jamming of multiple frequencies may be possible for noise jamming from different sectoral antennas or phased array antennas. When using deceptive jamming, the threat radar's frequency, PRF, and scan rate are adjusted to produce false targets. Thus, only one threat radar can be deceived each instant if only one RF memory unit is used in the SPEW scheme. The present SPJ systems apply the jamming process within the definition of PFM. Current PFMs are prepared according to the importance of the threats. The operational and technical staff define this importance level, which cannot be changed during missions. However, the importance of threat only partially relates to the pre-studies and past experiences but also to the situation of the operation.

There are two future trends in this context: one is for the electronics, and one is for the PFMs. The SPJ systems, generally platform-specific, have gradually increased their capabilities and reached an alpha-plus level, but there is much to do. Typically, the used subunits in the SPEW systems, specifically in SPJ, are limited due to the SWaP requirements. In the future, by developing more capable solid-state devices, the subunit dimensions and power requirements will be reduced, and more stages can be located in smaller volumes.

The current PFM preparation process depends on prioritizing the threats, and during the mission, the importance weights of the threats remain the same. In the preparation stage, machine learning methods may optimize the PFMs and augment their successes. In the preparation, much information exists for the threat radars, and many different jamming features can be defined. In this situation, supervised machine learning techniques are appropriate in the development and optimization process. After making the required development studies on MATLAB or Python environments, PFMs may be tested on a hot-mock-up or *hardware-in-the-loop* (HITL) system. Thus, optimal prioritizing may be reached for a PFM.

Nevertheless, sometimes the least prior system may be the most danger-ous according to the situation and must be strived first. This time, an adaptive algorithm is needed to change the threats' predefined priorities during the mission. For this reason, using machine learning techniques in the jamming process is inevitable. Unsupervised techniques will be more suitable this time for the continuously changing environment and situation. These machine-learning algorithms can be used for power management and threat priority in SPJ systems. Also, they will adjust the power output and the number of false targets produced by the RF memory units.

3. All the parameters of the victim radar must be known and programmed to the PFM for effective jamming. A threat without programming to the PFM cannot be jammed at the present systems. However, there are some reasons that the PFMs do not contain information on the threat radar systems. These factors can be summarized as threat systems having high mobility, launching missiles from different and unexpected platforms, such as UAVs and *uncrewed surface vehicles* (USVs), lack of SIGINT studies due to insuf-ficient planning or equipment, and developing counterintelligence. Another functional future trend for SPJ systems is the ability to apply countermeasures against unexpected threats, which does not exist in PFM. This property may be provided by two means. If the information on the unexpected threat is known, remote programming capability during the mission is a further step in the pacer ware. However, if the threat is unfamiliar, a machine-learning system may be used to decide the jamming method. For this purpose, a comparison may be conducted with the known threats, and using the simi-larities between the known and the unknown threats, a supervised method such as SVM, ensemble methods, neural networks, naive Bayes, or regres-sion methods. Thus, a suitable jamming method may be adapted using the data for the known threats.

The operational future projection for SPJ systems is:

1. Today, UAVs are entering the operational plans in many different aspects. The missions given to the UAVs from military aspects have been surveillance, reconnaissance, SIGINT, relay, mapping, SAR, early warning, SIJ, and kami-kaze drones since the beginning of their usage. However, recent operational requirements force us to use these invaluable and practical vehicles in dif-ferent roles. When pilots and crew are at significant risk, UAVs are the only solution for the missions. Thus, the fighter (both for attack and AI), combat service support, logistics, and target designation roles have been added to the UAVs' missions. Almost all the UAV missions require high-cost payloads and avionics. Furthermore, medium and large military UAVs generally have new technology, high-cost engines, and unique fuselage composite materials.

    Although no staff loss is expected from the UAV missions, costs are crucial for conducting operations and warfare. Thus, using self-protection systems for UAVs will soon become compulsory. Many countries study for onboard and podded SPEW systems for UAVs. In this context, the UAV producer countries change the game, shape the theater, and import the UAV roles into the operational plans. A weapon system that provides such radical alteration in the operational fields requires additional protective accessories and force multiplier equipment. The SPEW system is a critical force multiplier system for UAVs. The most challenging part of the UAVs' SPEW systems is SPJ since the SWaP opportunities are limited. Furthermore, some roles of the UAVs are ultimately dangerous and at risk. For example, hovering is required for target designating, and slow motions are needed for surveillance and reconnaissance. In these cases, UAVs are at hazard from radar threats, and SPJ systems are the most effective solution to mitigate risks. As a result, it will become imperative to develop SPEW systems for UAVs shortly, which require high capacity and low SWaP requirements.

2. Another critical operational future projection for SPJ usage has become green in recent decades. This time, let us change our focus to another valuable but single-use-only equipment: cruise missiles. Cruise missiles are high-technology, valuable equipment but disposable. This means that the targets of cruise missiles are priceless for military purposes and national interests. For this reason, the success of the missile mission is of the utmost importance.

The IADS systems are sophisticated and do not miss any unexpected flying object, even in low altitudes. Furthermore, cruise missiles travel a long path and pass from different regions with different IADS systems composed of several radar threats. Recently, some countries have used SPEW systems to increase the possibility of cruise missile success. In the future, cruise missiles will be equipped with SPEW systems, especially with robust SPJ systems. The trade-off between the system's capabilities and the cost must be considered since the system will be disposable.

6.8   Problems

Problem 6.1: Assume that an attack helicopter is cruising through the 6,000-ft altitude to avoid the *man-portable air defense* (MANPAD) systems. The helicopter has a fully equipped SPEW system, and the SPJ system has noise and false target generation capabilities. The SPJ system has four antennas, which cover 360° azimuth. Each antenna covers a 90° sector in the azimuth with 5-dB gain, and within the 3-dB points of the main lobes (assume that 90°), the gain level is considered constant.

After entering the highland hostile territory, where LoS is limited, two different radars from the first sector (0° to 90° according to the helicopter's heading) and one radar from the third sector (180° to 270°) detected the helicopter. The radars are monostatic; the first two are E/W, and the other one is a surveillance radar of a SAM system. All the radars of the IADS systems are connected via a C3 link. Tracking may be formed by any SAM tracking radars in the IADS after continuous detection. The detected radars technical specs are given in Table 6.1.

The preset jamming starts to operate to prevent track initiation. The Tx can be used from 2 to 18 GHz. The total jammer Tx power is 200W, and the total jammer loss is 4 dB. Calculate the jammer efficiency for these specifications.

Table 6.1 Threat Radar Specifications

Radar	Operational Frequency (GHz)	ERP (dBm)	Rx BW (MHz)	Ground Projection Distance (km)	Total Radar Loss (dB)	RCS of the Target (m²)	Polarization Loss Between Jammer and Radar (dB)	Angle with Respect to the Heading (Degrees)
E/W radar 1	3	115	80	8.2	5	5	0	35
E/W radar 2	3.7	120	100	7.6	4.5	7.5	3	46
SAM Surv. Radar	4.2	95	75	4.5	4	9.5	0	205

Problem 6.2: A SAM system's tracking radar operates at 6.5 GHz. It allows for manual tracking with an experienced operator tracking and launching air targets. The tracking radar's Tx peak power is 150 kW, and the antenna gain is 25 dB. The sensitivity of the radar is −92 dBm. This tracking radar locks on a fighter aircraft with 5-m² RCS from its aspect angle. The fighter is at 6 km with the radar and has an SPJ system with an adjustable process gain RF memory loop. The jammer process gain (G_J) can be adjusted between 10 and 60 dB. The tracking is conducted from the radar scope, and the brightness of the echoes is proportional to the return power. The operator can sense higher changes than ±20% in the brightness levels and eliminate the false targets. The repeater's Rx antenna gain is 1.25 dB, and the Tx antenna gain is 5 dB. If the signal at the repeater jammer's Rx is higher than −20 dBm, the jammer saturates and gives a 55-dBm saturated signal (or noncoherent noise signal). Assuming that no polarization loss exists between the radar and jammer antennas, the total jammer loss is 4.5 dB, and the radar loss is 5 dB. What is the required jammer process gain for the defined conditions if the jamming can be done in a coherent repeater zone? Otherwise, what is the jamming effectiveness? (*Note:* Please refer to Figure 6.3 when solving the problem.)

Problem 6.3: Repeat Problem 6.2 if the fighter approaches within 3 km of the SAM system. (All values will remain the same; the only value that changes is the distance.)

Problem 6.4: A simulation study will be conducted to jam the air-to-air semiactive radar-homing missile operating at 5 GHz. The missile is launched from a fighter aircraft. The target aircraft has a 10-m² RCS, and after the target aircraft detects the missile, it starts to apply jamming. For this situation, a MATLAB simulation will determine the *J/S* for different distances and find the crossover distance. Please use suitable antennas for the simulation.

Problem 6.5: A linear array antenna is used in a UAV's SPJ system and is planned to be mounted at the front edge of the wings. The steering requirement of the array is ±30° horizontally. In the linear array antenna design, antenna elements with a gain of 5 dB will be used. The planned highest frequency of the antenna array is 6 GHz. The overall gain of the antenna does not want to fall below 15 dB. How many antenna elements should be placed in this case, and at what intervals? What is the boresight gain for this design?

Solutions

Solution 6.1: First, all the values are turned into linear cases:

$$G_{Ja(dB)} = 5 \text{ dB} \rightarrow G_{Ja} = 3.16, \quad L_{J(dB)} = 4 \text{ dB} \rightarrow L_J = 2.5$$

$$ERP_{E/W1(dBm)} = \left(P_{E/W1}G_{E/W1}\right)_{dBm} = 115 \text{ dBm} \rightarrow P_{E/W1}G_{E/W1} = 316.23 \text{ MW}$$

$$L_{E/W1(dBm)} = 5 \text{ dB} \rightarrow L_{E/W1} = 3.16, \quad L_{P1} = 1$$

$$ERP_{E/W2(dBm)} = \left(P_{E/W2}G_{E/W2}\right)_{dBm} = 120 \text{ dBm} \rightarrow P_{E/W2}G_{E/W2} = 1 \text{ GW}$$

$$L_{E/W2(dBm)} = 4.5 \text{ dB} \rightarrow L_{E/W2} = 2.81, \quad L_{P2} = 2$$

$$ERP_{SAM(dBm)} = \left(P_{SAM}G_{SAM}\right)_{dBm} = 95 \text{ dBm} \rightarrow ERP_{SAM} = 3,162 \text{ MW}, \quad L_{P(SAM)} = 1$$

$$L_{SAM(dBm)} = 4 \text{ dB} \rightarrow L_{SAM} = 2.51, \quad L_{PSAM} = 1$$

Since two antennae are used for barrage jamming, the total power is divided into two, and the jammer power per antenna is 100W. First, let us determine the barrage jamming BW. For this purpose, it would be suitable to put a 500-MHz safety BW margin arbitrarily on the starting and ending sides of the band,

$$B_J = (4.2 + 0.5) - (3 - 0.5) = 2.2G \text{ Hz} = 2,200 \text{ MHz}$$

Using (6.1) to obtain the jamming efficiency (J/S) for each radar separately:
For radar E/W1:
We may determine the total losses and the slant range between the E/W1 radar and the helicopter ($R_{E/W1}$). Then substitute the numeric values in (6.1):

$$L_1 = \frac{L_J L_{P1}}{L_{E/W1}} = \frac{(2.5) \times (1)}{3.16} = 0.8 \quad h = 6,000 \text{ ft} = 1,829m$$

$$R_{E/W1} = \sqrt{(8200)^2 + (1829)^2} = 8,401.5m$$

$$\left(\frac{J}{S}\right)_{E/W1} = \frac{P_J G_{Ja} 4\pi R_{E/W1}^2 B_{R1}}{ERP_{E/W1}\sigma_1 L_{E/W1} B_J} = \frac{(100)(3.16)(4\pi)(8,401.5)^2(80)}{(316.23 \times 10^6)(5)(0.8)(2,200)} = 8$$

$$\left(\frac{J}{S}\right)_{E/W1(dB)} = 9 \text{ dB}$$

It would be effective if the jamming's effectiveness is 0 dB (or 1 in ratio, which means $J = S$).
For radar E/W2:

$$L_2 = \frac{L_J L_{P2}}{L_{E/W2}} = \frac{(2.5) \times (2)}{3.16} = 1.58 \quad h = 6{,}000 \text{ ft} = 1{,}829\text{m}$$

$$R_{E/W2} = \sqrt{(7{,}600)^2 + (1{,}829)^2} = 7{,}817\text{m}$$

$$\left(\frac{J}{S}\right)_{E/W2} = \frac{P_J G_{Ja} 4\pi R_{E/W2}^2 B_{R2}}{ERP_{E/W2}\sigma_1 L_2 B_J} = \frac{(100)(3.16)(4\neq)(7{,}817)^2(100)}{(1 \times 10^9)(7.5)(1.58)(2{,}200)} = 0.93$$

$$\left(\frac{J}{S}\right)_{E/W2(\text{dB})} = -0.31 \text{ dB}$$

As we can conclude, the jamming is ineffective. The 500-MHz safety BW margins may be reduced to overcome this issue, but the jamming BW may not cover all the signals this time.

For the SAM radar:

$$L_3 = \frac{L_J L_{P(SAM)}}{L_{SAM}} = \frac{(2.5) \times (1)}{2.51} = 1 \quad h = 6{,}000 \text{ ft} = 1829\text{m}$$

$$R_{SAM} = \sqrt{(4{,}500)^2 + (1{,}829)^2} = 4{,}857.5\text{m}$$

$$\left(\frac{J}{S}\right)_{SAM} = \frac{P_J G_{Ja} 4\pi R_{SAM}^2 B_{SAM}}{ERP_{SAM}\sigma_3 L_3 B_J} = \frac{(100)(3.16)(4\pi)(4{,}857.5)^2(75)}{(3.162 \times 10^6)(9.5)(1)(2{,}200)} = 106.3$$

$$\left(\frac{J}{S}\right)_{SAM(\text{dB})} = 20.26 \text{ dB}$$

Thus, the jamming is effective.

Solution 6.2: Let us write the given values in linear form:
For the tracking radar:

$$f_0 = 6.5 \text{ GHz}$$
$$(G_{Tx})_{\text{dB}} = (G_{Rx})_{\text{dB}} = G_{\text{dB}} = 25 \text{ dB} \rightarrow G = 316.23$$
$$P_{Tx} = 150 \text{ kW}$$
$$L_{R(\text{dB})} = 5 \text{ dB} \rightarrow L_R = 3.16$$

For the SPJ and its platform:

$$\sigma = 5 \text{ m}^2$$
$$G_{J(\text{dB})} = 10 - 60 \text{ dB} \rightarrow G = 10 - 10^6$$
$$P_{J(sat)} = 55 \text{ dBm} \rightarrow P_J = 316.22\text{W}$$
$$L_{J(\text{dB})} = 4.5 \text{ dB} \rightarrow L_R = 2.82$$
$$G_{Ja(Rx)(\text{dB})} = 1.25 \text{ dB} \rightarrow G_{Ja(Rx)} = 1.33$$
$$G_{Ja(Tx)(\text{dB})} = 5 \rightarrow G_{Ja(Tx)} = 3.16$$

The first part of the solution is validation. For this purpose, the target echo power at the radar Rx is obtained by using the two-way radar equation (2.33). The aim is to find whether the return echo power is higher than the sensitivity.

$$P_{Rx} = \frac{P_{Tx}G^2\sigma c^2}{(4\pi)^3 R^4 f^2 L_R} = \frac{(150 \times 10^3)(316.23)^2(5)(3 \times 10^8)^2}{(1981.4)(6,000)^4(6.5 \times 10^9)^2(3.16)} = 19.7 \text{ pW}$$

$$P_{Rx(dB)} = -107 \text{ dB} = -77 \text{ dBm}$$

As seen from the result, −77 dBm > −92 dBm, the return signal is higher than the sensitivity level and can be detected and tracked by the radar. In the second part of the solution, the radar signal power that comes to the repeater Rx is calculated to find the jammer operational mode. To that end, the one-way radar equation (2.29) is used.

$$P_{Ja(Rx)} = \frac{P_{Tx}G_{Tx}G_{Ja(Rx)}c^2}{(4\pi R f)^2 L_{R(dB)}} = \frac{(150 \times 10^3)(316.23)(1.33)(3 \times 10^8)^2}{(4\pi \times 6,000 \times 6.5 \times 10^9)^2(3.16)} = 7.48 \times 10^{-6}$$

$$P_{Ja(Rx)(dB)} = -51.25 \text{ dB} = -21.25 \text{ dBm}$$

Thus, $P_{Ja(Rx)} = -21.25$ dBm < -20 dBm, and the operation is in the linear or coherent jamming region. We will make the calculations for the jamming processing gain. The acceptable limits are defined as ±20% of the return signal power. Thus, the acceptable jammer power limits at the radar Rx are defined as:

$$P_{Rx} = 19.7 \times 10^{-12} \text{ W} \rightarrow \begin{cases} P_{Rx(max)} = 23.64 \times 10^{-12} \text{ W} \\ P_{Rx(min)} = 15.76 \times 10^{-12} \text{ W} \end{cases}$$

Sending back the min value or close to it with the repeater may create very low power values at the radar Rx and remain below the sensitivity level. For this reason, selecting values close to the max value will be suitable. Let us choose J/S as 1.1 arbitrarily. Using (6.5), the required jammer process gain (G_J) can be obtained as follows:

$$\frac{J}{S} = \frac{G_{Ja(Rx)}G_J G_{Ja(Tx)}c^2}{4\pi\sigma f^2 L_J L_{pol}}, \quad \frac{J}{S} = 1.1$$

$$1.1 = \frac{(1.33)G_J(3.16)(3 \times 10^8)^2}{4\pi(5)(6.5 \times 10^9)^2(2.82)} \rightarrow G_J = 21,759$$

$$G_{J(dB)} = 43.4 \text{ dB}$$

Solution 6.3: Since the distance gets closer, the sensitivity inspection is not required and is assumed to be valid. For this purpose, the radar signal power that comes to the repeater Rx is calculated to find the jammer's operational mode.

$$P_{Ja(Rx)} = \frac{P_{Tx}G_{Tx}G_{Ja(Rx)}c^2}{(4\pi Rf)^2 L_{R(dB)}} = \frac{(150 \times 10^3)(316.23)(1.33)(3 \times 10^8)^2}{(4\pi \times 3{,}000 \times 6.5 \times 10^9)^2 (3.16)} = 3 \times 10^{-5}\,\text{W}$$

$$P_{Ja(Rx)(dB)} = -45.23\,\text{dB} = -15.23\,\text{dBm}$$

30

Thus, $P_{Ja(Rx)} = -15.23$ dBm > -20 dBm, and the operation is in the repeater non-coherent jamming or saturation region. In this situation, the signal at the repeater jammer's Rx is higher than -20 dBm; the jammer saturates and gives a 55-dBm saturated signal.

Solution 6.4: The MATLAB code in Table 6.2 is prepared for the defined conditions.

Table 6.2   MATLAB Code for Jamming the Air-to-Air Semiactive Radar-Homing Missile

```
%% Jammer Simulation for Problem 6.4
clc, clear, clf
%% Parameters
AggressorSpiralGain = 3; %dB
VictimSpiralGain = 3; %dB
MissileHornGain = 7; %dB
AgressorSpiralPower = 40; %dBm
VictimSpiralPower = 40; %dBm
RCS = 10; %m^2
K1 = 92.45; %dB
%% K1 is the dB compensation coefficient when the frequency is used
%% in GHz and the range in km
K2 = 21.46; %dB
%% K2 is the dB compensation coefficient when the frequency is used
%% in GHz, and the RCS is in m2
Freq = 5; %GHz
v_delta = 60;
lambda = 3e8 / (Freq*1e9);
DopplerFreq = 2*v_delta/lambda;
S_mean_dB = zeros(1,10000);
J_mean_dB = zeros(1,10000);
%% RCS Gain
RCSGain = 10*log(RCS) + 20*log(DopplerFreq) + K2;
%% Jamming and Signal Powers Seen From Missile RX
index = 1;
for DistanceAggressorToVictim = 1:10000 %km
S_dB = 0;
J_dB = 0;
FSLAggressorToVictim = 20*log(Freq*DistanceAggressorToVictim) + K1;
innerloopcount = 0;
```

Table 6.2   (*continued*)

```
for DistanceVictimToMissile = flip(1:DistanceAggressorToVictim)
FSLAVictimToMissile = 20*log(DopplerFreq*DistanceVictimToMissile) + K1;
S_dB = S_dB + AgressorSpiralPower + AggressorSpiralGain + MissileHornGain +
RCSGain...
- FSLAggressorToVictim - FSLAVictimToMissile;
J_dB = J_dB + VictimSpiralPower + VictimSpiralGain + MissileHornGain ...
- FSLAVictimToMissile;
innerloopcount = innerloopcount + 1;
end
S_dB = S_dB/innerloopcount;
J_dB = J_dB/innerloopcount;
S_mean_dB(1,index) = S_dB;
J_mean_dB(1,index) = J_dB;
index = index + 1;
end
%% Plotting J & S
semilogx(1:10000,S_mean_dB, 'k')
hold on
semilogx(1:10000,J_mean_dB, 'k--')
grid on
legend('Echo', 'Jammer');
xlabel('Range from Radar to Target (km)');
ylabel('Power at Radar Rx (dBm)');
title('Jamming and Echo Powers at Rx vs.Range');
%% Crossover Distance
Rcs = (AgressorSpiralPower + AggressorSpiralGain + RCSGain...
- VictimSpiralPower - VictimSpiralGain - K1 - 20*log(Freq))/2;
```

When the simulation is run, the graph in Figure 6.39 is obtained. As seen in Figure 6.39, the jamming signal at the radar Rx is higher than the echo signal up to the crossover point. The crossover range is approximately 36 km. Also, note the slopes of the jamming and echo signals. While the jamming signal (the dashed line) has a –20-dB/decade slope, the echo signal has a –40-dB/decade slope. This is because of the jamming signal's one-way link and the radar signal's two-way link.

Solution 6.5: First, to solve the problem, the highest possible value of the space between the antenna elements must be found. For this purpose, we have to calculate the wavelength for the highest operating frequency.

$$\lambda = \frac{c}{f} = \frac{3 \times 10^8 \text{ m/s}}{6 \times 10^9 \text{ Hz}} = 0.05\text{m} = 5 \text{ cm}$$

The most significant possible spacing between the antenna elements is half the wavelength to avoid grading lobes. Figure 6.40 shows a possible linear array configuration in a UAV wing. In this case, the distance between antenna elements will be, at

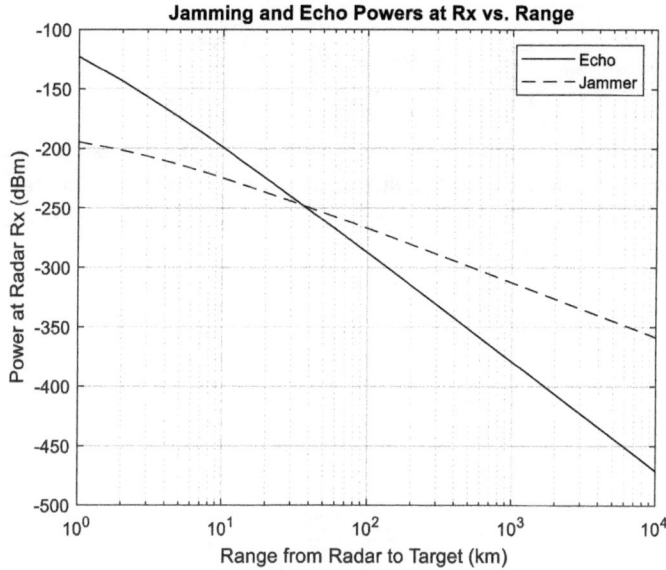

Figure 6.39 The simulation results for the air-to-air semiactive radar-homing missile.

most, 2.5 cm. Since the orientation is not desired above 45°, choosing a spacing of 2.5 cm would be appropriate.

Deviation from the boresight causes a reduction in gain with the cosine of the steering angle (φ_s). The number of antenna elements is obtained by using (6.16). The antenna array's gain is to remain at least 15 dB for ±30°. This equation is solved for the number of antennas (N) by choosing the angle value of 30°.

$$G_{\varphi_s} = 10\log_{10}(N) + G_e + 10\log_{10}(\cos\varphi_s)$$
$$15 = 10\log_{10}(N) + 5 + 10\log_{10}(\cos 30°) \rightarrow N = 11.54$$

The number of antennas is chosen as $N = 12$ so that the number of antennas is greater than 11.54 and the closest integer. Thus, using the boresight gain equation (6.14),

$$G_{BS} = 10\log_{10}(N) + G_e$$
$$G_{BS} = 10\log_{10}(12) + 5 \rightarrow G_{BS} = 15.8 \text{ dB}$$

The beamwidth is obtained for this condition using (6.12).

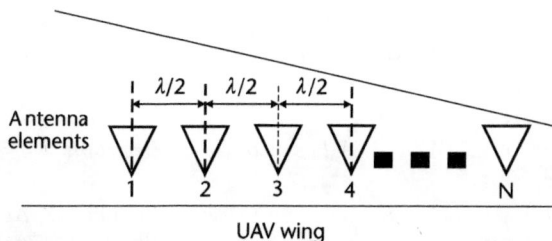

Figure 6.40 The linear antenna array localization in a UAV wing.

$$\theta_{3dB} = \frac{\left(\theta_{3dB}\right)_e}{N}$$

In this equation, the 3-dB beamwidth information is required. To calculate the antenna beamwidth, we assume that the antenna elements are horn antennas. Thus, we can use the following gain formula for an unsymmetrical 60% efficient horn antenna.

$$\left(\theta_{3dB}\right)_e = \frac{31,000}{\theta_{1d}\theta_{2d}}$$

and we can expand the assumption for a symmetrical usage as follows:

$$G = \frac{31,000}{\left(\theta_{3dB}\right)_{el}\left(\theta_{3dB}\right)_{az}} \rightarrow \left(\theta_{3dB}\right)_{el} = \left(\theta_{3dB}\right)_{az}$$

$$5 = \frac{31,000}{\left(\theta_{3dB}\right)^2} \Rightarrow \theta_{3dB} = \left(\theta_{3dB}\right)_e = 78.7°$$

We can revert to (6.12) and find the beamwidth as follows:

$$\theta_{3dB} = \frac{\left(\theta_{3dB}\right)_e}{N} = \frac{78.7}{12} = 6.56°$$

References

[1]   Joint Publication 3-85 (2020): Joint Electromagnetic Spectrum Operations, USA Joint Chiefs of Staff (CJCS), May 22, 2020.

[2]   Wiegand, R. J., *Radar Electronic Countermeasures System Design*, Norwood, MA: Artech House, 1991.

[3]   Schleher, D. C., *Electronic Warfare in the Information Age*, Norwood, MA: Artech House, 1999.

[4]   De Martino, A., *Introduction to Modern EW Systems*, 2nd ed., Norwood, MA: Artech House, 2018.

[5]   Golden, A., *Radar Electronic Warfare*, New York: American Institute of Aeronautics and Astronautics, 1987.

[6]   Burnside, W. D., et al., "Curved Edge Modification of Compact Range Reflector," *IEEE Transactions on Antennas and Propagation*, Vol. AP-35, No. 2, 1987, pp. 176–182.

[7]   Balanis, C. A., *Antenna Theory Analysis and Design*, 4th ed., New York: John Wiley & Sons, 2016.

[8]   Adamy, D. L., *EW 101: A First Course in Electronic Warfare*, Norwood, MA: Artech House, 2001.

[9]   Haupt, R. L., and P. Nayeri, "Pulse Dispersion in Phased Arrays," *International Journal of Antennas and Propagation*, Vol. 2017, Article ID 5717641, 2017.

[10]   Tang, G., et al., *Techniques and System Design of Radar Active Jamming*, Singapore: Springer, 2023.

[11]   Chrzanowski, E. J., *Active Radar Electronic Countermeasures*, Norwood, MA: Artech House, 1990.

[12]   Neri, F., *Introduction to Electronic Defense Systems*, 2nd ed., Norwood, MA: Artech House, 2001.

[13]   Air Combat Command Training Support Squadron (ACC TRSS), *Electronic Warfare Fundamentals*, Nellis AFB, NV, 2000.

CHAPTER 7

Effective Use of the Self-Protection Jammer

A self-protection jamming system's practical use is somehow different from the jamming effectiveness. The jamming effectiveness of an SPJ system is the measure of the jammer (J) to the echo signal (S) power ratio (J/S) at the radar Rx. However, the effective use of an SPJ system is relevant to the optimal usage of the whole system in any circumstance. The SPJ systems' effective use is related to different factors; sometimes, they are interwoven. These factors can be classified as follows:

- Target-specific jamming techniques;
- Special ECCM properties of the targets;
- Similarities between the threat systems;
- Operational fields' effects;
- EM density of the environment;
- SPJ system aspects.

While many sources inspect the technical properties of SPJ systems, the real secret is their practical application. This practical usage is not solely based on technical aspects but also tactical and operative foundations. Understanding the technical interpretation of the tactical requirements for SPJ systems is equally important.

In this chapter, we delve into the effective usage of SPJ systems from various operative aspects and factors. The chess-like nature of EW, with opponents applying tactical methods, underscores the importance of understanding these aspects. Operational requirements define the technical specifications for EW systems and lead to the development of state-of-the-art products. While operational requirements are critical for every EW system, our analysis focuses on SPJ systems. By examining the tactical and operative parts of SPJ in technical terms, we can gain a deeper understanding and effectively use these systems.

7.1 Developing Specific SPJ Techniques

SPJ is generally different than the support jamming techniques. This is because deceptive jamming occupies a vital part of SPJ compared to support jamming, composed mainly of noise jamming. Also, the geometry of the jamming and mathematical models are different for them. These factors were described in Chapter 3 in detail.

The development of target-specific SPJ techniques is not a solitary endeavor. If the threat radar type and specifications are known, developing a SPJ technique is possible using one or more methods in the correct sequence. Noise and deceptive SPJ techniques were detailed in Chapter 6. After analyzing the threat radar, a theoretical study can be conducted, and the findings are tested on a hot mock-up or a hardware-in-the-loop (HITL) system. These tests are typically performed in an *EW support center* (EWSC) or an equivalent facility. Developing a specific SPJ method is a collaborative effort involving engineers and various other parties.

One must understand the parties' roles to understand the development of the SPJ method. These parties are intelligence, operations, and communications-electronics staff. The parties' responsibilities for the jamming method development are sometimes specifically for self-protection and sometimes for general cases. The parties' roles in SPJ can be summarized as follows. However, countries and organizations may have various military insights, and the branches' roles and tasks can differ.

- Intelligence staff:
  - Tasking SIGINT units with the radar threat and C3 intelligence collection plan.
  - Providing radar systems' intelligence on locations, C3 structure, and technical capabilities.
  - Assisting in the preparation of the intelligence-related portion of the EW plans.
  - EW intelligence will be disseminated to the operational and the communicating-electronics branches to update the operation plans and plan the required SPEW system PFM preparations and SPJ developments.
  - Maintaining appropriate EW databases [1].
- Operations staff:
  - Determine the SPEW system requirements according to the operations plans and define the PFM preparation and SPJ development requirements.
  - Tasking the EWSC to develop SPJ techniques against required threat radars.
  - Make trials to exercise the developed jamming methods on SPJ performance.
  - Updated SPEW system performances are integrated into the EW appendix of the operation plans.
- Communications-electronics staff:
  - Keep up with the proper staff and equipment at the EWSC.
  - Planning to develop the PFMs and SPJ methods and keep them current according to the SIGINT reports and platform conditions.
  - Coordinate the development of the required SPJ methods with operational, intelligence staff, and scientists experienced in EW.
  - Make the tests for the prepared PFMs and the developed SPJ methods on hot-mock-up or HITL systems.

Under this information, a generic flow diagram for the SPJ development process from the requirement to test is given in Figure 7.1.

The SPJ process starts with detecting new radar-guided threats by SIGINT or ES systems. SIGINT units conduct threat detection activities according to intelligence

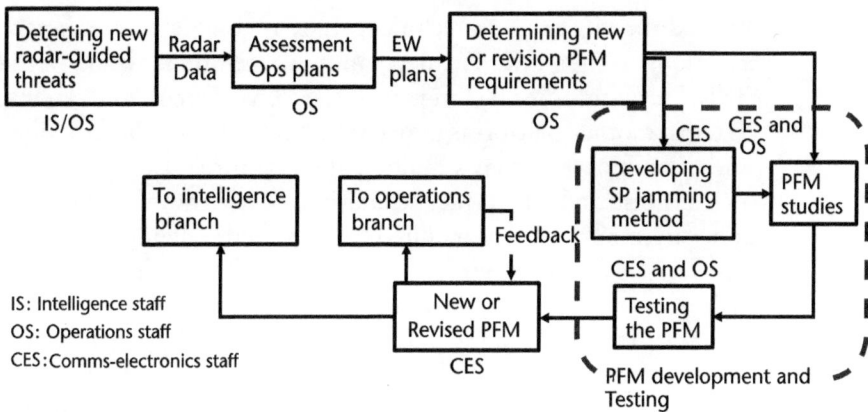

Figure 7.1   SPJ development process.

sense perception and intelligence plans. ES systems generally do not acquire data with the required resolution to analyze and characterize the RF sources but only detect and classify the sources stored in the EID database. However, most modern ES systems have additional property to obtain and store unexpected threat signals undefined in their EID list in an operation region. The unexpected signals can belong to an unknown threat or belong to a known one but could not have been precisely located before. For this reason, the assets may belong to the intelligence or the operations branch.

The SIGINT units can further analyze and inspect the signals, and operational units can analyze the threats by considering their location and operational roles. Thus, only the intelligence branch analyzes and obtains the radar data that will be sent to the operations branch. After taking the radar data, the operations branch inspects the effects of the new information on the operational plans. The changes will be fulfilled in the EW appendix of the operation plans. These amendments cause the preparation of new PFMs or revisions to them. The operations branch determines the PFM requirements and sends them to the EWSC. Studies of SPJ techniques occasionally require adding new radar threats to the PFMs.

The technical staff of the communications-electronic branch works to develop and test the jamming techniques. This development process is a subprocess of the PFM development and testing process. Developing a jamming technique against a radar system and adapting it to a PFM should include some phases.

- *Phase 1:* This phase aims to develop an efficient jamming technique for a specific radar system. This phase requires technical study, simulation study, effectiveness analysis, and field tests. The central part of the jamming technique development is conducted in this phase. However, more attention needs to be paid to coordinating the jamming method with the other jamming techniques in the PFM since it is an isolated study from PFM. This coordination is fulfilled in the second phase. The effectiveness test of the jamming method is also realized separately from the PFM. These primarily contain laboratory tests, such as those conducted on hot mock-ups and HITL. Also, they may cover the field trials in the test fields or theater.

- *Phase 2:* In this phase, a technical study is conducted to adapt the jamming technique to the PFM. The new jamming technique would not be used solitarily but with other techniques in the PFM. During operations, the SPEW systems' mother platforms may encounter different combinations of threats. These combinations may simultaneously require different jamming methods, such as noise and false target generation. Furthermore, false target generation or noise jamming may be required to be utilized contemporaneously for various targets. The threats would probably be at different distances and directions; their operating frequencies and operational methods differ.
- *Phase 3:* After developing and adapting a jamming technique to the PFM, the aim is to test the overall PFM in the laboratory and the field with various scenarios. These tests assess the overall PFM regarding jamming techniques' effectiveness and coordination.

The developed jamming technique is imported into a new or revised PFM. At the end of the process, the latest product is a new or revised PFM, passed to the operations branch for use in operations. The bugs and discrepancies of the PFMs detected during the missions are reported to the technical staff for revision. Moreover, the PFMs are sent to the intelligence branch to archive and compare with the changing threats in the related regions.

7.2   Operational Field Aspects

The theaters' landform structures affect not only the operational methods but also the RF behavior. The SPEW systems are used on airborne and shipborne platforms, and the RF-guided threats are launched against them from land-based, shipborne, and airborne platforms. Figure 7.2 demonstrates the possible scenarios for SPEW and corresponding threat radar platforms. It can be readily seen that the radar-guided systems are engaged in the mother platform of the SPEW systems with a discriminative background. For this purpose, the first assumption is that radars and jammers use high-gain antennas. Thus, they have limited main lobes spreading on both sides of the boresight. Naturally, the threat radar antenna gains are higher than the jammer antennas.

The effect of the SPJ systems' performance due to the operational fields can be classified according to the battlefield's environmental impact and EM density. Although environmental effects combined with the noise and false target deceptive jammer might differ, they can be calculated similarly. The environmental and EM density effects are dealt with separately. However, mentioning the clutter issue would be essential before determining the impacts on the operation area.

7.2.1   Clutter Effects on SPJ Operations

Clutter refers to any object that produces unwanted radar signals that can disrupt normal radar operations. When these signals enter the radar through the antenna's primary lobe, they are called main lobe clutter. Otherwise, they are known as sidelobe clutter. Here, we assume that only the main lobe clutter is considered in the

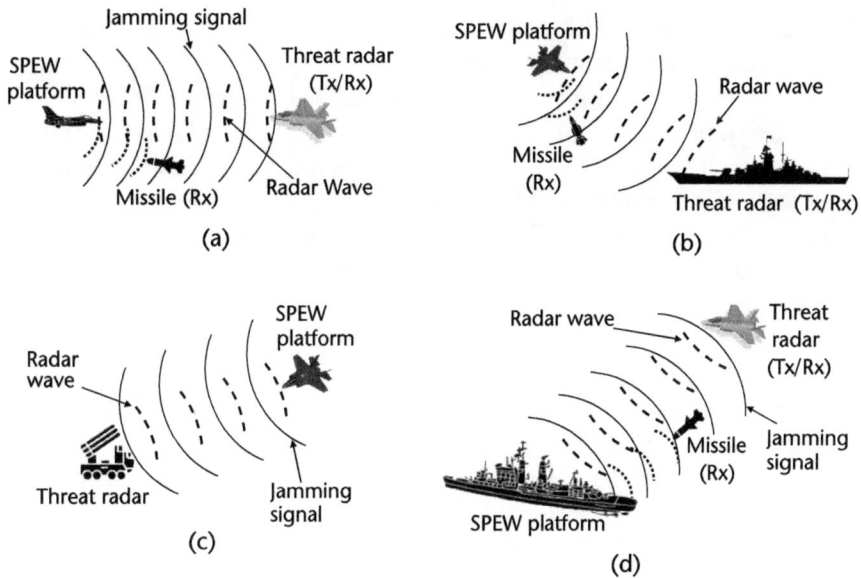

Figure 7.2   The possible scenarios for threat radar platforms versus SPEW platforms.

calculations. Furthermore, it is thought that no additional clutter-elimination process is utilized in the radar systems for simplicity. Clutter is categorized into surface (or area) and airborne (or volume) clutter [2].

The clutter signals measured at any region include a deterministic component dependent on the RF-illuminated area's physical features. For example, a scene's mountains, sea surface, rivers, lakes, or artificial structures do not move. When radar data is collected over the same scene on different days, these measurements give a common deterministic component. Additively, there will also be a random component, due to other variations such as the swaying of trees and waves on the water's surface.

Random parts of the clutter are signal-dependent and, more importantly, are not similar to thermal noise because spatial and temporal correlations characterize them. For this purpose, several families of distributions can be used to fit the observed amplitude statistics over a wide range of conditions, including the log-normal, the Weibull, and, especially, the compound Gaussian model.

The individual clutter components, known as scatterers, have random phases and amplitudes. In most cases, the level of clutter signal is much higher than that of the Rx noise level. Therefore, a radar's ability to detect targets hidden in a background of high clutter depends on the *signal-to-clutter ratio* (SCR) rather than the SNR. White Gaussian noise usually introduces the same noise power across all radar range bins. However, clutter power may vary within a single-range bin. Since clutter returns resemble target-like echoes, a radar can only differentiate target returns from clutter echoes based on the target RCS (σ_t) and the expected clutter RCS (σ_s) via a clutter map. The clutter map method is very compatible with designs made with CFAR architectures.

The *clutter map constant false alarm rate* (CMAP-CFAR) is a signal-processing technique to detect targets in cluttered environments. It is a CFAR algorithm that adjusts the threshold for detecting targets based on the level of clutter in the surrounding environment. Specifically, the detector output of each range-resolution cell is averaged over multiple scans to estimate the background level. The CMAP-CFAR processor uses digital filtering to update the background power estimate corresponding to each map cell in every scan. Studies have been conducted on CMAP-CFAR detection systems that explore various background distributions. The main advantage of using a CMAP-CFAR algorithm is that it can help to reduce the number of false alarms generated with the radar system. This situation is critical in scenarios where the radar operates in a cluttered environment [3].

Determining the J/S ratio is insufficient to describe the whole interference process. Generally, the noise + interference-to-signal ratio (N + I/S) refers to the general interference effects on the signal. The term interference includes intentional and unintentional disturbance. Ultimately, the total interference-to-signal ($\Sigma I/S$) ratio determines the total corruptive impact of the interference on the radar. The interference can be unintentional emissions by friend or foe, jamming, noise, or clutter. Also, some other EM interference stems from motors, generators, ignitions, or cell services. If the signal's power at the Rx is S_t, the area clutter power is S_C, the volume clutter is S_V, and the jamming noise effect is J. At this stage, let us assume that all the interference signals except clutter are clustered under J, and the noise effect occurs under the S_t. Thus, the total interference-to-signal ratio is defined as follows:

$$\frac{\sum I}{S_t} = \frac{S_C + S_V + J}{S_t} \qquad (7.1)$$

Area (surface) clutter: Surface clutter includes land and sea clutter, often called area clutter. Clutter RCS can be defined as the equivalent RCS attributed to reflections from a clutter area, A_C. The clutter RCS (on average) can be defined as:

$$\sigma_C = \sigma^0 A_C \rightarrow \left\{ \sigma^0 = \frac{m^2}{m^2} : \text{unitless} \right\} \qquad (7.2)$$

In (7.2), σ^0 is the unitless clutter scattering coefficient (or area reflectivity), generally defined in dB, and A_C is the clutter area. The area clutter emerges in airborne threat radars in the look-down mode, as shown in Figure 7.3. In the figure's scheme, the grazing angle (ψ) is the angle from the earth's surface to the boresight. ΔR, defined as $c\tau/2$, represents the range resolution of a pulse.

Area clutter reflectivity is characterized by its mean or median value of area reflectivity σ^0 (dimensionless), the probability density function (PDF) of the reflectivity variations, and their correlation in space and time. Many of the PDFs are applied to modeling σ as well. Examples include exponential, lognormal, Weibull, and K power distributions. The values of σ^0 of terrain observed by the radar are a function of the following:

- *Terrain type and condition:* surface roughness and moisture;
- *Weather:* wind speed and direction, as well as precipitation;

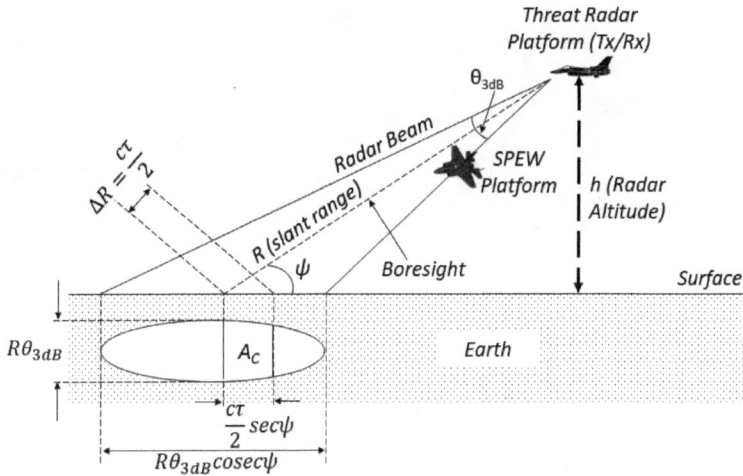

Figure 7.3   The surface (or area) clutter scheme for an airborne radar system.

- *Engagement geometry:* the engagement geometry, especially the grazing angle, results from the altitude of the airborne platform, the distance between the radar platform and the target, and the main lobe 3-dB beamwidth (directivity angle, θ_{3dB});
- *Radar parameters:* the operating frequency (wavelength) and polarization.

Thus, in the light of Figure 7.3, the clutter area is defined as follows:

$$A_{\text{C}} = R\theta_{3dB}\frac{c\tau}{2}\sec\psi \qquad (7.3)$$

where τ is the *pulse width* (PW) in seconds, c is the speed of light (3×10^8 m/s), R is the slant rate in meters, θ_{3dB} directivity angle, and ψ grazing angle. Both angles must be taken into calculations in degrees or radians.

Clutters only disturb a radar's operation if illuminated within the antenna's main lobe. The effective slant range extent of the resolution cell is less than the range resolution and the elevation beamwidth, as each is projected onto the scattering surface. The limiting factors could be the relative range values, resolution, and grazing angle. If the range resolution limits the effective extent, the resolution cell is considered *pulse-limited*, and if the main lobe extent is the limiting factor, it is called *beam-limited*.

Figure 7.4(a) shows the beam-limited case, and Figure 7.4(b) demonstrates the pulse-limited case. Let us assume ϕ_{3dB} and θ_{3dB} are the antenna's azimuth and elevation 3-dB beamwidths, respectively. Also, ψ is the grazing angle of the antenna boresight with the clutter surface. The range resolution is smaller than the projection extent in the pulse-limited case. The pulse-limited case ($A_{\text{P-L}}$) area is defined by the intersection of the beam and pulse width on the ground as follows [4]:

$$A_{\text{P-L}} = c\tau R\tan\left(\frac{\theta_{3dB}}{2}\right)\sec\psi \qquad (7.4)$$

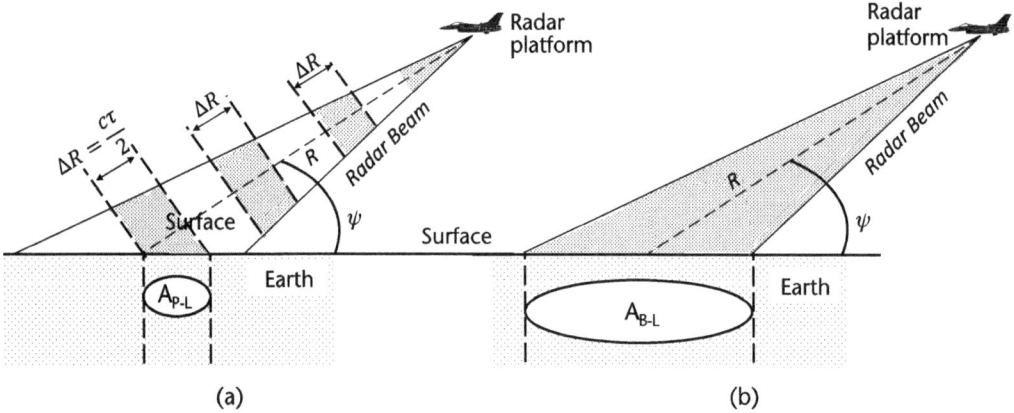

Figure 7.4 Relative geometry of (a) pulse-limited case and (b) beam-limited case.

In the beam-limited case, the area is defined by the union of the beam and pulse width on the ground. The range resolution is enormous compared with the projection of the vertical beam width onto the surface. In this case, the result for the beam area ($A_{\text{B-L}}$) on the ground is given by (7.5).

$$A_{\text{B-L}} = \pi R^2 \tan\left(\frac{\theta_{3dB}}{2}\right) \tan\left(\frac{\phi_{3dB}}{2}\right) \cosec \psi \qquad (7.5)$$

Generally, the area or clutter reflectivity (σ^0) varies significantly with grazing angle. It decreases rapidly at very low grazing angles and increases swiftly at very high grazing angles, with a milder variation in a middle flat or plateau region. Figure 7.5 is a theoretical diagram of this behavior.

The low grazing angle region ranges from zero to a critical angle (ψ_C). This critical angle is determined by the *root mean square* (RMS) height of surface irregularities in wavelengths. Rayleigh defines a surface as smooth if it is below the critical angle and rough if it is above. In other words, Rayleigh defines a surface as smooth

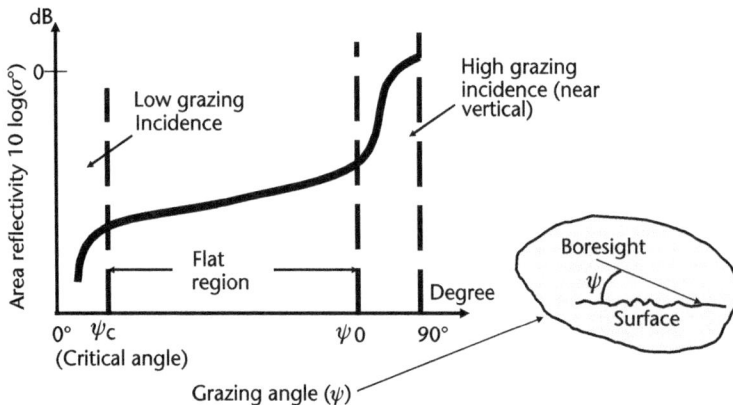

Figure 7.5 Relationship between the clutter scattering coefficient and the grazing angle.

if it is grazing at an angle below the critical angle [4]. The critical angle shown in Figure 7.5 is obtained using (7.6).

$$\frac{4\pi h_{RMS}}{\lambda}\sin\psi_C < \frac{\pi}{2} \Rightarrow h_{RMS}\sin\psi_C < \frac{\lambda}{8}$$

$$\psi_C < \sin^{-1}\frac{\lambda}{8\,h_{RMS}}$$

(7.6)

where h_{RMS} is the RMS height of the surface irregularities, and λ is the radar wavelength. In the low grazing incidence region, σ^0 increases rapidly with increasing incidence angle. Even rough terrain can appear smooth in this region if the Rayleigh roughness criterion is satisfied [5].

Example 7.1: Let us consider an airborne radar with an operating frequency of 3.2 GHz and looking to ground with a 5° grazing angle. For this situation, apply the Rayleigh roughness criterion and comment on the result.

Solution 7.1: The transmitted wavelength is 0.094m, and the Rayleigh roughness criterion is

$$h_{RMS} < \frac{\lambda/8}{\sin(5°)} = \frac{0.01175}{0.0087} \Rightarrow h_{RMS} < 1.35\text{m}$$

Therefore, an RMS surface roughness greater than 1.3m may still appear smooth to radar scattering.

When the critical angle exceeds, the flat region is encountered. The incident wave encounters surface irregularities in the flat region, so the dependence of the clutter reflectivity on the grazing angle is much less than at lower angles. In the flat region, the dependency on the grazing angle is minimal. Clutter in the high grazing angle region creates more coherent reflections than the lower grazing angles, and the diffuse clutter components disappear. In this region, the smooth surfaces are more significant than the rough surfaces, which is the opposite of the low grazing angle region. So the grazing angle is defined as the radar altitude (h) ratio to the slant range (R). As a result, the clutter reflectivity (σ^0) can be calculated using the following equation,

$$\sin\psi = \frac{h}{R} \rightarrow \sigma^0 = \gamma\sin\psi = \frac{\gamma h}{R}$$

(7.7)

The flat region of the area clutter can be inspected in two separate cases: the sea and the land. In the case of sea clutter, the clutter reflectivity (σ^0) according to the grazing angle (ψ) can be calculated as:

$$\sigma^0 = \gamma\sin\psi$$

(7.8)

In the above equation, γ represents the RCS characteristic of the illuminated area (or the clutter type) at the radar operating frequency and polarization. With another definition, γ is equivalent to the surface cross-sectional area intercepted by the beam as mapped onto the plane perpendicular to the line of sight between the radar and the ground. γ is thus approximately independent of the grazing angle for small angles. However, the tendency is in the direction of using σ^0 to describe the average radar reflectivity of surface clutter from the empirically obtained diagrams. The average clutter values depend on clutter type, grazing angle, frequency, and polarization.

Consider a wave incident on a rough sea surface, as shown in Figure 7.6. Due to the sea surface height irregularity, the rough path is longer than the smooth path by a distance of $2h_{RMS} \sin\psi$. This path difference can be represented as:

$$\Delta\psi = \frac{2\pi}{\lambda} 2h_{RMS} \sin\psi \qquad (7.9)$$

The critical grazing angle for sea clutter is computed when $\Delta\psi = \pi$, or the first null, as defined in (7.10). For this purpose, the RMS height irregularity of the sea surface ($h_{RMS(s)}$) is used.

$$\sin\psi_C \left(\frac{4\pi h_{RMS(s)}}{\lambda} \right) = \pi \rightarrow \psi_C = \sin^{-1}\left(\frac{\lambda}{4h_{RMS(s)}} \right)$$

$$h_{RMS(s)} = 0.025 + 0.046 S_n^{1.72} \rightarrow \{S_n: \text{sea state}\} \qquad (7.10)$$

where λ is the operating frequency of the radar.

The sea states, also called the international sea and swell scale, are defined in the Douglas Sea scale and are given in Table 7.1 [6].

The second case is land clutter, and when the radar's height is constant, $\sin\psi$ is proportional to the reciprocal of R ($1/R$). The constant gamma model represents clutter reflectivity [4].

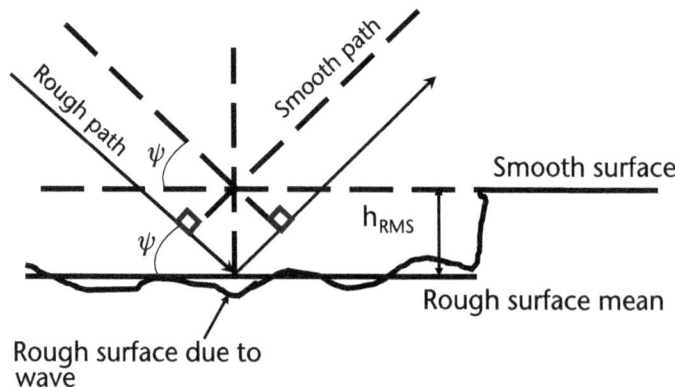

Figure 7.6   Rough sea surface description.

Table 7.1 Douglas Sea Scale Definitions

Sea State (S_n)	Height (m)	Description
0	No wave	Calm (Glassy)
1	0–0.1	Calm (rippled)
2	0.1–0.5	Smooth
3	0.5–1.25	Slight
4	1.25–2.50	Moderate
5	2.5–4	Rough
6	4–6	Very rough
7	6–9	High
8	9–14	Very high
9	>14	Phenomenal

$$\sigma^0 = \gamma \sin\psi = \frac{\gamma}{R} \tag{7.11}$$

This model predicts that σ^0 is maximum at normal incidence and becomes extremely small as the grazing angle tends to zero. However, as stated before, it does not account for the behavior of σ^0 when observed at zero to critical grazing angles. Furthermore, in the near-normal case or the high grazing angle region, the scattering becomes more directional and rapidly increases to a maximum value based on the reflectivity and smoothness of the clutter in a manner somewhat analogous to the behavior of the main lobe of a rough flat plate at near perpendicular incidence. In this region, specular reflection becomes dominant so that σ^0 increases rapidly with a grazing angle up to a maximum incidence of 90°. The magnitude of σ^0 at 90° incidence depends on the RMS surface roughness and the dielectric properties of the clutter.

The literature presents various predictive models for σ^0 as a function of critical parameters. One prevalent example with practical applications is the GTRI model from the Georgia Tech Research Institute [7].

$$\sigma^0 = A(\psi + C)^B \exp\left[\frac{-D}{1 + \left(\Delta h_{\mathrm{rms}}/10\lambda\right)}\right] \tag{7.12}$$

where ψ is the grazing angle in radians, and Δh_{RMS} is the RMS surface roughness. The parameters A, B, C, and D depend on the clutter type and radar frequency. Sample values for 10 GHz are given in Table 7.2 [5].

Considering the explanations for the area clutter, we can calculate the clutter power at the radar Rx as follows:

$$S_C = \frac{P_{\mathrm{Tx}} G^2 c^2 \sigma_C}{(4\pi)^3 R^4 f^2} \tag{7.13}$$

Table 7.2 GTRI Land Clutter Model Parameters for 10 GHz

Parameters	Soil/Sand	Grass	Trees	Urban	Dry Snow	Wet Snow
A	0.25	0.023	0.002	2	0.195	0.0246
B	0.83	1.5	0.64	1.8	1.7	1.7
C	0.0013	0.012	0.002	0.015	0.0016	0.0016
D	2.3	0	0	0	0	0

This equation is derived from (3.8) for a monostatic radar, and the Rx losses (L_R) are omitted.

Volume clutter: Volume clutter occurs in any radar and SPJ engagement. For this reason, this kind of clutter is consistently considered in the process. Figure 7.7 shows the resolution volume scheme for an airborne radar system. Volume clutter has large extents, including rain, chaff, birds, and insects. The volume clutter coefficient is typically expressed in square meters or RCS per resolution volume. Birds, insects, and other flying particles are often called angle clutter or biological clutter. Rain clutter can be suppressed by treating the droplets as perfect small spheres. Practically, using the Rayleigh approximation of an ideal sphere to estimate the rain droplets' RCS is common. An approximation for the volume RCS (σ_w) can be written without considering the propagation medium refraction index (m) as follows [2]:

$$\sigma_w = 9\pi r^2 (Kr)^4 : \text{for } r \ll \lambda \quad \left\{ K = \frac{2\pi}{\lambda} \right\} \tag{7.14}$$

where r is the radius of a rain droplet, λ is the radar's wavelength, and K is an approximation constant according to the radar's operating frequency. Even though (7.14) is an approximation, the volume clutter reflectivity is strictly related to the wavelength of a radar's operating frequency. Now let us obtain a more precise

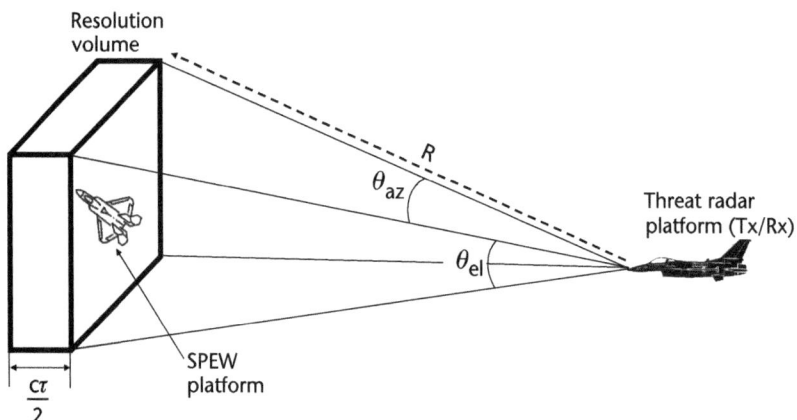

Figure 7.7 The resolution volume scheme for an airborne radar system.

volume RCS using. For this purpose, the total RCS of a single-resolution volume can be calculated as

$$\sigma_w = \sum_{n=1}^{N} \sigma_n V_w \rightarrow \left\{ V_w \simeq \frac{\pi}{8} \phi_{3\text{dB}} \theta_{3\text{dB}} R^2 c\tau \right\} \tag{7.15}$$

The equation is expressed as RCS per volume V_w with unit resolution. It is the sum of the RCS of all individual scatterers in the volume. Moreover, $\phi_{3\text{dB}}$ and $\theta_{3\text{dB}}$ are radar's antenna azimuth and elevation beamwidths in radians. N is the number of scatterers, and σ_n is their RCSs within the resolution volume. Additionally, R is the range between the radar and the volume, τ is the radar's pulse width in seconds, and c is the speed of light.

Consider a propagation medium with an index of refraction m. The nth rain droplet RCS approximation in this medium is as follows.

$$\sigma_n \simeq \frac{\pi^5}{\lambda^4} K^2 D_n^6 \rightarrow \left\{ K = \left| \frac{m^2 - 1}{m^2 + 2} \right| \right\} \tag{7.16}$$

where D_n is the nth droplet diameter. When the refraction index (m) enters the calculations, we can see that it is a function of both the temperature and wavelength. However, for radar operating frequencies between 3 and 10 GHz (wavelengths between 10 and 3 cm) and temperatures between 0°C and 20°C, the value of K^2 is approximately a relatively constant 0.93 for scatterers composed of water and 0.197 for ice [7].

Given the derivations and definitions for the volume clutter, we can calculate the clutter power at the radar Rx as follows:

$$S_V = \frac{P_{\text{Tx}} G^2 c^2 \sigma_w}{(4\pi)^3 R^4 f^2} \tag{7.17}$$

This equation is similar to the (7.13) derived for a monostatic radar. Also, the Rx losses (L_R) are omitted.

7.2.2   Effects of Operational Environments on SPJs

The operational environments directly affect the performance of the SPJ systems. This effect is added to the jamming effectiveness calculations using clutter effects. The clutter calculations are made according to the scenario. The probable scenarios for mutual operations of threat radar and SPJ platforms are given in Figure 7.2, and it is helpful to mention the situations separately. This way, the operational fields' effects on the SPJ system's performance can be discussed more clearly and elaborately.

1. *Threat radar and SPEW system are on air platforms:* Considering air-to-air engagement shown in Figure 7.2(a), the airborne platforms' altitudes are vital. When the guidance is semiactive, the missile (the threat radar Rx) is lower

than or at the same altitude as the target (or SPEW) platform, and the radar Rx's main lobe direction mainly covers the sky background. However, when command guidance is utilized, the radar's Rx is on the launcher platform, and the same situation exists when it is lower or at the same altitude as the target. Thus, no area (or surface) clutter (S_C) is expected, but volume clutter (S_V) is encountered. For this case, the jamming efficiency can be obtained by using (7.1), (7.17), (3.8), and (3.9) as follows:

$$\frac{\sum I}{S_t} = \frac{S_V + J}{S_t} = \frac{\left(\dfrac{P_{Tx}G^2c^2\sigma_w}{(4\pi)^3 R^4 f^2}\right) + \left(\dfrac{P_J G_{Ja} G c^2}{(4\pi R f)^2}\right)}{\dfrac{P_{Tx}G^2c^2\sigma_t}{(4\pi)^3 R^4 f^2}} = \frac{\sigma_w}{\sigma_t} + \frac{(ERP)_J (4\pi R^2)}{(ERP)_R \, \sigma_t} \qquad (7.18)$$

ERPs are the multiplication of power output and antenna gain. The J index shows the jammer, and the R represents the radar at the ERP. As seen from (7.18), an increment in the volume clutter increases the effect of the jammer efficiency. However, if the altitude of the threat radar Rx platform is higher than the target platform, the land or sea may enter the scenario in the background. Thus, the expected background clutter comprises both area and volume clutters. This time, the event is modeled by adding area clutter to the nominator of (7.18) as follows:

$$\frac{\sum I}{S_t} = \frac{S_C + S_V + J}{S_t} = \frac{\sigma_C + \sigma_w}{\sigma_t} + \frac{(ERP)_J (4\pi R^2)}{(ERP)_R \, \sigma_t} \qquad (7.19)$$

As seen from (7.17), the jamming effect at the threat radar Rx increases when more clutter enters the process. This result is natural since clutter and noise jammers cause a degradation in the radar Rx's operations. When their effects combine, increasing the jamming efficiency is not surprising.

2. *Threat radar is on a shipborne, and the SPEW system is on an air platform:* This scenario is given in Figure 7.2(b); the threat missile is firing from the sea surface to the airborne platform. When a target platform is flying from a low profile, entering the area clutter into the process is possible due to the land. However, this is a sporadic case for the open sea conditions. Thus, in this case, the radar is probably exposed to jamming and volume clutter, and the J/S calculation is as given in (7.18).

3. *Threat radar is land-based, and the SPEW system is on an air platform:* Figure 7.2(c) demonstrates this scenario, similar to the previous one. However, the engagement occurs on land now. The background comprises volume and area clutters for low-profile flying airborne platforms, and the J/S can be calculated using (7.19). However, for high-profile flights, the background is probably sky, and only volume clutter is considered for J/S calculations, as in (7.18).

4. *Threat radar is on an airborne platform, and the SPEW system is on a shipborne platform:* This situation is demonstrated in Figure 7.2(d). Whether

the threat radar's Rx is on the missile, as in semiactive guidance, or on the airborne launcher platform, as in the command guidance missile system, the background of the main lobe is the sea. Thus, the area clutter is added to the jammer to obtain J/S using the following equation:

$$\frac{\sum I}{S_t} = \frac{S_C + J}{S_t} = \frac{\sigma_C}{\sigma_t} + \frac{(ERP)_J (4\pi R^2)}{(ERP)_R \sigma_t} \tag{7.20}$$

To obtain σ_c using (7.2) in the above equation, σ^0 is calculated according to (7.7) by considering sea states in Table 7.1.

7.2.3   Electromagnetic Environment Survey

Two essential factors that affect the environmental effects of SPJ operations are the operational environment and the enemy and ally's RF threat systems' deployment to the field. The former relates to the environmental impact on radar performance, as discussed above. The latter is associated with forming the operational area's *electronic order of battle* (EOB). The EOB is a subset of the overall order of battle that consists of the identification, strength, command structure, disposition, and operating parameters of the RF-dependent systems. This includes radiating, receiving, and inactive systems within an operational area or those that could be readily deployed. The EOB is the identification of Tx and Rx in an area of interest, a linkage to the system and platform supported, a determination of their geographic location and range of mobility, and a characterization of their signals and parameters [8].

The EOB for an operation zone contains the radar threat systems as well. EOB forming is a mission dedicated to the intelligence units. They conduct this duty with the experienced staff and the assets attributed to them. They take support from the other units, such as operation and technical branches, when required. Furthermore, they seek to identify, catalog, and update the EOB of threats.

Their information can be extracted from the EOB when considering the radar systems in an enemy IADS for a probable operation region. This is the starting point of EW application planning. Thus, the PFMs and the required jamming techniques may be developed and prepared before an operation. However, the level of environmental RF density is critical. When focusing solely on the threat radar signal density, the operational limitations of the SPEW system will be an essential factor, both in identification and jamming purposes. The high number of radars in the operation area requires a complicated detection and identification process for RWRs and a sophisticated operation for the jammer systems. Thus, a dense radar environment is a challenging area for SPEW systems.

As discussed in Chapters 4 and 5, the RWR systems' transaction throughputs increase exponentially. Identifying radars with close technical specifications, such as operating frequency, PRF, and pulse width, will be an arising problem. Moreover, deinterleaving may be complicated if similar threat radars are in close directions. However, some critical issues listed below emerge for SPJ systems' performance in a dense threat radar environment.

- In the normal operation mode of the jammers, it is advantageous that the threat radars' operating frequencies are close together. A reasonable barrage jamming is possible without reducing the power to a predetermined level. However, this might be a challenging problem when these radars use different operation methods, such as pulsed, CW, FMCW, or pulsed Doppler.
- If the simultaneously required jamming methods are noise and false target generation, power management and resource allocation might be a problem both for close and distant frequencies. Furthermore, when noise jamming is used with methods that require continuity, such as RGPO and VGPO, conducting the process will be challenging whether the jamming frequencies are close or distant.

    When an SPJ jams different threats with different frequencies sequentially, a response time is required for each frequency shift. Thus, as the threat radar number increases, the jamming effectiveness and following the preprogramming sequence on time become more complex.

- A possible scenario for a high-density RF zone is distributed radars. Thus, in the region, a SPEW system detects the radars from different directions, and SPJ can jam them with a lower power simultaneously or sequentially. Simultaneous jamming from different directions causes lower power jamming. However, when jamming is conducted sequentially, the response time suffers due to switching directions to the other direction. Thus, a compromise emerges for jamming methods: using lower power without response time lagging or suffering response time lag without power reduction.

    Some critical issues are essential for SPJ systems to cope with problems in the dense radar signal environment. Some relate to the SPJs' architecture and some to tactical usage. The following items show the rules for effectively utilizing SPJ systems in a dense threat radar environment.

- An effective SPEW system suitable for dense RF environments should have an advanced multifunction capability through simultaneous threat identification and jamming operation. Fast and reliable RF emitter recognition in highly dense electromagnetic environments and of complex radar waveforms. Multithreat capability providing jamming effectiveness against simultaneous threats. An *active phased array* (APA)-based Rx/Tx with configurable ERP will be adaptable for this case.
- Multiple-threat jamming capability depends on the dimensions, weight, and power consumption of the SPEW system. Thus, a platform with high payload capability and power supply has higher multiple threat-jamming capabilities. This means that a warship has a higher ability than an airborne platform, and a cargo aircraft has a higher capability than a fighter aircraft.
- SPJ systems' capabilities have developed rapidly since the 1990s. Power amplifiers have a critical role in this development. TWT technology has vastly improved, increasing efficiency and decreasing volumes. Concerning volume, they have evolved from midi-TWTs (about 1 liter) to mini-TWT (about 0.2 liters) and finally to *microwave power modules* (MPMs), which several US firms have developed under a program started by the Naval Research Laboratory in 1990 [9]. Fighters' SPJ systems have been reached to jam up to 20

threat radars sequentially. However, fighters' simultaneous jamming capability is less than this number.

- When the number of radar threats increases, the SPJ systems must apply many different jamming methods simultaneously or sequentially. For an effective and successive jamming process, power management is crucial. For this purpose, optimal resource allocation methods have been used for many years and continue to develop new approaches.

- While developing jamming techniques, related application tactics should also be studied for potential multithreat scenarios. For this purpose, the support centers test the methods and tactical scenarios using a hot-mock-up set, and all the developed techniques and tactics must be tried at the military drills.

7.3 Threat Systems Aspects

More than threat radar systems' structures and operational methods, combinations of different radars affect the SPJ systems' performance. A radar may have a complex and state-of-the-art architecture, but there is a way to solve its weak points, and a suitable jamming technique can be developed. However, complications begin when more than one radar is encountered in the operation zone. Jamming two radars simultaneously will be possible if the effective jamming method is preset noise. The power level will decrease when their frequencies are discrete, but no additional complexity is added to the procedure. This situation will not change for more than two radar threats; however, the power level will decrease according to the frequency spectrum.

When noncoherent repeater jamming is the required jamming method exposed to more than one radar, it is impossible to jam radars simultaneously more than the number of repeaters. Deception techniques are utilized to jam more sophisticated radars effectively. Utilizing deceptive techniques in a threat radar system would be an ordinary process. However, applying deceptive jamming techniques to more than one radar will be a challenging problem. Say that a range deception, a velocity deception, and an angle detection method are to be applied simultaneously. This time, some performance problems arise from the SPJ, especially from the ECM generator and the power amplifier subunits.

Another area for improvement emerges from insufficient SIGINT information for the operation region. If not all the radar systems' technical and operational properties are determined clearly, it would not be a surprise to fail jamming. For this purpose, all SIGINT activities are to be conducted to obtain every specification and tactical usage of the radar systems at the operation zone.

7.3.1 Defining Potential Threat Systems

Detecting and identifying threat systems have a crucial role in SPEW systems. Chapters 4 and 5 explained the process from signal detection by an Rx to identifying a potential threat. These successive phases comprise a potential threat determination process. Furthermore, this process is essential for self-protection jamming. As stated before, all the identification and jamming processes are defined in PFMs. However, the success of a PFM depends on the threat system definition accuracy and details. For this purpose, the threat system technical specifications must be obtained by ELINT and ESM capabilities and further analyzed carefully to determine distinguishing characteristics. Thus, discriminative properties for exact identification can be obtained.

After determining the threat, its properties and weaknesses are exploited for jamming. For this reason, defining potential threat systems is a critical point in the identification and jamming processes. Although the capabilities and limitations of the SPEW systems are on one side of this event, the threat systems' specifications and technological levels are on the other side. The limitations generated by threat radar system features in SPJ can be grouped as follows:

- The technological level includes the enhanced matched filter, novel coherency methods, high efficiency, and Rx sensitivity.
- Adaptive Radar Resource Management within some physical and practical boundaries, such as energy, dwell time, and PRF.
- The EP (or ECCM) capabilities against jamming, such as jamming and clutter elimination methods and side-lobe blanking methods.
- The LPI capabilities to prevent detection by enemy ELINT and ESM facilities and jamming by enemy assets, such as frequency hopping, frequency chirp, and *direct-sequence spread spectrum* (DSSS).

7.3.2 Similarities in Threat Specifications

In a dense RF environment, multiple radar threats could likely engage with an SPEW mother platform. The complexity of the SPEW systems' detection and jamming functions increases with the growing number of radar threats. This situation is common for the air defense structures, such as a well-implemented IADS. Thus, the air attack packages and IADS are considered for this explanation. Assume that the number of radars in the operation area is very high, leading to RWR overload and delays in response. Moreover, threat identification would be a challenging problem when the radar parameters are similar. Even if a well-defined EID is utilized, there would be possible incorrect identifications due to the threats with similar properties.

This erroneous decision is dangerous in two aspects. The first one is operational since the reactions against each threat are definite, such as leaving all control to the SPEW system or maneuvering, manual chaff dispensing, and manual jamming. Thus, pilots' responses will be according to the RWR display, even though the threat is different. The second aspect is technical. If the emitter identification is wrong, the jamming will also be false, and the applied jamming will probably be ineffective. Moreover, unnecessary jamming will lead to redundant source usage.

When considering the sophistication of the technical properties of modern radar systems, encountering an exponential increment is not a surprise. In an operational region, both uncontemporary and modern systems may exist simultaneously. This case brings additive complications to the jamming process, applying straightforward and sophisticated techniques together. The straightforward jamming techniques include high-power preset noise, noncoherent repeater noise, and false target generation. The cross-eye, cross-polarization, RGPO, and VGPO are sophisticated jamming techniques. Applying two or more techniques simultaneously would always be challenging.

Thus, the SPEW systems' engagement of radars with different properties is challenging for RWR and SPJ systems. However, engaging multiple radars with similar properties is another problem. For this case, let us assume the IADS system consists of radar batches with similar properties, such as operating frequency and frequency agility, PRF, pulse amplitude, pulse width, and ECCM capabilities. Let us consider the air-attacking scenario given in Figure 7.8. In this mission, the attack group aims to devastate the IADS system radars and its C3 structure as much as possible. Assume that the attack group has SPJ systems, and each system is self-separate. Each radar type constitutes the defense system, and early warning (A), AAA (B), SAM (C), and AI have similar properties. The IADS system has a region control center, and all the air picture is fused here via a C3 system. Also, the accurate target information is distributed with the C3 to the defense weapon systems, such as AAA, SAM, and AI.

Similarities in defensive radars challenge the fighters' RWR systems to identify threats. Furthermore, increasing the number of threats means more errors and delays in the response time for the RWR systems. The degradation in RWR performance affects the SPJ systems' performance. However, if all the threats are identified accurately this time, simultaneous jamming for different radars will be a problem. Say that three same coherent radars with jamming noise-eliminating capability and two same radars that do not use any ECCM engage a fighter simultaneously. The SPJ system on the fighter is preprogrammed by a PFM to produce false echoes for the coherent threat radars and a noncoherent repeater jammer for the latter. If the PFM has been realized for multiple sequential or simultaneous jamming for false echo and repeater noise and prioritization has been defined, the jamming efficiency

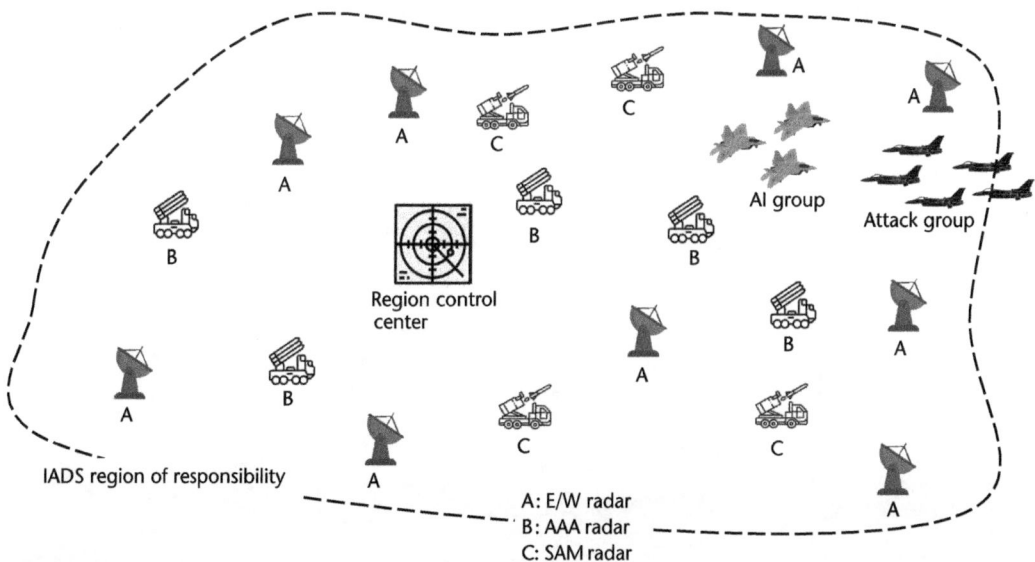

Figure 7.8 A possible scenario for an attack on a region with IADS.

may be high. The SPJ system must technically meet the PFM requirements for jamming success.

Another scenario, shown in Figure 7.9, is like the first one. An adversary's IADS system has an early warning (A), AAA (B), SAM (C), and AI, which have similar properties. The IADS system has a region control center, and all the air pictures are fused here via a C3 system. Also, the accurate target information is distributed with the C3 to the defense weapon systems, such as AAA, SAM, and AI. The difference between these scenarios is in the attacking force and its task. In this scenario, the attacking group aims to destroy a target, say a vital production facility for the defender's economy and defense industry. An SOJ aircraft supports the task group. With this appearance, the attack converges to a SEAD operation when the attacking force uses ARMs.

Let us focus on the SPEW system properties of the task group under the SOJ support. In this mission, the attacking group aims not to attack the air defense system but to reach a specific range and destroy the target. Thus, the defense system is a thwart rather than a target. The similar disadvantages of this scenario to the previous scenario can be counted as follows:

- The operation environment's RF structure is very dense, and one task force unit may be exposed to more than one threat simultaneously, making it challenging for the fighters' RWR systems to identify threats.
- Increasing the number of threats means more errors and delays in the response time for the RWR systems.
- The degradation in RWR performance affects the SPJ systems' performance. However, if all the threats are identified accurately this time, simultaneous jamming for different radars will be a problem.
- In a simultaneous multithreat engagement scenario, with coherent and jamming noise-eliminating capability and without any ECCM capability, the SPJ system on the fighter is preprogrammed by a PFM to produce false echoes and noncoherent repeater jamming simultaneously or sequentially. If the PFM has

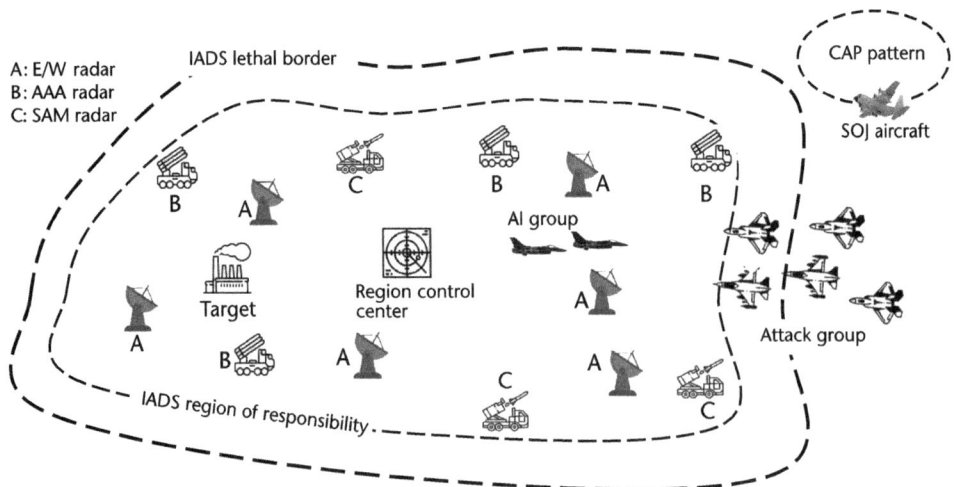

Figure 7.9 A possible scenario for an attack on a specific target by fighting down the IADS.

been prepared for multiple sequential or simultaneous jamming for false echo and repeater noise and prioritization has been defined, the jamming efficiency may be high. The SPJ system must be technically suitable to meet the PFM requirements for jamming success.

This scenario has similar handicaps and relatively more challenging mission objectives. However, the task force, such as SOJ aircraft, has necessary jamming support. This situation must be considered for operation planning, execution, and PFM preparation. In the condition of SOJ operation and the jamming plan being sufficient, a particular purpose of PFM can be prepared for fighters in the attack pack. Since SOJ applies a high-power wideband barrage jammer, the task force units do not have to produce noise jamming during the planned mission. This will degrade the detection of E/W and substantially degrade the track initiation and locking processes of SAM and AAM systems. Thus, the attacking group can reach the target through close-distance noise or deceptive jamming. However, the PFM's RWR properties should be the same as those of the regular case without SOJ support.

An essential point must be considered here. If the operational situation changes, such as ceasing the SOJ support, a backup PFM for the region must exist in the fighters' SPEW systems, and the pilots must switch to the PFMs to survive.

Threats are similar not only in the air-to-ground scenarios but also in the air-to-sea surface scenarios. This time, let us consider the scenario in Figure 7.10. In this scenario, the air task group attacks a carrier battle group. The main difference between this case and the others is that adversary forces comprise attack units. The carrier battle group is assembled to attack a location, and the owner of the related region forces attack to intercept them. Even though naval forces seem like the defenders, they are the invaders. Thus, the main task of the air group is in the air interception mission. In other words, the carrier battle group becomes hunted while as a hunter.

The air-attacking group aims to destroy the aircraft carrier capital ship and has escort jammer support. The geography is impressively different from the former scenarios. This time, only the horizon limits the LOS of the RF systems. Moreover, the warships have excellent hard-kill and soft-kill capabilities, probably because the

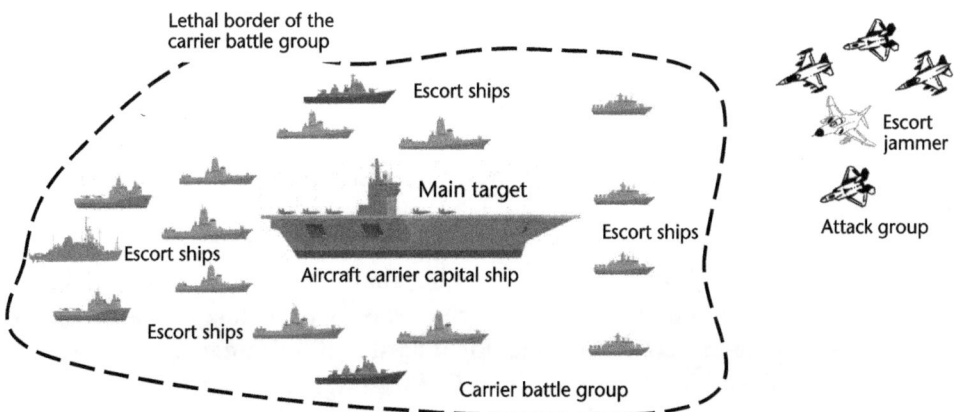

Figure 7.10 A possible scenario for an attack on a carrier battle group.

source of most EW principles emerged from the naval forces. Although the carrier battle group aims to reach the central operation region and has a smaller control airfield than an IADS, they have enormous destructive weapons, surveillance, and EW capabilities.

In this case, the air group must have well-studied, meticulously prepared, and repeatedly tested and studied PFMs in their SPEW system. Since the naval group's close cruise formation makes simultaneous engagement inevitable for the air task group, *escort jammer* (EJ) support would be a logical tactic. The EJ aircraft alleviates the load of the fighters' SPJ systems; however, they should still produce different types of noise and deception jamming simultaneously. Moreover, their RWR system capabilities must be as if they are alone.

7.3.3 ECCM Techniques in the Threat Systems

When discussing the threat system aspects for SPEW systems, ECCM is a critical matter to mention. ECCM covers all the techniques for reducing the effectiveness of an active or passive EW threat. So, this time, we changed our point of view from SPEW to radar systems. Some ECCM techniques aim to prevent radar systems from being detected by ESM systems, and some are useful in neutralizing the ECM, which degrades radar system operation. Hence, developing ECCM techniques requires intelligence about the active and passive EW systems. Weapon-related radar system designers have developed complex and effective ECCM techniques to prevent jammer systems from neutralizing radar-guided weapon systems. Furthermore, most of the ECCM techniques are generated against the SPEW systems.

Weapon system radars are primarily concerned with searching, acquiring, detecting, and tracking target returns in thermal noise, clutter, and interference. Thus, radars are potential targets of jamming by adversary forces, which can use active techniques to protect a platform from being detected and tracked by the radar. This is accomplished through two approaches: masking and deception [10].

Noise jammers are designed to hide targets by generating interference that blends in with the radar receiver's thermal noise. This decreases the radar sensitivity as the constant false alarm rate threshold increases to match the higher noise level. As a result, detecting the presence of jamming becomes more challenging.

However, deceptive jamming techniques aim to inject false information into the radar processor. Deception is the intentional and deliberate transmission or retransmission of amplitude, frequency, phase, or otherwise modulated intermittent or continuous wave signals to mislead the radar's interpretation or use of information [11]. Deception jammers receive, modify, amplify, and retransmit the radar's signal to generate false targets. Thus, they produce the mother platform's range, Doppler, and angle information at the radar's Rx far from the actual position.

As discussed in Chapter 6, deception-jamming techniques are enhanced by accurately replicating and reproducing the radar signal through DRFM. These techniques attempt to deceive radar systems by providing the radar Rx to process jamming signals coherently, making it hard to discriminate between true and false targets. RGPO/RGPI and VGPO/VGPI techniques are used in this context. These methods aim to mislead the radar in tracking mode, assuming that the target is tracked correctly. Thus, the tracker range or velocity gate must be removed from the target return. Given both RGPO/RGPI and VGPO/VGPI, proposing an effective ECCM method to suppress these jamming threats is crucial. Regardless of the ECCM technique, the radar must maintain regular operation under jamming conditions.

At the forefront of the defense against jamming threats, radar designers have developed crucial strategies known as ECCM. Depending on the main radar subsystem where they occur, these strategies can be categorized as antenna-related, Tx-related, Rx-related, or signal-processing-related [12]. Each of these strategies plays a vital role in ensuring the radar's resilience and effectiveness in the face of jamming. The ECCM effect is a radar system's signal processing gain (G_{SP}). For the SPJ (from the main lobe) of monostatic radars, the ECCM effect can be added to (3.10) with the $B_R/B_J = 1$ (radar Rx to jammer noise bandwidth ratio) assumption as given here:

$$\frac{J}{S} = \frac{P_J G_{Ja} 4\pi R^2}{P_{Tx} G_R G_{SP} \sigma L} \quad \left\{ L = \frac{L_J L_{pol}}{L_R} \right\} \tag{7.21}$$

The components of the equation can be defined as follows:

- P_{Tx}: peak radar Tx power (watts);
- P_J: peak jammer power (watts);
- G_{Ja}: jammer antenna gain in the direction of radar (ratio/unitless);
- G_R: radar antenna gain for Tx and Rx operation (ratio/unitless);
- R: range between radar to jammer (meters);
- L: total losses that enter the process, consisting of jammer losses (L_J), radar losses (L_R), and antenna polarization losses (L_{pol}) (unitless).

For the support jamming (from the main lobe) of monostatic radars, the ECCM effect is imported to (3.21) as given here:

$$\frac{J}{S} = \frac{P_J G_{Ja} 4\pi R_{R\text{-}T}^4 B_R}{P_{Tx} G_{Tx} G_{SP} \sigma R_{J\text{-}R}^2 B_J L} \tag{7.22}$$

The components in (7.22) that are not mutual with (7.21) are as follows:

- B_R = radar Rx bandwidth (MHz);
- B_J = Jammer noise bandwidth (MHz);
- $R_{R\text{-}T}$: radar to target range (meter);
- $R_{J\text{-}R}$: jammer to radar range (meter);
- G_{SP}: radar signal processing gain (unitless).

7.3.3.1   Antenna-Related ECCM

The radar antennas are the first component for protection against jamming. The antenna's directivity in the transmission and reception phases allows space discrimination to be used as an ECCM strategy. Techniques for space discrimination include antenna coverage and scan control, reduction of main beamwidth, low side lobes, *sidelobe blanking* (SL-B), *sidelobe cancelers* (SL-C), and adaptive array systems. Some techniques are proper during transmission, whereas others operate in the reception phase. Additionally, some are active against main-beam jammers, and others provide benefits against side lobe jammers [13]. The SL-B, SL-C, and adaptive array systems can be used jointly to face noise jammers and deception jammers contemporaneously impinging on the side lobes of the victim radar.

Antennas' beam patterns can be manipulated to eliminate false targets from side lobes or reduce the power of noise jamming entering from the antenna side lobes. With a well-designed antenna or incorporating specific auxiliary circuits, the antenna side lobes can be minimized to a level where jamming power is adequate only when injected into the radar's main lobe. The SPJ systems are advantageous in this design. However, the SOJ systems are less fortunate in terms of jamming, and they will be effective only in preventing detection in the small sector of the radar's azimuth beamwidth centered on the jammer. Thus, low antenna sidelobes or effective sidelobe control is of utmost importance as it restricts jamming and detection to the main lobe [14].

An effective technique is SL-B, an ECCM technique against low-duty cycle coherent pulsed interferences. The purpose of an SL-B system is to prevent the detection of unexpectedly strong target returns and interference pulses entering the radar receiver via the antenna sidelobes. The idea is to employ an auxiliary antenna coupled to a parallel receiving channel in addition to the main antenna to distinguish signals entering the sidelobes from those entering the main beam. The side lobe signals are aimed to be suppressed in some applications.

This technique compares the detected signal amplitude from the main channel with an auxiliary channel. Figure 7.11 shows the block diagram of SL-B. Specifically, when the auxiliary channel signal power is greater than the main channels, the radar is likely under attack by a deception jammer from the side lobes, and the detection is blanked. For the distinguishing process, the antenna gains must be selected appropriately. The possible presence of noise jamming makes the detection task even more challenging and reduces the capability of blanking deceptive jamming.

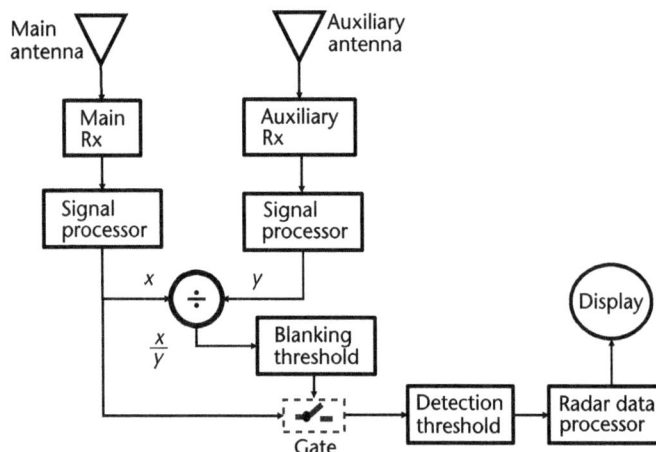

Figure 7.11   A generic block diagram of a sidelobe blanker.

SL-B becomes ineffective in the presence of CW or high-duty-cycle jammers since it usually inhibits the detection of actual targets. In this situation, the SL-C presents an efficient ECCM for noise jammers. The method places nulls in the Rx's sidelobes of the main lobe along the arrival directions of the noise jammers, which are adaptively estimated using auxiliary channels.

Modern radars use a digitally-based approach to implement the SL-C function. Specifically, digital samples from each channel of an electronically scanned array are weighted to shape the resulting beampattern adaptively. These techniques belong to the more general family of algorithms called *adaptive digital beamforming* (ADB). In cases where the dominant interference is jamming, and its level exceeds the target's, angle-of-arrival processing techniques, such as SL-C and ADB, are good selections [4].

The objective of the SL-C is to suppress the high-duty cycle and noise-like interferences received through the radar's side lobes. This is accomplished by equipping the radar with an array of auxiliary antennas used to adaptively estimate the direction of arrival and the power of the jammers and, subsequently, modify the radar antenna's receiving pattern to place nulls in the jammers' directions.

Side lobe cancellation uses the interference signal received by the auxiliary antenna to suppress the directional interference coming in through the antenna sidelobe direction. The basic idea is to send the interference signals received by the antenna's main lobe and side lobes to the adaptive processor simultaneously and calculate the optimal weight through a specific adaptive algorithm to minimize the total output power to achieve interference cancellation purposes.

Figure 7.12 gives a block diagram for a typical closed-loop adaptive sidelobe cancellation system consisting of a main channel and several auxiliary channels. The dashed lines in the figure represent the path for the processed signals within weighting calculations. However, solid lines show the transfer of the unprocessed

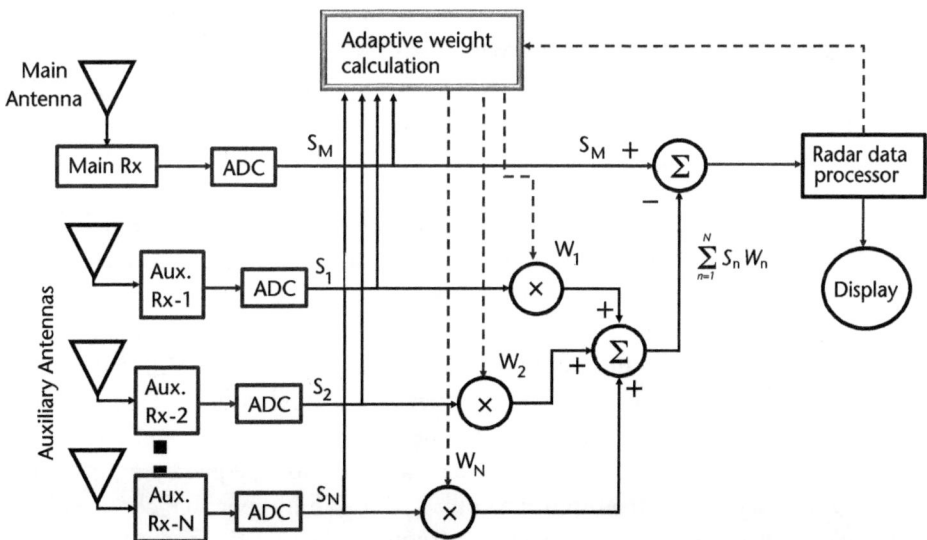

Figure 7.12   A generic block diagram of a closed-loop adaptive side lobe canceler.

signals. The main channel signal is weighted by the main antenna and synthesized to determine the target direction. Each auxiliary antenna feeds the auxiliary channel signal separately. The auxiliary antennas are usually omnidirectional antennas with low gain. Moreover, their gains are selected as equivalent to the sidelobe level of the main antenna.

When the desired signal and interference arrive simultaneously at the main and auxiliary antennas, the expected signal component in the auxiliary channel is far less than that in the main channel, while the interference amplitude is roughly the same. If the auxiliary channel is weighted in real time to compensate for the fixed phase difference caused by the wave path difference between the main channel and the auxiliary channel and then subtracted from the output of the main channel, the interference can be eliminated adaptively.

The structure of the SL-C system is mainly closed-loop and open-loop. The feedback between the radar data processor and adaptive weight calculator of Figure 7.12 is absent in the open-loop SL-C systems. The direct matrix inversion method is the commonly used open-loop adaptive side lobe cancellation weight algorithm. The *minimum mean square error* (MMSE) criterion calculates the optimal weight. The weight of the auxiliary channel can make the output power of the main and auxiliary channels after weighting is the minimum.

The closed-loop adaptive SL-C algorithm is a typical adaptive filtering algorithm based on the steepest descent method. It takes the MMSE as the criterion and continuously feeds back the calculation error to the weight estimation module to ensure the convergence iteration and obtain the optimal weight. The closed-loop SL-C method achieves the best interference cancellation effect by gradually changing the weighting coefficient. This method's advantage is simplicity, but the convergence speed is slow.

7.3.3.2 Tx-Related ECCM

The various ECCM strategies are intricately linked to the precise management and utilization of power, frequency, and the radiated signal waveform. One such strategic approach to counter noise jamming involves a power boost in the radar Tx. Combined with the targeted spotlighting of the radar antenna on the intended object, this tactic significantly expands the radar's detection range. However, this approach, known as spotlighting or burn-through mode, has limitations. As the radar focuses on a specific direction, it neglects other potential areas of interest. Moreover, the burn-through mode proves ineffective against chaff, decoys, repeaters, spoofers, and similar countermeasures.

More compelling is the use of complex, variable, and dissimilar transmitted signals that place a maximum burden on a SPEW system regarding RWR and jamming. Changing the operating frequency and PRI adds ECCM property to radar systems. In Chapter 5, different frequency and PRI agility methods are mentioned. If we consider the ECCM techniques achieved by frequency agility (chirp RF, RF agile (varying RF from pulse to pulse), and RF hopper (RF changes from batch to batch of pulses), these modes are frequency-diversity modes for the use of wide instantaneous bandwidth. The RF hopper (batch-to-batch) approach allows Doppler processing, which is incompatible with the RF agile (pulse-to-pulse) basis. In an RF hopper waveform, the center frequency of each transmitted pulse is moved between many center frequencies in either a random or a programmed schedule. Predicting the subsequent pulse frequency by the noncooperative Rx systems is almost impossible using the frequency of the current pulse. Frequency diversity in the environment may indicate that several radar threats exist or one radar operates at different frequencies. The frequency agility and diversity aim to increase the RWR processing time and load for emitter identification. Furthermore, they force the SPJs to spread their energy over the entire agile bandwidth of the radar. This reduces the jammer density and results in the effectiveness of the SPEW system. However, when we pay regard to the ECCM modes obtained by varying PRI, the following methods are encountered: staggered PRI, jittered PRI, and pulse group repetition interval (PGRI).

Varying PRI methods are beneficial for aggravating the RWR's deinterleaving and emitter identification processes. They are also valuable ECCM techniques for deceptive jamming but are not effective against noise jamming techniques. PRI diversity techniques make deception jamming or spoofing of the radar difficult since hostile SPEW systems do not know how to anticipate the transmitted waveform's fine structure. Therefore, they provide maximum range performance against such jammer types.

Typically, pulse compression, or intrapulse coding, aims to achieve higher range resolution in improving target detection capability without reducing average radar power. This method also does not require exceeding radars' peak power limitations. Thus, it reduces the jamming effect of the chaff returns and resolves targets to a higher degree.

Another ECCM method that Tx can conduct is the ability to examine the jammer signals, find nulls in their transmitted spectra, and select the radar operating frequency with the lowest level of jamming power. This approach is beneficial against pulsed spot noise and preset barrage noise. The method's effectiveness mainly depends on the extent of the radar agile bandwidth and the acquisition spend and frequency tracking of an intelligent jumper.

7.3.3.3   Rx-Related ECCM

Radar Rx systems are critical when applying ECCM methods since they are the main stage exposed to jamming. Jamming signals can saturate the radar processing chain if they are large and capable enough to pass the antenna's ECCM expedients. In radar systems, wide dynamic range Rx must be used to avoid saturation, which results in the virtual elimination of information about the target. One result of jamming is the possible saturation of the radar receiver and display. Depending on the intended application of the radar, special processing circuits may be included in the radar Rx to reduce the effect of clutter. These include *fast time constant* (FTC) Rx, a wide dynamic range logarithmic (log) Rx, and CFAR.

An FTC circuit produces extremely short-duration time constants (τ) compared to conventional circuits. FTCs can be implemented using a required combination of resistance and capacitance ($\tau_C = RC$) or inductance and resistance ($\tau_L = L/R$). If these circuit elements are of a proper magnitude and are appropriately connected about input and output, the resulting circuit output will be the time derivative of its input. In radar systems, FTCs reduce the effects of certain types of undesired signals, emphasize short-duration signals to discriminate against the low-frequency components of clutter, and protect against the interference of either FM or AM signals. FTCs are also used as ECCM devices against jammers in radar amplifier circuits. They differentiate incoming pulses so that only the leading edge of the pulses is used.

A log Rx's video output is proportional to the logarithm of the envelope of the RF input signal over a specified range. Since these processing circuits prevent clutter noise saturation and Rx false alarms, they also effectively reduce the effects of noise jamming. A log Rx might help against noise jamming but has detrimental effects against clutter when using Doppler processing. It may help to prevent Rx saturation in the presence of variable intensities of jamming noise. Compared with an Rx of low dynamic range, moderate jamming noise levels will typically cause the computer to saturate so that the target signal will not be detected. The primary drawback of using a log characteristic is that it can result in the spectral spreading of received echoes. Thus, maintaining clutter rejection in an MTI or pulse Doppler radar would only be possible if the spectrum of clutter echoes were spread into the spectral region where returns were expected.

Radars always work in a dense environment with noise, unintended interferences, jammers, and unwanted reflections, such as from the Earth and the sea. These unwanted signals and echo returns can fully occupy the radar processor and make target detection difficult. Noisy signals are nonstationary and must be used in a detector that has a feature that automatically adjusts its sensitivity to the intensity of interference variation. In this way, the false alarm probability is held constant. A detector with this feature is known as the CFAR detector, as mentioned in Chapter 2.

Dicke Fix circuit is another specific Rx antisaturation circuit used primarily to prevent swept jamming. For this purpose, nonlinear memoryless devices, called hard or soft limiters, are engaged in counter-jamming signals. These devices cut jamming signals with wide amplitudes. The Dicke-Fix Rx counters high rates of swept-frequency CW jamming and swept spot noise jammers. In a radar Rx, the Dicke-Fix uses a wideband IF amplifier and a limiter ahead of the narrow-bandwidth IF amplifier. The wideband amplifier allows a rapid recovery time from the effects of the swept jammer, and the limiter cuts the jamming signal. After transiting through a wideband amplifier and the limiter without remarkable degradation, the narrowband target signal is integrated with the narrowband filter matched to the signal. The Dicke-Fix Rx systems are not used in modern radars, especially those that employ Doppler processing.

7.3.3.4 Signal-Processing-Related ECCM

There are similarities between the Rx-related and signal-processing-related ECCM techniques. The main difference is in the aims rather than the methods. Fundamentally, Rx-related ECCM techniques aim to prevent saturation under chaff and jamming. However, signal-processing-related ECCM techniques intend to discriminate and detect the targets under a dense RF environment. These techniques are fixed and adaptive MTI, optimum pulse Doppler processing, and CFAR. MTI and pulse Doppler processing methods are coherent signal processing techniques that greatly alleviate clutter and chaff's effects. They provide effective radar ECCM capability, especially against chaff and various forms of noise jamming. Noncoherent tech-niques are also required since coherent devices can achieve only a limited degree of clutter, chaff, and jammer suppression. Therefore, the cancellation ruins may still be an essential source of false alarms. Among the noncoherent devices, the CFAR detector and the pulse-width discriminator are worth mentioning.

MTI and pulse Doppler processing techniques are based on Doppler filter-ing. Fixed and adaptive MTI techniques are coherent processing methods and provide the radar with the potential for detecting and tracking aircraft through chaff clouds. MTI-based techniques fulfill this function in much the same manner as they provide for the reduction of clutter returns from rain, clouds, and ground clutter. These methods employ a PRF that provides unambiguous range coverage while using a comb Doppler filter whose nulls are tuned to the average radial speed of the chaff [12].

Another coherent antichaff technique is pulse Doppler processing. This method uses a high PRF to provide unambiguous Doppler coverage in conjunction with a Doppler filter bank, allowing separation of the target from the chaff by exploit-ing their different motion characteristics. The characteristics of chaff are similar to those of weather clutter, except that the chaff-scattering elements are cut to respond to a broad spectrum of radar frequencies. Weather clutter and chaff differ from ground clutter in that wind velocity determines the mean Doppler shift and the spread. Chaff moves with the local wind, and there are tricks to make an MTI null out both moving and stationary unwanted echoes.

CFAR, mentioned in the antenna-related ECCM part, can also be considered a noncoherent signal-processing-based ECCM method. CFAR circuits reduce the effects of jamming by increasing the radar's detection threshold in the presence of jamming. This increment causes the Rx sensitivity of the radar to deteriorate at the same time. Furthermore, CFAR has a possible negative side effect. In the presence of severe jamming, it may automatically increase the radar's detection threshold to prevent excessive false alarms. Since the radar display appears normal, the operator may not detect the jamming. Unique displays sometimes warn an operator that the radar is being jammed and indicate an approximate bearing to the jammer.

Another noncoherent method is the pulse-width discrimination circuit, which measures the width of each received pulse. It is only accepted if the received pulse is approximately the same width as the transmitted pulse. A pulse-width discrimina-tion technique can also help to reject chaff. The echo returns from chaff corridors are much wider than the transmitted pulse. However, if a target is within the chaff corridor, the pulse width discriminator might also eliminate the target.

7.4   Self-Protection Jammer Aspects

For the case in Figure 7.8, the required PFM is prepared for jamming one and mul-tiple threats simultaneously or sequentially. Thus, the jamming sequence and threat priority plans are essential. However, the operational limits of the SPJ capabili-ties must be within the PFM definitions. Thus, handling multiple targets, optimal source allocation, and developing jamming techniques for the operational require-ments and compatibility with the SPJ system is vital. In this section, necessary SPJ systems' specifications are mentioned and related to the challenging points of the jamming techniques.

7.4.1 SPJ System Specifications

The SPJ system is a subunit of the SPEW system, and all the subsystems have their design, implemented hardware, and embedded software. This means each subsystem has different defined specifications. Even though the subsystems are separate systems, such as SPJ, RWR, CMDS, MWS, DIRCM, and LWR, they constitute the SPEW system. There have been some examples of using the subunits independently. However, using them as a part of a platform EW defense system is more common. Thus, SPEW is an integrated system that combines many subsystems and features. The general operating system runs on the overall system, and the PFMs are running on the operating system. However, subsystems have embedded software and input/output (I/O) protocols to exchange data among the subsystems.

In this context, as each subsystem, SPJ has its specifications. These specifications determine the capabilities of the SPJ system, and its technical competence limits determine the applicable jamming techniques and PFMs. Thus, after deciding the SPJ systems' operational requirements, the essential specifications are based on the production development and procurement projects. The main SPJ specifications and some generic values adaptable to the practical systems are listed here:

- Operating frequency range: 1–20 GHz.
- Repeater jammer Rx sensitivity: −65, −55 dBm.
- Maximum Tx power: minimum 45 dBm.
- Jamming type: The jammer's capability to produce different jamming techniques. These techniques can be listed, but not limited to, as follows:
  - Noise techniques: cover pulse, continuous noise;
  - Range deception techniques: noise methods, RGPO;
  - Velocity deception techniques: VGPO, adaptive initial phases;
  - Angle deception techniques: AM deception techniques, monopulse techniques, frequency offset techniques;
  - False target generation techniques.
- Jamming sector in forward and aft hemisphere: ±60° in azimuth and ±30° in elevation.
- Simultaneous operation: both in the front and rear hemispheres or just one hemisphere is possible simultaneously.
- Operation modes: automatic (the crew's attention is not distracted), semiautomatic (with the crew's attention), or manual.
- EMC/EMI with the platform's onboard avionics, navigation, or communication equipment. The SPJ is fully compatible with the platform's radar and other RF equipment.
- Operation time: The operation time is no less than 4 to 6 hours.
- Startup time after the power supply is provided: within 1 minute.
- Jamming response time: Once the ECM is within lethal range, the jamming starts (the response time), which is less than 1 second.

- Power consumption: The maximum allowable power consumption for SPJ systems is 2 to 4 kW for fighters, helicopters, and fighter UAVs and 4 to 6 kW for larger aircraft, such as assault, bomber, and cargo aircraft. For warships, the permitted power consumption levels for SPJ systems may be up to 10 to 15 kW.
- System volume: With all units (including control panel dimensions) of an SPJ system, the volume occupied is between 0.1 and 0.5 m^3 for airborne platforms and 1.5 to 3 m^3 for shipborne platforms. The volume allowance depends on the platform's capabilities. The dimensions are mainly related to the system's analog or digital structure and the Tx power level.
- System weight: The weight of all SPJ systems units (including control panel dimensions) depends on the platform type. The allowable weight is 50 to 100 kg for airborne platforms. For the shipboard platforms, the permittable weight can be higher than this.

7.4.2 Challenging Points for Developing Jamming Techniques

Developing a jamming technique is a technical issue after obtaining the technical specifications of a threat radar system. However, the technical requirements are captured by intelligence-gathering means and studied for extracting technical information by engineering studies. When looking at the past developments of radar systems, it is not a fantastic expectation to say that further developments will exist. Each technical novelty in the radar systems can be neutralized with a newly developed jamming technique.

This can be extended for a practical application with a suitable example. In the first part of the 2010s, *multiple-input multiple-output* (MIMO) radars were developed [15]. A MIMO radar with a *frequency diverse array* (FDA) produces a range-angle-dependent steering vector, which can suppress the deceptive jamming in the joint range-angle domain. By combining FDA and MIMO technology, the FDA-MIMO system has the result of associating a waveform to each point range/angle of the space, with the possibility of recovering this information in the receiver after appropriate processing. Thus, the range-angle-dependent transmit steering vector is obtained. The FDA-MIMO radar was investigated to suppress the deceptive jamming in the joint Tx–Rx domain [16]. A time-varying delay modulation and phase modulation technique has been developed for this novelty in radar technology to ensure that FDA-MIMO radar cannot accurately obtain target information through range gate analysis and spatial spectrum searching [17].

Thus, when one knows the novelty and the related technique for a radar system, one can find a way to jam it. The challenging point is not to develop a jamming technique for only one threat radar and use it solitarily in the operational field while using different jamming techniques against many radars in actual combat conditions without degrading any jamming efficiency. For this reason, the following studies in the given order or by changing the sorting according to the requirements would reduce the challenge level of the jamming techniques development:

- Study each threat radar separately and determine their weak sides and possible effective jamming methods.

- Develop the most effective jamming technique for each radar system.
- Study other techniques for determining the effects of each threat radar. Thus, the alternative jamming methods and their impact on the radar systems should be specified. Say that the RGPO is the most effective way to jam a specific radar after starting to track. However, examining the effects of noise jamming on this radar during and before tracking is crucial. Thus, noise-jamming effectivity can be measured on the radar operation, and preventing track initiation probability may be detected.
- For each PFM, the probable threat radar combination scenarios are established, and the jamming techniques' efficiencies are assessed. The studies must be realized first in simulation and then in a laboratory environment, such as a hot mock-up or hardware-in-the-loop. These assessments are to be conducted for the worst cases.
- These studies determine the best results for different scenarios, and the trials are conducted in the field. The feedback of the operational personnel, pilots, and crew is essential, and they are evaluated and required amendments on the jamming techniques are realized. In this stage, the compatibility of multiple jamming techniques simultaneously gains importance. Thus, rather than using the most effective jamming technique for individual radar, coherent operation is more critical without degrading the overall effectiveness.
- Besides compatibility, the adverse mutual effects of the simultaneous or sequential jamming for different radars must be eliminated or reduced. For this purpose, the neutralization phases of the threat radars are defined in the PFMs. Let us consider two threat radars. Suppose that, in the regular case, one is applied VGPO, and the other is injected with a false target. When engaged to a SPEW system, the PFM is defined as one is applied VGPO, as usual, but preset noise is applied to the other before locking.
- Jamming technique development is a dynamic and continuous process. After using the jamming techniques in a real mission, the feedback on the advantages and drawbacks is vital and gives direction for further improvement.

7.4.3 Preparation of SPJ Techniques

Preparing an SPJ technique is a multidisciplinary process that brings together intelligence, operations, and engineering staff. Furthermore, the preparation process depends on different actions and contributions from various parties. For this purpose, the intelligence staff gathers and analyzes the operation field and the threats in the theater. The engineers analyze the radar threats in detail. Additionally, the effectiveness of the technical capabilities of SPJ systems against the threats is examined. Operations personnel prepare EW plans based on operational needs, intelligence, and technical analysis.

Some preparations should be conducted to develop the self-protection jamming technique in this context. These preparations can be summarized as the following action items:

- Technical and locational intelligence is collected to extract the threat and friendly forces' EOB in the potential mission area.

- A detailed field analysis is realized to describe the EM operational environment. For this purpose, the EM properties and the physical and environmental characteristics of the potential mission area are extracted.
- Operational factors are reviewed within the theater to identify risks to the mission and identify the possible threats to the SPJ platforms.
- A detailed technical and operational analysis is conducted for each threat radar.
- The existing PFMs are checked to the above requirements, and the required amendments for the self-protection jamming are specified. Also, the new self-protection jamming requirements are defined.
- Requirements for aligning self-protection jamming techniques with *rules of engagement* (ROE) are determined by considering threats' guidance methods and operational estimates.

7.5   Future Trends in Developing Jamming Techniques

The SPJ systems are to be successful, or the threat radar will be successful, and the SPJ mother platform crew cannot return home. The responsibility of the SPJ systems and the working staff in the background is arduous. One side of the SPJ systems' accomplishment depends on the technical properties and technological level discussed in Chapter 6. The second side is the jamming techniques' effectiveness. Both are vital and complete each other. A future projection for SPJ systems is given before, and the future trends in SPJ techniques are mentioned here.

Future trends in developing SPJ techniques can be clustered into two main groups. The first concerns the jamming technique's development process and adaptation to PFMs. The second concerns jammer control methods for optimizing the process during missions.

1. *Future trends for the jamming development process:* Experienced engineers and operation staff generally conduct the SPJ development process. However, developing a jamming technique against some threat radars may be complex. When a new jamming method is designed for a threat, adding it to a PFM may be challenging. Providing compatibility between the techniques without reducing jamming efficiency is a complex, vital, and time-consuming study. The development time for SPJ techniques is generally limited, and they should be imported to the PFM in the required time. The jamming technique development process and adding it into a PFM must be studied theoretically and tried on simulation and hot mock-up or HITL systems. For this reason, a more practical way must be focused on as a future projection.

   When a threat radar system signal structure is known, it can be simulated, and some features may be extracted from the radars' operating frequency, frequency agilities, modulation bandwidth, pulse width, antenna scan type, PRF, and PRF agility. The most effective jamming technique may be developed by using machine learning techniques. Supervised methods are good candidates for developing SPJ techniques. Although the development process is time-limited in some situations, it does not have to be in real time. The main aim of this process is to teach the SPJ system via PFM and how to react to different threat combinations. Thus, SVM, decision trees, neural networks, or their variations are compatible with developing jamming techniques for a specific radar system.

As stated before, the PFM preparation process becomes more complex when multiple threat radars are jammed simultaneously. This time, jamming sequence adjustment methods and adaptive resource allocation processes should be developed to obtain the optimal PFM structure. One must consider the RWR functions for the SPJ systems when solving the resource allocation problem. The objective is to maximize the jammer on time and minimize proper sensing time. The two issues are interconnected because sensing the environment informs the tailoring and focusing of jammer techniques to improve overall jamming effectiveness. For this reason, RWR and SPJ functions must be optimized together.

With the increasing number of variables, features, and combinations, the process complexity will increase exponentially, and the solution will shift from machine learning to deep learning methods. The standard deep learning methods may be good candidates for developing optimal PFMs with multithreat jamming, such as convolutional neural networks, long and short-term memory networks, and recurrent neural networks.

2. *Trends in jammer control optimization:* SPJ systems operate in dense RF environments in modern operation fields. In the future, an increasing number of radar threats are expected, and the RF density of the theater will gradually increase. For this reason, the systems' capabilities and the power sources must be used most effectively. Thus, an optimization is necessary for power management and source allocation. In the standard approach, the PFMs are prepared for determined scenarios close to the hostile or potential enemies' EOB structures obtained by intelligence units. Optimization studies have also been conducted for these scenarios, and they are not adaptive to unexpected conditions. These unpredictable situations may be the relocation of threat radar systems or the addition of new threat systems to IADS. The optimum PFM structures are corrupted in these variances, and the most influential power usage and source allocation cannot be reached. Thus, not only in the development phase but also in the operations, PFMs should be adaptive.

If only one threat radar with static behavior (a single work mode) is considered, it can easily result in a local descent optimal jamming strategy for source allocation. Local descent optimization algorithms are intended for optimization problems with multiple input variables and a single global optimum, such as line search or its variation Brent-Dekker algorithms. Thus, facing one threat at an instant does not require additional adaptivity in the PFM structure.

When simultaneously confronting many radars, some of which have adaptive behaviors (multimodes) and the ability of parameter agility, ample jamming action space is usually needed to ensure that the correct actions are included. This situation dramatically increases the complexity of the jamming process and requires high-level optimization methods for power management and resource allocation. High complexity will lead to a long processing time, and the jammer can hardly find the optimal power management and resource allocation strategy in a limited time, which is improper for the quick-response requirements of SPJ. In this situation, SPJ systems require PFMs with adaptive optimization methods during the missions.

To achieve adaptive structured PFMs, they must be changed from standard database structure to adaptively changing source allocation and power utilization construction. Moreover, the SPEW systems' central operation systems are to be compatible with the adaptive control of power management and resource allocation by PFMs. These studies have been started for the last two decades with different adaptive optimization methods and machine learning techniques at the academic level. This property would probably be the operational requirement for the next-generation SPJ systems.

The SPJ resource-allocation problem can be considered a nonpolynomial (NP)-hard problem due to its exponentially increasing nature. The NP-hard problems cannot be solved in polynomial time in a deterministic way [19]. Traditional NP-hard problem solution algorithms, such as the Hungarian algorithm, zero-one programming, fuzzy multi-attribute dynamic programming, and close degree, are suitable for solving small-scale jamming resource allocation problems [20]. Along with the fast development of cognitive EW, mosaic warfare, multidomain warfare, and other tactics, the traditional optimization algorithms cannot meet the calculation speed requirement of the battlefield.

For this reason, unsupervised learning ML techniques may be a solution for the adaptive optimization of SPJ applications' power management and resource allocation. Thus, during missions, PFMs can adaptively optimize power management and resource allocation. According to the operation field difficulty level, unsupervised ML algorithms such as K-Means, Fuzzy C-Means, Hierarchical, Gaussian Mixture, neural networks, and hidden Markov are used.

7.6 Problems

Problem 7.1: Assume that a command-guided missile is launching from a warship to a fighter aircraft, as shown in Figure 7.13. When launching, the ship changes its radar operating frequency from 4.2 to 6.4 GHz, and the SPEW system on the fighter senses this change. Immediately, SPJ starts an aggressive repeater noise jamming with a Tx power of 5 kW, and its antenna gain is 7 dB. The radar's antenna is a phased array antenna; its beamwidth angle is 2° in elevation and 3° in azimuth, and the RCS of the fighter through the engagement direction is 5.8 m². The radar's peak power is 800 kW, the pulse width is 1 μs, the slant distance between the radar and the target is 15 km, and the sum of all individual scatterers RCS within the volume is 2.3×10^{-6} m²/m³. The figure shows that the missile does not have an Rx, and a data link provides the steering. The jammer applies spot noise jamming, and the radar's and jammer's bandwidths are close together, and B_R/B_J is assumed to be unity. Calculate the total jamming efficiency for this case.

Problem 7.2: Assume that an ASM is launching from a fighter to a vessel, as shown in Figure 7.14. In the first stage, the missile cruises with inertial guidance in the sea-skimming mode, and in the terminal phase, it climbs to a specific altitude and uses the RF active guidance mode. The operating frequency of the missile's radar

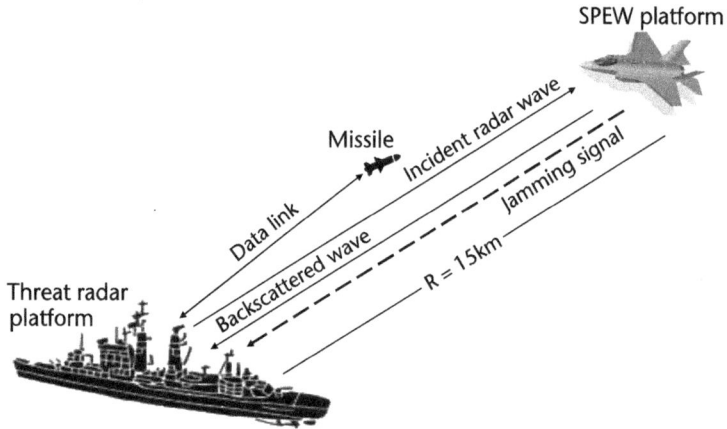

Figure 7.13   A command-guided RF missile launching from a warship to a fighter.

is 4.6 GHz, the peak power is 3 kW, its antenna is a plane array with a gain of 15 dB with equal beamwidths for azimuth and elevation, and the pulse width is 0.5 μs. The vessel's RCS is 45 m^2, and the sea state is 4 from Table 7.1 (Douglas Sea Scale Definitions). The RCS characteristic of the illuminated area (γ) is assumed to be 0.1 m^2. The vessel's ESM system detects the missile at a slope range of 5.3 km, and its SPJ system immediately starts to apply spot noise jamming with all power. The full power of the SPJ is 25 kW, and the antenna gain is 10 dB. The radar and jammer's bandwidths are close, and B_R/B_J is assumed to be unity. Calculate the jamming efficiency when the missile to vessel range is 5 km, and the grazing angle (ψ) is 20°. Also, the critical grazing angle for the condition must be found.

Problem 7.3: Write a MATLAB code for determining sea state properties for wave height, wave slope, and wing velocity. Determine the properties for all sea states from 1 to 8 and draw them.

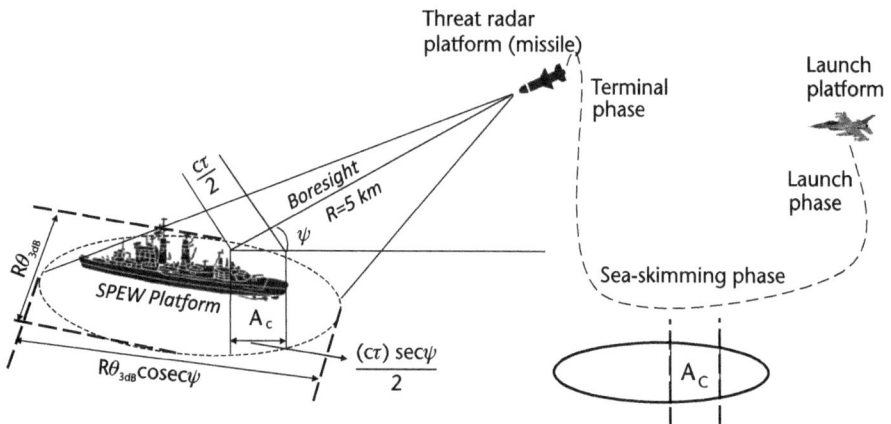

Figure 7.14   An ASM missile launching from a fighter to a vessel.

Solutions

Solution 7.1: Let us define the given information:

$$\sum \sigma_i = 2.3 \times 10^{-6} \text{ m}^2/\text{m}^3 \quad \sigma_T = 5.8 \text{ m}^2$$

$$f_0 = 6.4 \text{ GHz} \quad P_J = 5 \times 10^3 \text{ W} \quad P_{Tx} = 800 \times 10^3 \text{ W}$$

$$\left(G_J\right)_{dB} = 7 \text{ dB} \rightarrow G_J = 5 \quad \tau = 1 \times 10^{-6} \text{s}$$

To find the radar antenna gain, first find the directivity of the radar antenna by using the following general equation for phased array antennas as follows [18]:

$$D_0 = \frac{32,400}{\theta_{el} \times \theta_{az}} \text{ (unitless)} \tag{7.23}$$

where θ_{el} and θ_{az} are elevation and azimuth beamwidth angles in degrees, respectively. So the directivity is

$$D_0 = \frac{32,400}{\theta_{el} \times \theta_{az}} = \frac{32,400}{2 \times 3} = 5,400$$

The gain is obtained from the following general gain to directivity relationship,

$$G_0 = \eta D_0 \rightarrow \{\eta: \text{ total antenna efficiency}\} \tag{7.24}$$

If we do not have any information about the antenna efficiency, we may assume it is 55%. Thus,

$$G_0 = \eta D_0 = 0.55 \times 5,400 = 2,970$$

Now, we can calculate the total RCS of a single-resolution volume using (7.15) as follows:

$$\theta_{el} = 2° \rightarrow \theta_{el} = 0.035 \text{ rad} \quad \theta_{az} = 3° \rightarrow \theta_{az} = 0.0523$$

$$V_w \simeq \frac{\pi}{8}\theta_{az}\theta_{el}R^2 c\tau = \frac{3.14}{8}(0.035)(0.0523)\left(15 \times 10^3\right)^2\left(3 \times 10^8\right)\left(1 \times 10^{-6}\right)$$

$$V_w \simeq 48.5 \times 10^6 \text{ m}^3$$

$$\sigma_w = \sum_{n=1}^{N} \sigma_n V_w \rightarrow \sigma_w = 2.3 \times 10^{-6} \times 48.5 \times 10^6 = 111.55 \text{ m}^2$$

Using (7.17), the volume clutter return effect on the radar Rx is calculated as follows:

$$S_V = \frac{P_{Tx}G^2c^2\sigma_w}{(4\pi)^3 R^4 f^2} = \frac{(800 \times 10^3)(2,970)^2(3 \times 10^8)^2(111.55)}{(1981.4)(15 \times 10^3)^4(6.4 \times 10^9)^2} \approx 1.73 \times 10^{-8}\,W$$

Let us find the jammer and target echo return effects on the radar returns,

$$J = \frac{P_J G_{Ja}Gc^2}{(4\pi R f)^2} = \frac{(5 \times 10^3)(5)(2,970)(3 \times 10^8)^2}{(157.75)(15 \times 10^3)^2(6.4 \times 10^9)^2} = 4.6 \times 10^{-6}\,W$$

Since $G_{Tx} = G_{Rx} = G$ for the radar:

$$S_t = \frac{P_{Tx}G^2c^2\sigma_t}{(4\pi)^3 R^4 f^2} = \frac{(800 \times 10^3)(2,970)^2(3 \times 10^8)^2(5.8)}{(1,981.4)(15 \times 10^3)^4(6.4 \times 10^9)^2} = 9 \times 10^{-10}\,W$$

Finally, using (7.18), the total jamming efficiency (J/S) can be calculated by taking the volume clutter into account:

$$\left(\frac{J}{S}\right)_{Total} = \frac{\sum I}{S_t} = \frac{S_V + J}{S_t} = \frac{1.73 \times 10^{-8} + 4.6 \times 10^{-6}}{9 \times 10^{-10}} = 5,130$$

$$\left(\frac{J}{S}\right)_{Total\_dB} = 37.1\ dB$$

Solution 7.2: Let us define the given information:

$$\gamma = 0.1\ m^2 \quad \psi = 20° \quad \sigma_T = 45\ m^2$$
$$f_0 = 4.6\ GHz \quad P_{Tx} = 3 \times 10^3\,W \quad \tau = 0.5 \times 10^{-6}\,s$$
$$(G_{Tx})_{dB} = 15\ dB \rightarrow G_{Tx} = G_{Rx} = G = 31.62$$
$$P_J = 25 \times 10^3\,W \quad (G_{Ja})_{dB} = 10\ dB \rightarrow G_{Ja} = 10$$
$$(radar\text{-}target\ distance)\ R = 5\ km$$

We must find the directivity for the missile antenna's 3-dB angles. We may assume the efficiency to be 55%.

$$G_0 = \eta D_0 \Rightarrow D_0 = \frac{31.62}{0.55} = 57.5$$

Then we obtain the 3-dB angles, since they are equal,

$$D_0 = \frac{32,400}{\theta_{el} \times \theta_{az}} \quad \& \quad \theta_{el} \times \theta_{az} = \theta_{3dB}^2$$

$$\theta_{3dB} = \sqrt{\frac{32,400}{57.5}} = 23.73° \rightarrow \theta_{3dB} = 0.414 \text{ rad}$$

The clutter area A_c is calculated by using (7.3)

$$A_C = R\theta_{3dB}\frac{c\tau}{2}\sec\psi = (5 \times 10^3)(0.414)\frac{(3 \times 10^8)(0.5 \times 10^{-6})}{2}(1.06)$$
$$A_C = 174.44 \text{ m}^2$$

The clutter reflectivity (σ^0) according to the grazing angle (ψ) can be calculated using (7.8),

$$\sigma^0 = \gamma\sin\psi = 0.1 \times \sin(20°) = 0.0342 \text{ m}^2$$

The area clutter is calculated as

$$\sigma_C = \sigma^0 A_C = (3.42 \times 10^{-2})(174.44) = 6 \text{ m}^2$$

The clutter power at the radar Rx is calculated by (7.13).

$$S_C = \frac{P_{Tx}G^2c^2\sigma_C}{(4\pi)^3 R^4 f_0^2} = \frac{(3 \times 10^3)(31.62)^2(3 \times 10^8)^2(6)}{(1,981.4)(5 \times 10^3)^4(4.6 \times 10^9)^2} = 6.2 \times 10^{-14} \text{ W}$$

Let us find the jammer and target echo return effects on the radar returns,

$$J = \frac{P_J G_{Ja}Gc^2}{(4\pi Rf)^2} = \frac{(25 \times 10^3)(10)(31.62)(3 \times 10^8)^2}{(157.75)(5 \times 10^3)^2(4.6 \times 10^9)^2} = 3.41 \times 10^{-13} \text{ W}$$

Since $G_{Tx} = G_{Rx} = G$ for the radar:

$$S_t = \frac{P_{Tx}G^2c^2\sigma_T}{(4\pi)^3 R^4 f^2} = \frac{(3 \times 10^3)(31.62)^2(3 \times 10^8)^2(45)}{(1,981.4)(5 \times 10^3)^4(4.6 \times 10^9)^2} = 4.64 \times 10^{-13} \text{ W}$$

Finally, using (7.20), the total jamming efficiency (J/S) can be calculated.

$$\left(\frac{J}{S}\right)_{\text{Total}} = \frac{\sum I}{S_t} = \frac{S_C + J}{S_t} = \frac{6.2 \times 10^{-14} + 3.41 \times 10^{-13}}{4.64 \times 10^{-13}} = 0.87$$

$$\left(\frac{J}{S}\right)_{\text{Total\_dB}} = -0.6 \text{ dB}$$

The Sea State 4 is moderate (from Table 7.1). We can obtain the critical grazing angle (ψ_C) for the condition using (7.10), as follows:

$$h_{\text{RMS(s)}} = 0.025 + 0.046 S_n^{1.72} \rightarrow \{S_n = 4\}$$

$$h_{\text{RMS(s)}} = 0.524\text{m}$$

$$\lambda = \frac{3 \times 10^8}{4.6 \times 10^9} = 0.065\text{m}$$

$$\psi_C = \sin^{-1}\left(\frac{\lambda}{4h_{\text{RMS(s)}}}\right) = \sin^{-1}\left(\frac{0.065}{4 \times 0.524}\right) = 1.78 \text{ rad} = 101.8°$$

Solution 7.3: The MATLAB code is shown in Table 7.3.

Figure 7.15(a) shows the wave height (in meters), the wave slope (in degrees) in Figure 7.15 (b), and the wind velocity in Figure 7.15(c).

Table 7.3 Sea State Analyze for Problem 7.3

```
% Sea State Analyze for Problem 7.3.
clc, clear all

SS = 1:8; % Sea states

% Initialize outputs
numSeaStates = numel(SS);
WH = zeros(1,numSeaStates);
WS = zeros(1,numSeaStates);
WiV= zeros(1,numSeaStates);

% Sea state properties
for y = 1:numSeaStates
        [WH(y),WS(y),WiV(y)] = searoughness(SS(y));
end

% Plot results separately
plot(SS,WH), xlabel('Wave Height(m)'), ylabel('Sea State')
figure(2)
plot(SS,WS), xlabel('Wave Slope(deg)'), ylabel('Sea State')
figure(3)
plot(SS,WiV), xlabel('Wind Velocity(deg)'), ylabel('Sea State')
```

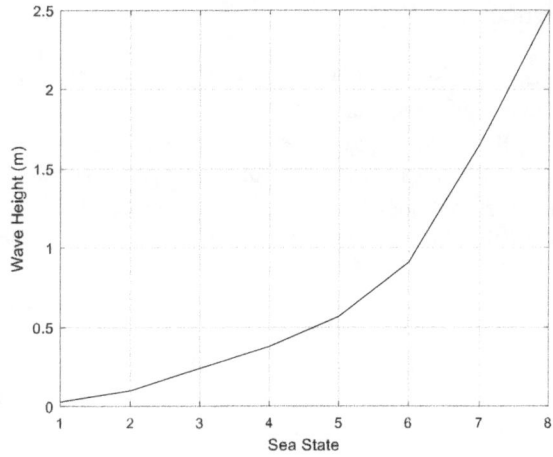

(a)

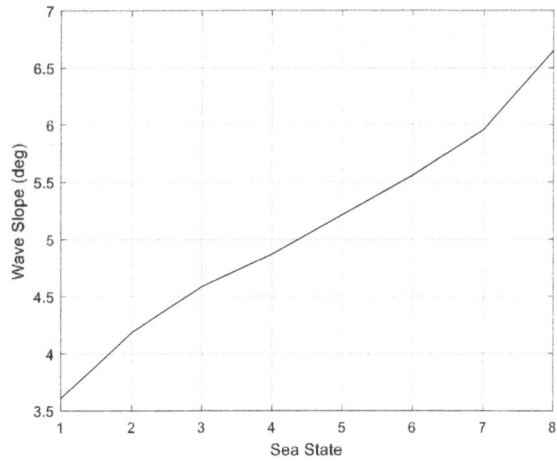

(b)

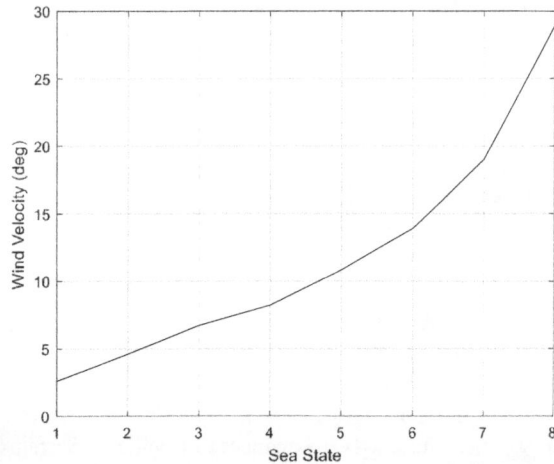

(c)

Figure 7.15 Sea state analysis drawings: (a) the wave height (in meters), (b) the wave slope (in degrees), and (c) and the wind velocity.

References

[1]   US Marine Corps, MCRP 3-32D.1, *Electronic Warfare*, Department of the Navy, Washington, D.C., May 2, 2016.

[2]   Mahafza, B. R., *Radar Systems Analysis and Design Using MATLAB*, 3rd ed., Boca Raton, FL: CRC Press/Taylor & Francis Group, 2013.

[3]   Mbouombouo Mboungam, A. H., Y. Zhi, and C. K. Fonzeu Monguen, "Clutter Map Constant False Alarm Rate Mixed with the Gabor Transform for Target Detection Via Monte Carlo Simulation," *Applied Sciences*, Vol. 14, No. 7, 2024, p. 2967.

[4]   Richards, M. A., J. A. Scheer, and W. A. Holm, *Principles of Modern Radar*, Raleigh, NC: SciTech Publishing, 2010.

[5]   Currie, N. C., J. L. Eaves, and E. K. Reedy, (eds.), "Clutter Characteristics and Effects," in *Principles of Modern Radar*, New York: Chapman & Hall, 1987.

[6]   Owens, E. H., and M. Schwartz, (ed.), "Beaches and Coastal Geology," in *Encyclopedia of Earth Sciences Series*, New York: Springer, 1984, p. 722.

[7]   Richards, M. A., *Fundamentals of Radar Signal Processing*, 2nd ed., New York: McGraw-Hill Education, 2014.

[8]   Joint Publication 3-85 (2020): Joint Electromagnetic Spectrum Operations, USA Joint Chiefs of Staff (CJCS), May 22, 2020.

[9]   De Martino, A., *Introduction to Modern EW Systems*, 2nd ed., Norwood, MA: Artech House, 2018.

[10]   Yan, L., et al., "New ECCM Techniques Against Noise Like and/or Coherent Interferers," *IEEE Transactions on Aerospace and Electronic Systems*, Vol. 56, No. 2, 2019, pp. 1172–1188.

[11]   Bandiera, F., et al., "Detection Algorithms to Discriminate Between Radar Targets and ECM Signals," *IEEE Transactions on Signal Processing*, Vol. 58, No. 12, 2010, pp. 5984–5993.

[12]   Farina, A., "ECCM Techniques," in *Radar Handbook*, 3rd ed., M. I. Skolnik, (ed.), New York: McGraw-Hill, 2008, pp. 24.1–24.67.

[13]   Farina, A., *Antenna-Based Signal Processing Techniques for Radar Systems*, Norwood, MA: Artech House, 1992.

[14]   Reedy, E. K., "Radar ECCM Considerations and Techniques," in *Principles of Modern Radar*, J. L. Eaves and E. K. Reedy, (eds.), New York: Chapman & Hall, 1987, pp. 680–699.

[15]   Davis, M. S., G. A. Showman, and A. D. Lanterman, "Coherent MIMO Radar: The Phased Array and Orthogonal Waveforms," *IEEE Aerospace and Electronic Systems Magazine*, Vol. 29, No. 8, 2014, pp. 76–91.

[16]   Tan, M., C. Wang, and Z. Li, "Correction Analysis of Frequency Diverse Array Radar About Time," *IEEE Transactions on Antennas and Propagation*, Vol. 69, No. 2, 2020, pp. 834–847.

[17]   Zhao, Y., et al., "Research on Main-Lobe Deceptive Jamming Against FDA-MIMO Radar," *IET Radar, Sonar & Navigation*, Vol. 15, 2021, pp. 641–654.

[18]   Balanis, C. A., *Antenna Theory Analysis and Design*, 4th ed., New York: John Wiley & Sons, 2016.

[19]   Paschos, V. T., "An Overview on Polynomial Approximation of NP-Hard Problems," *Yugoslav Journal of Operations Research*, Vol. 19, No. 1, 2016.

[20]   Gao, Y., and D. Li, "Electronic Countermeasures Jamming Resource Optimal Distribution," in *Information Technology and Intelligent Transportation Systems*, Vol. 455, V. Balas, L. Jain, and X. Zhao, (eds.), Advances in Intelligent Systems and Computing, New York: Springer, 2017.

CHAPTER 8

Use of Self-Protection Jammer Systems for Airborne Platforms

Airborne platforms are the main striking power and primary intelligence means for all the armed forces. However, they are vulnerable to surface and air threats without using SPEW systems. The main types of threats use RF guidance for medium and long ranges. The primary countermeasure systems for these kinds of threats are self-protection jammer systems. The airborne platforms that use SPJ systems are bombers, cargo aircraft, helicopters, UAVs, and cruise missiles. Each platform type has a different role and tactical usage. Their SPEW systems and SPJ subsystems have similar structures in hardware and software. However, the operational requirements of the platforms for the jammer response time, simultaneous jamming, and jamming switch time from one threat to another are different. These properties determine the dimensions, weight, power requirements, and capabilities of the SPJ systems. However, each platform provides various SWaP capabilities, and the designers should compromise the operational requirements and the platform properties. In military operations, where needs are endless, and resources are limited, a trade-off must be made, as in every aspect of life.

For an SPJ system's mother platform, SPJ is not the only payload; the ammunition, intelligence sensors, and fuel tanks are also loaded according to the mission type. However, the primary payloads of these airborne platforms are the mission-centric ones, not the SPJs. This underscores the nature of SPJ systems as force multipliers pivotal in mission success and the safe return of the airborne platform and crew to the air base. This chapter delves into the comprehensive usage of the SPJ for different air platforms, focusing on fighters, helicopters, and UAVs primarily for close engagement and penetrating the enemy's lethal border. The inspection details potential threats, SPJ system structure, and reprogramming properties for each platform type, underscoring the critical role of the SPJ systems in the operations.

8.1 SPJ Systems in Fighters

At present, fighters are the main striking power for attack operations and essential units of the active air defense structures. The fighters can reach very high speeds and maneuvering capabilities that increase the self-protection jamming effectiveness with the required evasive maneuvers from radar-guided threats. The fastest fighter jet recorded is the MiG-25 (Foxbat), with 3,524 km/h (~1,903 knots). However, the production of fast-speed fighters has long been stopped by aircraft manufacturers, and the new generation aircraft are significantly slower on average than their pre-decessors. This is because speed stopped being as important in air combat and was sacrificed in favor of maneuverability, stealth, and fuel efficiency, among other factors. The fourth and fifth-generation fighters are generally designed up to 2,172 km/h (~1,173 knots).

Fighter aircraft can carry both attacking and defensive weapon systems. They can penetrate the enemy air defense area and make dogfights throughout the home-land borders or transfrontier regions. Furthermore, they conduct AI missions on the homeland or ally airfields to protect the area from enemy attacks. All these fighter missions have been realized since the beginning of air combat, and the systems and tactics have been developed in parallel with these requirements. Also, the SPJ systems on the fighters emerged and have been improved to meet these needs. Here, we will discuss the fighters' potential radar-guided threats, the SPJ system structures of fighters, and their reprogramming requirements.

8.1.1   Potential Radar-Guided Threats for Fighters

The present strategic environment is uncertain and complex and changes rapidly. The uncertainty, complexity, and evolution were the situation in the past and will be in the future. However, the present environment is more dynamic than the past; in the future, it will be more emotional than today. Since the past changes stemmed from political issues and military powers, the present and future rapid changes emerge from political issues, military power, and developments in weapon technologies. While the fundamental character of war has not changed, the essence of conflict has evolved. The military environment and its threats are increasingly transregional, multidomain, and multifunctional [1]. In this context, modern fighter aircraft face SAMs, AAAs, and AAMs in almost any scenario.

SAM and AAM occasionally use EO-guidance in short-range engagements, such as IR, video, and UV. However, in medium and long ranges, they use radar guid-ance. Some AAA systems utilize radar guidance in all ranges. Thus, the fighters' main threats are SAM and AAM at the medium and long ranges and AAA systems. Detailed technical information about radar guidance was presented in Chapter 2. From a tactical point of view, we can involve radar-guided missile methods such as command, beam-rider, and homing guidance. This classification comprises the conventional technical classification such as active, passive, semiactive, and com-mand guidance. Let us explain them briefly.

- *Command guidance:* A command-guided system accurately determines the missile position according to the control station, the target, and the desired trajectory. A computer is usually used to determine the error between the actual missile position and the desired position. The control point contains a Tx for command signal transmission, and the missile has an Rx. The Rx activates the missile control circuits when it receives command signals from the Tx. This equipment makes it possible to follow the missile's flight and correct for errors that would cause a miss.

    Middle-range and long-range SAM systems occasionally use com-mand guidance and are the primary threats to fighters. Fighters have high maneuvering and speed capability, but their success rates are excellent. The drawing for the command guidance operation is given in Figure 2.38.

- *Beam-rider guidance:* There are similarities in the operation of the beam-rider and command guidance systems. Guidance information is not generated within the missile but collected and analyzed by devices at the launch platform or other control points. However, beam riding is not considered a form of command guidance because the signals it transmits are not specific commands in a command system. Instead, the Tx transmission of a beam-rider guidance system sends only information, not commands. To do this, the control station projects a narrow beam of radar energy that indicates the direction of the target or, in some systems, the direction of a calculated intercept point. The missile's guidance system must interpret the information contained in the radar beam and then formulate its steering commands. These commands operate to keep the missile as close as possible to the center of the beam. As a result, the missile can be said to ride the beam to its intended target.

  The beam-rider system is highly effective for use with short-range and medium-range SAM and AAM systems. For missiles of more extended range, a beam-riding system may be used during the midcourse phase of light while the missile is still within effective range of the beam-transmitting radar. Beam-rider guidance systems are highly fatal for fighters at short and medium ranges. As it approaches the limit of the beam-riding range, the missile may switch over to some other form of guidance, such as homing. Figure 8.1 shows the beam rider guidance method.

- *Homing guidance:* The guidance process for medium-range and long-range missiles involves three phases: launching, intermediate or midcourse guidance, and terminal guidance. The most crucial phase is terminal guidance, which ensures the missile hits its target. The guidance system must function correctly during this phase. For radar-guided systems, guidance is provided by homing signals. The missile or the victim platform may produce these

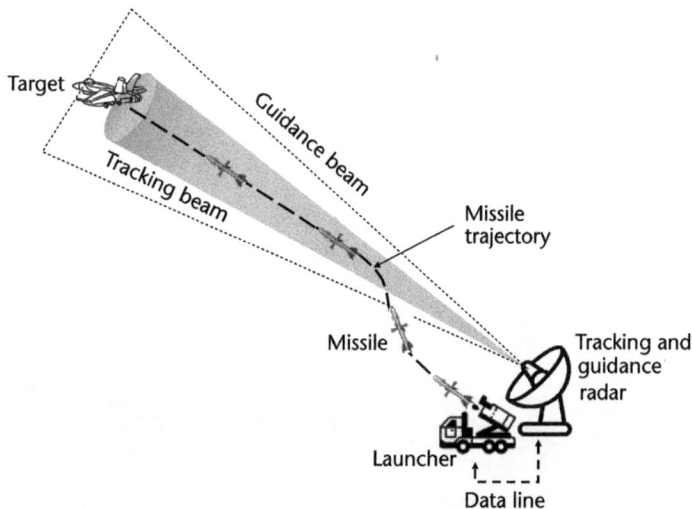

Figure 8.1 Beam rider guidance method.

signals. In the active guidance method, the missile, a complete radar system, emits the homing signal.

The victim platforms emit specific signals that can be used for the homing stimulus, such as radar, navigational, and even jamming signals. The missiles that use these signals to track the targets are passive homing guidance systems. If the victim is illuminated by some source other than the missile's equipment, the system is known as a semiactive homing system. The source may be the launching platform or another platform that provides LOS between the illuminating and the victim platforms. Figure 8.2 presents the two different homing-guidance methods: (1) passive homing guidance, the missile uses the RF emission by the target, and (2) semiactive guidance, the missile uses the RF illumination from the launch platform.

8.1.2   SPJ System Structure for Fighters

The SPJ configuration meets the operational needs of fighters in internal or pod configuration. They are also part of SPEW systems in most practical applications or operate alone in some configurations. The general configuration of an internal SPJ system for a generic fighter is given in Figure 8.3, in which the other SPEW subsystems are neglected, and only the SPJ components and the SPEW CMU are considered. The SPJ systems generally have active electronic antenna arrays in the 4.5-generation and fifth-generation fighters. Thus, the given configuration is supposed to have phased array structures. The bands are defined in Section 4.6.2; the frequency band definitions are shown in Table 2.3. According to the definitions, Band 1 covers E and F bands, Band 2 covers G and H bands, and Band 3 contains I and J bands.

In the design of the SPJ systems, 360° azimuth coverage is aimed. For this purpose, the antennas are located to achieve full coverage. The typical antenna locations in contemporary fighters are at the front spars, on the stabilizers, and in their vicinity. For conventional antennas, such as spiral, blade, and horn, the antenna number is selected to divide the full azimuth into suitable sectors for the operational requirements. So the required gain can be achieved. In this case, additional Rx antennas are needed or will be operated in a time-sharing between Tx and Rx. The elevation coverage should not be 360°, but it should provide the operational

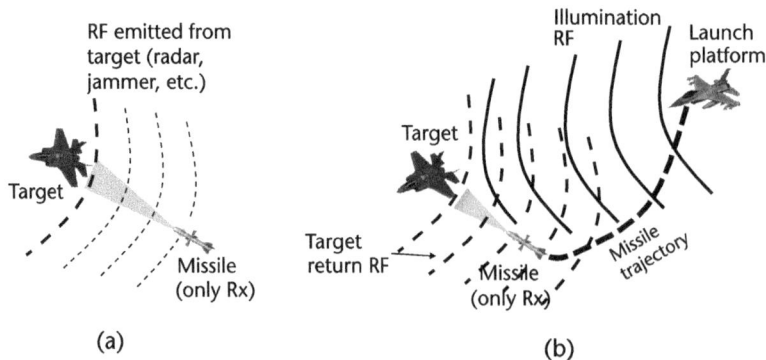

Figure 8.2   Homing guidance methods: (a) passive, and (b) semiactive.

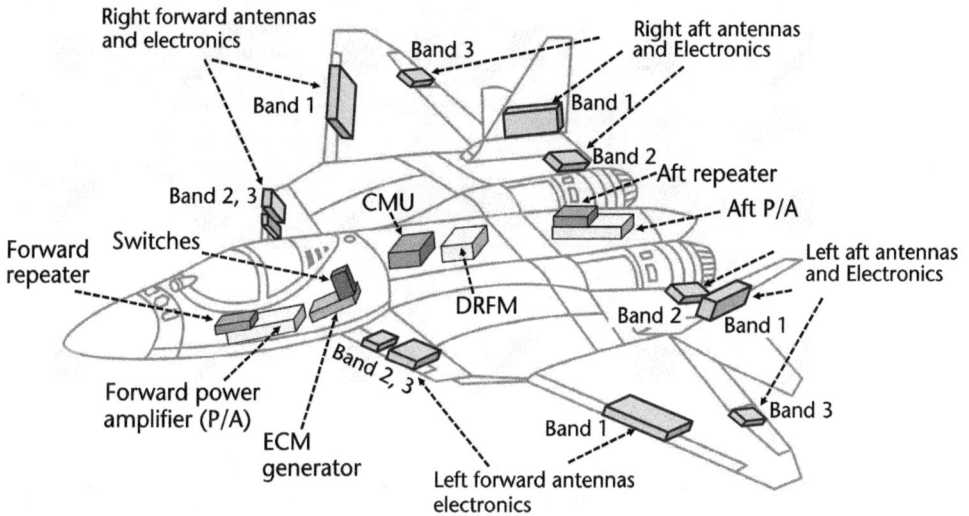

Figure 8.3   The general configuration of an internal SPJ for fighters.

requirements. The configuration given in Figure 8.3 is sufficient for array antennas in terms of azimuth and elevation coverage and gain. Furthermore, simultaneous Rx and Tx operations will be possible.

The internal SPJ systems have a distributed structure at the mother platform. In contemporary SPJ systems, the interfaces between the subsystems are digital, and their SWaP requirements are compatible with the platforms' capacities. Figure 8.3 demonstrates the subsystems of an SPJ system; however, some additions and removals may occur. Since the most crucial part of the design is remaining within the SWaP limits, saving from components is essential. For this reason, some components are single, such as DRFM, ECM generator, and switches, and some others are double, such as aft power amplifiers (P/As) and repeaters.

An alternative to the internal SPJ architecture is a podded structure for the fighter aircraft. Podded SPJ structures are generally more capable than their internal counterparts, and adaptation to different platforms is possible. However, their flexibility for using various platforms and modification capabilities may need improvement. Figure 8.4 shows a generic configuration for a podded SPJ that can be engaged with fighters. In this configuration, the SPJ system does not have an RWR. For an integrated system, it can be assumed that an internal RWR system supports SPJ. Moreover, the antennas are planar arrays for jamming Tx and repeater Rx.

The podded systems' power levels may be much higher than the internal ones if they have power generators. Their other advantages are antenna gains and processing capabilities compared to the internal systems. The pods' input/output (I/O) interfaces are Mil-STD-1553 and ethernet to facilitate communication with other onboard systems. This property provides the podded SPJ adapting for many different platforms. However, from this advantage, two disadvantages emerge:

- The podded SPJ may only be compatible with some platforms' functional and operational requirements.
- The pod's SWaP needs may exceed the providing limits of some platforms.

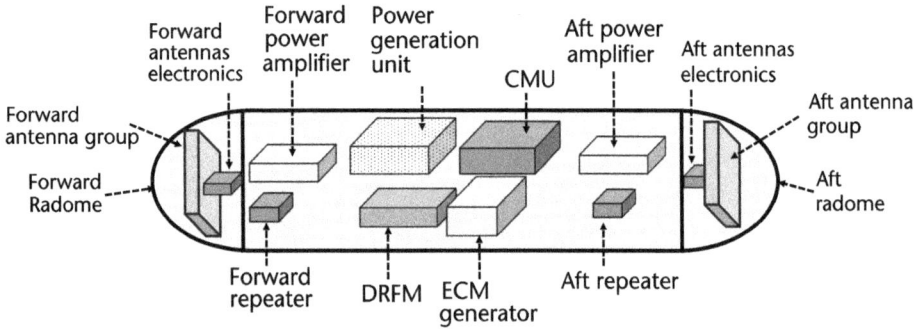

Figure 8.4 The general configuration of a podded SPJ for fighters.

The main disadvantage of using podded SPJ, related to the second item, is their higher weight than the internal structures. As separate, podded SPJ systems, they are independent structures interacting commensurately with the other systems in data processing. Also, their relatively independent structure reduces the EMI/EMC effects up to a level. However, these advantages produce an additive power requirement for the independent podded system. Since these systems can be adapted and mounted to different platforms, they are self-sustained, and their power requirement may be higher than the internally mounted systems. This causes more power to be drawn from the host platform.

At this stage, comparing the fighters' SWaP needs with the podded and internally mounted SPJ systems would be beneficial. First, the size comparison between the two structures is meaningless since the internal architecture allows mounting the subcomponents in the proper spaces of the fighters. However, subcomponents should be placed in a separate outer shell in the podded architecture. Furthermore, the outer shell must be ruggedized and have a suitable structure for the fighters' aerodynamics. For these reasons, it would be unnecessary to compare the two structures, and internally mounted SPJ systems are more advantageous than podded ones.

When considering the weight and power consumption of the podded and internal architectures, it would be suitable to consolidate the comparison by using the actual SPJ systems in Table 8.1. In addition to weight and power consumption specifications, Table 8.1 presents the function, used platform, and type of some SPJ systems. As seen from the table, the power consumption and weight of the podded systems are higher than those of their internal counterparts with higher capabilities. Weight and power consumption will increase with technological improvement while they are reduced.

The podded SPJ's structural design is intended for fighter aircraft that will probably fly at supersonic speeds. The outer shell geometry of the pod is designed to resemble the structure of the host platform's external centerline fuel tank. The pod cases require lightweight and resilient materials. For this reason, the podded structures, aluminum alloys, and composite materials are used in the outer shell structure. Even though they have relatively low weights, the internal structures do not have such a shell, and they are better in weight. Also, as mentioned before and can be seen from the actual systems, internally mounted systems are more fortunate than podded ones in power consumption.

A podded SPJ system also causes heating problems since the jammer and its potential heat source subcomponents, such as RF power amplifiers and power generators, are in a closed body. The increment in the pod system may temporarily render the electronic components inoperable or damage them permanently. Proper precautions should be considered during the design to prevent this unwanted situation—a cooling structure utilizing prevalent to solve this issue. The generally used parts of

Table 8.1   SPJ System Specifications for Fighters [2, 3]

System Type	Function	Entering Service	Frequency Band (GHz)	Fighter Platform	Type	Weight (kg)	Power Consumption
AN/ALQ-131 (V)	SPJ-EJ	1983	2–20	F-16, F-15, F-111, F-4, A-7, A-10, Harrier	POD	299	~3 kW
AN/ALQ-184 (3 bands)	SPJ	1988	2–10	F-15, F-16, A-10	POD	288	~9 kW
EL/L-8222	SPJ	~2000s	2–18	A-4, F-5, F-18, F-16, F-111, Jaguar, Mig-21, Mig-27, Mig-29 Tornado	POD	100	~3 kW
PAJ-FA (new version)	SPJ	1990s	6–20	Mirage-III, Jaguar	POD	85	~1 kW
AN/ALQ-165	SPJ	1987	1–35	F-18, F-16, F-14, F-16, AV-8B	Internally mounted	91–150	~3 kW
AN/ALQ-126B	SPJ	1984	2–18	A-4, A-7, F-4, F-14, F-16, F-18	Internally mounted	86	~3 kW
AN/ALQ-214	SPJ	1999	1–35	F-15, F-16, F-18	Internally mounted	49	~1 kW
AN/ALQ-178	SPJ	1986	0.5–20	F-16, Mirage V	Internally mounted	121	~3 kW
AN/ALQ-187	SPJ	1988	6.5–18	A-7, F-4, F-16	Internally mounted	7	~1 kW

Note: ~ is used for approximate/expected values.

the cooling system are the turbo compressor, exchanger, pump, accumulator, pipe system, unions, and fittings.

The cooling systems aim to guarantee the performance of the podded SPJ systems under all fighter flight conditions. The podded systems generate a lot of heat, and conventional cooling systems are often too big and heavy to cool sensitive electronics on-air platforms where size and weight matter. The contemporary podded SPJ design includes a *ram air turbine* (RAT) on the front. Liquid cooling technology is used in the active heat control system of the podded SPJ system. During the flight, the podded systems with liquid and air cooling and natural heating receive the air as ram air through the air inlet near the front of the hull and spread it within the pod through the turbo compressor.

A RAT is also used as a power source. It is a small wind turbine installed in an aircraft connected to a hydraulic pump or electrical generator. Due to the aircraft's speed, the RAT generates power from the airstream by ram pressure. It may be called an air-driven generator on some aircraft. A RAT is utilized only for cooling purposes in the pods if the host fighter platforms can supply the power required for the system. In some cases, the existing power supply in the aircraft is sufficient and efficient, and there will be no need for the RAT-type cooling technology. Thus, eliminating the use of additive RAT components may be possible.

Another critical disadvantage of podded structures is that occupying a pylon reduces the additive fuel tank or weapon-carrying capability. It can be inferred that podded and internal SPJ structures have advantages and disadvantages. Podded systems are ready to mount to the platform. However, an internal system is designed for a specific platform, or an additional study is made to adapt an existing SPJ system to the platform. More precisely, an adaptation project is conducted for component localization, electrical connections, data exchange, cabling, and compatibility with the other EW, avionics, and radar systems. Furthermore, specific technical properties are added to the original SPJ, and a new version arises.

8.1.3   Fighters' RCS Effects on SPJ Systems

Inspecting the host systems' RCS would benefit the SPJ systems' effectiveness. Moreover, the type of the SPJ system, such as podded or internally mounted, affects the RCS of the mother platform. The RCS depends on the air platforms' structure, size, and aspect angle. Moreover, it strictly relates to a threat radar's operating frequency. The data obtained in a series of experiments for various aircraft, aspect angles, and radar bands shows that, in many cases, the Rayleigh distribution is a good approximation for aircraft RCS statistics. However, there were exceptions, especially for smaller and all aircraft at broadside [4]. We can classify all the fighters into the roles of attack and bomber, and both in Rayleigh distribution RCS statistics. Figure 8.5 shows the RCS values of different fighter types.

Figure 8.3 gives the RCS versus angle drawings for different fighter generations' *x-y* or yaw planes. The charts in Figure 8.5's unit is dBm^2, defined as decibels relative to 1 m^2 or 10log(RCS/1 m^2). The aircraft's heading is taken as 0°. Although RCS takes different values for different frequency bands, it shows similar behavior for each aspect of different frequency bands. However, this may be invalid when the *radar-absorbent materials* (RAM) are considered since these materials behave

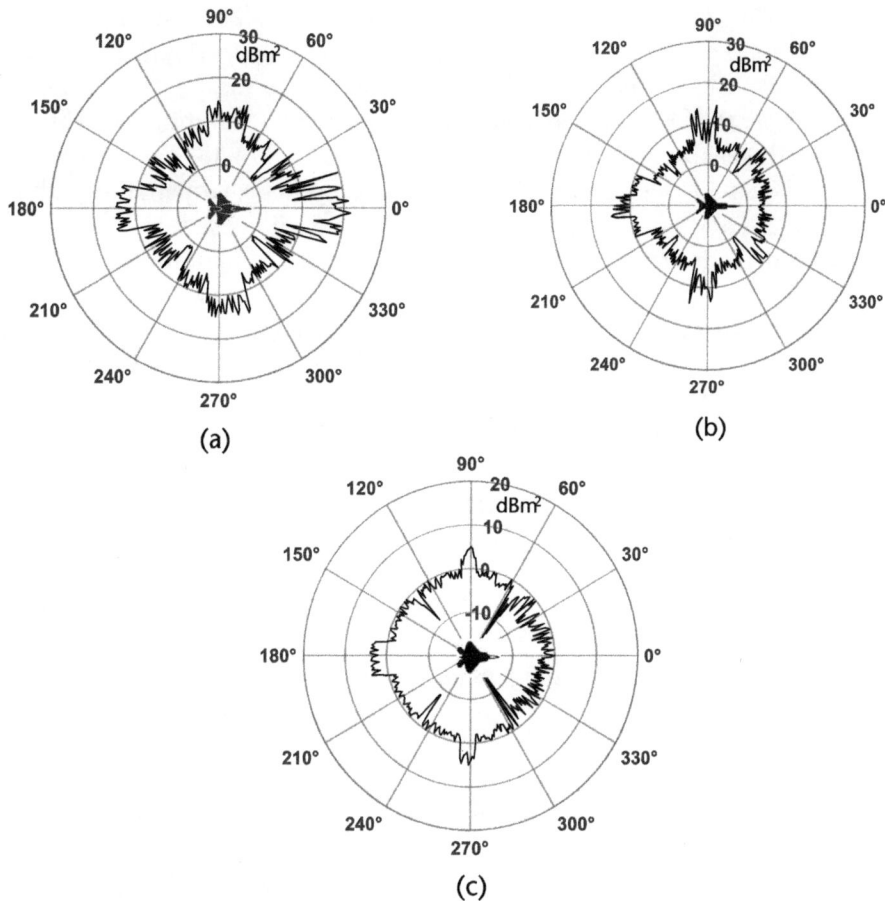

Figure 8.5   RCS polar curve of: (a) a typical third-generation fighter, (b) a typical fourth-generation fighter, and (c) a typical fifth-generation fighter.

differently for various frequencies. For this reason, focusing on the general behavior of RCS from different aspects is more important than the frequency effect on the RCS [5].

Figure 8.5(a) gives the general RCS behavior of third-generation fighters, which covers 1960–1970. RCS reduction applications are rare in this generation and previous generations. The figure shows that the RCS is higher from the front and rear because the radar can "see" the engine parts. Also, it is more extensive from the sides because of the larger cross-section of the fuselage and the angles between the wings and the fuselage. The general characteristics of third-generation fighters are that they are multirole, use advanced avionic systems, and have the first precision ammunition.

Some examples of third-generation fighters are the F-111, F-4, F-5, MIG-17, MIG-21, MIG-23, Mirage F1, and J-8. Some properties may exhibit differences from fighter to fighter, and some of their RCS capabilities may not be as good as the others. Still, Figure 8.5(a) shows a generic model for third-generation fighters.

The general characteristics of fourth-generation fighters are advanced avionics and improved sensitivity. In addition, they have advanced radar systems and enhanced maneuverability. The F-117, the first low-observable fighter aircraft no longer in service, also belongs to this generation. The fourth generation covers a very long period, from 1970 to 2000. Continuing developments in this prolonged period pushed some fighters into a group known as the 4.5 generation. Here, we will consider the fourth generation as a whole without separating it into two groups. The fourth-generation fighters are F-14, F-15, F-16, F/A-18, F-117, MIG-29, MIG-31, Su-27, Su-33, Su-35, Mirage 2000, Tornado, Rafale, Gripen, Eurofighter-Typhoon, J-9, J10, and FC-1.

In the fourth-generation fighters, some of the undesirable RCS effects of the third-generation fighters were reduced by additional improvements. These improvements can be summarized as using RAM, changing the structural design of the fighters, and redesigning jet engine cowlings. As a result of these improvements, the RCS characteristics of fourth-generation fighters have some advantages over third-generation fighters. The RCS values of fourth-generation fighters drastically reduce in heading direction and slightly mitigate in all the other directions. Figure 8.5(b) shows the RCS plot versus the angle in the x-y plane of a typical fourth-generation fighter.

The fifth-generation fighters have fully integrated avionics and sensors, greater speed and maneuverability, and noticeable improvements in RCS reduction. The fighters classified in the fifth generation are the F-22, F-35, J-20, J-31, and Su-57. Their RCS reduction techniques include structural design, shaping, and using RAM.

The experiences gained from the fourth-generation fighters have been used for structural designing. The knowledge gained from the F-117 was used to shape the fighters. The designers tried to avoid any possible surface or edge whose normal vectors would look at a direction where a possible enemy radar might be found, especially for the frontal aspect. Also, all bumps and curves have been avoided in the designs. In the same way, any external loads, such as pylons, missiles, fuel tanks, and pods, are avoided and carried internally as much as possible instead. Fifth-generation fighters use a high percentage of composite materials, decreasing the weight and reducing the RCS in their structure. Thus, a generic RCS plot versus angle in the x-y plane for a fifth-generation fighter is given in Figure 8.5(c).

As seen from the above analysis, the central echo regions that characterize the RCS signature for fighters are the nose, broadside, and tail. A simple approximation for the nose-on aircraft RCS is

$$\sigma = 0.01L^2 \tag{8.1}$$

where σ is the RCS and L is the aircraft length [4].

For using SPJ against monostatic radar, the effect of RCS on the signal-to-jamming ratio can be defined as follows, which is derived from (3.10),

$$\frac{S}{J} = \frac{P_{\text{Tx}} G_R G_{\text{SP}} \sigma B_J L}{P_J G_{\text{Ja}} B_R 4\pi R^2} \quad \left\{ L = \frac{L_J L_{\text{pol}}}{L_R} \right\} \tag{8.2}$$

As seen from the equation, the increment of the RCS benefits the radar system. The components of (8.2) can be defined as follows:

- P_{Tx}: peak radar Tx power (watts);
- P_J: peak jammer power (watts);
- G_{Ja}: jammer antenna gain in the direction of radar (ratio/unitless);
- G_R: radar antenna gain for Tx and Rx operation (ratio/unitless);
- G_{SP}: radar signal processing gain (unitless);
- R: range between radar to jammer (meters);
- B_R: radar Rx bandwidth (MHz);
- B_J: jammer noise bandwidth (MHz);
- L: total losses that enter the process, consisting of jammer losses (L_J), radar losses (L_R), and antenna polarization losses (L_{pol}) (unitless).

Generally, the mean, median, or nose-on RCS is used in (8.2). For detailed and high-resolution analysis, the polar curve may be used for the instant radar boresight aspect. However, this analysis can only be conducted for specific and detailed calculations, such as determining the vulnerabilities of a platform. The median RCS is the middle RCS value of all RCS data. This means that 50% of the RCS spikes inside the arcs will be higher than the median value, and the other 50% will be smaller than the median value. However, mean RCS is the average of all RCS data. Practically, the median and mean RCS values do not give all the details about the RCS. Moreover, two air platforms with the same median and mean RCS are not equally visible to radar when viewed from the same direction. Thus, we also need the contour map to see the arrangement of the reflection lobes.

Also, the RAM effect on the RCS of air vehicles must be considered in this discussion. The RCS is defined by the air vehicles' structure, size, and radars' aspect angle and operating frequency. The air vehicles' structure consists of shape and material. If the materials are the same for the aircraft, the same or similar shapes will not substantially change the behavior of the air platform. Furthermore, different operating frequencies will change the magnitudes of the RCS, but the polar curves' shape will be similar for various frequencies.

When radar-absorbing coating material is used in the air vehicle's body, the polar curves' behavior for different frequencies will not be similar. Because these materials have absorbent properties in specific frequencies, their utilization will spoil the polar curves' shape similarities for various properties. These coating materials may be silicone rubber composites, polypropylene-based composites, honeycomb sandwich structures, or carbon fiber epoxy composites [6]. Also, their thickness gives them different radar-absorbent properties.

For example, MF-500 Urethane broadband RAM coating is a carbonyl iron-based paint coated with a proprietary urethane acrylic resin. When this material is applied to the platform 2.03 mm, it provides maximum absorption at 8 GHz. If applied 1.70 mm, the maximum absorption will exist at 10 GHz. The level of absorption is higher than 20 dB in both cases. However, the absorption level in the other frequencies converges to nonabsorbent materials. Another example of RAM coating materials is the MR11-0010-00 tuned frequency absorber. It is a silicone-based material, and when applied with a 1.7-mm thickness, it provides 25-dB absorption at 10 GHz. The stealth design adjusts RAM material usage for the most dangerous frequency bands.

Table 8.2 gives various fighters' mean yaw (or x-y) plane RCS. The mean RCS values for the optimal design frequencies regarding their shapes and materials are taken. These values are also for the standard or no-load aircraft. The mean RCS may increase if the payloads, such as ammunition, fuel tank, EW, and EO pods, are not carried in internal weapon bays.

Even though the values in Table 8.2 should be used in rough calculations, they give approximate results that may not be ignorable. However, providing the RCS from different aspects would be beneficial for understanding. The problematic areas for fighters and piloted aircraft are their cockpits, wings, and tail (or nozzle). The cockpit and the wings are physical reflectors due to their shape. However, the tail has a different RCS source since the exhaust plume from a jet motor contains radar-reflective ionized gases. Thus, focusing on these parts is essential to increase the comprehension level of RCS behavior. In this context, Table 8.3 presents median RCS values for different aspects of two general-class fighters. Also, Table 8.3 presents the RCS values for two different operating frequency bands. Only generic fourth-generation and fifth-generation fighters are considered for a realistic perspective. Table 8.3 shows that the primary threat frequency is X-band, and the maximum RCS reduction studies aim to achieve in this band. Thus, a second tracking radar system, generally in an unexpected band, would be a good solution for detecting stealth aircraft. Albeit, this solution will increase the hardware.

Ammunition and payload effects on RCS values must also be considered in operational research and planning to ensure the success of the missions and reduce possible casualties. Special measures can considerably reduce aircraft RCS when operational requirements are defined. These measures must be taken into consideration from the design phase.

All the EW and EO systems must be carried internally on interchangeable pallets to reduce the RCS increment effect of the external payloads. Ammunition must

Table 8.2 Mean RCS Values of Fighters in Yaw Plane

Fighter Type	F-4	MiG-21	F-16 A	Tornado	MiG-29	F-16 C	Rafale	F-117	F-35	F-22
Mean RCS (m^2)	6	4	5	8	3	1.2	2	0.003	0.005	0.0001

Table 8.3 Median RCS Values of Generic Fighters from Different Aspects

		L-Band		X-Band	
Fighter Class	Aspect	Fourth Generation	Fifth Generation	Fourth Generation	Fifth Generation
Bomber and combat fighters	Nose (m^2)	5–8	1–3	2–4	0.1–0.5
	Wings (m^2)	10–15	2–5	4–8	1–3
	Tail(m^2)	18	7	5–10	2–4
Combat fighters	Nose (m^2)	2–5	0.5–1	1–3	0.001–0.01
	Wings (m^2)	5–8	1–3	2–5	0.1–1
	Tail (m^2)	6–9	2–4	3–6	0.1–1

be taken in internal weapon bays or semi-buried conformal locations. A convenient method is to place the ammunition in a body covered by expendable radar-absorbing fairings that can be dropped immediately before the weapon is released [7]. If any other type of underwing store is needed, it must be carried in a radar-absorbent container. It must be noted that high RCS reduces the effect of jamming and increases the possibility of detection by the threat radar.

8.1.4   Reprogramming Fighter SPJ Systems

The preprogramming activities include planning, preparing, and modifying the PFMs. The PFM planning and organizing processes were defined in Chapter 7. In this part, we will discuss the modifications in the PFMs, and let us assume that the modification process is reprogramming. The reprogramming of an SPEW system is to maintain or enhance the effectiveness of RWR and SPJ subsystems maintained by technical and operational units. The SPJ and RWR subsystems' requirements during crises or hostilities dominate reprogramming. Even though this part focuses on reprogramming SPJ systems, it is not feasible to think of the reprogramming of SPJ and RWR subunits separately. For this reason, considering reprogramming SPEW systems containing both active and passive parts would be beneficial. The SPEW reprogramming includes changes to the systems' hardware, software, threat weapons, and intelligence data [8]. The middle-term and long-term reprogramming requirements of the fighter aircraft's SPJ system stem from the following issues:

- *Tactical changes:* Fighters are central to the attack package and air interception force. Thus, fighters' tactical requirements are highly dynamic and change according to regional and global threats and technological developments in defense technology. The tactical changes consist of modifications in fighter tactics, equipment settings, and EW systems mission-planning data.
- *SPEW system software adaptation:* The software changes include the changes to the SPJ systems' software. These changes include the system's software altering and developing programmed look-up tables, threat libraries, signal-sorting routines, and jamming capabilities.
- *SPEW system hardware adaptation:* Hardware changes and long-term system development are necessary when tactics or software changes are insufficient to meet the operational requirements due to equipment deficiencies. These changes usually occur when the complex nature of a change leads to a system modification. Hardware changes typically require depot-level support. Both hardware and software can increase the jamming capabilities. The most important of these capabilities are the number of threats that can be jammed simultaneously and the switching time in sequential jamming.

SPEW systems' short-term reprogramming activities provide operational units with timely equipment corrections, tailored solutions for specific theater or mission requirements, and effective responses to changes in adversary threat systems. These activities emerged from short-term reprogramming requirements covering threat changes and theater and mission tailoring.

One of the requirements for prompt reprogramming is threat changes in the potential foes. The SPEW reprogramming support software and hardware are primarily designed to respond to adversary threat changes affecting the system's operational effectiveness. The second is theater tailoring, which reprograms SPEW systems for operations in a specific region or theater. Regional tailoring usually reduces the number of threats in system memory, decreasing processing time and reducing system display ambiguities. The last requirement is mission tailoring, reprogramming the SPEW system for the host platform's mission. Mission tailoring is required to improve system response to the host platform's priority threats.

All air platforms have similar short-term, medium-term, and long-term requirements for reprogramming SPEW systems. However, some slight changes may occur due to their tactical use, maneuverability, and being a priority target.

8.2   SPJ Systems in Helicopters

Helicopters are crucial to many operations, but their structures are unsuitable for high maneuverability and speed. Compared to fixed-wing aircraft, helicopters are less efficient and have limited range. Helicopters are inherently unstable and challenging to fly, and choppers with tail rotors are particularly sensitive to crosswinds. The average cruising speed of a helicopter is generally less than 300 km/h (~160 knots). The maximum flight speed in a helicopter (V-22 Osprey) is about 508 km/h (~275 knots).

Helicopters have excellent maneuverability due to the acceleration limit of up to 4G, which occurs at low speeds. However, they are sensitive to additional weight, which limits the amount of armor and protective equipment that can be carried. The rotors are particularly vulnerable and almost impossible to protect. In low-visibility conditions, helicopters may need to fly at higher altitudes, which increases their susceptibility to radar-guided threats and direct fire [9].

Helicopters have different operational requirements than fighters. Due to their structures and flight principles, their velocity and maneuvering capabilities are less than those of fixed-wing aircraft. Their fundamental advantages are vertical take-off and landing and low-profile flight capabilities. The primary role of helicopters is to support land battles by providing mobility and firepower. Helicopters are unique vehicles that link air advantages with land forces. Moreover, they are compatible with search and rescue missions even in tough landforms like dense forests, cliffs, and mountain chains.

Attack (or armed) helicopters offer a range of advantages over their fixed-wing counterparts. For one, they are less expensive and do not require an airfield. Additionally, they can be deployed forward with land forces and are dedicated to supporting combined-arms ground battles. In urban warfare, attack helicopters excel at using precision engagements to take out specific targets with minimal collateral damage. They can be relied upon to support ground troops whenever and wherever needed.

The increased use of transport (or general purpose) helicopters on the battlefield is closely related to air mobility, a concept founded on using helicopters to provide increased mobility for ground combat forces on and around the battlefield. A safer air assault operation in terms of radar-guided threats can be conducted by transport helicopters rather than their fixed-wing aircraft counterparts. Since they can fly low-profile while penetrating the enemy zone and approaching the drop zone. Moreover, in an assault operation with helicopters, the requirement for parachute usage vanishes. However, they may be vulnerable targets for short-range EO and radar-guided AAA systems this time. Similar situations are valid for search and rescue operations. Helicopters are the unique solution for rescuing pilots downed behind enemy lines. The search may be conducted by different means besides helicopters, such as satellites, UAVs, and reconnaissance aircraft.

As can be extracted, the operation requirements for helicopters are critical and, in some conditions, essential. According to these requirements, helicopters have been imported into operational plans and tactics developed. Also, using the SPJ systems on helicopters has increased in parallel with the requirements. Here, we will discuss the helicopters' potential radar-guided threats, the SPJ system structures used in the helicopters, and their reprogramming requirements.

8.2.1   Potential Radar Guided Threats for Helicopters

The main threats to helicopters are SHORAD and MANPAD systems due to their operational roles and low-flying profiles. From a classification point of view, air defense weapons, tank main armament, antitank guided missiles, field artillery, fighter aircraft, and armed helicopters are potential threats to helicopters. Many threat systems in this class are video or IR-guided systems. Only short-range and medium-range AAA systems can pose a threat to helicopters at a low profile. However, when flying high-profile or over the sea, the radar-guided weapon systems given in Section 8.1.1 threaten helicopters.

They do not have a jet engine like plume exhaust, and the limited heat relative to the fighters may be further reduced with an IR suppressor. Furthermore, using DIRCM and flare gives satisfactory results for protection against IR-guided missiles. However, these countermeasure systems can only partially protect against IR-guided missiles due to helicopters' low maneuverability and speed.

The situation is even worse for helicopters when they become targets of radar-guided threats, which can be directed from land, sea surface, or air. Significantly, the range may extend to the middle-range and long-range when the radar-guided missile comes from a seaborne or airborne platform. Due to the helicopters' maneuvering and speed weaknesses, radar-guided threats generate critical problems even though they have SPJ systems.

8.2.2   SPJ System Structure for Helicopters

This section focuses on the radar-guided missiles used against helicopters, and the structures of the SPJ systems by including the RWR systems are explained. The launching platform may be a land-based, surface-based, or airborne platform. Furthermore, in a mission area, adversaries may use all kinds of platforms, which means short-range to long-range weapon systems may be the threats. RWR and SPJ systems used in fighters may be used in helicopters for radar-guided missile threats. However, this situation causes some drawbacks due to the helicopters' structures and maneuvering capabilities. RWR and SPJ systems must have the following features to protect helicopters from radar-guided threats when operating together.

- *Short detection time:* Helicopters' flight profiles are relatively low compared to fixed-wing aircraft. This is both an advantage and disadvantage for them. The low profile gives the helicopters a short LOS range to the threat systems and the threat radars can detect them at a short distance. However, they have a very short time to react after detection, such as jamming or evasive maneuvers. For this reason, their RWR systems must immediately identify the threat and warn the pilot or crew. Thus, the Rx of the RWR and repeater jammer must have a high dynamic range. Another factor for an immediate warning is the RWR processing time, which must be short enough. The computer hardware must have a high capability for a short processing time, or the PFM structure must be simple for quick detection.
- *Short reaction time:* After detecting and identifying one or multiple radar-guided threats by helicopter, an appropriate countermeasure method or a sequence of methods is applied according to the PFM. A fighter can use jamming, chaff, evasive maneuvers, or multiple countermeasures. However, utilizing all these countermeasures is not valid for helicopters. Generally, using chaff with proper evasive maneuvers will be sufficient. Sharp maneuvers and instant accelerations are not possible with helicopters. Thus, helicopters' usage of chaff needs to be improved in tactical aspects, and evasive maneuvers utilized with other countermeasures are impossible.

  Another challenging problem with chaff usage in helicopters is how to dispense chaff. The airflow of the primary and tail rotor and the helicopter's slow (even zero) speed. When a helicopter hovers at low altitude, a chaff burst drawn into the rotor vortex can make the chopper visible to radar. One solution to this issue is to dispense chaff at the back into the airflow of the tail rotor. This technique has several benefits, including quick blooming and Doppler broadening due to the chaff filaments' rapid movement in the airflow. However, one drawback is that chaff bursts will always appear on one side of the helicopter, which may be better for engagement purposes.

  As a result, jamming is the most essential type of countermeasure that helicopters can use against radar-guided threats. For this reason, SPJ is the most critical part of the helicopters' SPEW systems when they confront radar-guided threats. As stated, helicopters have a short time to evade radar-guided missiles. Helicopters' SPJ systems must react immediately to avoid the fatal zone. The first requirement of the short reaction time requires identifying the threat quickly, as described above. The second requirement is a fast-responding jammer system.

  The identification and jamming processes must be accurate and generate highly effective jamming on the threat radars. These requirements depend on the hardware and the programming capabilities of the SPJ systems. Thus, the hardware used in helicopters should be up-to-date technology, and the PFM data structure must be flexible to ensure reductions and extensions in the number of threats. Power amplifiers are preferred for their solid-state, highly efficient, and quick-switching capability. The antennas should be electronically steering planar arrays capable of handling multiple threats simultaneously or instantly switching from one direction to another.

- *Helicopter-compatible PFM structures:* The requirements differ from those of fixed-wing aircraft. The first reason is that helicopters' low-flying altitudes cause them to be exposed to fewer threats than fixed-wing aircraft. Furthermore, in different missions, different landforms are encountered. In a highland theater area, LOS ranges are probably very short. However, an operation over the sea may require handling more threats from far distances. Thus, the threat numbers in the PFMs must be effectively adjusted from region to region by supporting vital intelligence. When the number of threats is reduced, the threat identification and jamming switching for different threat periods shorten. Furthermore, power management will be more adaptable, and jamming efficiency will be higher.

The general configuration of an internal SPJ system for a generic helicopter is given in Figure 8.6. In the figure, the other SPEW subsystems of the chopper are neglected, and only the SPJ components and the SPEW CMU are considered. The SPJ system consists of elements similar to those of the fighter counterpart. However, the properties and specifications must meet the operational requirements of the helicopter. The properties of the SPJ subsystems also depend on the type of helicopter, such as whether it is armed or for general purposes. Podded SPJ systems are unsuitable for general-purpose helicopters, but armed helicopters may use podded SPJ via their pylons instead of carrying a missile payload.

In the design of the SPJ systems, 360° azimuth coverage is aimed. However, this requirement cannot be met due to the limited number of appropriate helicopter locations. This case is valid not only for internal systems but also for podded systems at helicopters. The required antenna bands are given in Section 8.1.2: Band 1 covers E and F bands, Band 2 covers G and H bands, and Band 3 contains I and J bands. The typical forward antenna locations in helicopters are at the nose and the front of the main rotor transmission casing. The aft antennas are at the back of the main rotor transmission casing, as well as the right and left stabilizers.

Conventional antennas such as spiral, blade, and horn can be used in helicopter SPJ systems. The antenna number is limited due to the mounting constraints of the helicopters. Mounting one set of antennas for the front and one or two for the rear

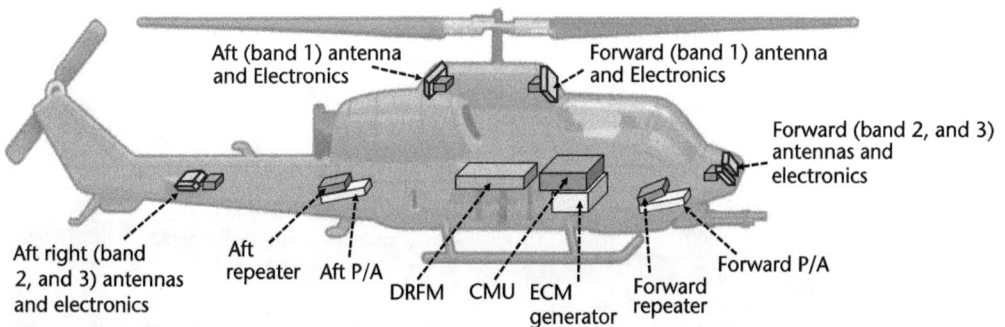

Figure 8.6 The general configuration of an internal SPJ for helicopters.

may be possible. For this reason, the azimuth coverage is generally less than 360° for SPJ systems in helicopters. Furthermore, additional Rx antennas are needed in conventional antenna architecture, or they will be operated in time-sharing between Tx and Rx. Using additive antennas can be eliminated using phased array antennas, and the configuration given in Figure 8.6 is fictionalized for this case.

An internal or podded SPEW or SPJ system may be possible in choppers and fighters. However, the podded structures have more disadvantages when used with helicopters than with fighters since the choppers have a low speed. As mentioned in the fighters' podded SPJ, the contemporary system design includes a RAT on the front. However, this is not the case for choppers since the intake air will probably be insufficient. This situation results in the power being drawn from the host platform, which adds power consumption.

Also, when the SWaP requirements are entered into the process, the podded systems lose their chances in the face of the internally mounted counterparts. In terms of size comparison, internal structures consistently surpass the podded ones. For this reason, the analysis will focus on the weight and power consumption.

For weight and power consumption, actual systems are considered for comparison purposes. Table 8.4 gives some podded and internally mounted SPJ systems used in choppers. One must know that the systems are potentially suitable for fighter or UAV platforms. As Table 8.4 shows, internal structures' weight and power consumption properties are far better than those of podded structures.

Another issue emerges when podded structures are utilized in helicopters—some areas need improvement in identifying and jamming top-attack weapons. Since the Rx cannot detect them, RWR and SPJ systems will not provide the required protection. This situation is probably encountered when a fighter or another armed helicopter

Table 8.4 SPJ System Specifications for Helicopters [2, 3]

System Type	Function	Entering Service	Frequency Band (GHz)	Helicopter Platforms	Type	Weight (kg)	Power Consumption
Intrepid Tiger II	SPJ	2012	~0.5–18	UH-1Y Huey, MV-22B Osprey	POD	100	~4 kW
AN/ALQ-167 (V)	SPJ	~1990s	0.5–18	Lynx HAS Mk 3, Sea King	POD	107	~5 kW
AN/ALQ-162 (V)	SPJ	1998	6–20	MH-60, AH-64, MH-53J, MH-60G SOF	Internally mounted	~19	~1 kW
AN/ALQ-211	SPJ	2002	2–18	AH-64D, MH-47, MH-60	Internally mounted	44	~3 kW
SPJ-20	SPJ	2002	6–17.5	Attack and general purpose	Internally mounted	42	~3 kW

Note: ~ is used for approximate/expected values.

threatens the mother helicopter of the SPEW system. The coverage has blind sectors due to rotors, tail boom, external stores, and other protruding structures.

8.2.3 Helicopters' RCS Effects on SPJ Systems

Section 8.1.3 provided a detailed analysis of the RCS effects on fighters. Similarly, helicopters' RCS values depend on their shape, absorber material, size, threat radars' aspect angle, and operating frequency. Furthermore, when helicopters utilize SPJ against monostatic radar, the effect of RCS on the signal-to-jamming ratio can be calculated as (8.2). The helicopter's RCS solution for calculating static electromagnetic scattering characteristics is the same as that of fighters or combat UAVs. However, due to the rotor's high-speed rotation, the static calculation obviously cannot reflect or meet the helicopter's dynamic scattering characteristics [10].

Although the rotors (or blades) are the primary source of the difference between helicopters' RCS and their fixed-wing counterparts, the contribution of the helicopters' airframes is also essential to the RCS. When helicopters and their blades are manufactured from composite materials, the RCS values of the blades and airframes made of metal may be reduced by 7 to 10 dB. However, the helicopter airframe will still have metal components like the engine.

The rotation speed of the helicopter blades is an important point and a characteristic signature in determining the helicopter's RCS. Contemporary helicopters generally include single-rotor and tandem-rotor. The tandem rotor can be divided into coaxial, longitudinal, and transverse schemes. The helicopter's speed is limited by the extreme operating conditions of the helicopter's rotor blades.

The carrier and the tail blades can be regarded as rotating with the corresponding dipole frequencies. If the length of the blade ℓ is equal to half the wavelength of the irradiated area, $\ell = \lambda/2$, the blade will scatter as a resonant half-wave dipole with the maximum RCS value. In the quasi-resonant region, the resonant increase of the RCS is possible when the wavelength of the irradiation field is commensurate with the size of the entire structure of the helicopter or its components [11]. This increment in the RCS causes vital problems near the resonance frequencies at tens of megahertz. However, these frequencies are much lower than the radar operating frequencies, and their RCS values are acceptable in the radar operating frequencies. The designers and producers consider operating in the resonance frequencies in the design phases of helicopters.

Table 8.5 shows the RCS for various types of helicopters in the design frequencies and threat systems operating bands. The shape of the helicopters and the use of absorber materials to reduce RCS aim to optimize the design in the threat frequency bands, mostly the X-band. Table 8.5 presents the mean RCS values for the yaw (or x-y) plane.

As in fighter aircraft, the effects of ammunition and payload on RCS values must also be considered in operational research and helicopter planning. This way, helicopter casualties might be reduced, and the mission's success will increase. Special measures can considerably reduce helicopter RCS according to the operational requirements. These measures start from the design phase and continue throughout their service lives.

Table 8.5 Mean RCS Values of Different Helicopters in the Yaw Plane

Helicopter Type	Function	Entering Service	Mean RCS (m²)
Sikorsky S-92/USA	Transport	1998	~12
Kamov Ka-62/RUS	Multipurpose	2014	~6
Bell 429/USA	Multipurpose	2006	~5.5
Eurocopter EC.725 Cougar/FRA	Multipurpose	2000	~4
Agusta Westland AW.189/ITA	Multipurpose	2011	~10
Kamov Ka-50 Erdogan/RUS	Attack	1999	~5.5
Agusta A.129 CBT/ITA	Attack	1997	~5

Note: ~ is used for approximate/expected values.

Due to their structural differences, helicopters' RCS is substantially higher than fighters' RCS. Any external payload or ammunition mounting results in an additional RCS increment to the total RCS, which sometimes has undesirable consequences. From an operational point of view, the helicopter aims to carry the EW and EO systems internally on interchangeable pallets to reduce the RCS increment effect of the external payloads. However, this is not as possible as the fighters, especially the EO systems. Furthermore, putting ammunition in internal weapon bays would be impossible for helicopters, but semi-buried conformal locations may be a solution.

A convenient method is to place the ammunition in a body covered by expendable radar-absorbing fairings that can be dropped immediately before the weapon is released. If any other type of external payload is needed, it must be carried in a radar-absorbent container. As in the fighters, high RCS reduces the effect of jamming and increases the possibility of detection by the threat radar. However, this situation results in catastrophic outcomes for low-maneuverable RCS helicopters.

8.2.4 Reprogramming of Helicopter SPJ Systems

The preprogramming activities of helicopters' SPJ systems cannot be separated from SPEW preprogramming. Thus, SPJ and RWR are considered together for preprogramming RF parts of the SPEW systems. As stated, preprogramming includes planning, preparing, and modifying the PFMs. The modification process in the PFMs is reprogramming. The middle-term and long-term reprogramming requirements of the helicopters' SPJ are similar to the fighter aircraft's SPJ systems' requirements, which were given in Section 8.1.3. The exception may occur in tactical changes since the helicopters' tactical usage and tasks differ for fighters.

Helicopters' structures and flight principles differ from those of fixed-wing aircraft. Their velocity and maneuvering capabilities are their weakness. However, their fundamental advantages are vertical take-off and landing, hovering, and low-profile flight capabilities. The operational role of helicopters is to support land forces for mobility and firepower. Additionally, they conduct search and rescue missions in allies' and adversaries' areas. Thus, the tactical requirements of helicopters are highly dynamic and change according to regional and global threats and technological developments in defense technology. The tactical changes consist of modifications in helicopter tactics, equipment settings, and EW systems mission-planning data.

Short-term reprogramming of helicopters' SPEW systems provides operational units with timely equipment corrections, customized solutions reconciled to specific operational or regional requirements, and effective responses to changes in adversary threat systems. These activities have arisen due to short-term reprogramming needs, which address threat changes and the need for theater and mission-specific tailoring.

One of the requirements for prompt reprogramming is threat changes in the potential operations. The SPEW reprogramming support software and hardware are primarily designed to respond to adversary threat changes affecting the system's operational effectiveness. The second is theater tailoring, which reprograms SPEW systems for operations in a specific region or theater. With reliable intelligence, regional tailoring usually reduces the number of threats in system memory, decreasing processing time and reducing system display ambiguities. This reconciling comes into prominence for helicopter SPEW systems. Furthermore, mission tailoring shortens the system response time, increases threat identification accuracy and jamming efficiency, and prioritizes threats.

8.3 SPJ Concept for UAVs

UAVs have been indispensable in supporting operations for many years due to the situational awareness provided by imagery, real-time video, and signal intelligence. However, new missions have been added to their surveillance and reconnaissance roles for several decades. The current missions for the UAVs can be listed as follows:

- Airborne surveillance;
- Search and rescue;
- Third-party targeting and designation;
- Artillery correction;
- Monitoring chemical, biological, and radiation attacks and spreads;
- Battlefield damage assessment;
- Support jamming: stand-in jamming and SEAD for mobile targets;
- Air-to-ground attack.

UAV usage in combat is expected to increase, and UAVs will be substituted for conventional fighters as a future projection. Many countries' UAV doctrines and operational plans are changing in this direction. The combat UAVs engaged in fighter roles are air-to-air combat, both for attacking and air interception.

UAVs fall into two distinct groups: remotely piloted and autonomous vehicles. First, the differences between piloted or conventional aircraft and the general characteristics of UAVs are mentioned. The speed, range, altitude, and payload properties are critical when comparing the differences between piloted and unpiloted vehicles. The most fundamental characteristic is that piloted air vehicles rely on the presence of humans to detect and respond to changes in the vehicle's operation. Human operators have limitations and shortcomings when used instead of sensors; however, the current technology level of nonhuman sensors needs to be sufficiently developed to replace their human counterparts. Aircraft development history has relied on pilots because technology can only assess the aircraft's operation with a human presence. Humans can sense and make provisions for the aircraft's condition, including unusual vibrations indicating structural damage or impending engine failure. Even though artificial intelligence technology has demonstrated stunning development, more is needed to replace pilots and crew with machines. However, piloted aircraft are designed for longer lifetimes than their unpiloted counterparts, principally because the human payload is intrinsically valuable.

The piloted and autonomous UAVs must be evaluated differently when compared with conventional aircraft. Since pilots enter the process, the piloted UAVs converge to their traditional counterparts. Let us compare the distinctions between the piloted UAVs and conventional aircraft. The most drastic difference between them is their ability to sense events within and outside the vehicle, known as situational awareness. In traditional aircraft, the electronic and EO sensors and cockpit view are directly taken by the pilot, and steering them for decision-making is conducted from the same place, the cockpit. However, the piloted UAV pilots are at the operation centers and use the UAVs from there. All the sensor data, except the cockpit view, is conveyed to the operation center, and the pilots try to make critical decisions about flight and use lethal force remotely.

To understand military autonomous UAV systems, let us consider the definitions of automated and autonomous systems. Automated systems are physical systems that function with no or limited human operator involvement, typically in structured and unchanging environments, and whose performance is limited to the specific set of actions it has been designed to accomplish. Normally, these are well-defined tasks with predetermined responses, or their behaviors are "scripted" according to simple rule-based prescriptions. However, autonomous systems are intelligence systems that can independently compose and select among alternative courses of action to accomplish goals based on their knowledge and understanding of the world, themselves, and the local, dynamic context. Unlike automated systems, autonomous systems are designed to respond to situations that are not preprogrammed or anticipated before system deployment [4].

Autonomous weapon systems are weapon systems that, once activated, can select and engage targets without further intervention by an operator. This includes but is not limited to operator-supervised autonomous weapon systems designed to allow operators to override the weapon system's operation but can select and engage targets without further operator input after activation [5]. There are three broad classes of autonomy as it pertains specifically to the role it plays in the use of weapons, delineated by the degree of control that humans can exert on the weapon deployment: human-controlled (semi-autonomous), human-supervised, and fully autonomous [12].

- *Semi-autonomous (human in the loop):* Once the weapon system is activated, it only engages those targets a human operator has selected. Specific functions may include acquiring, tracking, identifying, cueing, and prioritizing potential targets.
- *Supervised autonomous (human on the loop):* Once activated, the weapon system operates under human supervision. The human operator can intervene and terminate engagements, including in the event of a weapon system failure.
- *Fully autonomous (human out of the loop):* Once activated, the weapon system may select and engage targets without further human intervention.

No human supervises the task's operation or can intervene in the event of a system failure [13].

Here, autonomous is used in place of fully autonomous. Thus, autonomous UAVs can perform their functions without a human operator. Conventional missions like surveillance and signal intelligence tend to be smaller and less costly than piloted vehicles. However, this case is different for the combat or fighter UAVs, and their dimensions are similar to those of their piloted counterparts. They require robust hardware and software capabilities for many decision-making functions conducted by pilots in conventional fighters. These functions are indispensable needs of fighters and given as follows but not limited to these only: situational awareness, navigation, communication, controlling avionics and radar, detecting targets, discriminating enemy and ally units, air interception, evasive maneuvers, using missiles, bombing, SPJ, RWR, CMDS, and collaborating with the other ally units.

However, more autonomous control and artificial intelligence technology development is needed for UAVs to conduct all the above functions autonomously. Their capabilities are insufficient to analyze the events or display the freedom of action essential for success in military operations. As with other autonomous vehicles, UAVs cannot replicate the human ability to understand the nuances that distinguish between success and failure in war. Autonomous UAVs cannot demonstrate initiative but must rely on expert systems or artificial intelligence, a product of lists of explicit rules and contingencies.

These systems will lack the human ability to adapt and behave in unpredictable ways, which will increase their vulnerability to enemy actions. One of the reasons for this limitation is that expert systems need to improve at dealing with information that operates at the edge of their knowledge, which increases their vulnerability to enemies who can use deception or ambiguity to confuse the vehicle. Given the sheer number of objects that highly automated systems may encounter and the number of decisions that must be made correctly and rapidly, the problem is the relative ease with which an adversary could exploit autonomous vehicles.

Autonomous vehicles lack the human ability to adapt and behave in unpredictable ways, which makes them more vulnerable to enemy actions. One of the reasons for this limitation is that expert systems need to improve at dealing with information that operates at the edge of their knowledge, which increases their vulnerability to enemies who can use deception or ambiguity to confuse the vehicle. This brittleness problem is a significant concern, given the large number of objects highly automated systems may encounter and the number of decisions that must be made correctly and rapidly. As a result, adversaries could easily exploit autonomous vehicles.

The most crucial property of reliance on humans is that no two pilots react similarly in every situation. Military operations involve how humans identify threats and targets, make decisions in unfamiliar and ambiguous situations, and function analytically and creatively. The advantage of human pilots is that they can adapt to new and different circumstances, make decisions based on incomplete or ambiguous information, and deal with unexpected situations, such as damage or malfunction. One condition distinguishing piloted aircraft from their unmanned counterparts is that uncrewed vehicles do not have these human qualities. While piloted vehicles use electronic devices, such as radio, radar, radar warning, and electronic countermeasures, to supplement the pilot's senses, the most fundamental characteristic of piloted vehicles is that humans make critical decisions about using lethal force.

Conventional aircraft are planned to be substituted by the piloted UAVs shortly, and a mixed form of piloted and autonomous, dominantly autonomous, UAVs in the long term. This is because UAVs offer some advantages over conventional aircraft:

1. UAVs eliminate the risk to a pilot's life.
2. UAVs' aeronautical capabilities, such as endurance, are not bound by human limitations.
3. Improving low observable technology for conventional aircraft may use inherently unstable designs that might be too dangerous for humans. However, these designs can be applied to the UAVs.
4. UAVs can conduct dull, dirty, or dangerous missions that do not require a pilot in the cockpit [14]. The long endurance sortie is an example of a dull mission. Flying through nuclear clouds to collect radioactive samples is an example of a dirty mission. Intelligence surveillance and reconnaissance sorties were flown in the presence of active threats, such as man-portable air defenses or integrated air defense systems, which can be an advantage for dangerous missions.

Using self-protection in UAVs is a relatively new concept that emerged from the new operational concepts of UAVs. Here, it is intended to give information about the usage of the SPJ for UAV platforms, potential radar-guided threats for them, the SPJ system structures of UAVs, and their reprogramming requirements.

8.3.1 UAVs in the Operational Field

Human-crewed aircraft are the core airpower presently and have dominated the twentieth century. They will probably stay as the primary air vehicle for military operations soon. However, technological advances have led to the development of UAVs that can perform some military missions. They fulfill these missions separately or with piloted aircraft. Over the past decades, military forces have used UAVs to perform various tasks, including the following.

- Intelligence, surveillance, and reconnaissance;
- Cargo and resupply for relatively small and light assets;
- Third-party targeting and designation;
- Stand-in jamming (SIJ);
- Close air support;
- Communications relay.

Technological advances are leading to the development of UAVs that can perform military missions that once were reserved for piloted aircraft. Many countries revise their operational plans and try to consider the future operational environment. According to these predictions, UAVs are good candidates for different roles in future military operations. Furthermore, they may replace piloted aircraft for several missions. These missions are listed here:

- Transportation;
- Aerial refueling;
- Air-to-air combat;
- Combat support (including tactical bombing and providing firepower in land battles);
- Strategic bombing;
- Battle management and command and control (BMC2);
- Suppression of enemy air defense (stand-off-jammer);
- Destruction of enemy air defense (DEAD);
- Escort jamming.

Several experimental concepts have recently been developed, such as aircraft system-of-systems (SoS), swarming, and lethal autonomous weapons. These concepts aim to define new roles for the future generations of UAVs in parallel with the latest military doctrines. These concepts can be summarized as follows [7]:

- *System-of-systems (SoS)* refers to a collection of systems, each capable of independent operation, that interoperates to achieve additional desired capabilities. These systems may be deployed on separate aircraft, including crewed, optionally crewed, and uncrewed aircraft.
- *Swarming* refers to cooperative behavior, generally enabled by artificial intelligence and networked communications, in which a group of UAVs autonomously coordinates to accomplish a mission [15–16]. Notional swarming concepts range from large formations of low-cost UAVs that could overwhelm adversary defensive systems to smaller, more tailored formations that execute electronic attacks or ISR missions.
- There is no international agreement on the definition of *lethal autonomous weapon systems* (LAWSs). LAWSs are a class of weapon systems capable of independently identifying a target and employing an onboard weapon to engage and destroy it without manual human control [13]. This concept is known as human out-of-the-loop or fully autonomous.

When considering the future roles of UAVs, it may examine a range of issues, including lethal autonomous weapons and arms control, cost comparisons of future UAS with piloted aircraft, personnel and skills implications, concepts of operation and employment, and the proliferation of unpiloted technologies. The technological trends, military doctrines, and requirements force the study of combat UAV usage to substitute for fighter roles. Combat UAV usage has been started in the adversary theater as the third-party targeting and designation. Furthermore, for future military operation requirements of UAVs, they will be engaged in air-to-air and air-to-ground combat for offensive, defensive, and support missions. Some are related to fighter missions, such as dogfighting, DEAD, combat support, bombing, and escort jamming. Some are related to cargo aircraft, such as transportation and aerial refueling. In each case, the UAVs conduct their missions in the adversary regions, requiring SPEW systems and their inseparable part SPJ systems.

8.3.2 Potential Radar-Guided Threats for UAVs

Military UAVs' nominal operating altitudes can range from 1,200 feet above ground level to over 18,000 feet at flight level, while their service ceilings can be between 50 and 65,000 feet. Chapter 1 gave the UAV classification in Table 1.2, and to inspect the potential radar-guided threats against UAVs, we will consider Class II and III here. Furthermore, the propulsion systems are assumed to be similar to their piloted counterparts.

Here, the radar-guided threats to UAVs are examined for their existence and future military roles. These roles are in summary, cargo and resupply, third-party targeting and designation, SIJ, close air support, communications relay, air-to-air, and air-to-ground combat for offensive and defensive purposes. Fighters, cargo aircraft, and helicopters conduct these missions. However, some of them, such as SIJ and communications relay, are special for UAVs.

The threats for low-profile, medium-profile, and high-profile flight UAVs are considered, using the above assumptions, as follows:

- *Low-profile flights:* The main threats to the low-profile UAVs are SHORAD and MANPAD systems. Moreover, air defense weapons, tank main armament, antitank-guided missiles, field artillery, fighter aircraft, and armed helicopters are potential threats to UAVs. Many threat systems in this class are video or IR-guided systems. Radar and EO-guided AAA systems are ultimately fatal threats for them at low profiles. Furthermore, radar-guided weapon systems threaten UAVs when flying over the sea.
- *Medium and high-profile flights:* Since the threats are almost identical, it would be appropriate to consider medium and high-profile UAV flights together. Radar-guided SAM, AAM weapons, and piloted fighters are the main threats in these altitudes. Radar-guided AAA systems are another crucial jeopardy for UAVs in the middle-profile altitudes. Also, dogfights can be considered in medium and high-profile categories, and this time, the threats are radar and EO-guided AAM systems.

As mentioned, UAVs are mainly used for dull, dirty, or dangerous missions. Another crucial difference between UAVs and conventional piloted aircraft and helicopters is the variety of operations on the same platform. Thus, a Class-III UAV can be used for third-party targeting and designation, BMC2, SIJ, and supporting land battles by providing firepower, intelligence, surveillance, reconnaissance, and communications relay. The critical point is engaging different payloads and ammunition for each duty to fulfill all these missions using the same platform. However, their operational altitudes will change according to the mission. The common UAV properties of these missions are their propulsion systems. They might use piston or turboprop engines, with lower maneuvering capabilities than fighters. UAVs may substitute different conventional airborne platforms, such as cargo aircraft and helicopters.

Some missions require structural variations in the UAVs in addition to the payloads. These missions are strategic bombing, aerial refueling, stand-off-jamming, and transport. The UAVs in these roles generally utilize turboprop engines. These missions are conducted by cargo aircraft. When UAVs are used in fighter roles, their engines are turbojet and turbofan. These roles are air-to-air combat, DEAD, tactical bombing, and escort jamming. The fighter roles require high maneuvering and fast-changing situation capabilities. Also, different roles need various payloads, missiles, and ammunition.

Radar-guided threats against helicopters and fighters also threaten UAVs. Thus, UAV platforms require RWR and SPJ systems as a subsystem of SPEW systems. However, a tradeoff study is needed for a cost-effective solution.

8.3.3   SPJ Operational Requirements for UAVs

From the threat analysis conducted above, the present and future missions of the UAVs require SPJ systems. For this purpose, prerequirement studies should be undertaken to obtain a cost-effective system. The most essential part of this study is to decide the protection level of the platform. Since the loss of the UAV does not mean the loss of a pilot, the decision about the SPEW system protection level depends on the technology of the UAV platform, payloads, and mission importance. Two concepts may represent the protection level; the first is the number of the SPEW system's subcomponents. A typical SPEW system comprises seven subsystems: RWR, SPJ, CMDS, DIRCM, IR suppressor, MWS, and LWS. In the first approach, the SPEW system includes some of the subsystems. For example, it is composed of just the IR part or just the RF part. As another approach, just the passive parts (RWR, MWS, and LWR) are in the SPEW system.

The second concept uses a complete SPEW system with different specifications, such as jammer power, Rx sensitivity, Rx/Tx processing capabilities, operating frequency coverage, simultaneous detection, identification, and jamming capabilities. When the mission's success is critical, the payloads are very valuable (due to their physical value or confidential technology), or the UAV platform uses state-of-the-art technology in structural or engine design.

The SPEW system usage on the UAVs is still conceptual, and some tests have been realized for the last decade. Some experiences may have been discovered in military operations, but their usage and operational success level have not been reported in any sources. Also, all the reports about the tested SPEW systems were about the podded structures. As stated above, the basic structures of the UAVs are similar for a group of missions, and the difference occurs in the payloads and ammunition. Thus, the pylons are limited and crucial for the payloads. It must be noted that the SPEW system is a force multiplier for UAVs and not the central part of the mission. Thus, using one of the pylons for the SPEW system results in one deficient payload or ammunition. A modular onboard SPEW system would be more practical and suitable for UAVs. Furthermore, if the system consists of subcomponents in a line-replaceable unit (LRU) structure, which is easily mounted and removed, it will increase operational and logistical effectiveness.

UAVs' operational role and technical properties continuously change according to operational requirements and technological developments. Being a UAV pilot is ordinary today, even in demanding and challenging missions. However, it had only been a dream for a few decades before. With the rise of UAVs in the operational area, their usage and roles have increased, and this increment will continue. Future military concepts include system-of-systems, artificial intelligence-enabled manned-unmanned teaming, swarming, and lethal autonomous weapon systems [15].

The extractions from these concepts are using artificial intelligence in UAVs and increasing autonomy. The most exciting result of these concepts is autonomous dogfighting. Dreaming of a fully autonomous UAV in a dogfight is heartwarming. The future operational area will contain piloted and autonomous UAVs at different levels. The SPJ requirements for piloted and autonomous UAVs are to be inspected separately. The following sections mention these needs.

8.3.3.1   Piloted UAVs' SPJ System Requirements

A well-accepted definition for autonomous and piloted UAVs is given in [17]: "UAV is a powered, aerial vehicle that does not carry a human operator, uses aerodynamic forces to provide vehicle lift, can fly autonomously or be piloted remotely, can be expendable or recoverable, and can carry a lethal or nonlethal payload." The piloted UAV properties can be summarized as follows:

- They do not carry a pilot or crew.
- Their aerodynamic structures must be like that of conventional aircraft.
- The UAVs are remotely controlled for navigation and payloads.

These properties affect the SPEW systems and, consequently, SPJ systems. In addition to the fuel tanks, engine, engine electronics, and battery block, navigation devices, sensors, payloads, and datalink communication hardware are on the UAV. An operator or pilot on the ground control station conducts the control units and the decision-making, thus saving not only the crew but also the automation hardware and software. This saving provides a substantial advantage to piloted UAVs since they do not have to carry humans or use advanced technologies for automation. However, some disadvantages may emerge due to the structure. In the case of a remote pilot, control may be continuous or episodic. When the control is continuous, the only controller of the navigation and payloads is the pilot. However, in episodic control, some automation is required, depending on the type and duration of the automation. Here, we will focus only on continuous remote pilot control and Class-II and III types of UAVs.

The most dominant disadvantage of a piloted UAV is the sensors' instantaneous field-of-view (FOV) limitation. During the flight, situational awareness is obtained from daylight cameras, FLIR, and radar systems. Typically, airborne radars are large systems mounted in conventional human-crewed aircraft. However, the radar that has been integrated into a UAV must be a competent radar system that is compact and lightweight. The control operators at the ground station can obtain situational awareness according to the sensor data sent to them. All the sensors have their instantaneous FOV, and the operators must make the navigational and payload usage decisions based on this data. However, in conventional piloted aircraft, pilots are in the cockpit and act like another sensor. They can change their focus to different directions and immediately switch the following sensor according to the situation.

Another sensor type for increasing situational awareness is SPEW systems. Both RF and EO threats can be noticed and neutralized by the system. Again, let us change our focus to the RF part of the SPJ, which consists of RWR, SPJ, and chaff dispenser. Similar to the radar systems, the RWR systems that have been integrated into UAVs must be competent systems that are compact and lightweight when compared to their aircraft counterparts. The UAVs with piston or turboprop engines have low maneuvering capabilities and high-flying profiles. This situation makes them easy targets, and their RWR sensitivities and data processing capabilities should be high. Moreover, the chaff usage may sometimes be insufficient for UAVs due to their maneuvering limits. The SPJ system capabilities must compensate for the UAVs' low maneuvering and SWaP properties. Thus, the SPJ systems' ERPs and instantaneous frequency coverage must be high. Furthermore, the response and switching times are to be as low as possible. This claim is valid when considering piston and turboprop engines. The UAVs in this propulsion type are generally engaged in surveillance, intelligence, and third-party designation missions.

However, the situation has changed for UAVs with turbojet and turbofan motors. Their maneuvering and SWaP capabilities are high, and the technical properties of RWR, SPJ, and chaff dispenser systems converge with those of their conventional fighter aircraft counterparts. In case of transferring required data to the controller at the ground station in the shortest time, this operation method would be the optimal case for logistical, technical, and security aspects. The most dangerous operations are conducted by fighters, such as attacking the inside parts of adversary regions, using close air support, AI, and especially dogfights. Thus, using UAVs in place of fighters will save many human lives. Even in podded or onboard architecture, maintenance, repair, and LRU/SRU replacement would be at the level of conventional fighters, which the logistics and technical staff have conducted for many years.

The use of UAVs in place of fighters is a recent concept. There are some gaps in the idea that must be filled. The first issue about this subject is the control method. The question is: "Should it be fully automated, half-automated, or piloted remotely?" When we consider the answer to be remotely controlled by a pilot, we can assume that the SWaP-providing capability of the UAV will increase. Since there will not be a crew in the air vehicle, the decision-making will be conducted in the ground area. However, the challenging problem is that fighter UAVs have not been tested sufficiently in the operation fields. Furthermore, the psychomotor performances and behaviors of the pilots engaged in such UAV duties must be studied. In a conventional fighter, a pilot is concerned about his or her life and the precious airborne platform with all high-technology avionics and weapon systems. However, when substituting the fighters with remote-piloted UAVs, the pilots worry about only the valuable UAVs with precious electronics and weapons. The possible behavior of the pilots in trials and actual operations, their training processes, and their moods under such pressures must be studied.

When considering advantages and disadvantages, the SPJ and RWR systems are to be studied separately, even if it is not our main subject. Although the SPJ and RWR systems are for situational awareness, they are generally utilized when the mother platform is threatened. As stated, using UAVs in fighter roles is a new concept; equipping them with SPJ and RWR is conceptual. Some trials were made to adapt different podded SPEW systems to the Class-III UAVs in fighter roles. The podded systems are smaller in dimensions and weight than the conventional fighter counterparts. Some podded systems are too small to protect the mother platform. The trial results of the success of SPJ and RWR systems for UAVs have not been shared as a scientific report. Furthermore, no official usage of an SPJ and RWR system in operation has been reported yet. Thus, the requirements for the conceptual approaches are mentioned here. The UAVs in the fighter roles are considered first.

- For piloted fighter UAVs, SPJ and RWR systems must be at least as capable as conventional fighters' systems. However, fighter UAVs' SPJ and RWR systems' response, processing, and switching times must be shorter than traditional fighters. This is because the situational awareness data is sent to the ground station, the pilot decides, and the control command is returned to the UAV.
  For piloted fighter UAVs, onboard SPJ and RWR systems would be more valuable than the podded systems. Thus, a platform-coherent system would provide the needs, and the pylons could be utilized for payloads and weapon systems.
- The indicators and command interfaces in the remote-control station for the SPJ and RWR systems should be more detailed, user-friendly, and accurate than those in conventional fighters. Operations experiments will also be conducted to improve the indicators and controllers.
- The experienced pilot can tolerate EMI and EMC discrepancies during flight with conventional aircraft. However, this is not possible for UAVs. Thus, the SPJ and RWR systems must have low EMI and high EMC compared to other UAV electronic systems. This is another essential factor in the onboard design of SPJ and RWR systems.
- The control signals are vital; thus, more than one link should be established between the ground control station and the UAVs if interrupted and jammed. The link signals have ECCM properties, such as frequency agile or DSSS. Also, the command signals must be encrypted to prevent deception by the adversary units.

Piston and turboprop Class-II and III UAVs, such as surveillance, intelligence, and third-party designation missions, generally conduct the other UAV missions. The above fighter requirements are also valid for SPJ and RWR properties for these kinds of UAVs. However, at this time, some additional operational and technical properties should be considered.

These types of UAVs need a proper application to be the most cost-effective. Cost overrun, while considering the technology level of the UAV, may not be an acceptable solution. In each specific UAV system design, the trade-offs, usually based on the costs of alternative systems, must be made to ensure that the selected self-protection capabilities are cost-effective. Here, cost-effectiveness means that some of the technical properties of RWR and SPJ systems may be reduced according to the mission type and flight profile. Say that the sensitivity, dynamic range, processing capabilities, and instant frequency band of the RWR may worsen to reduce the price. However, the jammer ERP, frequency coverage, switching time, and response time may deteriorate similarly. Thus, in this case, the understanding turned that developing an all-around survivable UAV could not have been easy.

8.3.3.2   Autonomous UAVs' SPJ System Requirements

As mentioned, there are three types of autonomy concerning weapon deployment: human-controlled (semiautonomous), human-supervised, and fully autonomous or autonomous. Let us consider the navigation in addition to the weapon deployment and make the autonomous UAV definition again. In the autonomous operation of UAVs, navigation and weapon/payload usage are conducted fully autonomously. The mission's target is defined and given to the UAV. Then the UAV fully autonomously determines the probable routes, selects the required weapon systems, and engages the potential threats at the route without further human intervention.

The main disadvantages of the piloted UAVs are time delay during taking data and sending the control command, determining sensor precedency, sensors' FOV, pilots' decision-making from a ground control center, and the systems' links vulnerability to intervention and jamming. However, the advantages can be counted as human operator controls, especially in extreme cases, the extra hardware of the systems for navigation and payload control, and extra software and artificial intelligence support for automation and decision making. Also, for UAVs with no state-of-the-art technology and relatively dispensable payloads, cost-effectiveness should be considered, and further study for determining the technical specifications of the SPJ and RWR system.

The advantages and disadvantages of autonomous UAVs differ from those of piloted ones. The data stream delay is no longer a disadvantage for autonomous UAVs since the data taken from the sensors is used on the UAV computer. Moreover, the ground control center and the systems' links vulnerability to intervention and jamming is meaningless for autonomous UAV systems. Maybe a mission or target-changing order can be taken from the ground control center via a data link in an extreme case, but autonomous UAV systems do not require them at all during missions.

Determining sensor precedency, changing sensors' focal points to the required direction, frequency, and decision-making are the main properties of autonomous UAV systems. These essential technological improvements include machine and deep learning, adaptive control, resource allocation, and power management techniques. These properties include state-of-the-art technology for hardware and software and cannot be considered dispensable. The pilot and crew are entirely extracted from the processes, and the orders given by decision-makers to the tactical commanders directly make plans on the machines rather than staff. However, be aware that the machines are precious for price, technology level, and conducting the mission. The tactical commanders depend on them to fulfill their missions.

The most important expectation from them is to be standard and accurate. These are their tactical properties, and they seem like advantages. These properties may have advantages or disadvantages, depending on the technical properties. A well-designed, equipped with state-of-the-art hardware and software, carefully programmed systems without missing any detail, and applied high-technology artificial intelligence systems might meet standard and accurate mission conducting. An autonomous UAV system with these properties deserves top-level EW protection. Autonomous UAVs have less space than piloted UAVs since they have additional hardware to control the navigation, payloads, mission, and SPEW system. Although they do not require a link system to send data and take control orders, they still have a link for communication purposes.

Autonomous UAV usage for military purposes is still in trials, and missions are limited. However, almost all countries plan to use autonomous UAVs dominantly in the future and prepare their military doctrines in this direction. Some of the planned future roles for autonomous UAVs are related to fighter missions, such as dogfighting, DEAD, combat support, tactical bombing, and escort jamming. Moreover, the plans include strategic bombing, transporting, air assault, stand-off-jamming, and third-party designating.

The tactical plans and technical properties of autonomous UAVs dictate that they use a capable SPEW system. These UAVs probably may be the highest priority targets. Thus, the SPJ and RWR should have high-level performance properties. Like the piloted UAVs, no official usage of an SPJ and RWR system on the autonomous UAVs has been reported yet. Thus, the requirements for the conceptual approaches are mentioned here. Autonomous UAVs are considered entirely for all roles and engine types since they are state-of-the-art structures, and cost-effectiveness is not conceived in the first stage. The conceptual requirements are summarized as follows:

- The autonomous UAVs' SPJ and RWR systems must be at least as capable as conventional fighters. The main properties, such as sensitivity, jammer ERP, frequency coverage for Rx, and jamming, should parallel recent technology.
- The autonomous system uses the information taken from the sensors. For this reason, the sensors must be very accurate. The general architecture of the UAV system may be in two types: The SPJ and RWR systems are expected to take orders from the general autonomous system, or the central autonomy enters the decision process like a pilot. The other architecture is UAVs' SPJ and RWR systems, which operate independently and do not take any input from the central autonomy. The former is a central management system, and the latter is a distributed structure. In both cases, the Rx sensors and jammer processes must be accurate.
- The response, processing, and switching times can be at the level of conventional fighters.
- Onboard SPJ and RWR systems would be more valuable than the podded systems for piloted fighter UAVs. Thus, a platform-coherent system provides the needs, and the pylons can be utilized for payloads and weapon systems. Furthermore, the central autonomy can be applied more effectively.
- The SPJ and RWR systems must have low EMI and high EMC compared to the other electronic systems of the UAVs. As stated in piloted UAVs, this is another essential factor in the onboard design of SPJ and RWR systems.

8.3.4   SPJ System Structure for UAVs

As stated above, engaging the podded system at the UAV systems has some disadvantages. However, the onboard SPEW system usage for UAVs is a relatively novel concept. A conceptual layout of SPJ subunits for a generic UAV is given in Figure 8.7. Figure 8.7(a) shows the side view layout, and Figure 8.7(b) shows the top view layout. In the conceptual design, Band 1, 2, and 3 antennas are planar arrays, and they are located under the wings. These antennas can also be utilized for RWR purposes.

Repeater jamming subcomponents, repeater, and power amplifier (P/A) are usually in two sets, front and rear, for fighter jets and helicopters. Mechanical tolerance for sharp maneuvers with high G forces, possible maximum fuel tanks for long missions, and RCS reduction make the UAV structure a small space for internal avionics, additive electronics, and cabling. For this reason, wings are appropriate places for mounting antennas and their electronics. Thus, UAVs can utilize the repeater jamming subcomponents in left and right sets. Each vacant space can be evaluated using the subcomponents of the SPJ system.

In the configuration in Figure 8.7, the space between the fuel cell assembly is suitable for mounting left and right repeaters and P/A electronics. Moreover, the space in the accessory bay is appropriate for DRFM, ECM generator, and CMU units. Also, the other subunits of the SPEW system can be mounted in the same

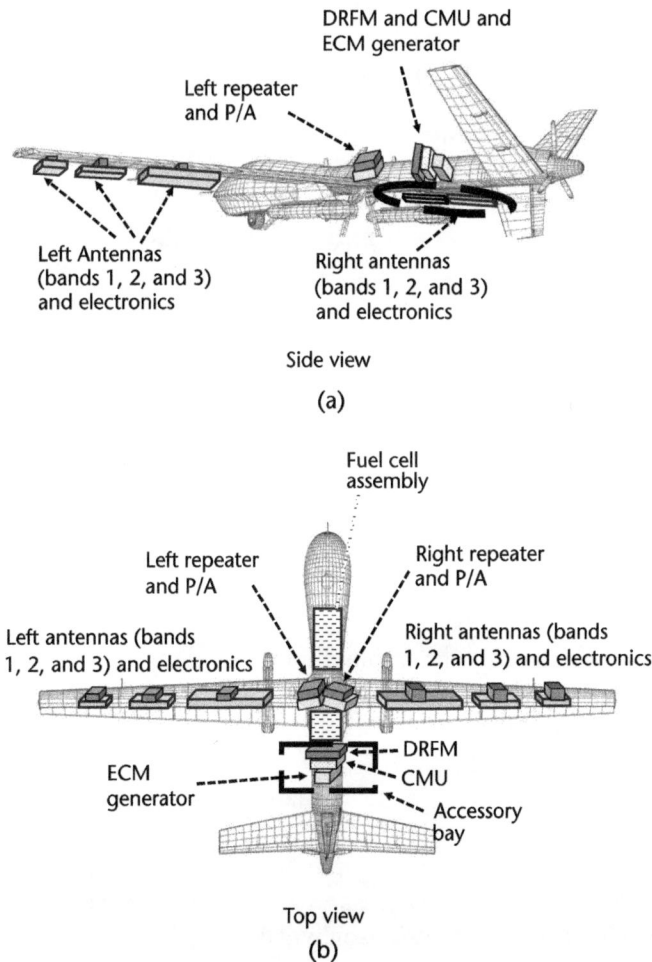

Figure 8.7 The conceptual configuration of an internal SPJ for a generic UAV: (a) side view, and (b) top view.

places, or a further design study is made to fit them into the UAV body. As seen from the example, each UAV design should be considered separately, and different studies should be conducted to adapt the SPJ (also SPEW) systems.

In the design phase of the onboard SPJ systems, some critical issues are considered to determine the required protection level of the UAVs. For this reason, both the monetary and technological values of the UAVs and their payloads become prominent. UAVs do not carry pilots and crew; however, Class-II and III UAVs are valuable. The cost becomes nonignorable when the payload values are added to the UAVs. The SPEW system, including SPJ, at conventional airborne vehicles is the insurance for pilots and crews to return home. They will be the returning insurance for the airborne robots soon.

Presently, most of the UAV operations are conducted remotely by human operators. However, technological trends and military doctrines are directed at autonomous UAVs. Nothing can compare with human life, but the technology behind fully autonomous aerial vehicles is precious. Moreover, their mission may protect many lives or provide a significant advantage in a battle or even the result of a war. Thus, using the SPEW systems will be inevitable soon.

Let us narrow our scope again with the RF part of the SPEW subsystems, which are SPJ and RWR. The RF part of the SPEW system properties can be determined according to the following criteria:

- The SWaP capabilities of the UAV;
- The monetary and technological values of the UAV;
- The mission importance level;
- Payload's monetary and technological values;
- Confidential state-of-the-art technologies of the UAVs and payloads.

According to the criteria, the protection level and the subcomponent types of the RF part of the SPEW systems are of particular importance. More protection requires more SWaP demands. The technical specifications that affect the SWaP demands are as follows:

- Rx's sensitivity;
- Instantaneous Rx bandwidth;
- Data processing capabilities;
- Operating frequency band;
- Simultaneously threat-handling capability;
- Jamming power level;
- Applicable jamming types;
- DRFM capabilities;
- Simultaneous and sequential jamming capabilities.

The SWaP comparison for the podded and internal SPJ systems is similar to that of the fighters if the UAV has a jet engine. However, if the engine is an internal combustion type, the comparison will converge to the helicopters. The UAVs' usage concepts, operational roles, and related doctrines are in the rearrangement phase.

Thus, they are improving rapidly, and the roles of the UAVs have increased at the same speed. We probably must discuss many things far beyond our imaginations in a few decades. However, using the existing data, Table 8.6 can be obtained.

Table 8.6 demonstrates the weight and power consumption for some used or potential candidates for podded and internal structures. No internally mounted SPJ system has been reported for UAVs, but the potential ones have been selected. For this reason, the balance between the podded and internal systems seems close to each other.

However, it must be noted that the internal systems given in Table 8.6 are designed for helicopters and fighters. Say that AN/ALQ-167 has been tested on MQ-9, but it is a system for choppers and fighters. AN/ALQ-211 has been used in different helicopters and F-16 fighters. Moreover, the main aim of SPJ-20 systems is helicopters. Up until now, only the podded systems have been tested or operated with UAVs. Shortly, the concept will inevitably extend to internally mounted SPJ systems.

8.3.5   UAVs' RCS Effects on SPJ Systems

As mentioned, there are various UAVs, both in size and mission. Here, we focus on combat UAVs's RCS structures that can utilize the SPJ systems given in Table 8.6. Also, these are assumed to be fixed-wing ones. The most incredible advantage of UAVs regarding the RCS structure is that they do not carry a crew, so they do not have to consider aerodynamics and people's physical tolerances simultaneously. The design of UAVs is primarily aimed at optimizing aerodynamics while reducing RCS. In the design, the shape and materials used might differ from those of fighters and helicopters, and they are more suitable for RCS reduction than their piloted counterparts.

Another critical advantage of UAVs compared to piloted air vehicles is that they do not require a cockpit, an important RCS source for fighters and helicopters.

Table 8.6   SPJ System Specifications for UAVs [2, 3]

System Type	Function	Entering Service	Frequency Band (GHz)	Fighter Platform	Type	Weight (kg)	Power Consumption
Tactical radar jammer (TRJ)	SPJ-EJ	1996	0.5–18	Combat UAVs	POD	46.3	~1 kW
AN/ALQ-167 (V)	SPJ	~1990s	0.5–18	MQ-9, and potentially all the combat UAVs	POD	107	~5 kW
AN/ALQ-211	SPJ	2002	2–18	Potential for combat UAVs	Internally mounted	44	~3 kW
SPJ-20	SPJ	2002	6–17.5	Potential for combat UAVs	Internally mounted	42	~3 kW

Note: ~ is used for approximate values.

The primary RCS sources for fixed-wing air vehicles are the cockpits, wings, and nozzles. However, the rotors (or blades) are an essential component in addition to the airframe for rotary-wing vehicles. Although fixed-wing UAVs do not have cockpits, they still have wings and nozzles.

However, the main objective in the operational design of combatted aircraft is to reduce the RCS effects on the cockpit. This is because the threat radar and the attack aircraft engage in nose-to-nose or air-to-ground engagement in the attack and AI missions. Thus, the operational feedback from the operation staff to the designers aims to detect before being detected in the nose-to-nose and air-to-ground engagements. For this reason, the contemporary aircraft designs aim to reduce the RCS signature of the cockpits. This is a natural process in combat UAVs since they do not require a conventional cockpit.

Even though combat UAVs are new in terms of fighters and helicopters, they use the same technology as their piloted counterparts. So the same materials as the fighters are used in the combat UAV design. The fighters' experiences are obtained in the shape design, and the newest technological software tools are utilized. Moreover, many research and development studies are conducted for them. In these studies, researchers, designers, and producers focus on the structural design, ammunition, and payload layouts. As stated in fighter and helicopter RCS discussions, their effects on RCS values cannot be underestimated.

The podded payloads and ammunition usage are very common in the combat UAVs. This is because of the operational flexibility and the recently determined operational roles of the combat UAVs. As in the fighters, EW and EO systems must be carried internally on interchangeable pallets to reduce the RCS increment effect of the external payloads of the UAVs. Moreover, ammunition must be stored in internal weapon bays or semi-buried conformal locations to ensure RCS reduction. An effective way to do this is to place the ammunition in a body covered by expendable radar-absorbing fairings. These fairings can be dropped immediately before the weapon is released. If any other type of underwing store is required, it must be carried in a radar-absorbent container.

Although UAVs are engaged in challenging and dirty missions nowadays, they will be the backbone of the airspace in the future. The SPEW systems will be their integral part. Thus, a reduced RCS would increase the jamming efficiency of the SPJ systems and decrease the possibility of detection by threat radars.

Table 8.7 gives the mean of approximate yaw (or x-y) plane RCS for some fixed-wing UAVs. This table provides some ideas for the medium and large fixed-wing UAVs. Although currently, the example UAVs given in the table are not utilized in combat roles directly, they have close shapes and dimensions to them. Also, some physical dimensions affecting the RCS values are given in Table 8.7. In this context, the UAVs' fuselage (body) lengths and wing spans are given in meters.

8.3.6 Reprogramming of UAV SPJ Systems

The preprogramming activities of UAVs' SPJ systems cannot be separated from the entire SPEW system preprogramming. Thus, SPJ and RWR are considered together for preprogramming RF parts of the SPEW systems. As stated, preprogramming

Table 8.7   Mean RCS Values of UAVs in Yaw Plane

UAV Type	MQ-1 Predator	KHAI-112	Heron	Barracuda
Function	Multipurpose	Multipurpose	Reconnaissance	Reconnaissance
Body length (m)	8.23	2.7	8.5	8.25
Wing span (m)	14.84	3.5	16	7.22
Mean RCS (m²)	<~4	<~6	<~6	<~2

Note: ~ is used for approximate values.

includes planning, preparing, and modifying the PFMs. The modification process in the PFMs is reprogramming. The SPJ and RWR systems' usage at UAVs is conceptual at present. UAVs have engaged in operational fields and have had many different roles for years. Also, they are planning to substitute with manned airborne vehicles soon. Even though manned air vehicles will not fall into disuse in the next few decades, UAVs are expected to be used increasingly.

Thus, procedures and standards must be defined and developed for preprogramming and reprogramming activities of UAVs' SPJ and RWR systems. The preprogramming activities' PFM planning and preparing processes will be similar to those of the fighters and helicopters defined in Chapter 7. Here, a projection for long-term, middle-term, and short-term reprogramming requirements is given.

The middle-term and long-term reprogramming requirements of the UAVs' SPJ should be similar to those of the fighter aircraft and helicopters' SPJ systems. The exception may occur in tactical changes since the UAVs' tactical usage and tasks differ regarding their types. Suppose that the engine of a UAV is a piston or turboprop. In that case, its flying profile and maneuvers are similar to helicopters in low altitudes and cargo aircraft in high altitudes. However, if a UAV's engine is turbojet or turbofan, its flying profile and maneuvers are like that of fighters.

Short-term reprogramming of UAVs' SPJ and RWR systems provides operational units with timely equipment corrections, customized solutions reconciled to specific operational or regional requirements, and effective responses to changes in adversary threat systems. These activities will emerge from the short-term reprogramming needs, which address threat changes and the need for theater and mission-specific tailoring. These needs are summarized as follows:

- *Threat changes in the potential foes:* The SPJ and RWR reprogramming support software and hardware are primarily designed to respond to adversary threat changes affecting the UAVs' SPJ and RWR systems' operational effectiveness.
- *Theater tailoring:* PFMs are prepared for a specific region or theater. With the change in the theater situation, reprogramming is conducted for regional tailoring. This process reduces the number of threats in system memory, decreasing processing time and reducing system display ambiguities.
- *Mission tailoring:* This requirement involves reprogramming the SPJ and RWR systems for the UAV's mission. Mission tailoring improves system response up to the mother platform's threat priority.

In general understanding, the UAV reprogramming requirements and proce-dures are similar. However, an essential property of UAVs, the endurance or long flight times, becomes prominent. In conventional aircraft, pilots or crew are not only in the process but also on the airborne platform. Thus, the mission's duration depends on the pilot's situation and fuel capability. One of the properties of UAVs is that they are suitable for long hours of operation. Their limits depend on only fuel capability, and by refueling, they may be continuously in operation. During this dense program, they may change operational field and mission type. Say that a UAV may be sent to close air support in a closed region while patrolling a bor-derline. Even though the area is close, the threats are different. Moreover, suppose that the region's threats must be updated due to the added threats. Conventional air vehicles must return to the base, and pacer-ware should be applied immediately for this condition.

Pacer-ware is used for emergency reprogramming actions, such as adding a new threat that appears not part of the current EID or removing the threats from the EID. It is a process of immediately changing PFMs in the line and loading them to the SPEW system's central operating system. The pacer-ware procedure can be applied during flight using the data link for the UAV systems. The SPEW system usage on the UAVs is a new concept, and using some novelties in their procedures is inevi-table. Changing or adding PFM during the mission is a new procedure beyond the pacer ware, and it would be appropriate and increase the operational effectiveness.

8.4   Future Projection of the SPJ for Airborne Platforms

The SPJ systems can be inspected in three main parts: the hardware structure and capabilities, software capabilities, and preprogramming capabilities. These are the main elements of the SPJ architectures. Moreover, they may be examined according to their platform types. The SPJ system's structures and general usage on different platforms are elaborately given. Furthermore, today's and near-future trends and requirements of the airborne SPJ systems and the roles of the airborne platforms using SPJ systems have been described conceptually. In this part, a future projection is carried out regarding past experiences, present conditions, and future expecta-tions for the operational areas and potential threats.

- *Future projection for hardware structure and capabilities of SPJ systems:* The SPJ systems have turned to an all-digital structure for the last decades. Digiti-zation reduces size, weight, and power (SWaP) requirements and more incred-ible speed and accuracy. Since utilizing SPEW and, consequently, SPJ at UAVs is a new concept, the structure of the SPEW system must be entirely digital.

  Most of the SWaP capacity of an airborne platform is dedicated to the SPJ systems. The physical dimensions of some subcomponents of the jammer systems, such as power amplifiers and DRFM systems, are substantial. The main disadvantage of the SPJ systems compared to the radar transmitters, which operate in a relatively small bandwidth, is that they must cover an extensive frequency range of 0.5 to 20 GHz. The semiconductor technology limits the dimensions of these subcomponents.

SPJ systems with fewer SWaP requirements and higher jammer ERP and efficiency are expected to improve semiconductor power amplifiers in the future. The developments in semiconductors and nanotechnology are the push factors for the required improvements. Also, the main central computer hardware must always be up to date with the technology since the hardware limits the operating system software capability.

- *Future projection for software capabilities of SPJ systems:* The operating system is the central determiner of the capacity of the SPEW system. The operating system manages the subsystems of the SPEW system and, relatedly, SPJ and RWR. The future projection for the operating system is expected for SPJ and RWR together. In each airborne platform, response time must be as short as possible, which requires optimal operating systems and developed hardware. Some features provided by software for SPJ and RWR on different platforms stand out. For fighters, automatic operation without overloading the pilot and processing accuracy is essential. The processing speed and accuracy came forward for helicopters and piloted UAVs. Full autonomy is required for autonomous UAVs.

  Developing decision-making and resource allocation algorithms using adaptive control and machine learning methods has been important for decades. Academic development in this field cannot be ignored, and some studies have been adapted to the actual SPJ and RWR systems. Shortly, these studies will increase exponentially, and more artificial intelligence studies will be utilized on the SPJ and RWR systems for decision-making and resource allocation. Thus, the limited power and sensor capabilities of the SPJ and RWR systems will be used optimally.

  The data streaming on the SPJ and RWR exponentially increases during the missions. Applying conventional data pruning and adaptive control techniques remains insufficient. Thus, using machine learning and deep learning algorithms for these systems has significantly been considered for the last decade. These studies are expected to continue increasingly, and application to the systems is inevitable.

- *Future projection for preprogramming capabilities of SPJ systems:* The SPJ and RWR systems are considered together to predict the future of the preprogramming process. The limited EID database capacity of the PFM will reach a sufficient level to store the properties of all the known threat systems. Developments in hardware and software capabilities will provide this property. Adaptive PFM structures can be beneficial in this case. One PFM will include all the EID and required jamming techniques, which is a good dream.

Preparing a PFM covering all the treats and jamming methods is a far objective. Until this aim is reached, an in-flight PFM loading and changing via data link will be a suitable solution. Machine learning and deep learning techniques are adaptable for threat identification and jamming processes and have been studied for a few years in academic fields. Furthermore, they have been tested on hot-mock-up and HITL environments. Soon, they will be engaged in real systems and become widespread for SPJs and RWRs.

The most challenging part of preparing a PFM is developing jamming techniques. When multiple jamming methods are applied simultaneously, many inputs and variables exist. Their effects on each other and the optimal use of sources are complex problems. Artificial intelligence methods should be considered to solve this problem.

References

[1] Joint Publication 3-01 (2018): Countering Air and Missile Threats, USA Joint Chiefs of Staff (CJCS), May 2, 2018.

[2] Streetly, M., (ed.), *Jane's Radar and Electronic Warfare Systems 2002–2003*, 2002–2003 edition, Janes Information Group, June 2002.

[3] Donaldson, P., *Electronic Warfare Handbook 2008: The Concise Global Industry Guide*, Shephard, U.K., 2008.

[4] Barton, D. K., and S. A. Leonov, *Radar Technology Encyclopedia (Electronic Edition)*, Norwood, MA, Artech House, 1998.

[5] Pakfiliz, A. G., "Increasing Self-Protection Jammer Efficiency Using Radar Cross Section Adaptation," *Computers & Electrical Engineering*, Vol. 98, 2022, p. 107635.

[6] Kim, S., et al., "Carbon-Based Radar Absorbing Materials Toward Stealth Technologies," *Adv. Sci.*, Vol. 10, No. 32, 2023.

[7] Richardson, D., *Stealth Warplanes: Deception, Evasion, and Concealment*, London, UK: MBI Publishing Company, 2001.

[8] Joint Publication 3-51 (2000): Joint Doctrine for Electronic Warfare, USA Joint Chiefs of Staff (CJCS), April 7, 2000.

[9] Heikell, J., *Electronic Warfare Self-Protection of Battlefield Helicopters: A Holistic View*, Helsinki University of Technology Department of Electrical and Communications Engineering Applied Electronics Laboratory, Finland, 2005.

[10] Zhou, Z., and J. Huang, "Influence of Rotor Dynamic Scattering on Helicopter Radar Cross-Section," *Sensors (Basel)*, Vol. 20, No. 7, 2020, p. 2097.

[11] Lutsenko, V., et al., "The Acousto-Electromagnetic Portrait Signatures for the Aerodynamic and Ground Technology Objects," *Telecommunications and Radio Engineering*, Vol. 77, No. 11, 2018, pp. 971–993.

[12] DoD Directive 3000.09, *Autonomy in Weapon Systems*, January 25, 2023, as amended.

[13] Ilachinski, A., *AI, Robots, and Swarms: Issues, Questions, and Recommended Studies*, Virginia: Center for Naval Analysis, January 2017.

[14] US Department of Defense: Unmanned Systems Roadmap 2007-2032, Office of the Secretary of Defense, December 2007.

[15] Hoehn, J. R., K. Sayler, and M. DeVine, *Unmanned Aircraft Systems: Roles, Missions, and Future Concepts*, Washington, D.C.: Congressional Research Service, 2022.

[16] Scharre, P., *Robotics on the Battlefield Part II: The Coming Swarm*, Washington: Center for a New American Security, October 2014.

[17] Joint Publication 1-20, Department of Defense Dictionary of Military and Associated Terms, March 23, 1994, as amended through April 15, 1998.

CHAPTER 9

Decoys for Self-Protection Jamming

Every military operation aims to fulfill the primary mission and to protect military assets, staff, and facilities from various threats. The success of an operation depends on protecting the military assets, staff, and facilities through physical security and different countermeasures to deter and detect intruders. Throughout this book, we have studied protection and countermeasure techniques, concepts, roles, and their planning and implementation. These include the various electronic countermeasures to deter, delay, detect, or prevent threats. This chapter will focus on the decoys used against radar-guided threats for self-protection jamming.

The passive decoys against radars are chaff for the air and ship platforms, and corner reflectors are generally for shipborne platforms. The corner reflectors are also deployed from an airborne platform with a parachute. Even though their structures are simple, passive decoys are very effective against conventional radar systems with limited ECCM techniques. The active decoys provide more flexibility against an approaching radar-guided missile and can imitate the mother platform's RF emissions. The active decoys' transmitted signals can be modulated to produce more realistic or confusing target parameters for the victim radar [1]. They can be classified as expandable and towed decoys.

When ejected from a mother platform, the expandable active decoys break off their connection with the platform. This type of decoy is versatile and used in airborne and shipborne platforms. Towed decoys, the other kind of active ones, are equally adaptable. They are used to defend both shipborne and airborne platforms. Sea surface towed decoys can be either floating above water long distances from the protected ship or airborne, and they can be of the repeater type or fed with low-power RF via a fiber-optic cable. The flexibility of these active decoys is a testament to their effectiveness in self-protection jamming.

9.1 Concept of Using Decoys for Self-Protection Jamming

Decoys are active or passive countermeasures to imitate the mother platforms, saturate the enemy radars, or stimulate them. They fulfill these countermeasures by saturating or deceptive jamming. Their usage for self-protection EW purposes has gained importance with the technological developments of defense radars and missile systems. According to their mission, the types of decoys are summarized as saturation, seduction, or detection of radar-guided threat systems.

Saturation decoys aim to prevent the enemy's defense from detecting the platforms. These decoys are usually an expendable vehicle designed to emulate a platform. The three main characteristics of saturation decoys are their electronic signature, flight program, and mission type. Saturation decoys must present an electronic signature or radar return indistinguishable from the aircraft that they protect. Decoys can do this by either passive or active measures or combining both. A passive decoy is a flying radar reflector. The size, shape, and materials used in the decoy are optimized to ensure that the proper amount of radar energy is returned to the enemy radars. Active decoys utilize radar repeater systems to receive, amplify, and send back to the threat radar signal to confuse the adversary units.

Seduction decoys attract the attention of a radar that has established track on a target, causing the radar to change its track to the decoy. Then the decoy moves away from the target. Tracking radars considers only narrow segments of azimuth (and sometimes elevation), range, and return-signal frequency using angle, range, and frequency gates. If the decoy can move any or all those gates far away from the real target, the radar's tracking lock on the target will be broken. Thus, seduction decoys could also be called break-lock decoys.

Another use of radar decoys is to trick a defensive system that uses radar, such as an air-defense network, into activating its radars. These types are called *detection decoys*, making the system vulnerable to detection and attack. Independent maneuver decoys are typically used to achieve this. The decoys are designed to mimic real targets, so the acquisition radars or other acquisition sensors will mistake them for genuine targets and hand them off to tracking radars. Once the tracking radars turn on, they become vulnerable to attack by antiradiation missiles fired from aircraft outside the lethal range of the enemy weapon.

Regardless of their mission types, decoys can be classified as passive and active decoys based on their structure. Their natures and features vary depending on the platforms used, such as naval and air. However, decoys will not be inspected separately for platforms; they will be analyzed for passive and active types. Moreover, active systems are explained in further detail under towed and expandable decoys.

9.1.1   Passive Decoys

Two types of commonly used passive decoys are chaff and corner reflectors. Their structures are simple and full of theoretical background. They do not emit RF signals but behave as reflecting antennae in different manners. Chaff is a volumetric radar clutter constituted by a vast number, several hundreds of thousands, of small metalized, shorted tuned dipole antennas, which are dispensed into the atmosphere to confuse or disrupt radar operation. However, a corner reflector is a structure used as a radar target, sometimes in passive decoy roles and, on occasion, calibrating test equipment such as in an anechoic chamber. Corner reflectors are used for many reasons: they have very high RSC for a small size, the high RCS is maintained over a wide incidence angle, and an exact solution is known for their RCS.

9.1.1.1   Chaff

Chaff is a passive defensive decoy that airborne and shipborne forces employ to avoid detection. The radar reflections off the chaff may cause a tracking radar to break the lock on the target. Chaff is also utilized to support other countermeasures with proper maneuvers, contributing to the general SPEW systems. Chaff consists of many reflectors, usually in metallic foil strips packaged in a bundle, and they are cut at about $\lambda/2$ to achieve the maximum ratio of RCS. The length of the foil determines the optimum reflected frequency. Chaff bundles usually consist of various strip lengths for multiple-frequency coverage. The chaff effectively reflects radar signals in different bands depending on the size of the chaff fibers. The effective frequency bandwidth of a single chaff length is varied by ±5%, but various chaff length is often mixed with increasing chaff effective bandwidth.

Immediately after deploying chaff, the chaff cloud obscures the aircraft from radar detection, breaking the radar lock. Pilots often perform beam maneuvers to maximize the effectiveness of chaff before releasing them. The objective of chaff is to reduce the effectiveness of the radar by providing a substantial, bright target return that masks the returns from targets of interest. When dispensing large quantities from aircraft or vessels, it forms a cloud that temporarily hides the platform from radar detection. The two significant types of chaff are aluminum foil and aluminum-coated glass fibers. The aluminum foil type is the older one but is still in operation. Chaff is used either to create a corridor in the airspace controlled by a surveillance radar to conceal some intruder aircraft penetrating enemy territory or to provide screening or self-protection to air platforms when subject to illumination by AAA, SAM, or AAM systems' tracking radars.

When the radar waves interact with the chaff, they resonate, and the waves are reflected to the radar. The ideal length of a strip is half the length of the enemy radar wavelength. Being this length, chaff acts as a resonant dipole and reflects much of the energy to the radar. Since the frequency used by the enemy radar is unknown, a batch of chaff often contains different lengths to counter a more extensive range of radars using various wavelengths. Chaff is released from containers usually located under an airplane's wings and is dispersed by the turbulence and wing-tip vortices. Due to their tiny weight, the chaff quickly reacts to the turbulent winds and creates a large cloud. This cloud can then act as a false target to cover the airplane, jam the enemy radar, and potentially break the lock of a radar-guided missile.

A chaff bundle's RCS depends on the threat radar's frequency and the launching platform's relative position or aspect. Figure 9.1 shows a single chaff cartridge's normalized RCS (RCS/RCSmax) based on normalized frequency. The frequency normalization is conducted according to the resonant frequency (f/f_r). The largest chaff RCS occurs at about f_r resonant frequency, adjusted to the radar's operating frequency. Moreover, chaff produces echoes at multiples of the fundamental frequency.

Say that the resonant frequency is 4 GHz, and the maximum RCS exists in this frequency. However, RCS peaks with decreasing values at the harmonics of the resonant frequencies, such as 8, 12, 16, 20, … GHz, are observed. The first harmonic takes approximately half the maximum RCS value obtained in the resonant frequency. Thus, it can be concluded that the chaff echoes affect most of the EW spectrum between 2 and 18 GHz. The chaff cartridge should provide a sufficient RCS to mask the mother platform. For this purpose, the mother platform's maximum RCS and threat radar's operating frequency must be considered.

Chaff usage in airborne platforms should be inspected separately because of the various air platforms. Chaff is used by different air platforms, such as fighters, bombers, cargo aircraft, helicopters, and UAVs, over a wide range of altitudes and

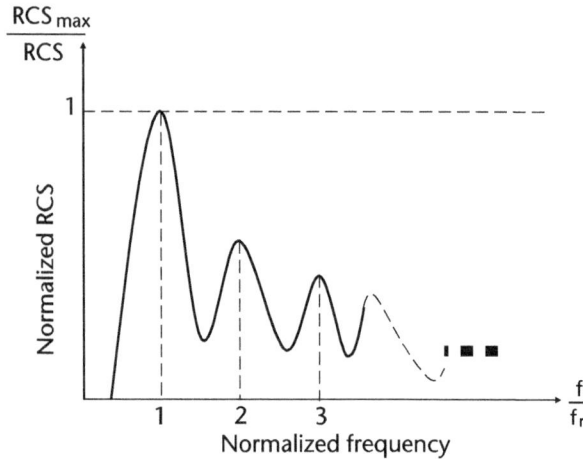

Figure 9.1 Chaff frequency response.

flight maneuvers or tactics. The deployment of chaff does not interfere with the flight characteristics of the dispensing aircraft. Fighters can drop chaff or flares at approved altitudes during flight maneuvers, such as turns, climbs, descents, airspeed, and G-loading. Although less maneuverable than fighters, bombers, cargo aircraft, helicopters, and UAVs can drop chaff or flares at any approved altitude while in a turn, climb, or descent. Specific descriptions of how chaff is employed in training for a combat situation are not releasable.

The purpose of chaff is to divert the tracking radar off the target and onto the chaff cloud. However, when deployed from an aircraft, the chaff's rate of slowing down and moving out of the Doppler filter passband of MTI and pulsed Doppler systems is exceptionally rapid. Therefore, chaff is typically used in the terminal phase of a missile engagement and is required only to break the lock sufficiently so the missile cannot recover. The chaff should have a significantly higher amplitude than the target's RCS to pull the track centroid to ensure maximum effectiveness as far away from the target and maintain the chaff RCS advantage in the Doppler passband for as long as possible [2].

The following two figures show the tactics for chaff usage on-air platforms. Figure 9.2 shows the corridor chaff utilization. The fighters in the main strike package conducting the operation say they aim to hit an important target and must be protected. An RF-guided SAM track radar from the enemy air defense system tries to initiate tracking. The chaff bombers, which fly in front of the main strike package, dispense chaff steadily over a long period to form a corridor that conceals the following aircraft. Chaff bombers release a series of partially overlapping chaff clouds to create chaff corridors through which the main strike package will pass. Thus, the chaff corridor screens the main strike package from the threat radar system.

Figure 9.3 shows the self-screening chaff utilization. The target for this scenario is the fighter, which utilizes self-screening chaff. There are two different threats for the target: one is a SAM system, and the other one is an AI aircraft. The angular aspect between the fighter aircraft and the chaff cloud, or the chaff/target line in the figure, affects the RCS presented to the tracking (or the threat) radar. Chaff RCS is

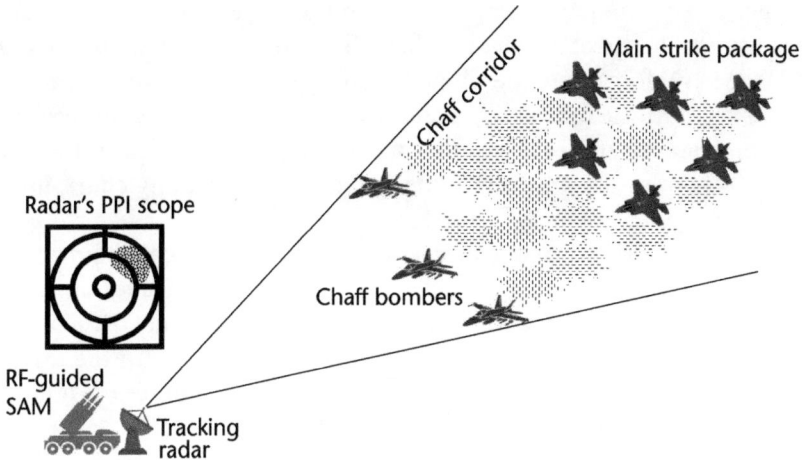

Figure 9.2   A conceptual chaff corridor scenario.

the greatest according to the target RCS when the chaff/target line is perpendicular to the threat radar's main beam. Chaff RCS gets closer to the angle between the chaff/target line, and the threat radar's boresight is approaching 0° and 180°. The target's RCS receives the highest value according to the chaff's RCS when the angle is 0° or 180°. Thus, the aspect is essential when developing self-protection maneuvering and chaff dispensing tactics against threat radars.

Chaff can be dispensed from packets in front of or behind an aircraft. The effectiveness of using chaff can be improved by combining it with maneuvers and self-protection jamming. The targeted aircraft can use self-protection jamming techniques to deceive the seeker's range and velocity gates. These techniques include RGPO, VGPO, and coordinated range-VGPO. Cooperative target maneuvers that introduce changes in the update rate of angular tracking or reductions in radial Doppler can be done simultaneously with the ejection of chaff to improve the effectiveness of these tactics.

The chaff utilization contributes to the self-protecting of ships against radar-guided *antiship cruise missiles* (ASCM) with sea-skimming capabilities. The

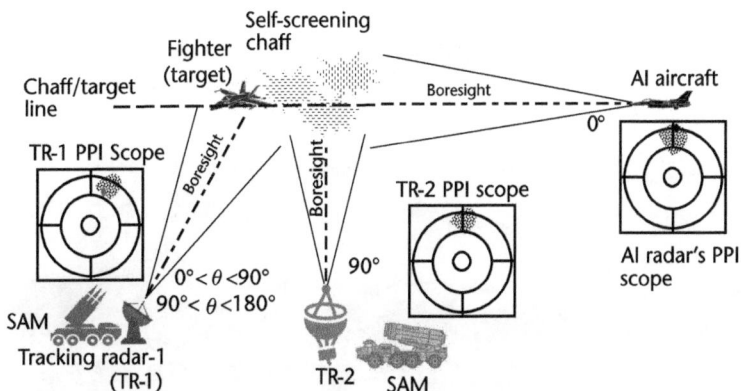

Figure 9.3   A conceptual self-screening chaff corridor scenario.

emerging technological capabilities and proliferation of modern ASCM present a considerable threat to naval surface ships and their missions. As stated in Chapter 1, the naval forces have developed an array of hard-kill, such as missile interceptors, and soft-kill, such as decoys, chaff, and jammers, EW counter-ASCM systems to defend against this threat. Solely using chaff is insufficient for ASCM threats. The hard-kill and soft-kill countermeasures (CM), including chaff, used on shipborne platforms are shown in Figure 9.4.

With the disadvantageous nature of the ships according to the airborne platforms, in terms of their massive body and low maneuvering capability, chaff is an additive CM to the others. For optimal effectiveness, each CM system must be employed quickly, correctly, and judiciously relative to the threat scenario. For example, a single decoy may be very effective at countering simultaneous threats from one direction but could be more effective at countering threats from multiple directions. In the latter case, several decoys or a combination of hard- and soft-kill options must be used. As seen in Figure 9.4, integrating individual counter-ASCM systems into a single defensive strategy increases the complexity of the combined systems' capabilities and limitations. In this scenario, chaff usage is both essential and complementary.

At this point, mention the RCS calculation of chaff clouds. Due to the incoherence of the waves reflected from each chaff, the average RCS of a chaff cloud can be calculated as the average RCS of a single chaff times the number of operational chaffs in the cloud [3].

$$\bar{\sigma}_{cloud} = \left(N\bar{\sigma}_{chaff}\right)\mu_{ops} \qquad (9.1)$$

The expressions in the equation can be defined as follows:

- $\bar{\sigma}_{cloud}$: average RCS of a chaff cloud (m^2).
- N: total number of chaffs in the cloud (unitless).

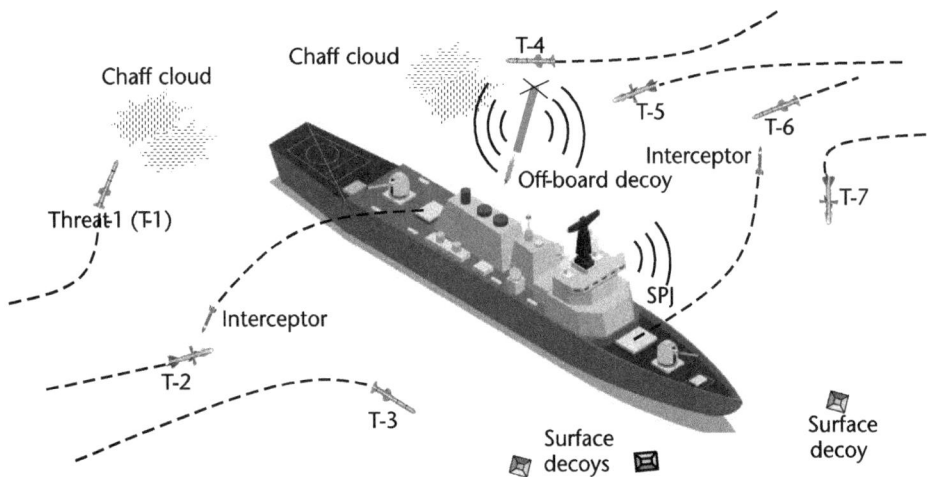

Figure 9.4 Conceptual shipboard chaff usage scenario, including hard-kill and soft-kill CM.

- $\bar{\sigma}_{chaff}$: average RCS of a single chaff (m²).
- μ_{ops}: percentage of operational chaff (unitless).

The operational chaff ratio is typically between 50% and 65% of the total number since the chaff particles tend to stick together and break apart [4]. Another phenomenon affecting a cloud's RCS is mutual coupling, an undesirable electromagnetic interaction that occurs when antennas are closely spaced. When this happens, the other antennas partly absorb the energy transmitted, which reduces the overall antenna efficiency. Similarly, a reduced RCS occurs between different scattering chaff particles in a chaff cloud. This effect is negligible if the average spacing between chaff particles is more significant than two wavelengths and is often disregarded in chaff models.

Determining the directions of chaff in a cloud is a challenging process due to the chaotic nature of the aerodynamics behind an airplane. Therefore, it is expected to assume a uniform distribution of orientations. The maximum RCS of chaff occurs when the length of the chaff dipole is half the radar operating frequency wavelength. The average RCS of a single dipole, when viewed broadside, is defined as [5]:

$$\bar{\sigma}_{chaff} = 0.88\lambda^2 \tag{9.2}$$

where λ is the wavelength of the radar in meters. The average RCS of a single chaff with a length equal to half the radar wavelength for an average aspect angle is as follows:

$$\bar{\sigma}_{chaff} = 0.18\lambda^2 \tag{9.3}$$

Using (9.3) rather than (9.2) in calculating (9.1) would be appropriate. The total chaff average RCS ($\bar{\sigma}_R$) within a radar resolution volume is defined as:

$$\bar{\sigma}_R = \frac{\bar{\sigma}_{chaff}NV_{CS}}{L_bV_R} \xrightarrow{\bar{\sigma}_{chaff}=0.18\lambda^2} \bar{\sigma}_R = \frac{(0.18\lambda^2)NV_{CS}}{L_bV_R} \tag{9.4}$$

The expressions in the equation can be defined as follows:

- N: total number of chaffs in the cloud (unitless).
- $\bar{\sigma}_{chaff}$: average RCS of a single chaff (m²).
- λ: radar operating frequency wavelength (m).
- V_R: radar resolution cell volume (m³).
- V_{CS}: chaff scattering volume (m³).
- L_b: radar antenna beam shape loss for the chaff cloud (unitless).

9.1.1.2 Corner Reflector

A typical corner reflector is composed of several metal planes. It creates a radar false target effect by strongly reflecting incident radar echo in an extensive incident angle range. Thus, corner reflectors are passive decoys that reflect radio waves to the source without active electronic components. It is helpful for passive jammer applications. Corner reflectors are made up of perpendicular plates that intersect with each other. There are two types of corner reflectors: dihedral and trihedral. Dihedral corner reflectors have two plates as surfaces that are on orthogonal planes, as shown in Figure 9.5. These reflectors require precise mechanical alignment to function correctly.

The maximum RCS for a dihedral corner reflector is given in (9.5). In the equation, d and h represent the lengths of the edges of the dihedral corner reflector, as defined in Figure 9.5. Moreover, λ is the wavelength of the radar's operating frequency.

$$\sigma = \frac{8\pi d^2 h^2}{\lambda^2}$$

$$(9.5)$$

The trihedral corner reflector has three right-angle plates, and the trihedral corner reflector is highly tolerant to misalignment. This structure offers a convenient way to set up a quick field. Figure 9.6 shows two types of trihedral structures: orthogonal square plates and orthogonal triangular plates. In trihedral corner reflectors, the waves that hit the corner reflector are bounced by each surface three times, resulting in reversed direction waves sending back towards the source. Therefore, the trihedral corner reflectors provide a high RCS target for radar system testing and characterizations.

The maximum RCS for a square trihedral corner reflector is given in (9.6), and that for a triangular one is given in (9.7) [6].

$$\sigma = \frac{12\pi d^4}{\lambda^2}$$

$$(9.6)$$

$$\sigma = \frac{4\pi d^4}{3\lambda^2}$$

$$(9.7)$$

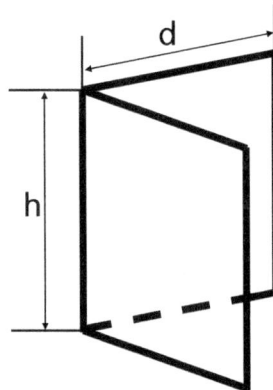

Figure 9.5 A conceptual dihedral corner reflector.

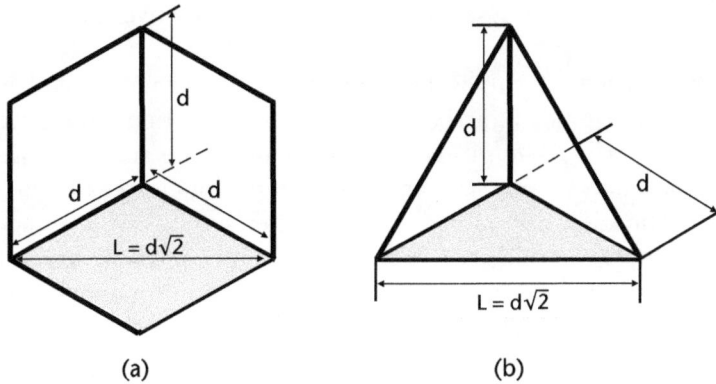

Figure 9.6   Two conceptual trihedral corner reflectors: (a) square and (b) triangular.

In the equations, d represents the trihedral corner reflectors' edge lengths as defined in Figure 9.6, and λ is the radar's operating frequency wavelength. The RCS is proportional to the fourth power of the size of d coherently reflecting target. This calculation shows that doubling the reflector's size will result in 16 times larger RCS. The primary trihedral corner reflectors are square and triangular designs, as in Figure 9.6 (a, b). Furthermore, circular and more complicated designs with four, five, and six triangular trihedral structures are possible.

Trihedral corner reflectors are easy to produce and equip; they can interfere with the radar of different systems and all horizontal directions. They are composed of three metal plates that intersect vertically. In an extensive incident angle range, the incident electromagnetic wave will reflect three times and be reflected in the original direction. Its maximum reflection direction is the center axis of the corner reflector.

9.1.2   Active Decoys

Active decoys have their own Tx and, up to the mission type, their own Rx. They are ejected from a mother platform, providing more flexibility against an approaching radar-guided missile than their passive counterparts. As an advantage, their transmitted signal can be modulated to produce more realistic or confusing target parameters for the threat radar.

The active decoys are designed to lure the tracking gates of a threat radar away from the aircraft. They enhance the self-protection capability of naval and airborne platforms by providing them the ability to counter radar-based threats. According to the mother platform and the threat radar system, an active decoy must be in a suitable position during the operations. It radiates a signal similar to the threat radar's RF, PW, and PRI, leading the missile to track the decoy instead of the radar. Generally, it has an omnidirectional antenna and transmits the same amplitude signal in all directions.

The difference between active decoy utilization of shipborne and airborne platforms is in the operational roles. The air platforms typically engage active decoys on an attacking mission, such as strategic bombing, airstrike, forward air control, and close air support. However, naval ships use active decoys in any mission, such as strategic deterrence, sea control, projection of power ashore, and maritime presence missions.

In airborne and shipborne applications, towed and expendable decoys are used against radar-guided missiles in saturation, seduction, and detection roles for different scenarios. The saturation decoys might be active or passive. The passive saturation decoys are chaff and corner reflectors, discussed in Section 9.1.1. The active saturation decoys use radar repeater systems to intercept the enemy radar signal, amplify it, and send back a radar return of the appropriate size to confuse the enemy. Transmitting the proper signal at an optimal power is crucial for active decoys. If the radar return is too large or too small, the enemy radar operator can distinguish between decoy and real aircraft signals, rendering the decoys useless [7].

The seduction decoy family is another active decoy type. They lure the threat radars' tracking mechanisms away from the mother platform. The purpose of a seduction decoy is to capture the tracking mechanism of the threat radar and break its lock on the target. Their operation is similar to that of deceptive jammers, such as RGPO and VGPO. However, decoys are more powerful in that they hold the attention of the tracking radar. Seduction decoys may be towed or expandable and are utilized in airborne and shipboard platforms. They do not only use jamming but also imitate the RF signals of the relevant platforms. Thus, the seduction decoys are very effective against RF homing threats. Another advantage of seduction decoys is that their signals are transmitted from a location away from the target. This distance provides an additive protection property against HOJ missiles.

The detection decoys, generally used in airborne platforms, are the other active decoy types. However, the operational purposes of their usage differ from the other decoy types. The lethal threats of the EADS are sometimes not activated up to a predetermined situation due to the applied tactical concepts. They take the required surveillance and situational awareness information from the other sensors of the EADS. In this case, SPEW system usage and evasive maneuvers would not help the air platforms to survive. The purpose of detection decoys is neither to confuse a threat radar nor to attract missiles by imitating a platform. Their purpose is to reveal the hidden lethal radar-guided threats. For this purpose, a target-like system with similar RF transmission, like imitation, is used. However, this time, the decoy imitates the platform and follows the operational procedures as the target platform.

The conceptual operation of the hidden threats depends on the surprise effect. They are activated when it is too late for the victim aircraft. These lethal threats should be eliminated for the mission's success and safe return. However, they cannot be neutralized or destroyed without detection. This dilemma can be overcome using a detection decoy, such as a *miniature air-launched decoy* (MALD) or MALD-jammer (MALD-J).

The MALD systems are small, low-cost, expendable, air-launched vehicles replicating how fighter, attack, and bomber aircraft appear to enemy radar operators. MALD-J systems are airborne close-in jammers for electronic attacks that can loiter on the radar sites. They have propellant systems, and their flight paths are preprogrammable. The MALD-J systems jam preprogrammed E/W and fire control radars while retaining the capabilities of the MALD. These decoys aim to stimulate and degrade an enemy's IADSs.

The MALD systems can imitate several aircraft simultaneously and allow an airborne strike force to accomplish its mission by deceiving enemy radars and air defense systems to treat MALD as a viable target. However, MALD-J systems allow an airborne strike force to accomplish its mission by jamming specific enemy radars and air defense systems to degrade or deny detection of friendly aircraft or munitions. The aircraft equipped with MALD-J can stimulate an enemy's integrated air defense system, enabling allied forces to target and engage enemy components. Because of its small size, the MALD can be carried into the target area before it is launched. Once launched, it uses its propellent system, preprogrammed or independent flight path, and electronically manipulated radar signature to make acquisition radars and target tracking radars mistake it for one of the attacking aircraft.

A general scenario containing saturation, seduction, and detection decoys for airborne platforms is given in Figure 9.7. In the given figure, the attack pack comprises four fighters to strike the target. Assume that the EADS in the region is composed of three *fire control* (FC) radars; one is hidden. Moreover, one E/W radar provides long-range target detections to the FC radars. In the scenario, fighter (F)-1 launches an expandable saturation decoy to suppress the EADS by jamming. F-2 launches a detection decoy to reveal the hidden threats by imitating the attacking fighters. F-3 launches an expandable seduction decoy to lure the threats on it. However, F-4 launches a towed one for the same purpose.

Active decoys used on shipboard platforms can be of two types: expendable or towed. The expendable or off-board active decoy is launched by an onboard decoy launcher with a ballistic trajectory in the direction of the incoming missile but at some angular offset concerning the ship-to-missile direction. This way, the decoys can lure the missile to a leading-edge range-tracking loop. To increase their operational life, they are suspended in the air by a parachute or a rocket and have a long endurance battery. However, the ship-towed decoys typically float above water long distances from the defended vessel and are of the repeater jammer type [1].

The naval active decoy system is integral to the ship's self-protection against active RF antiship missile attacks. The characteristics of the decoys, such as the

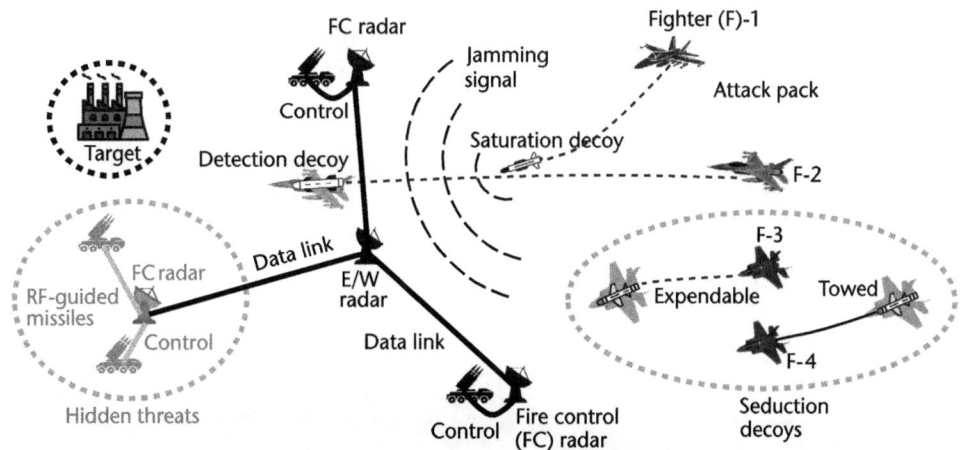

Figure 9.7   An attack pack's usage of saturation, seduction, and detection decoys.

reaction time and their effectiveness, are becoming critical for the new generation ASM and ASCM systems. The systems must cover a full 360° around the defended ship. The effectiveness of the decoy is wholly independent of ship maneuvers, even in the most extreme environmental conditions.

The decoys utilized by naval forces are saturation and seduction types. A typical scenario containing saturation and seduction decoys for shipborne platforms is given in Figure 9.8. In the given figure, the protected off-board decoys with parachutes and rockets simulate a radar return from a large ship overlapping the target signal. The off-board decoys with rockets provide a larger, more attractive target to the missile, consistent with the ASM and ASCM's range and angle tracking, and move slowly away from the ship, thus defeating the threat. However, off-board decoys with parachutes aim to launch to an acceptable height and range and to suspend as long as possible. The towed decoys generate radar return like a ship and lure the treats. They are more capable and powerful than off-board decoys and can be reutilized.

9.2   Towed Decoys for Self-Protection Jamming

The towed decoy protects ships and aircraft in saturation (repeater) and seduction (deceptive) jamming mode. Naval-towed jammer decoys typically float above water long distances from the defended ship and are repeater types. However, airborne towed jammer decoys have deceptive and repeater jamming capabilities. It is difficult for onboard SPJ systems to create angle-tracking errors against monopulse radars. In the towed decoy case, if the radar or missile seeker tracks the towed decoy, it is not tracking the targeted aircraft, and there is an inherent angle tracking error.

There are two basic types of towed decoys. The first is a simple repeater that retransmits the targeting radar waveform at a higher signal level to seduce the track away from the target aircraft; it is essentially a beacon. Once deployed, the system only requires power and control from the host platform. When the system receives an RF signal via the towed decoy onboard receiver that meets the threat criteria, it amplifies and retransmits the signal to seduce the threat track.

The second type is fiber-optic towed decoys. This type of decoy employs sophisticated Rxs and ECM technique generators onboard the host aircraft. The onboard receiver passes threat information, such as the SPJ operation, to the technique generator. It differs from the SPJ case in that it converts the RF technique to optical

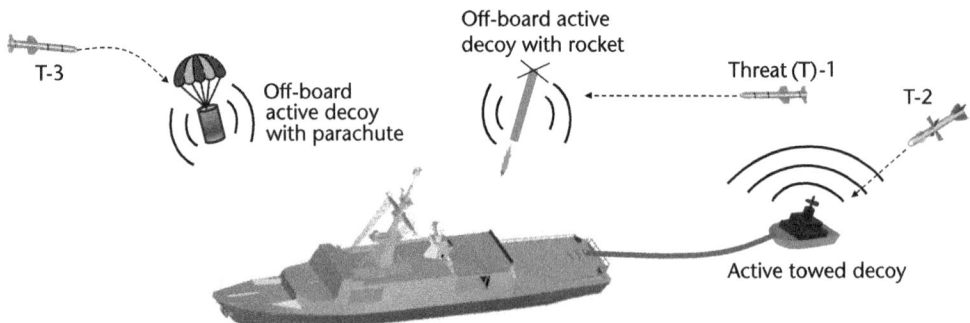

Figure 9.8   A warship's usage of off-board and towed decoys.

wavelengths and transmits it via fiber-optic cable to the decoy, where it is converted back to RF, amplified, and retransmitted [8].

Inspecting the towed decoys separately for airborne and shipborne platforms would be appropriate. As stated for active decoys, the difference between towed decoy employment of air and sea surface vehicles is in the operational roles. Air platforms typically engage active decoys on an attacking mission. In contrast, warships use active decoys in every kind of mission.

9.2.1   Towed Decoys for Airborne Platforms

Towed decoys are small jammers physically attached to the airborne platforms, as shown in Figure 9.7. They are designed to help individual aircraft survive and evade threat missiles in the terminal stages of an engagement. While towed decoys are primarily designed to provide sufficient miss distance between an attacking semi-active radar missile and the protected aircraft; they may also be effective against pulse Doppler radars and monopulse radars.

The primary role of the towed decoys is to protect the fighter by luring threat missiles away from the aircraft. The towed decoys combine techniques that disrupt adversaries' radars, preventing missile launches from occurring. They can also operate independently, enhancing their effectiveness against current and future threats. Suppose the aircraft is coming or approaching an area that is guarded by air defenses. The towed decoy will help the aircraft to dodge the incoming missiles. These decoys are typically released whenever the MWS or RWR systems detect the RF incoming missiles (SAM or AAM). These decoys will lure the incoming attacking SAM or AAM missiles away from the protected platform by generating an attractive false target signal. The missile will divert from homing the mother platform.

Towed decoys are employed against not only conventional threats but also sophisticated missiles. Due to their retractable nature, they can also be retracted back to the aircraft after successfully evading or jamming the incoming threat. After that, if another threat comes, they can be redeployed. The second-generation towed decoy systems are linked to the aircraft's SPEW system through a fiber-optic cable; they operate over a wide frequency range to counter various radars and missiles. It would be appropriate to consider the first and second-generation towed decoys separately.

9.2.1.1   First-Generation Towed Decoys

The first generation of towed decoys contains a simple repeater jammer that enhances any signal in the proper frequency range. The amplified signal is stronger than the aircraft's backscattered signal, so the missile is lured toward the decoy. These decoys are stand-alone units containing all the electronics, processors, Rx, and Tx. The only tie to the aircraft is for power and status. One of the significant advantages of these simple repeater devices is that they do not require the exact frequency of the enemy radar systems to be effective. They will enhance any signal coming at them.

An area of concern with towed decoys is a possible conflict between the onboard jamming system and the towed decoy. The onboard system could overpower the decoy, causing the attacking missile to ignore the decoy and track the aircraft. The towed decoys of the first generation comprise two main components: the launch controller subsystem and the towed decoys themselves. The launch controller subsystem stores the decoy before launch, supplies it with power, and monitors its electronics. However, the decoy body is a self-contained, factory-sealed unit with everything except power. The power is supplied through the tether from the host aircraft, while the decoy sends its operational status to the plane through the same rope.

The advantages of the first-generation towed decoys are as follows:

- They are towed behind aircraft with a cable, moving at the same speed as the host aircraft. Thus, the Doppler effect does not help distinguish decoys from the real target.
- They do not require an additional power unit on the decoy and utilize the power from the host platform.
- The towed decoys are often stored within wing pylons, wingtip pods, or aircraft fuselages and thus do not affect aircraft weapons loads.

However, the first-generation towed decoys have some nonignorable disadvantages:

- After deployment, decoys stay at a distance and are connected to the aircraft by a cable, thus limiting aircraft agility to low-G maneuvers.
- Decoys are towed, so they will always stay behind the real platform. Thus missiles with a two-way datalink or command guide can render towed decoys ineffective. In this case, the control is at the SAM operators so they can choose the target and lead the missile.
- These decoys rely totally on their internal components, such as processor, battery, and antenna. Thus, it lacks the processing, jamming power, and directivity of the SPJ or the aircraft's internal SPJ system.

9.2.1.2 Second-Generation Towed Decoys

The second-generation towed decoys are connected to the host platform via a fiber-optic cable. The decoy uses different jamming modulations through this cable. The second-generation towed decoys are utilized in the deceptive mode; however, they may change the mode to repeater jamming to protect the host platform from HOJ threats. These towed decoys only contain the Txs, usually mini TWT power amplifiers. The remaining units are in the mother platform or the pod. Thus, they can transmit RF-modulating signals via fiber-optic cables and high-voltage power supply via an inner core of the cable generated internally to the towing aircraft. They can exploit various jamming programs from the onboard ECM technique generator [7].

Since the second-generation towed decoys are tied to the platform, they will have the same speed and Doppler. Therefore, it aims to generate the desired deception or noise signal with a sufficient ERP to capture the threat's tracking gates. The decoy generates the required ERP value by itself. Thus, the TWT or semiconductor PA structures must be strong enough to provide the necessary ERP.

Figure 9.9 shows the aircraft employed towed decoy. The separation required between the decoy and the host platform is a primary consideration in developing a towed decoy system, and the separation limits are given in the operational procedures. The towed decoy must be positioned far enough behind the aircraft to prevent warhead fragments from missiles guided by the decoy from hitting the aircraft. However, this technique presents two conical regions where an approaching missile can fuse on and destroy the aircraft. These conical regions occur because the missile, while tracking the decoy signal, passes close enough to the aircraft to fuse on and destroy it before or after passing the decoy. Both cones have the vertex on the decoy and an aperture angle α [1]. The aperture angle consists of two symmetrical half-aperture (HA) angles with $\alpha/2$, and each angle is defined between the vertex line and the edges of the conic.

$$\frac{\alpha}{2} = \tan^{-1}\left(\frac{d_{l(m)} + W_f}{L}\right) \tag{9.8}$$

In the equation:

- $d_{l(m)}$: lethality range of the fuse or the radius of the fuse's lethality circle (m).
- W_f: the aircraft's fuselage width (m).
- L: the towing cable length (m).

An essential requirement for a towed decoy system is to achieve 360° coverage. When an aircraft equipped with a towed decoy is abeam a threat radar, the radar may be able to discriminate between the aircraft and the decoy. This is a function of the resolution cell of the threat radar. Figure 9.9 demonstrates different scenarios for various threat missiles' approach angles. The approach angle is defined between the heading of the missile and the vertex line. When the missile approaches from a low HA angle ($<\alpha/2$) and above the aircraft, such as T-1, it probably fuses on the aircraft while guiding the decoy. This situation continues until the approach angle equals the HA angle ($=\alpha/2$).

The hosting aircraft platform can avoid the threat of missiles when the missile's approach angle exceeds the HA angle ($>\alpha/2$). Figure 9.9 shows missile T-3 detonating far outside the aircraft's safety circle. Let us assume the missile is approaching from the decoy side, and the approach angle is lower than the aperture angle, like T-4. In this case, the fusing occurs on the decoy. From a technical standpoint, it is preferable to have a longer towing rope length (L) for the decoy to provide better

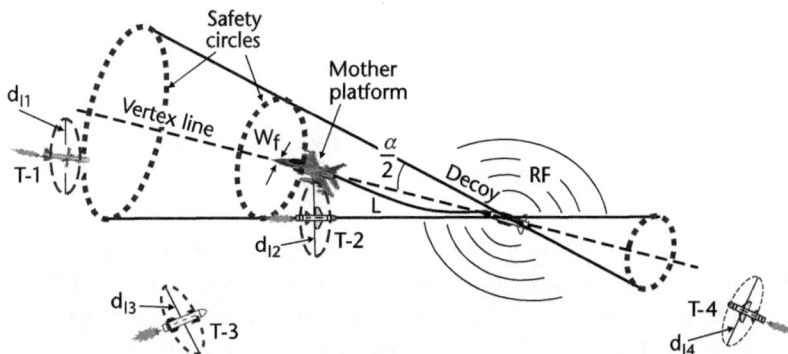

Figure 9.9 Air platform's safety circles in towed decoy usage.

protection. However, from an operational perspective, the restrictions on aircraft maneuvers due to the towed decoy can be significant and undesirable.

9.2.2 Towed Decoys for Naval Surface Platforms

The naval forces started the conceptual emergence of decoys; however, the advances and conceptual developments of active towed decoys have occurred in air platforms. Passive towed decoys with massive RCS, such as corner reflectors, are preferable to active ones for naval ship applications. However, active towed decoys are also possible on the shipborne platforms. The two main advantages of airborne platforms over shipborne ones are the body dimensions and maneuverability. Since the dimensions of the ships are vast, according to the air platforms, the RCS values of the vessels are very high compared to those of their airborne counterparts. Moreover, the airborne platforms have high maneuver capability, which makes it easy to develop active decoy utilization concepts, which is almost impossible for naval ships.

To provide the best protection, the towed decoy should be as far away from the defending ship as possible. Thus, once the RF-guided missile is locked on the towed decoy host platform, relocking back to the host platform is a weak possibility. However, the decoy should be in the RF-guided missile's main beam. At this point, finding the decoy's deployment angle and distance according to the mother platform becomes prominent. Thus, an optimized result of the distance of the decoy from the ship and the distance inside the main beam of the missile's RF seeker is an operational and technical problem to be solved.

For the best angle and range of the decoy deployment, the decoy must be a more attractive target than the mother ship or the decoy must be seduction type. For this purpose, a repeater-type active towed decoy with high gain will be an appropriate solution for increasing the self-protection jamming effectiveness. Seduction can be performed in two cases. The first case is if the host ship can maintain a good track of the incoming missile with its radar, allowing the missile turn-on time to be estimated. The decoy should be deployed before the missile turns on its seeker. When the seeker turns on and starts its search, it will lock on the stronger target echo, which, if correctly positioned, is the decoy.

In the second case, the host shipboard radar does not track the missile, or if the mother platform is unaware of the threat of the missile before its seeker turns on, then the decoy can still contribute to protecting the ship. When the seeker locks on to the vessel, it provides a very steady RF signal to the target ship. Even though the constant signal means the ship is being tracked, a towed decoy can be adequately launched at this point and may still lure the missile away from the ship.

Figure 9.10 shows the seduction towed decoy operation for the shipborne applications. According to the scenario, the towed decoy type is a repeater, and the threat missile has been engaged in the decoy. As discussed in Chapter 3, the repeater jammer systems use linear-constant gain and saturated power jamming. Let us assume that the repeater jammer operates in the linear-constant gain mode and that the transmitted signal is the accurate replica of the signal intercepted.

The main challenge is generating a signal with sufficient ERP to mask the ship's high RCS. One can utilize an antenna gain or RF power generated by a TWT to achieve the necessary ERP. In the former case, it is unlikely that a limited-size decoy

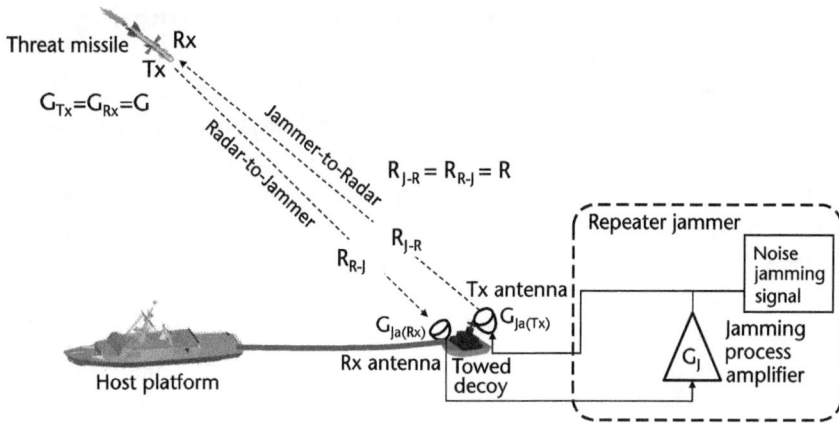

Figure 9.10   Warship towed decoy usage scenario with repeater jammer.

can carry a very directive antenna and an onboard pointing system. In the latter case, the decoy needs to have a rather heavy payload that includes electronic circuitry, a high-voltage power supply, and a TWT with an adequate cooling system.

In this type of operation, the output of the Tx is directed at the intercepted signal at the Rx. This system is referred to as a constant gain system. The output of the Tx is not necessarily the maximum output of the Tx. Still, it depends on the intercepted signal level multiplied by the amplifying system's gain. The jammer-to-signal ratio (J/S) is obtained using (3.18), derived in Chapter 3. The equation is written again as follows:

$$\frac{J}{S} = \frac{G_{Ja(Rx)}G_J G_{Ja(Tx)}c^2}{4\pi\sigma f^2 L} \quad , \quad \left\{ L = \frac{L_J L_{pol}}{L_R} \right\} \tag{9.9}$$

where L is the total losses that enter the process, consisting of jammer losses (L_J), radar losses (L_R), and antenna polarization losses (L_{pol}).

The power of the transmitter systems is not unlimited; the gain of the system eventually will drive the signal level to the point that demands maximum power out of the transmitter. Having reached this power level, any further increase in intercepted signal will result in the same maximum power level out of the transmitter. Since the threat system Tx output is constant, the Rx power level at the repeater jammer depends on the distance between the threat and the decoy. At ranges less than that point, the system operates as a constant power or saturated power jammer [9]. With a constant power output repeater, the jam-to-signal ratio is obtained using (3.10), adapted as follows for the situation.

$$\frac{J}{S} = \frac{P_{J(sat)}G_{Ja}4\pi R^2}{P_{Tx}G\sigma L} \tag{9.10}$$

In the equation, $P_{J(sat)}$ is the saturated or maximum repeater power output, and L represents the total loss defined in (9.9).

9.3   Air Expendable Decoys for Self-Protection Jamming

Towed decoys employed on an airborne platform restrict the ability to maneuver. In the present operational concepts, towed decoys are utilized with fighter aircraft and require a particular missile-aircraft orientation to be effective. The fighters and bombers can only carry a limited number of single-use towed decoys. Helicopters do not employ towed decoys because of the possibility of line entanglement in the primary or tail rotor. Transport and other nonattacking aircraft do not use towed decoys because they do not operate in threat environments that require radar-guided threat defense. Unfortunately, the missile engagement zones of many radar-guided threats are constantly expanding, which will likely put these aircraft in range. Furthermore, UAVs have been involved in the operations of various roles, and they may be as valuable as manned counterparts when they have dense payloads. Thus, towed decoys on UAVs will engage in the operational regions shortly. Their engagement will affect the UAVs' maneuvering capabilities like the fighters.

General expandable decoy information is given at the beginning of this chapter. In this context, active and passive expandable decoys are explained elaborately. This part will focus on the *expendable active decoy* (EAD) systems. EADs are employed to divert the tracking gates of an enemy's radar away from the aircraft. They provide naval and airborne platforms with the ability to counter radar-based threats, enhancing their self-protection capability. However, a decoy must do more than give the proper-sized radar return. To effectively deceive an enemy *integrated air defense system* (IADS) for a prolonged period, a decoy must possess flight characteristics similar to the aircraft it is protecting. Modern decoys can be powered with rockets or miniature engines, or they can glide for long distances based on the altitude and airspeed of the jet that releases them. Furthermore, their flight paths can be preprogrammed into an onboard autopilot, allowing the decoy to fly an independent ground track and increase their appearance as an attack aircraft worth tracking.

EADs are employed to defend aircraft against incoming missiles. These decoys can use noise or deception jamming techniques to disrupt the missile's guidance system. Noise jamming is the most common technique to saturate the missile's sensors. Deception jamming is used to confuse the missile's pulse Doppler radar, making it less effective. Multiple decoys can be released at a predetermined rate to increase the chances of evading the missile. The primary components of an EAD include the transmitting and receiving antennas, a technique generator, an amplifier, and a power supply. However, there are some limitations to the effectiveness of EADs. One of the main challenges is their size, which limits the amount and variety of jammer components that can be included in the decoy. Their size determines the power and frequency coverage of the jammer and the amount of time it can be effective. The unpredictable geometry and engagement scenarios also mean a wide antenna beam is necessary to ensure a disruptive effect.

The EAD systems in operations are *tactical air-launched decoy* (TALD) (United States), *miniature air-launched decoy* (MALD) (United States), the *BriteCloud* system (United Kingdom), *advanced self-protection decoys* (ASPD) (Israel), and the *president-S* system (Russia). The EAD systems fit into the physical space occupied by traditional chaff cartridges and are expended much the same way. However, the EAD's onboard DRFM technology can sample the EM spectrum produced by a threat radar, record the signal, and replay it with a more substantial return than the target aircraft. This process enables an EAD to capture a missile signal in its terminal phase and walk it off the target.

Two examples of seduction and saturation decoys are the TALD and MALD. Both decoys can work actively or passively and have preprogrammable flight paths. They are used in offensive operations against enemy air defense systems by diluting and confusing surface-based and airborne defenses with realistic tactical target characteristics. The TALD is an air-launched, aerodynamic vehicle whose purpose is to minimize the effectiveness of an enemy IADS. The TALD is a preprogrammed glide vehicle used to increase the survivability of strike aircraft. The improved TALD (ITALD) is a TALD that incorporates a propulsion unit. Both systems operate as expendable vehicles with no recovery capabilities. Launch platforms include the F-14, F/A-18, EA-6B Prowler, AV-8B Harrier II, and the P-3 Orion.

The approximate dimensions of a TALD system are 2.34m in length, 1.55m in wingspan, and 180 kg in weight. Their speed is up to 0.8 Mach, and their propulsion is the turbojet. The TALD system intends to confuse and saturate enemy air defenses as part of an overall SEAD strategy, thus allowing attacking aircraft and weapons a higher probability of penetrating the target. The TALD system can be launched from 40,000 ft, at which height it has a range of up to 126 km. However, a low-altitude range reduces this to approximately 30 km. With these specifications, TALD systems can perform similarly to the podded SPJ systems in terms of noise jamming power and deception jamming capabilities.

The MALD system could provide airborne platforms with an affordable, tactically effective radar decoy to accomplish preemptive and reactive lethal SEAD operations. The approximate dimensions of a MALD system are 2.38m in length, 0.65m in wingspan, 25 cm in body diameter, and 45 kg in weight. Their speed is up to 0.8 Mach, and their propulsion is the turbojet. The missile can be launched from 30,000 ft, at which height it has a range of up to 460 km. The MALD's small size, light weight, and low cost intend to enable missions with tactically significant decoys.

MALD-J is the radar jammer variant of the MALD system. This variant of the MALD decoy will be able to operate in both decoy and jammer modes. MALD-J is an airborne close-in jammer for electronic attacks that can loiter on a station. MALD-J aims to jam specific E/W, surveillance, and acquisition radars while retaining the capabilities of the MALD. MALD-J will stimulate and degrade an enemy's integrated air defense system.

The approximate dimensions of a MALD system are 2.84m in length, 1.71m in wingspan, 25 cm in body diameter, and 115 kg in weight. Their speed is up to 0.91 Mach, and their propulsion is the turbojet. The missile can be launched from 40,000 ft, at which height it has a range of up to 920 km. MALD-J systems' performances can reach those of the podded SPJ systems in terms of noise jamming power and deception jamming capabilities, as well as their impressive specifications.

Different kinds of aircraft that have used and potentially will employ MALD and MALD-J are listed here:

- Fighters: F-15, F-16, F-22, F-35, Eurofighter, MIG-29, AV-8B Harrier II, Gripen E;
- Bombers: A-10, B-1B, B-52, P-8A Poseidon;
- UAVs: MQ-1 Predator, MQ-1C Gray Eagle, MQ-9 Reaper.

MALD or MALD-J equipped systems are imported to the operational plans to improve the ability of airborne strike forces to access the battlespace by deceiving, distracting, or saturating enemy radar operators and IADS. The MALD system enables an airborne strike force to execute its mission by tricking enemy radars and air defense systems into believing the MALD module is a legitimate target. However, the MALD-J system is intended to allow an airborne strike force to achieve its objective by jamming specific enemy radars and air defense systems. This jamming degrades or prevents the detection of friendly aircraft or munitions. With MALD-J-equipped forces, it is possible to provoke an enemy's integrated air defense system, which enables friendly forces to target and engage the adversary's components.

The step model of the MALD platform was created and transferred to the ANSYS HFSS program. The materials of the aircraft body, wings, and tail sections were chosen as aluminum. Figure 9.11 shows the side view of the MALD-J platform obtained on the ANSYS HFSS program. The technical specifications of the modeled MALD-J system are listed here:

- Dimensions: Wing span 1.71m;
- Length: 2.84m;
- Weight: 115 kg;
- Power Plant: Hamilton Sundstrand TJ-150 turbojet, 337-lb thrust;
- Guidance: GPS/INS.

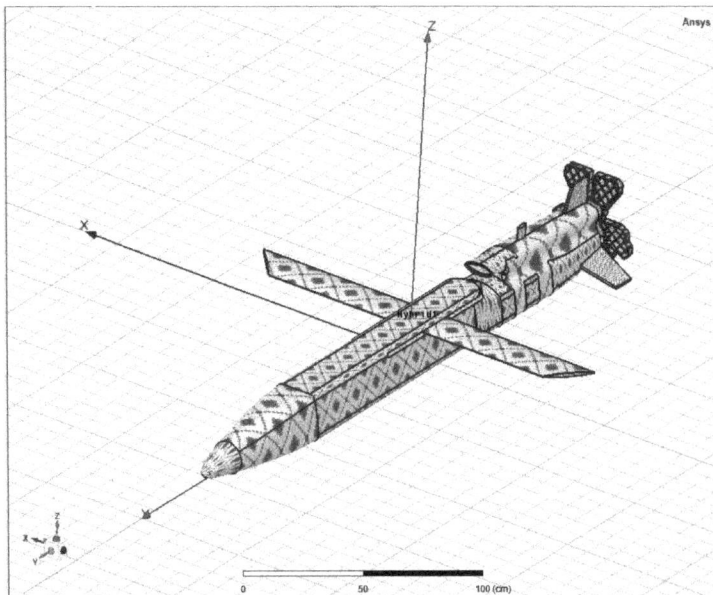

Figure 9.11 Side view of the MALD platform.

9.4   Future Trends in Self-Protection Jamming Decoys

In today's technology, airborne and shipborne platforms are not only weapon systems but also systems of systems. They contain many state-of-the-art systems, such as high-technology radars, EO sensors, communication systems, weapon systems, EW systems, and a significantly developed computer substructure. For the previous generation, it is said that airborne and shipborne platforms have considerably improved computer systems. However, for the current generation platforms, the concept of these platforms is called flying computers for the airborne platforms and shipborne platform floating computers. Thus, they are both precious and confidential from a technological point of view.

The operational capabilities have also improved with the computerization, and the conceptual changes followed them. The technological developments in the air and ship platforms compel us to utilize self-protection systems. However, using onboard or podded self-protection systems mitigates the threat risk at a level. Self-protection off-board systems, such as active and passive decoys, are employed for further protection. The additional development expectations for the passive decoys can be only a tiny amount. Indeed, chaff and corner reflectors are mechanical devices, and the improvement can be in the shape and material.

The leading technological development will probably occur in the active decoys, such as expendable and towed ones. We put future projections for shipborne and airborne *expendable active decoys* (EADs) and then continue with the towed counterparts.

The shipborne launch EADs hang in the air via rocket or parachute. The decoys operate like a repeater jammer and have a complex electronic structure with Rx/Tx, antennas, and processors. Moreover, adding rockets increases their complexity and value, but they are disposable. Thus, using capable drones that are not disposable for EAD systems would be possible. This time, the decoys are not expendable, and they cannot be called EAD but are still active decoys. Since they are not towed, calling them towed decoys is unsuitable. Thus, they can be used, and unless the threat missiles hit them, they can return to the mother platform and be used again. For this reason, the capabilities of the active decoys on drones can be increased. Say that their Rx sensitivities and processing capabilities, DRFM qualification and jamming power, response time, and simultaneous or sequential jamming capabilities can be improved. They can be preprogrammed to increase their effectiveness.

However, the design, production, and development of airborne EAD systems are hot topics today. Even though they have been employed in the last few decades, the future is full of developments on this topic. The EAD systems generally use the aircraft's existing dispensing system and, once launched, act independently to lure away hostile radar-guided missiles from the host platform. Typically, they have an advanced digital Rx, DRFM, and a sophisticated ECM technique generator. Their propulsion system is a rear-mounted pusher propeller, and their electronic components are battery-powered. Today's technology allows it to handle the most advanced radar threats. They can be preprogrammed to jam or replicate how the mother platforms (fighter, attack, and bomber aircraft) appear to enemy radar operators.

The most prevalent EAD systems are the MALD decoys, and future projections for the airborne EAD systems are carried out through these systems. For future predictions, their operation will become prominent. When they are used as saturation or repeater jammers, their Rx sensitivities and processing capabilities, DRFM qualification and jamming power, response time, and simultaneous or sequential jamming capabilities are aimed to increase. Fulfilling these properties without changing their size, weight, and power consumption is essential. Battery technology is also critical for these applications.

A significant operational deficit is encountered when MALD systems are employed as seduction or detection decoys. The MALD systems are used as repeaters to imitate the mother platforms on the enemy radars in the operational concepts. However, modern SIGINT/ES aircraft, AEW&C systems, and IADSs have ES Rx to detect the approaching system's emissions. Say that an allied fighter is flying in the close part of the adversary's territory border. The territory is controlled by an AEW&C aircraft connected to the IADS. Before passing the border, the fighter launches a MALD system to penetrate the territory. The MALD system can show itself to the radar systems in the IADS as the host fighter. However, the ES Rx at SIGINT/ES and AEW&C aircraft can discriminate the difference between the fighter and the MALD system due to the fighter's lack of radar and avionics emissions.

To solve this problem, the MALD system can imitate the fighter's radar, avionics, and communications emissions. In this case, the ES systems assume to detect the fighter when they encounter the MALD system. However, the radar systems in the IADS can see from the MALD system's RCS that the object is not a fighter, which causes a new problem. To handle this problem, a reverse RCS effect may be utilized, which aims to increase the RCS of the MALD system using 2-D antenna arrays [10]. The design for this purpose is to adjust the RCS with the mother platform.

Another future projection for EADs can be done for the UAVs. In future air combat concepts, UAVs occupy a critical place for various roles. These roles include fighter, bomber, and surveillance in the adversary territories. So UAVs as the host platforms of the MALD systems should be considered in this context.

The towed decoys are an essential self-protection tool; the developments and usage concepts have become prominent for airborne platforms. Today's towed decoy technology is rather sophisticated. They use fiber-optic cable for data transfer and apply high-power jamming response. Thus, missiles are guided safely from the mother platform to the towed decoy. The modern towed decoys work in concert with the aircraft's onboard or podded SPEW system, using various countermeasure techniques to suppress, deflect, and seduce pulsed and CW RF threats. The usage of towed decoys at the UAV should be considered a future projection. Moreover, it would be appropriate to investigate possible methods for using towed traps in cruise missiles.

9.5   Problems

Problem 9.1: Assume that a fire control radar, operating at 8.4 GHz, starts tracking a fighter with the scenario in Figure 9.12. The fighter's RCS is 9 m^2 from the aspect of the radar boresight, and the range between the radar and the fighter is 12 km. At this point, the fighter's RWR system detects the threat and dispenses chaff as the first countermeasure. The fighter's backscatter must be below the chaff cloud

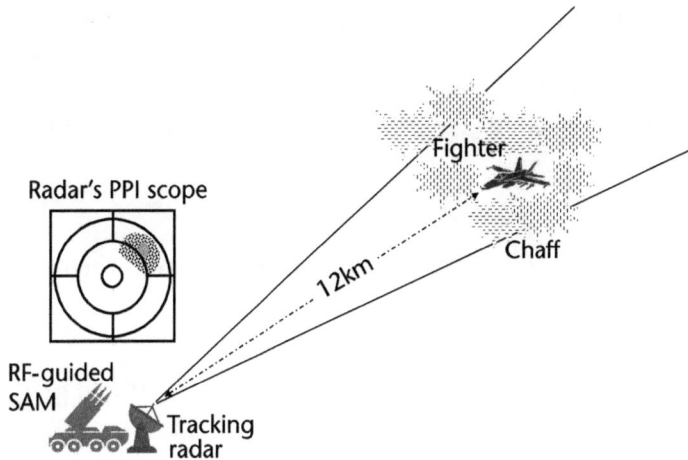

Figure 9.12 Chaff usage scenario for Problem 9.1.

backscatter on the radar Rx to perform an evasive maneuver that can break the lock. If the chaff dipole dimension is optimal, calculate the approximate number of chaffs in the cloud to break the lock. (*Note:* The operational chaff ratio is 60%.)

Problem 9.2: A warship's ESM system detects an RF activation at 15 GHz. When it compares the signal properties with the EID, it detects that this is an ASM system that is an active radar terminal homing missile. The ship deploys two trihedral corner reflectors; one is square, and the other is triangular, as shown in Figure 9.13. The ship's RCS, according to the ASM radar boresight, is 25,000 m^2, and the corner reflectors' edge length (d) is 1m. The warship and the corner reflectors are in the main lobe of the missile, and the distance difference according to the missile is negligible. When the RCS of any corner reflector is higher than the ship's RCS, the vessel is likely to survive. What is the chance of the ship's survivability? (*Note:* All the surface clutters are negligible.)

Problem 9.3: A bomber's RWR detects an activation at 10.4 GHz, and after the processing, it identifies the threat as an AAM system with active radar guidance. The SPEW system automatically launches the towed decoy. The scheme is presented

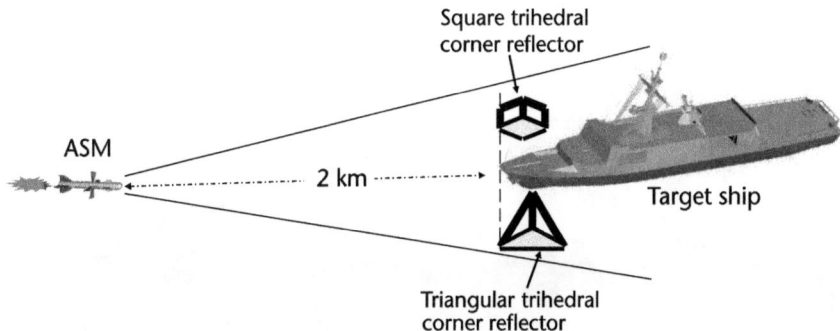

Figure 9.13 Corner reflector usage scenario for Problem 9.2.

in Figure 9.14. The towing cable length is 30m, the lethality range of the threat missile's fuse is 20m, and the bomber's fuselage width is 5m. Assume that the vertex line and the bomber's heading are the same. If the approach angle of the missile is 15° according to the bomber aircraft's heading, what is the bomber's chance to survive?

Solutions

Solution 9.1: If the chaff dipole dimension is optimal, then the dipole length is half of the wavelength:

$$\lambda = \frac{c}{f} = \frac{3 \times 10^8}{8.4 \times 10^9} = 0.036\text{m} \rightarrow \ell_{\text{dipole}} = \frac{\lambda}{2} = 0.018\text{m} = 1.8 \text{ cm}$$

The average RCS of a single chaff for an average aspect angle is calculated by using (9.3):

$$\bar{\sigma}_{\text{chaff}} = 0.18\lambda^2 = 0.18(0.018)^2 = 5.832 \times 10^{-5} \text{ m}^2$$

The aim is to find the chaff cloud and the fighter backscatter signal powers. The evasive maneuver can break the lock when the chaff cloud's backscatter signal power is higher than the fighter's.

$$S_{\text{cloud}} > S_t$$

Inspired from (7.18),

$$\frac{\sum I}{S_t} = \frac{S_V + J}{S_t} = \frac{\left(\dfrac{P_{\text{Tx}}G^2c^2\sigma_w}{(4\pi)^3 R^4 f^2}\right) + \left(\dfrac{P_J G_{\text{Ja}} Gc^2}{(4\pi R f)^2}\right)}{\dfrac{P_{\text{Tx}}G^2c^2\sigma_t}{(4\pi)^3 R^4 f^2}} = \frac{\sigma_w}{\sigma_t} + \frac{(ERP)_J(4\pi R^2)}{(ERP)_R \sigma_t}$$

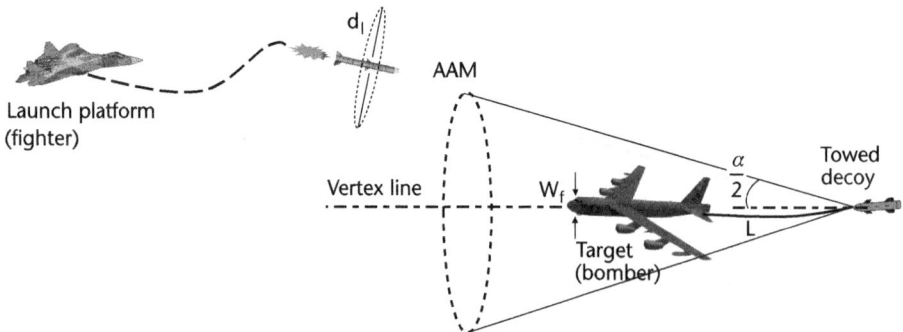

Figure 9.14 The scheme for Problem 9.3.

No jamming information is given in the question so that we can eliminate the jamming effect and the dominance of the chaff cloud in the volume; we can take the volume clutter as chaff cloud RCS,

$$\frac{\sum I}{S_t} = \frac{S_V}{S_t} = \frac{S_{cloud}}{S_t} = \frac{\left(\dfrac{P_{Tx}G^2c^2\bar{\sigma}_{cloud}}{(4\pi)^3 R^4 f^2}\right)}{\dfrac{P_{Tx}G^2c^2\sigma_t}{(4\pi)^3 R^4 f^2}} = \frac{\bar{\sigma}_{cloud}}{\sigma_t}$$

Thus, we can write

$$\bar{\sigma}_{cloud} > \sigma_t$$

Now, (9.1) is considered to obtain the average RCS of a chaff cloud as:

$$\bar{\sigma}_{cloud} = \left(N\bar{\sigma}_{chaff}\right)\mu_{ops} = \left(N \times 5.832 \times 10^{-5}\right) \times 0.6 > 9$$
$$N > 257,201.64 \Rightarrow N \simeq 260,000$$

Solution 9.2: The RCSs of the trihedral corner reflectors are calculated and compared with the ship's RCS, 25,000 m², to estimate the surviving possibility of the vessel. For this purpose, first, the wavelength of the ASM radar is calculated.

$$\lambda = \frac{c}{f} = \frac{3 \times 10^8}{15 \times 10^9} = 0.02\text{m}$$

The maximum RCS of the square trihedral corner reflector is calculated using (9.6).

$$\sigma = \frac{12\pi d^4}{\lambda^2} = \frac{12 \times \pi \times 1^4}{0.02^2} = 94,200 \text{ m}^2$$

The maximum RCS for the triangular trihedral corner reflector is obtained using (9.7).

$$\sigma = \frac{4\pi d^4}{3\lambda^2} = \frac{4 \times \pi \times 1^4}{3 \times 0.02^2} = 10,467 \text{ m}^2$$

As seen from the results, the possible target will be the square trihedral corner reflector, which has more than three times higher RCS than the ship's RCS. However, the triangular trihedral corner reflector's RCS is lower than the ship's RCS and probably does not provide any protection against the missile.

Solution 9.3: The question wants us to calculate the HA angle ($\alpha/2$) and compare it with the missile's approach angle according to the vertex line or bomber's heading. For this purpose, (9.8) is used.

$$\frac{\alpha}{2} = \tan^{-1}\left(\frac{d_{1(m)} + W_f}{L}\right)$$

The values are given as:
- $d_1 = 20\text{m}$ (lethality range of the fuse or the radius of the fuse's lethality circle);
- $W_f = 5\text{m}$ (the aircraft's fuselage width);
- $L = 30\text{m}$ (the towing cable length).

When the numeric values are substituted in the equation,

$$\frac{\alpha}{2} = \tan^{-1}\left(\frac{d_1 + W_f}{L}\right) = \tan^{-1}\left(\frac{20 + 5}{30}\right) = 39.8°$$

When we compare the obtained HA angle, 39.8°, with the missile's approach angle, 15°, we can say that the missile probably damaged the aircraft. If the crew wants to increase the bomber's chance of survival, they may increase the towing cable or change the heading angle.

References

[1]   De Martino, A., *Introduction to Modern EW Systems*, 2nd ed., Norwood, MA: Artech House, 2018.

[2]   Knott, E. F., J. F. Shaeffer, and M. T. Tuley, *Radar Cross Section*, 2nd ed., Raleigh, NC: SciTech Publishing, 2004.

[3]   Vakin, S. A., L. N. Shustov, and R. H. Durrwell, *Fundamentals of Electronic Warfare*, Norwood, MA, Artech House, 2001.

[4]   Nathanson, F. E., J. P. Reilly, and M. N. Cohen, *Radar Design Principles*, 2nd ed., Mendham, NC, SciTech Publishing, 1991.

[5]   Mahafza, B. R., and A. Z. Elsherbeni, *MATLAB Simulations for Radar Systems Design*, Boca Raton, FL: CRC Press, 2004.

[6]   Avionics Department, *Electronic Warfare and Radar Systems Engineering Handbook*, Point Mugu, CA: Naval Air Warfare Center, 2013.

[7]   Air Combat Command Training Support Squadron (ACC TRSS), *Electronic Warfare Fundamentals*, Nellis AFB, NV, 2000.

[8]   AGARDOGRAPH, R. T. O.; Series–Volume, *Flight Test Techniques. Electronic Warfare Test and Evaluation*, 2000.

[9]   Chrzanowski, E. J., *Active Radar Electronic Countermeasures*, Norwood, MA: Artech House, 1990.

[10]   Otenkaya, O. B., and A. G. Pakfiliz, "Increasing the Radar Cross Section Using Phase Array Antenna in the Body Structure of Miniature Air-Launched Decoy Platforms," *Review of Computer Engineering Research, Conscientia Beam*, Vol. 10, No. 1, 2023, pp. 1–15.

List of Acronyms and Abbreviations

A	Amplitude
AAA	Anti-aircraft artillery
AAM	Air-to-air missile
ACM	Air combat maneuvering
ADC	Analog-to-digital converter
ADT	Automatic detection and tracking
AEF	Active emitter file
AGC	Automatic gain control
AGL	Above ground level
AEW	Airborne Early Warning
AEW&C	Airborne Early Warning and Control
AF	Ambiguity function
AFC	Automatic frequency control
AI	Air interception
AOA	Angle of arrival
APA	Active phased array
ARM	Antiradiation missile
ASCM	Antiship cruise missiles
ASM	Air-to-surface missile
ASTE	Advanced strategic/tactical expendable
ATC	Air Traffic Control
ATM	Air Traffic Management
BIF	Intermediate frequency filter bandwidth
BLOS	Beyond line of sight
BMC2	Battle Management and Command and Control
BNJ	Barrage noise jamming
BP	Bandpass
BPF	Bandpass filter
BRF	Radio frequency filter bandwidth
BSM	Battlespace Spectrum Management

BW	Bandwidth
B/W	Beamwidth
C2	Command and Control
C3	Command, Control, and Communications
C4I	Command, Control, Communications, Computers, and Intelligence
C4ISR	Command, Control, Communications, Computers, Intelligence, Surveillance, and Reconnaissance
CFAR	Constant false alarm rate
CM	Countermeasures
CMAP-CFAR	Clutter Map Constant False Alarm Rate
CMDS	Countermeasures Dispenser System
CMU	Central Management Unit
CNN	Convolutional neural network
CNR	Carrier-to-noise ratio
COHO	Coherent oscillator
COMINT	Communication intelligence
COSRO	Conical scan on receive only
CP	Cover pulse
CSAR	Combat search and rescue
CVR	Crystal video receiver
CW	Continuous wave
CWD	Continuous-wave Doppler
DA	Differential amplifier
DAC	Digital-to-analog converter
DAS	Defensive Aids Suite
DDS	Direct digital synthesizer
DE	Directed energy
DEAD	Destruction of Enemy Air Defense
DEW	Directed energy weapon
DF	Direction finding
DIRCM	Directional infrared countermeasures
DL	Deep learning
DME	Distance-measuring equipment
DOA	Direction of arrival
DRFM	Digital radio frequency memory
DSP	Digital signal processing
DSSS	Direct sequence spread spectrum
EA	Electronic attack

EAD	Expendable active decoy
EADS	Enemy Air Defense System
ECM	Electronic countermeasures
ECCM	Electronic countercountermeasures
EDW	Emitter Descriptor Word
EID	Emitter identification
EIDD	Emitter Identification Database
EJ	Escort jammer
EKF	Extended Kalman filter
ELINT	Electronic intelligence
EM	Electromagnetic
E/M	Expectation maximization
EMC	Electromagnetic compatibility
EMI	Electromagnetic interference
EO	Electro-optical
EOB	Electronic Order of Battle
EP	Electronic protection
EPL	Emitter Program Library
ERP	Effective radiated power
ES	Electronic warfare support
ESM	Electronic support measures
ESR	Eclipsed signal return
EW	Electronic warfare
E/W	Early warning
EWO	Electronic weapons officer
F/Conv	Frequency converter
FC	Fire control
FFT	Fast Fourier transform
FLIR	Forward-looking infrared
FM	Frequency modulation
FMC	Flight and Mission Computer
FMCW	Frequency-modulated continuous wave
FML	Frequency memory loop
FOV	Field of view
FSK	Frequency shift keying
Ft	Foot
FTC	Fast time constant
FTSR	Faraway target signal return

GaN	Gallium arsenide
GaN-on-SiC	Gallium nitride on silicon-carbide
GCA	Ground control approach
GCS	Ground control station
GLM	Generalized linear model
GMM	Gaussian mixture model
GNC	Guidance, navigation, and control
GPB-MM	Generalized pseudo-Bayesian multimodel
GPR	Ground-penetrating radar
HA	High aperture
HALE	High-altitude long endurance
HC	Hybrid coupler
HARM	High-speed antiradiation missile
HF	High frequency
HITL	Hardware-in-the-loop
HMM	Hidden Markov model
HOJ	Home-on-jam
IADS	Integrated Air Defense System
IEEE	Institute of Electrical and Electronics Engineers
IF	Intermediate frequency
IFA	Intermediate frequency amplifier
IFF	Identification friend or foe
I/M	Input matching
IMM	Interacting multimodel
IMMPDAF	Interacting multimodel probabilistic data association filter
IR	Infrared
I/O	Input/output
IRST	Infrared search and track
ISAR	Inverse synthetic aperture radar
ISR	Intelligence, surveillance, and reconnaissance
ISTAR	Intelligence, surveillance, target acquisition, and reconnaissance
ITALD	Improved tactical air-launched decoy
ITU	International Telecommunication Union
J/S	Jamming-to-signal ratio
KF	Kalman filter
km	Kilometer
KNN	K-nearest neighbors
LAN	Local area network

LAU	LAN access unit
LAWS	Lethal autonomous weapon systems
LBR	Laser beam rider
LCMS	Laser countermeasure system
LFM	Linear frequency modulation
LHC	Left-hand-circular
LIDAR	Light detection and ranging
LNA	Low noise amplifier
LO	Local oscillator
LORO	Lobe on receive only
LOS	Line of sight
LPF	Lowpass filter
LPI	Low probability of intercept
LRF	Laser range finder
LRU	Line replaceable units
LTD	Laser target designator
LWR	Laser warning receiver
LWS	Laser warning system
m	Meter
MALD	Miniature air-launched decoy
MALD-J	Miniature air-launched decoy-jammer
MALE	Medium-altitude long endurance
MANPAD	Man-portable air defense
MBC	Main beam clutter
MCS	Mission control station
MCDU	Multifunction control and display unit
MCLOS	Manual command to line of sight
MHT	Multiple hypothesis tracking
MEADS	Medium extended air defense system
MF	Medium frequency
MIMO	Multiple-input multiple-output
ML	Machine learning
MLS	Microwave landing system
MM	Multimodel
MMIC	Monolithic microwave integrated circuits
MMSE	Minimum mean square error
MPM	Microwave power module
MTI	Moving target indicator

Mux	Multiplexer
MWR	Missile warning receiver
MWS	Missile warning system
NB	Naive Bayes
NBC	Nuclear, biological, or chemical
NBF	Narrowband filter
NF	Noise figure
NM	Nautical mile
NN	Nearest neighbor
NNF	Nearest neighbor filter
NAVWAR	Navigation warfare
OFP	Operational Flight Program
O/M	Output matching
OOK	On-off keyed
OOM	On-off modulation
OTH	Over-the-horizon
P	Polarization
PA	Pulse amplitude
P/A	Power amplifier
PAF	Periodic ambiguity function
PAR	Precision approach radar
PD	Probability of detection
PDF	Probability density function
PDR	Pulsed Doppler radar
PDA	Probabilistic Data Association
PDAF	Probabilistic Data Association filter
PDR	Pulsed Doppler radar
PDW	Pulse descriptor word
PFA	Probability of false alarms
PFM	Preflight message
PGPS	Pulse groups per second
PGRI	Pulse group repetition intervals
PLF	Polarization loss factor
PLL	Phase-locked loop
PMHT	Probabilistic multihypothesis tracking
POI	Probability of intercept
PoP	Pulse-on-pulse
PPI	Plan position indicator

PPG	Pulses per group
PRI	Pulse repetition interval
PW	Pulse width
PWSP	Platform and weapon system pair
RAM	Radar-absorbent materials
RAT	Ram air turbine
RCS	Radar cross-section
R&D	Research and development
RF	Radio frequency
RFA	Radio frequency amplifier
RFCM	Radio frequency countermeasures
RFiRx	Radio frequency illuminator or launch platform
RF-LOS	Radio frequency line of sight
RFPT	Radio frequency power transistor
RF SNR	Radio frequency signal-to-noise ratio
RGPI	Range gate pull-in
RGPO	Range gate pull-off
RHC	Right-hand-circular
RMS	Root mean square
RNN	Recurrent neural network
ROE	Rules of engagement
ROF	Radio operating frequency
RWR	Radar warning receiver
Rx	Receiver
SABR	Scalable agile beam radar
SAM	Surface-to-air missile
SAP	Signal analysis processor
SAR	Synthetic aperture radar
SCR	Signal-to-clutter ratio
SDW	Signal descriptor word
SEAD	Suppression of enemy air defenses
SFW	Stepped frequency waveform
SHR	Superheterodyne (superhet) receiver
SHORAD	Short-Range Air Defense
Si	Silicon
SIGINT	Signals intelligence
SIJ	Stand-in jammer
Si-LDMOS	Silicon-based laterally diffused metal oxide silicon

SIRFC	Suite of Integrated Radio Frequency Countermeasures
SL-B	Sidelobe blanking
SLC	Sidelobe clutter
SL-C	Sidelobe cancelers
SMM	Static multiple model
SMR	Surface movement radar
SNF	Strongest neighbor filter
SNR	Signal-to-noise ratio
SOJ	Stand-off jammer
SoS	System of systems
SPAF	Single-period ambiguity function
SPEW	Self-protection electronic warfare
SPJ	Self-protection jammer
SRW	Swept rectangular wave
STFT	Short-time Fourier transform
STOVL	Short-takeoff/vertical landing
SSDS	Ship self-defense system
STALO	Stable local oscillator
STT	Single target tracking
SVM	Support vector machine
SWaP	Size, weight, and power
TALD	Tactical air-launched decoy
TAS	Track and scan
TEWS	Tactical electronic warfare system
TDOA	Time difference of arrival
TNR	Threshold-to-noise ratio
TOA	Time of arrival
TRF	Tuned radio frequency
TTR	Target tracking radar
Tx	Transmitter
TWS	Track-while-scan
TWT	Traveling-wave tube
UAV	Unmanned air vehicle
USR	Unambiguous signal return
UV	Ultraviolet
VCO	Voltage-controlled oscillator
VGPI	Velocity gate pull-in
VGPO	Velocity gate pull-off

VS-IMM	Variable structure interacting multiple model
VSWR	Voltage standing wave ratio
WGN	White Gaussian noise
WN	White noise
WTA	Weapon-target assignment
YIG	Yttrium iron garnet

Glossary

Airborne   Adjective that describes items forming an integral part of an aircraft.

Airborne Early Warning and Control (AEW&C)   Air surveillance and control provided by airborne early warning aircraft equipped with search and height-finding radar and communication equipment for controlling weapon systems.

Aircraft   Any vehicle (such as an airplane, glider, or helicopter) that can travel through the air is supported either by its buoyancy or by the action of the air against its surfaces.

Air defense   All measures taken to nullify or reduce the effectiveness of hostile air action.

Air interception (AI)   An operation by which aircraft affect visual or electronic contact with other aircraft.

Air superiority   Air superiority is the degree of dominance in the air battle of one force over another that permits the conduct of operations by the former and its related land, sea, and air forces at a given time and place without prohibitive interference by the opposing force.

Air-to-air missile (AAM)   An air-launched missile against air targets.

Air-to-surface missile (ASM)   An air-launched missile against surface targets.

Amphibious assault   The principal type of amphibious operation involves establishing a force on a hostile or potentially hostile shore.

Autonomous   A system property that decides and acts to accomplish desired goals within defined parameters, based on acquired knowledge and evolving situational awareness, following an optimal but potentially unpredictable course of action.

Barrage jamming   Simultaneous electronic jamming over a broad band of frequencies.

Beam rider   A missile guided by radar or radio beam.

Chaff   Strips of frequency-cut metal foil, wire, or metalized glass fiber used to reflect electromagnetic energy, usually dropped from aircraft or expelled from shells or rockets as a radar countermeasure.

Communication intelligence (COMINT)   A subset of signal intelligence gained through the interception of foreign communications, excluding open radio and television broadcasts.

Command and control (C2)   Military commanders' authority, responsibilities, and activities in directing and coordinating military forces and implementing orders related to executing operations.

Countermeasures (CM)   CM involves employing devices and/or techniques to impair the enemy's operational effectiveness.

Direction finding (DF)   A procedure for obtaining bearings of RF emitters using a highly directional antenna and a display unit on an intercept receiver or ancillary equipment.

Doppler effect   The phenomenon evidenced by the change in the observed frequency of a sound or radio wave caused by a time rate of change in the effective length of the path of travel between the source and the point of observation.

Doppler radar   Doppler radar is any radar that detects motion relative to a reflecting surface by measuring the frequency shift of reflected radio energy due to the observer's motion or the reflecting surface.

Early warning (E/W)   This is the same as air defense early warning. It involves early notification of the launch or approach of unknown weapons or weapons carriers.

Electromagnetic compatibility (EMC)   EMC is the ability of equipment or a system to function in its electromagnetic environment without causing intolerable electromagnetic disturbances to anything in that environment.

Electromagnetic (EM) deception   EM deception is the deliberate radiation, reradiation, alteration, suppression, absorption, denial, enhancement, or reflection of EM energy in a manner intended to convey misleading information to an enemy or to enemy EM-dependent weapons, thereby degrading or neutralizing the enemy's combat capability.

Electromagnetic interference (EMI)   The EMI is any EM disturbance, whether intentional or not, that interrupts, obstructs, or otherwise degrades or limits the effective performance of electronic or electrical equipment.

Electromagnetic operations (EMO)   All operations that shape or exploit the EM environment or use it for attack or defense, including using the EM environment to support operations in all other operational environments. EM operations include (but are not limited to) *electronic warfare* (EW), *signals intelligence* (SIGINT), *intelligence, surveillance, target acquisition and reconnaissance* (ISTAR), *navigation warfare* (NAVWAR), and *battlespace spectrum management* (BSM).

Electromagnetic order of battle (EOB)   A subset of the order of battle that details the systems that operate in the EM spectrum within a specific area of interest. Note(s): The EM order of battle should include EM details such as system characteristics, parameters, functions, locations, and operation times.

Electromagnetic (EM) spectrum   The entire and orderly distribution of EM waves according to their frequency or wavelength. The EM spectrum includes radio waves, microwaves, heat radiation, visible light, ultraviolet radiation, x-rays, and EM cosmic and gamma rays.

Electronic attack (EA)   That division of EW involves using EM energy for offensive purposes. It uses EM energy, *directed energy* (DE), or antiradiation weapons to attack personnel, facilities, or equipment to degrade, neutralize, or destroy enemy combat capability.

Electronic countermeasures (ECM)   That division of EW involves actions taken to prevent or reduce an enemy's effective use of the EM spectrum through EM energy. ECM have three subdivisions: electronic jamming, deception, and neutralization.

Electronic intelligence (ELINT)   Intelligence derived from EM, noncommunication transmissions.

Electronic jamming This is the deliberate radiation, reradiation, or reflection of EM energy that impairs the effectiveness of hostile electronic devices, equipment, or systems.

Electro-optics (EO) The technology associated with those components, devices, and systems that are designed to interact between the EM (optical) and the electric (electronic) state.

Electronic protection (EP) The division of electronic warfare involves protecting personnel, facilities, and equipment from any effects of friendly or enemy use of the EM spectrum that degrade, neutralize, or destroy friendly combat capability.

Electronic warfare (EW) EW is the science that includes activities carried out to ensure the highest level of use of the EM spectrum by friendly units while preventing or making it difficult for enemy units to use it.

Escort jammer (EJ) EJs use a jamming platform to accompany maneuver forces. They are defensive and protect maneuver forces from threat weapons systems that use RF triggers. Successful EJs require precise intelligence regarding threat frequency use. EJs usually do not require high power levels or large antennas like SOJs. EJs use similar vehicle configurations to maneuver and screen vehicles for visual identification.

EW reprogramming EW reprogramming is the deliberate alteration or modification of EW or target sensing systems or the tactics and procedures that employ them in response to validated changes in equipment, tactics, or the EM environment.

Electronic warfare support (ES) or electronic support measures (ESM) ES or ESM is the division of EW involving actions taken to search for, intercept, and identify EM emissions and to locate their sources for immediate threat recognition and situational awareness.

Homing This is the technique whereby a mobile station directs itself or is directed towards a source of primary or reflected energy or to a specified point.

Identification, friend or foe (IFF) A system using EM transmissions to which equipment carried by friendly forces automatically responds, for example, by emitting pulses, thereby distinguishing themselves from enemy forces.

Jamming Jamming is deliberate interference caused by emissions intended to render unintelligible or falsify the whole or part of a desired signal.

Laser target designator (LTD) A system that directs laser energy at a target. The system consists of the laser designator or target marker with its display and control components necessary to acquire the target and direct the beam of laser energy thereon.

Launcher A device designed to support and hold a grenade, rocket, or missile in position for firing.

Missile guidance system This system evaluates flight information, correlates it with target data, determines a missile's desired flight path, and communicates the necessary commands to the missile flight control system.

Radar horizon (RF horizon) The locus of points at which the rays from a transmitter antenna are tangential to the Earth's surface. On the open sea, this locus is horizontal, but on land, it varies according to the topographical features of the terrain.

Repeater-jammer A receiver-transmitter device that amplifies, multiplies, and retransmits the signals received for deception or jamming.

Sea skimmer A missile designed to transit at less than 50 ft (or 15m) above the sea's surface.

Situational awareness The knowledge of the elements in the battlespace necessary to make well-informed decisions.

Stand-in jammer (SIJ) A SIJ is an aircraft that jams enemy receivers from close distances. It is a form of support jamming commonly performed by UAVs operating within the engagement range of hostile air defense systems.

Stand-off jammer (SOJ) SOJ is a jamming aircraft that jams the enemy receivers from long distances. The SOJ normally stations itself on its side of the border but, in any case, well outside the effective ranges of the enemy's air defense systems. SOJ system installed a special mission airborne platform on board to reduce long-range radar coverage on friendly intruders, consequently delaying their engagement and allowing the intruder package to enter the enemy area.

Suppression of enemy air defenses (SEAD) SEAD is any activity that neutralizes, destroys, or temporarily degrades surface-based enemy air defenses by destructive and/or disruptive means.

Surface-to-air missile (SAM) A surface-launched missile against air targets.

Surface-to-surface missile (SSM) A surface-launched missile against surface targets.

Target designation The act of assigning a target to a weapon system.

Target discrimination The ability of a surveillance or guidance system to identify or engage any one target when multiple targets are present.

Terminal guidance The guidance applied to a missile between midcourse guidance and its arrival near the target.

Weapons system A combination of one or more weapons with all related equipment, materials, services, personnel, and means of delivery and deployment, if applicable, required for self-sufficiency.

Wild Weasel An aircraft specially modified to identify, locate, and physically suppress or destroy ground-based enemy air defense systems that employ sensors radiating EM energy.

About the Author

Ahmet Güngör Pakfiliz graduated with a bachelor's degree in electrical engineering from Yildiz Technical University in 1991. The same year, he was appointed lieutenant as an engineer officer in the Turkish Air Force. During the first years of his military life, he gained experience in radar and avionics system maintenance and procurement projects. He earned his MSc in electrical and electronics engineering from Middle East Technical University in 1997. That year, he was appointed to the Electronic Warfare (EW) Department of the Turkish General Staff. From this point, he worked as a system engineer on radar and EW systems in procurement projects, preprogramming of the systems, and operational support. He received his PhD in statistical signal processing in electronics engineering from Ankara University in 2004. From 2004 to 2013, he worked as a technical director and engineering management in different parts of the Turkish Armed Forces. From 2009 to 2011, he was the national representative of the NATO Science and Technology Organization Board's Systems Concepts and Integration Panel. He retired from military service in 2013 as a colonel.

Then he worked as a faculty member at different universities in Turkey. During this period, Dr. Pakfiliz focused his studies on EW, avionics, radar, and signal processing. His other interests include applying EW in unmanned air vehicles and using artificial intelligence (AI) in self-protection EW system processes.

Dr. Pakfiliz remains active in military and academic circles, sharing his knowledge and insights through seminars and lectures on EW. His contributions extend beyond the classroom, with numerous technical papers on target tracking, EW, and signal processing published in international and national journals and presented at various conferences. He has also authored a comprehensive book on EW in Turkish. Dr. Pakfiliz holds the associate professor position at OSTIM Technical University in the Department of Electrical and Electronics Engineering.

Index

For further information on these and other Artech House titles, including previously considered out-of-print books now available through our In-Print-Forever® (IPF®) program, contact:

Artech House	Artech House
685 Canton Street	16 Sussex Street
Norwood, MA 02062	London SW1V 4RW UK
Phone: 781-769-9750	Phone: +44 (0)20-7596-8750
Fax: 781-769-6334	Fax: +44 (0)20-7630-0166
e-mail: artech@artechhouse.com	e-mail: artech-uk@artechhouse.com

Find us on the World Wide Web at: www.artechhouse.com